Advances in Polymer Science

Volume 295

Aims and Scope

The series Advances in Polymer Science presents critical reviews of the present and future trends in polymer and biopolymer science. It covers all areas of research in polymer and biopolymer science including chemistry, physical chemistry, physics, and material science.

The thematic volumes are addressed to scientists, whether at universities or in industry, who wish to keep abreast of the important advances in the covered topics.

Advances in Polymer Science enjoys a longstanding tradition and good reputation in its community. Each volume is dedicated to a current topic, and each review critically surveys one aspect of that topic, to place it within the context of the volume. The volumes typically summarize the significant developments of the last 5 to 10 years and discuss them critically, presenting selected examples, explaining and illustrating the important principles, and bringing together many important references of primary literature. On that basis, future research directions in the area can be discussed. Advances in Polymer Science volumes thus are important references for every polymer scientist, as well as for other scientists interested in polymer science - as an introduction to a neighboring field, or as a compilation of detailed information for the specialist.

Review articles for the individual volumes are invited by the volume editors. Single contributions can be specially commissioned.

Readership: Polymer scientists, or scientists in related fields interested in polymer and biopolymer science, at universities or in industry, graduate students.

R. Jayakumar
Editor

Chitosan for Biomaterials V

Insight into Pharmaceutical Uses

With contributions by

A. K. I. · M. Anraku · M. S. J. Anthraper ·
M. Araneda Cisternas · V. V. Arya Lakshmi ·
R. B. Chandrashekarappa · A. Dinesan · S. Gupta ·
J. Haldar · V. Hmingthansanga · R. Jayakumar · J. M. Joju ·
K. Joy · I. Kar · L. Kumar · F. Laffleur · H. Madhyastha ·
S. Maity · R. Malayandi · A. A. Mathew · M. Mohandas ·
S. Mukherjee · S. Natesan · Neha · M. Otagiri · D. Patra ·
M. Prabaharan · A. Priyadarshini · A. Ramprasad ·
S. M. Ravishankar · Sabitha M. · O. Şahin · P. R. Sahoo ·
D. Sathya Seeli · S. Şenel · M. Shanmuganathan ·
G. Shreeman Raj · A. Sundar · S. Sunil · T. Suresh ·
W. Tachaboonyakiat · S. B. Varadarajan · K. Varaprasad ·
M. Visakhan

Springer

Editor
R. Jayakumar
Polymeric Biomaterials Lab, School of Nanosciences and Molecular Medicine
Amrita Vishwa Vidyapeetham
Kochi, Kerala, India

ISSN 0065-3195 ISSN 1436-5030 (electronic)
Advances in Polymer Science
ISBN 978-3-031-84934-3 ISBN 978-3-031-84935-0 (eBook)
https://doi.org/10.1007/978-3-031-84935-0

This Springer imprint is published by the registered company Springer Nature Switzerland AG
The registered company address is: Gewerbestrasse 11, 6330 Cham, Switzerland

Preface

The goal of this particular volume, Chitosan for Biomaterials-V: Insight into Pharmaceutical Uses, is to offer a general overview of the therapeutic application of chitosan in the biomedical field. Chitosan is a well-known biocompatible and biodegradable polymer, which makes it the preferred material in a variety of biomedical areas including wound healing, tissue engineering, hemostasis, and drug delivery. This book volume is specifically focused on pharmaceutical/medicinal uses of chitosan. This book begins with a chapter on introducing chitosan as a multifaceted biomaterial and goes on to describe how the same has been used alongside drugs in different ways. The role of chitosan in the discovery and utilization of novel antimicrobial agents has been discussed in detail in the third chapter. The fourth chapter deals with the current trends of combining chitosan beads or films for drug delivery systems. The inherent mucoadhesive property of chitosan and its chemical modifications to enhance the same in the drug delivery have been explored in the subsequent chapter. The next few chapters focus on the delivery of different forms of chitosan such as films/membranes, scaffolds, and hydrogels and its efficient use for sustained drug release as well as its potential for use in peptide and protein drugs. The succeeding chapter discusses the role of chitosan nanoparticles as carriers in anticancer drug delivery, allowing for targeted and sustained delivery and retention at tumor sites. The formulations offered in the following chapters describe topical, transdermal, ocular, and oral drug delivery, which uses chitosan as a vehicle to improve drug delivery to specific target tissues. Chapter 13 explores the potential of chitosan as a nutraceutical, which includes dietary supplements and functional foods. The next chapter discusses the increasing involvement of chitosan in medicine with special regard to its use in hemostasis and wound healing. The penultimate chapter analyzes the prospects of biomolecule delivery utilizing different types of chitosan formulations such as nano/microparticles, scaffolds, and gels in tissue engineering, thus indicating the paradigm shift of the technology development in the field of regenerative medicine. The final chapter explores chitosan-based emulsions as versatile vesicular carriers, highlighting their potential in pharmaceutical applications. It is quite apparent in these profound chapters that chitosan is no longer a simple biopolymer, but a revolutionary promoter with potential for the pharmaceutical practice to be transformative. This compilation displays a worldwide commitment in terms of enhancing research and development efforts in this field. It is our hope that this volume will be of aid to researchers, practitioners, and students in particular as it will inspire them to explore the untapped potential of chitosan in promoting the betterment of human health.

Kochi, India R. Jayakumar

Contents

Chitosan: The Versatile Biomaterial 1
Arya Lakshmi V. V. and R. Jayakumar

Chitosan Biomaterials: Applications and Prospects in the Pharmaceutical Sector 35
Joel Maria Joju and R. Jayakumar

Harnessing Chitosan and Its Derivatives for Innovative Antimicrobial Biomaterials 63
Sudip Mukherjee, Subhankar Maity, Dipanjana Patra, and Jayanta Haldar

Chitosan Beads for Drug Delivery 117
Sevda Şenel and Oğuz Şahin

Mucoadhesive Chitosan in Drug Delivery 143
Flavia Laffleur

Chitosan Films and Membranes in Drug Delivery 161
V. V. Arya Lakshmi, Madhumitha Shanmuganathan, G. Shreeman Raj, and R. Jayakumar

Chitosan Scaffolds for Drug Delivery 183
Meghna Mohandas, Anushikaa Ramprasad, Tanushree Suresh, Sri Brindaa Varadarajan, Sri Manjari Ravishankar, and R. Jayakumar

Applications of Chitosan Hydrogel in Drug Delivery 213
Joel Maria Joju, Sona Sunil, Megha Visakhan, and R. Jayakumar

Chitosan and Its Derivatives in the Delivery of Peptides and Proteins 235
Pravas Ranjan Sahoo, Ipsita Kar, Ankita Priyadarshini, Aravindh Sundar, Revanasiddappa Bistuvalli Chandrashekarappa, and Harishkumar Madhyastha

Chitosan Nanoparticles: Drug Delivery Carriers in Cancer Therapy . . . 261
Kiran Joy, D. Sathya Seeli, and M. Prabaharan

Chitosan-Based Formulations for Enhanced Topical, Transdermal, and Ocular Drug Delivery . . . 285
Victor Hmingthansanga, Lalit Kumar, Neha, Rajkumar Malayandi, and Subramanian Natesan

Chitosan in Oral Drug Delivery . . . 319
Makoto Anraku and Masaki Otagiri

Chitosan as Nutraceuticals: Transforming Pharmaceutical Research . . . 339
Anjali K. I., Aparna Ann Mathew, Swati Gupta, and Sabitha M.

Chitosan-Based Biomaterials for Hemostasis and Wound Healing . . . 367
Matías Araneda Cisternas and Kokkarachedu Varaprasad

Delivery of Biomolecules Using Chitosan for Tissue Engineering . . . 403
Abhirami Dinesan, Mary Susan J. Anthraper, and R. Jayakumar

Chitosan-based Emulsions: Versatile Vesicular Capsules in Pharmaceutical Applications . . . 429
Wanpen Tachaboonyakiat

Adv Polym Sci (2025) 295: 1–34
https://doi.org/10.1007/12_2024_181

Published online: 10 December 2024

Chitosan: The Versatile Biomaterial

Arya Lakshmi V. V. and R. Jayakumar

Contents

1 Introduction ... 2
1.1 History ... 2
2 Chitosan: Structure and Derivatives ... 3
2.1 Chemical Structure of Chitin ... 3
2.2 Chemical Structure of Chitosan ... 5
2.3 Derivatives of Chitosan ... 5
3 Main Sources and Preparation of Chitosan ... 9
3.1 Sources of Chitosan Extraction ... 9
3.2 Processing of Chitosan ... 9
4 Chitosan Characteristics ... 11
4.1 Molecular Weight ... 11
4.2 Deacetylation Degree ... 12
5 Chitosan Properties ... 12
5.1 Physicochemical Properties ... 13
5.2 Biological Properties ... 17
6 Chitosan-Based Materials and Their Biomedical Applications ... 22
6.1 Wound Healing ... 22
6.2 Drug Delivery ... 24
6.3 Tissue Engineering ... 25
6.4 Hemostasis ... 27
7 Conclusion ... 28
References ... 28

Abstract Chitosan is a biodegradable linear polymer with randomly distributed β-(1-4)-linked D-glucosamine and N-acetyl-D-glucosamine. It is a prominent cationic natural polymer and is prepared by performing alkaline deacetylation on chitin, which is sourced from shrimp shells and other crustaceans. The net cationicity and presence of multiple reactive functional groups such as primary amine, and a

Arya Lakshmi V. V. and R. Jayakumar (✉)
Polymeric Biomaterials Lab, School of Nanosciences and Molecular Medicine, Amrita Vishwa Vidyapeetham, Kochi, India
e-mail: rjayakumar@aims.amrita.edu

primary and secondary hydroxyl group in its monomeric structure make chitosan a promising biomolecule. As a biopolymer, it has a multitude of potential applications and it can be enhanced by modification of its various properties. It is an excellent biopolymer having remarkable characteristics, making it suitable for applications in drug delivery, wound healing, antibacterial, hemostatic agents, food industry, agriculture, pharmacy, medicine, dentistry, ophthalmology, biomedicine, and bio-imaging techniques. This chapter examines the versatility of chitosan as a biomaterial exploring its traits, characteristics, properties, various applications, and wide range of uses.

Keywords Biomaterials · Biomedical applications · Characteristics · Chitosan · Preparation · Properties · Sources · Structure

1 Introduction

1.1 History

Chitosan is a remarkable biopolymer derived from chitin, which is a key structural element in the fungi cell wall and exoskeletons of arthropods, such as crustaceans. In recent decades there has been an increasing focus on developing natural alternatives to synthetic polymers, known as biopolymers. Chitosan and its derivatives are utilized in various industries such as food and nutrition, agriculture, aquaculture, pharmacy, medicine, and biomedicine industry [1]. Since the late 1970s, there has been growing interest in biopolymer chitosan within both basic scientific investigation and applied research due to its notable molecular structure, physical and chemical characteristics, and biological functions, which distinguish it from synthetic polymers [2]. Chitosan is a copolymer that has received FDA approval, has a lower molecular weight (MW) and a more crystalline nature than chitin (MW $>$ 100 kDa) and is prepared by the deacetylation of chitin [3].

In 1799 English scientist Charles Hachett reported chitin as the first stable chemical material [4]. Chitin and chitosan have a historical beginning that traces back to 1811 in France. Chemist Henri Braconnot discovered a non-soluble residue by the treatment of extraction of fungi with water, alcohol, and dilute alkali, and he named the alkaline-insoluble fraction as fongine/fungine [5]. In 1823, Odier discovered the same alkaline-insoluble residue from the elytra of insects through multiple reactions using hot solutions of hydroxide. He named it chitine/chitin, deriving from the Greek word chitos, which denotes a tunic or envelope [1]. Moreover, in 1859, Charles Rouget modified the previously reported chitin by treating it with concentrated caustic potash under reflux conditions, resulting in the formation of a new compound called "chitine modifiee" [5]. In the 1870s, Ledderhose conducted hydrolysis on arthropod chitin to elucidate the structure of the resulting compound. He treated shells of lobster with hot concentrated hydrochloric and observed that the dissolved shells upon evaporation yielded distinctive crystalline form. In 1894, Felix Hoppe-Seyler coined the term "chitosan" and during the mid-1930s, the characterization of both chitin and chitosan was done using X-ray and further by enzymatic analysis and infrared spectroscopy [1]. During the 1960s, chitosan gained prominence because of its capacity to combine with red blood cells. With the increasing popularity of chitosan applications, the first international conference solely dedicated to this material convened in 1977. The versatility and amenability of polysaccharide chitosan became a new window to expand and explore various applications because of its ease of processing and adaptability for modifications [5].

Chitin's extensive hydrogen bond network makes it entirely soluble in water and other organic solvents. This property renders the characterization difficult and limited its suitability for industrial application. Whereas, chitosan formed by the deacetylation of chitin has distinct properties because of its free amino groups capable of undergoing protonation. This feature influences various physicochemical properties such as solubility, hydrogen bonding, and reactivity [4]. The presence of hydroxyl and amine groups accounts for various characteristics such as polyelectrolyte behavior, antimicrobial activity, antioxidant activity, gel-forming ability, biocompatibility, metal chelation, and easy processability. These remarkable properties exhibited by chitosan in various unexplored streams are now recognized as a crucial raw material for producing diverse products utilized in food, pharmaceutical, healthcare, agriculture, industry, and environmental pollution protection (Fig. 1) [6].

Recent research has shown that processing chitosan for use in scaffolds, nanofibers, membranes, hydrogels, composites, and nanoparticles offers a range of advantages. Due to its polycationic nature, it has been paid much attention in chitosan-based biomedical materials. In addition, chitosan has also a wide range of applications in wound healing since it can stimulate hemostasis and accelerate tissue regeneration [7]. In brief, this review chapter highlights the multifaceted nature of chitosan exploring its characteristics, properties, and applications in biomedical field.

2 Chitosan: Structure and Derivatives

2.1 Chemical Structure of Chitin

Chitin serves as a vital structural element of the invertebrate exoskeleton of insects, crustaceans, other terrestrial arthropods [8] and it is the second most important natural polymer in the world. The chemical structure of chitin is composed of (1-4)-linked 2-acetamido-2-deoxy-b-D-glucopyranose units, and its structure is similar to cellulose, with an acetamide group (-$NHCOCH_3$) at the C2 position (Fig. 2) [9]. Chitin was first identified in 1884 and occurs in nature as ordered crystalline microfibrils [10]. Chitin and its deacetylated derivative chitosan have attracted significant interest due to their potential for innovative and novel applications [11]. In industrial processing, chitin is extracted from crustaceans through a method involving acid treatment by dissolving calcium carbonate followed by alkaline

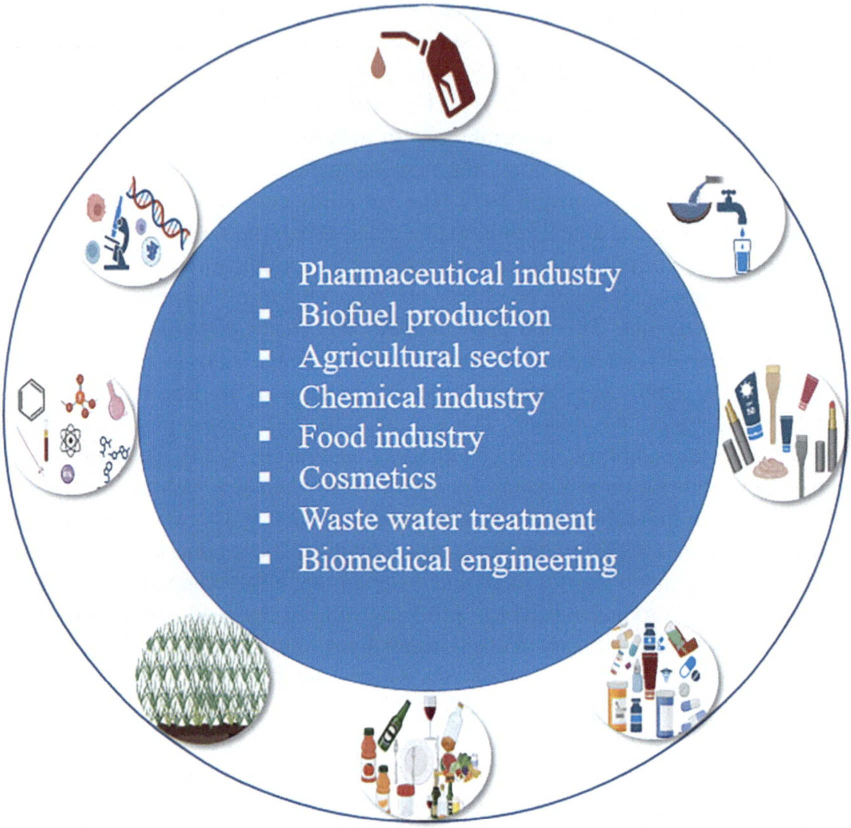

Fig. 1 Applications of chitosan in various fields

Fig. 2 Chemical Structure of Chitin

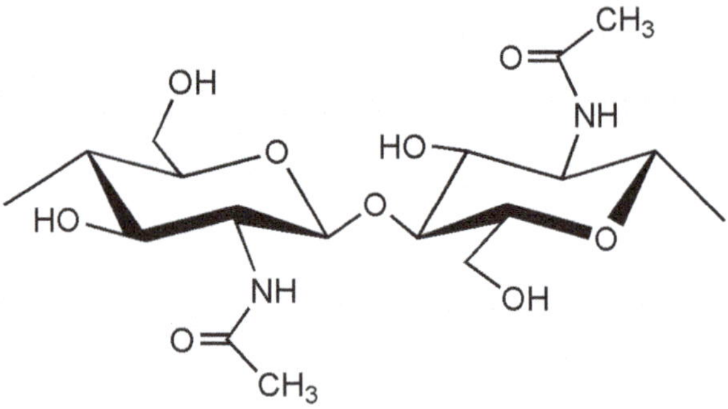

extraction to solubilize proteins [10]. Crustacean shell waste primarily comprises protein (30–40%), calcium carbonate and calcium phosphate (30–50%), and chitin (20–30%) [11]. In the solid state, chitin can exist in three different crystalline polymorph structures. α-chitin consists of anti-parallel N-acetyl glucosamine chains, β-chitin has parallel chains of N-acetyl glucosamine and γ-chitin is formed by two

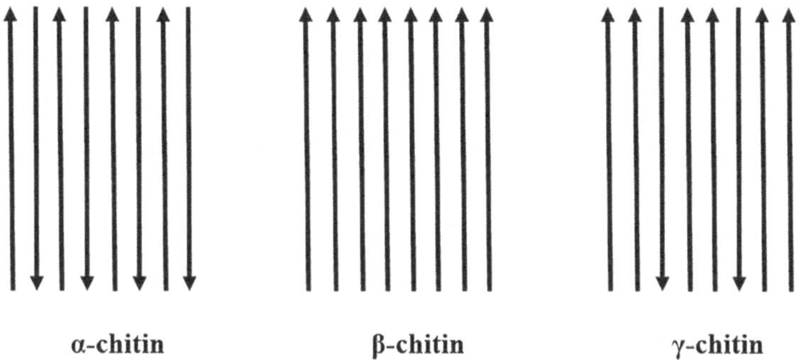

Fig. 3 Three polymorphic forms of chitin

Fig. 4 Chemical structure of chitosan

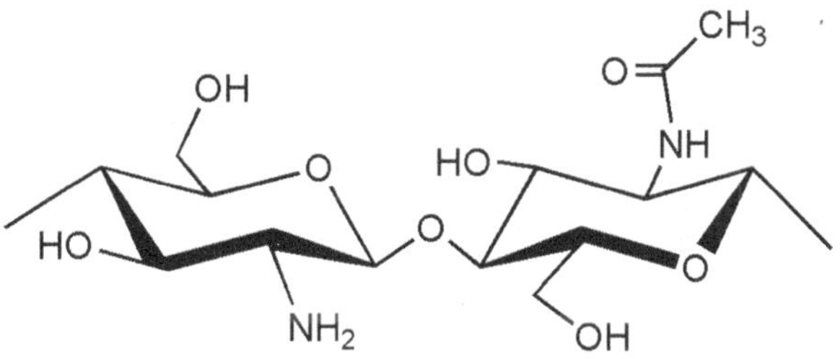

parallel chains alternating with one antiparallel chain. The α-chitin and β-chitin are the most common forms in nature with good mechanical properties [8]. A diagram showing three polymorphic forms of chitin is depicted in Fig. 3.

2.2 *Chemical Structure of Chitosan*

Chitosan's structure is composed of β (1-4)-linked glucosamine and N-acetyl-glucosamine units, which belong to the family of glycosaminoglycans, a group of polysaccharides known for their bioactivity (Fig. 4) [12]. The functional groups present in chitosan include 1° amino group (at C2), 1° & 2° hydroxyl groups (at C6, C3), glycosidic bonds & acetamide group. A large number of modifications to these functional groups produce derivatives with different properties and behaviors, such as altered solubility, biodegradability, and antibacterial activity [13].

2.3 *Derivatives of Chitosan*

Enhancing the properties of chitosan through modification is a key research focus resulting in enhanced solubility, pH-sensitive targeting, and a greater diversity of

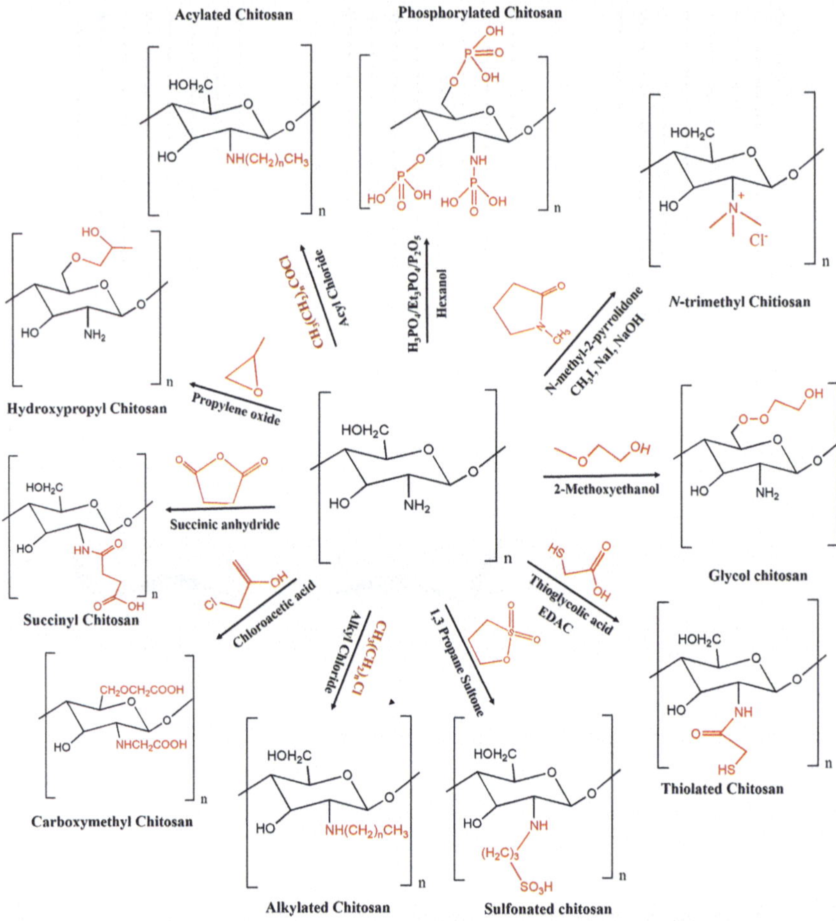

Fig. 5 Derivatives of Chitosan-I

delivery systems. Chitosan modification can be achieved through processes such as acylation, alkylation, carboxylation, phosphorylation, thiolation sulfonation, hydroxypropylation, and quaternization (Fig. 5) [14].

Carboxymethyl chitosan (CMC) is a well-known chitosan derivative with amphoteric properties. The reaction of chitosan under controlled conditions with sodium monochloroacetate and NaOH results in O- and N-carboxymethylation [15]. CMC can be prepared via direct alkylation, reductive alkylation, and Michael addition methods [16]. CMC can be used in a variety of biomedical applications because of its biological properties, which include water solubility, biodegradability, metal ion sorption capacity, antibacterial efficacy, antioxidant property, and increased moisture retention ability [17].

Thiolated chitosan is mostly prepared in the pharmaceutical and biomedical sector using thiolating agents such as thioglycolic acid and glutathione. Thiolation increases the mucoadhesive property of chitosan and enhances its permeation capability [18]. Acylated chitosan is another common derivative. In the process of acylation, aromatic/aliphatic acyl groups are introduced into the chitosan structure by treating with organic acids. N-acylated chitosan is produced when the acylation occurs between the C-2 position carbon and NH_2 group to form an amide whereas O-acylation occurs when C-6 position carbon reacts with OH functional group to form an ester [18]. Glycol chitosan can be synthesized by treating ethylene glycol with chitosan. It has hydrophilic character and has a greater solubility in water thus can be used in synthesizing of nanoparticles with insoluble drugs [18]. Alkylated chitosan can be developed by introducing alkyl groups at C2-NH_2, C3-OH, and C6-OH positions. N-alkylation refers to the introduction of alkyl groups at the C2-NH_2 position, whereas alkylation at C3-OH/C6-OH positions is termed O-alkylation [18].

Another significant derivative of chitosan is the N-succinyl chitosan (SCS). Under mild conditions, it is fully water soluble, and is prepared by the addition of succinyl groups to the N-terminal of the glucosamine units of chitosan. Owing to its versatile properties, including solubility, high hydrophilicity, biocompatibility, biodegradability and antibacterial activity, SCS is widely used in the biomedical field for drug delivery applications [19]. Quarternization is the best strategy for elevating chitosan's solubility and charge density. The quarternization of the free amine groups of chitosan produces a highly charged polyelectrolyte which can maintain its water solubility in the entire pH range and it showed enhanced mucoadhesive properties. It also shows enhanced antibacterial activity which depends on the degree of quaternization, the quaternary chemical structure, and the molecular mass [20]. Trimethyl chitosan can be synthesized by the dissolution of chitosan in N-methyl pyrrolidone with methyl iodide using NaOH, followed by reductive alkylation and additional methylation reaction with methyl iodide. It exhibits excellent mucoadhesive properties and is soluble in alkaline and neutral pH conditions [18]. Hydroxypropyl chitosan (HPCS) is another quarternized derivative of chitosan. It is prepared by treating chitosan with propylene epoxide under basic conditions [21]. It has a higher water solubility, plasticity, and high affinity for keratin. It also exhibits wound-healing properties, and enhanced compatibility with human tissues, making it a suitable film-forming agent. Additionally, HPCS shows anti-coagulant activity. HPCS is used as a drug formulation technology in nail treatments [22]. Sulfonated chitosan has distinct properties such as antiviral, anticoagulant, antimicrobial, and osteogenic activity. It can be produced by attaching sulfonate groups to free amino groups in chitosan, introducing compounds with sulfonate groups, or sulfonating the hydroxyl groups, resulting in sulfamate, sulfonated, and sulfated products, respectively. It has the ability to bind with growth factors and aid in direct neural differentiation, tissue repair, and regeneration. Hence, sulfonated chitosan is a highly promising candidate for various biomedical applications [23]. Phosphorylated chitosan is a water-soluble chitosan derivative having various promising wound healing properties such as hemostatic properties, metal chelating ability, antioxidant, anti-inflammatory, bactericidal, and angiogenic property [24]. Phosphorylated chitosan is produced by adding phosphonic acid or phosphate

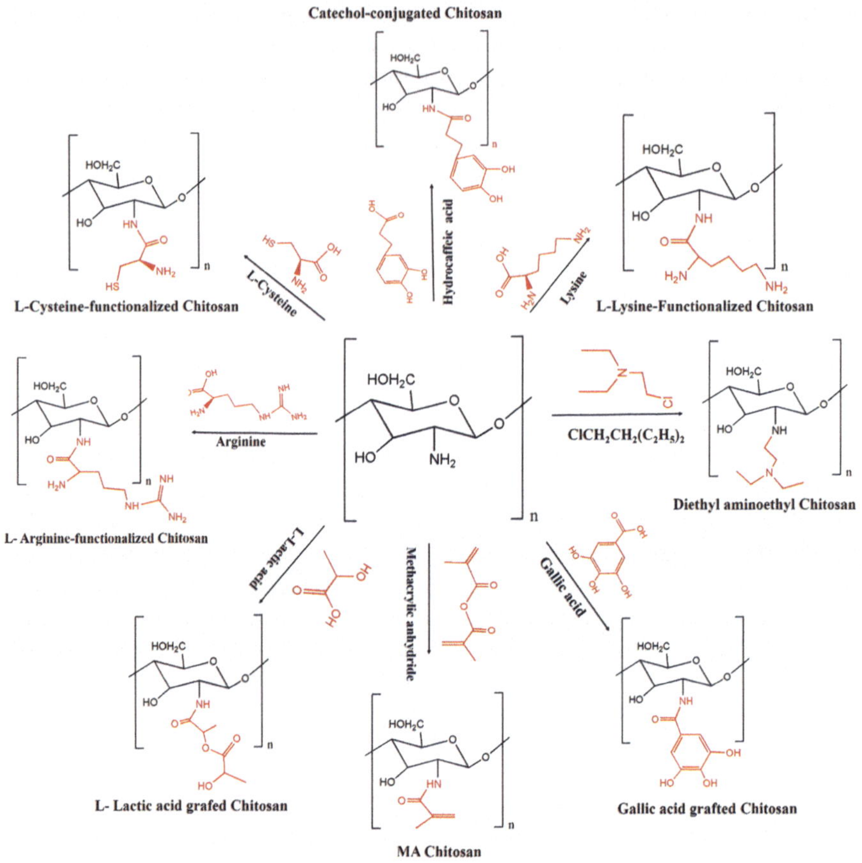

Fig. 6 Derivatives of Chitosan-II

groups onto chitosan's amino or hydroxyl groups. Additionally, phosphate derivatives of chitosan can be prepared through the interpolymer linkage of chitosan with tripolyphosphate or polyphosphate [25]. Moreover, the anionic parts in phosphorylated chitosan have a greater potential to modulate fibroblasts and endothelial cells. The amphoteric behavior of phosphorylated chitosan helps in cell adhesion and proliferation within extracellular matrix. Additionally, they can immobilize various growth factors near the cells. Thus, phosphorylated chitosan can play a significant role in promoting wound healing properties, as well as contributing to various biomedical applications [24]. There are many other derivatives of chemically modified chitosan that hold significant importance in the biomedical field (Fig. 6). Chitosan derivatives have great importance since they offer potential applications in antibacterial activity, sustained slow release, targeting, and delivery systems which are crucial for drug development in biomedicine [14].

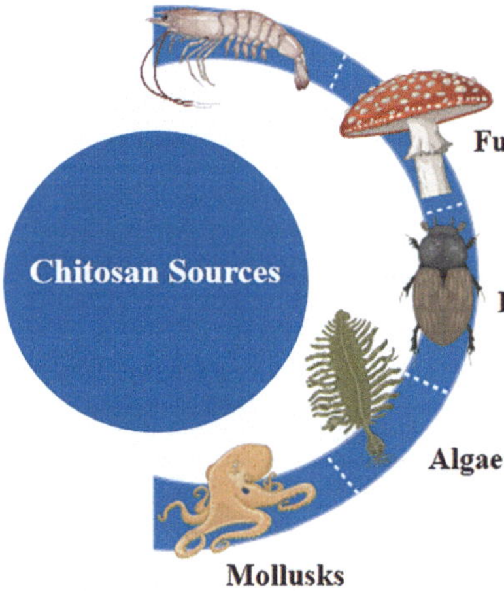

Fig. 7 Sources of chitosan

3 Main Sources and Preparation of Chitosan

3.1 Sources of Chitosan Extraction

Chitosan is a copolymer of glucosamine and *N*-acetyl glucosamine, derived from chitin [6], which is the second most abundant biopolymer in nature after cellulose [26]. Chitin is present in a variety of eukaryotes (Fig. 7) including exoskeletons of crustaceans, mollusks, algae, insects, and in the fungi cell wall. Chitin's presence in these species varies: 15–30% in crab cuticles, 20–30% in crustaceans, 30–40% in shrimps, 5–25% in insects, and 2–44% in cell walls of fungi. Currently, chitosan derived from crustaceans is used for industrial purposes due to the availability of crustacean waste from the food processing industry [6]. It is approximated about 2.0×10^{10} kg of chitin is produced annually by freshwater arthropods, and about 1.3 × 1,012 kg by marine arthropods [27].

3.2 Processing of Chitosan

Chitosan can be extracted from the crustacean by-products using chemical methods, or by biological methods (enzymatic and fermentation) [26]. The preparation of chitosan involves a series of chemical steps, including decalcification, deproteinization, decolorization, and deacetylation (Fig. 8). The shells of crustaceans are demineralized using weak acids like dilute HCl at room temperature as the first step, though alternatives like HNO_3, H_2SO_4, or CH_3COOH can also be used. Depending upon the source of chitosan, the concentration of acid and treatment conditions may vary. Deproteinization is carried out over a duration of 0.5 to 72 h at a temperature between 65 and 100°C using a mild NaOH solution. The deacetylation

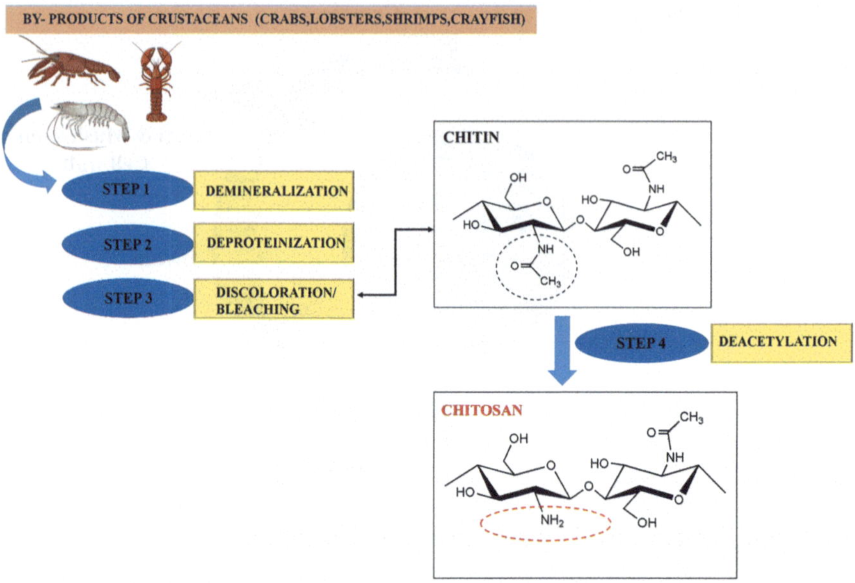

Fig. 8 Processing of chitosan from chitin

process is achieved by acetamide group hydrolysis using conc. NaOH or KOH at a temperature range of 100°C under heterogeneous conditions. Complete deacetylation of chitin produces chitosan with a deacetylation degree ranging between 40% and 13%, and chitosan's molecular weight is found within a range of 2×10^5 to 1×10^6 Da [28].

Enzymatic methods and fermentation methods also can be employed to produce chitosan from crustaceans. Enzymatic methods involve demineralization process similar to that used in chemical methods. The decalcification is carried out by the treatment with acid to remove the $CaCO_3$ in the shell. In this method, basic conditions and high reaction temperatures are replaced with enzymes for the deproteinization and deacetylation processes, which occur at milder temperatures ranging from 25 to 59°C [26]. Enzymes like papain, trypsin, alcalase, and pepsin are used during the deproteinization and demineralization stages [29]. Proteinases for enzymatic deproteinization are often isolated using microbes or fish entrails. Similarly, the enzymes for deacetylases are obtained from fish intestines or microbes. Nowadays modified microorganisms are also used as a source of enzymes for the extraction process [26]. Following demineralization and deproteinization, further steps are required, such as enzyme inactivation, followed by centrifugation. During centrifugation, the chitin is obtained in the form of a pellet. This pellet is then subjected to sequential washing with water, ethanol, and acetone, to obtain pure form of chitin. Chitin, upon deacetylation, yields chitosan [29]. However, enzymatic methods are less commonly used than chemical methods because the degree of deacetylation achieved via enzymatic processes is significantly lower, and they are also less effective at removing residual proteins from the shells during the deproteinization process [26].

The drawbacks of enzymatic methods are outdone by the fermentation process. They are lactic acid fermentation and non-lactic acid fermentation procedure. In the fermentation process using lactic acid, the microbial strains secrete lactic acid for the demineralization process. In the non-lactic acid fermentation process, other organic acids are used for the demineralization process [26]. Microorganisms like *Lactobacillus plantarum, Pseudomonas aeruginosa,* and *Bacillus subtilis* are commonly used in the fermentation method. The primary step of the fermentation process is the collection and grinding of shrimp. The ground mixture is combined with water and a suitable source of carbon followed by incubation at 37°C temperature with a suitable microbe required for the fermentation process. After 2–3 days of incubation, the filtrate is separated using filtration. The residue, which primarily consists of chitin, undergoes deacetylation process to produce chitosan. During the process of fermentation, the substances produced by the fermenting microorganisms play a key role in the demineralization steps [30]. For instance, in fermentation using lactic acid, lactic acid fermenting bacteria treat calcium carbonate which is found in the shrimp waste leading to the formation of calcium lactate. The formed calcium lactate is then removed through a washing process. The removal of proteins from chitin waste occurs through a process known as autolysis, which is commonly observed in fish and shrimp waste treatment [29].

4 Chitosan Characteristics

4.1 Molecular Weight

The molecular weight (MW) of chitosan is another significant factor that affects its bioactivity [26]. Chitosan is a glycosaminoglycan consisting of a number of sugar units per polymer molecule (n) that helps in the determination of its MW and plays a significant role in its properties. It also influences the crystal size and morphological character [31]. Prior studies have shown different MW cut-off values to differentiate between high, medium & low MW chitosan and oligosaccharide of chitosan. Regardless of the specific MW cut-off values, the bioactivity of chitosan typically exhibits greater efficacy with low MW (<20 kDa) compared to higher MW (>120 kDa) [32]. The deacetylation process of chitosan can result in modifications to its MW, which may vary depending on the source and preparation procedure [31]. The relationship between MW and solubility of chitosan in water follows a similar pattern to the relationship between MW and biological properties, with lower MW generally leading to greater solubility of the carbohydrate molecule [33].

4.2 Deacetylation Degree

Degree of deacetylation (DDA) quantifies the acetyl content within chitosan. The versatility of chitosan is due to the presence of the amino group in its structure. To enhance the presence of amine content and to achieve greater deacetylation levels, chitosan (DDA > 90%) undergoes multiple alkaline treatments [31]. Studies showed that the DDA of chitosan affects its solubility in acidic solutions and its crystalline behavior. An increase in DDA leads to higher charge density along the chain. As the DDA increases, the chitosan chain becomes more flexible, favoring the formation of a coil structure with intramolecular hydrogen bonding [34]. DDA has a crucial part in cell adhesion and cell proliferation, even though it cannot alter the biocompatible properties of chitosan. Lower DDA levels support cell adhesion and cell growth [35].

5 Chitosan Properties

Chitosan possesses multiple physiochemical and biological properties (Fig. 9). The versatile nature of chitosan arises by its cationic property and distinct behavior in solutions, particularly its solubility in acidic conditions. At low pH levels, typically below 6.3, the amine groups of chitosan become protonated, imparting a polycationic behavior to the polymer, while at higher pH levels (above 6.3), amine groups of chitosan are deprotonated and increase the nucleophilicity of the amine groups, enhancing their ability to participate in chemical reactions by readily donating pairs of electrons to form new bonds with other molecules [1].

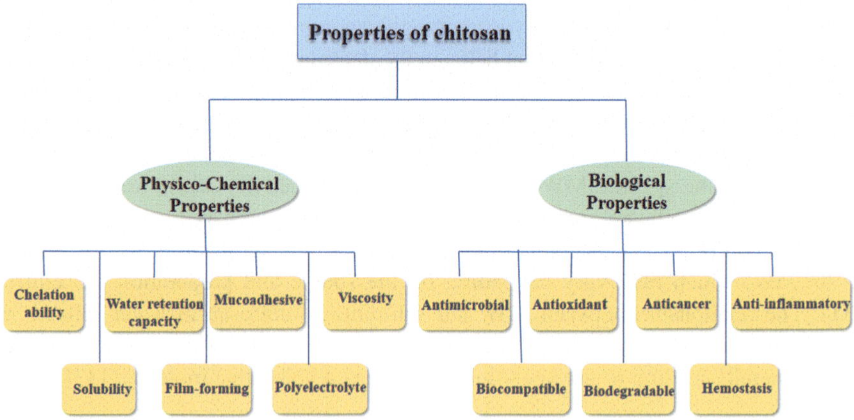

Fig. 9 Properties of chitosan

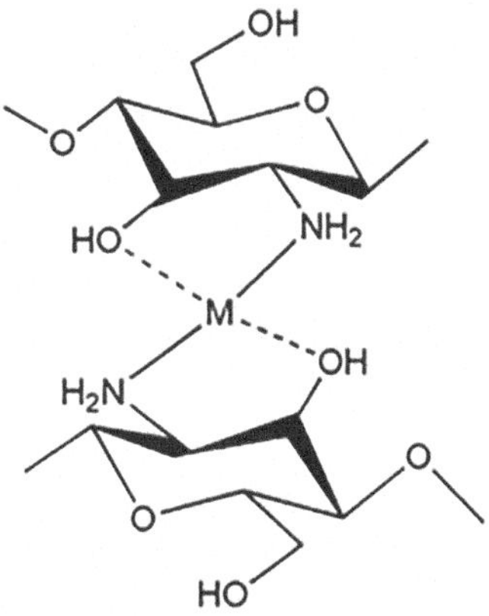

Fig. 10 Metal chelation interaction of chitosan with metal via inter or intramolecular complexation

5.1 Physicochemical Properties

Chitosan is renowned for its various unique properties including chelating ability, solubility, viscosity, film-formation, mucoadhesive property, and polyelectrolyte behavior [3].

5.1.1 Chelation

Chitosan has strong chelation ability [33]. In acidic pH, chitosan can exhibit chelating of several metal ions [3] and also it can form chelates with proteins, amino acids, cholesterol, and fats [36]. At near-neutral pH, chelation occurs through the free amino groups of chitosan or through electrostatic interactions on protonated amino groups in acidic solutions (Fig. 10) [3]. Chitosan also has a greater commercial interest because of its high percentage of nitrogen, 6–7%, which renders it an efficient chelating and complexing agent [1].

5.1.2 Solubility

Another significant property of chitosan is its solubility (Table 1). Chitosan dissolves in all aqueous acidic solutions when its average DDA falls below 50% and is soluble in dilute inorganic acids including HCl, HBr, HNO_3, and $HClO_4$, as well as in conc. H_2SO_4. It is also soluble in organic acids like acetic, citric, lactic, formic and in organic solvents such as tetrahydrofuran, ethyl acetate, and 1,2-dichloroethane. However, formic acid and acetic acids are most commonly used in the solubilization process of chitosan [1]. Moreover, in its crystalline form, chitosan is typically insoluble above pH 7 in aqueous solutions [37]. The chitosan's solubility is influenced by the pH-responsive amine groups. In lower pH, these amino groups of chitosan become protonated and permit to form cationic polysaccharides that are

Table 1 Solubility data of chitosan in various solvents

Solvent	Chitosan
H_2O	Insoluble
NaOH (1%)	Insoluble
Acetic acid (1%)	Soluble
HCl (1%)	Soluble
Dichloromethane	Insoluble
Pyridine	Insoluble
Ethanol	Insoluble
Dimethyl acetamide	Insoluble
Dimethylsulfoxide	Insoluble
Dimethylformamide	Insoluble
Acetone	Insoluble
Formic acid (1%)	Soluble
Lactic acid (1%)	Soluble
Citric acid	Soluble
Tetrahydrofuran	Soluble
Ethyl acetate	Soluble
1,2-Dichloroethane	Soluble

soluble. Whereas, at a pH above 6, the amino group becomes deprotonated, making chitosan insoluble [3]. Controlling the solubility of chitosan is challenging since it depends on various factors including DDA, the ionic concentration, the pH of the solution, the type of acid used for protonation, distribution of acetyl groups along the macromolecular chain, and also production conditions of chitosan [1].

5.1.3 Viscosity

The unbranched and linear form of chitosan was found to have a remarkable viscosity [3] and it can be used to determine polymer stability in solution [13]. Chitosan's viscosity in solution is impacted by several factors including DDA, concentration, ionic strength, pH, MW, and temperature. Typically, as the temperature rises, the viscosity of the polymer solution tends to decrease [38] and it also decreases with a reduction in the MW of chitosan [13]. Viscosity can vary depending on the strength of acids, in acetic acid, chitosan's viscosity increases, whereas in hydrochloric acid (HCl), it decreases [38]. Chitosan can be used as a viscosity-enhancing agent due to its high MW and linear, unbranched structure [28]. It can also act as a viscosity-controlling agent in cosmetics since it can alter product viscosity by changing polymer concentration and MW [39].

5.1.4 Film-Forming

Chitosan also exhibits film-forming properties, making it an ideal candidate for developing food coatings and packaging [40]. The films developed by the acid dissolution of chitosan can exhibit various beneficial properties. These properties can vary widely concerning the degree of acetylation, solution viscosity, drying temperature, and the concentration of acid used in the dissolution process [41]. Chitosan film is extensively used to prolong the shelf life of food and the incorporation of Ca^{2+} ions adjusts the permeation rate of CO_2 and O_2 through the chitosan membranes, enhancing the shelf life of fruits [42]. The film formation capability of chitosan on the surface of carbon steel creates a protective layer that hinders the initial bacterial attachment, thereby preventing biofilm formation and reducing corrosion [43].

5.1.5 Mucoadhesive

Chitosan, a positively charged polysaccharide is known for its mucoadhesive properties and has been broadly utilized as mucoadhesive drug delivery agents [44]. Chitosan contains –OH and –NH_2 groups, allowing it to form hydrogen and covalent bonds that are crucial in the mucoadhesion property of chitosan. The NH_2 groups in the structure of chitosan have a positive charge at pH < 6.5. This cationic nature facilitates greater electrostatic interaction with negatively charged components of mucus, such as sialic acid and epithelial surfaces (Fig. 11) [45]. Additionally, the mucoadhesive nature of chitosan is significantly affected by hydrophobic and hydrogen bonding interaction [18]. Many commercially available chitosans exhibit strong adhesions in vitro and these interactions between chitosan with mucus and its mucoadhesive properties are influenced by both physiological factors and the physicochemical characteristics of chitosan [46]. Chitosan derivatives have enhanced physicochemical and pharmaceutical properties. For instance, thiolated

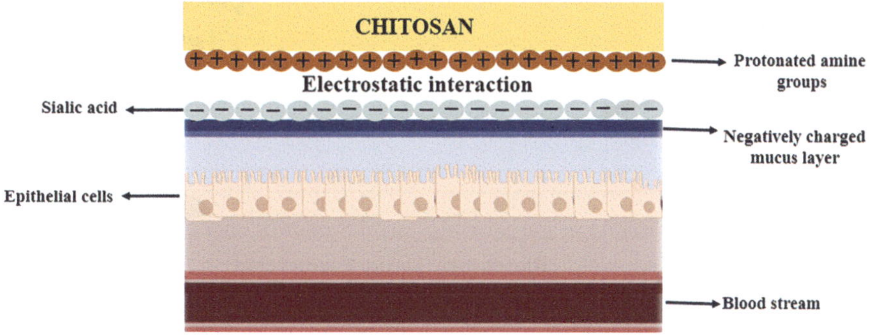

Fig. 11 Mucoadhesive behavior of chitosan

chitosan exhibited greater mucoadhesive strength as compared to unmodified chitosan due to its ability to form disulfide bridges with mucus glycoproteins [44].

5.1.6 Polyelectrolyte Behavior

A notable physicochemical feature of chitosan is the polyelectrolyte behavior. The inherent ionic property of chitosan is utilized to create different complexes in biomedical field. Due to the cationic nature of chitosan, it can form many polyelectrolyte complexes with many polyanionic polymers, such as chitosan-hyaluronic acid, chitosan-alginate, and chitosan-pectin complexes [47]. The positively charged amino groups at the C2 position of the repeating glucopyranose units within chitosan can interact electrostatically with the anionic groups often carboxylic acid groups of other polyions resulting in the development of polyelectrolyte complexes (Fig. 12) [48]. The amine groups of chitosan are protonated in solution, causing the formation of a positively charged polyelectrolyte. Its behavior in solution is significantly influenced by factors such as pH, ionic strength, salt addition, and DDA [47].

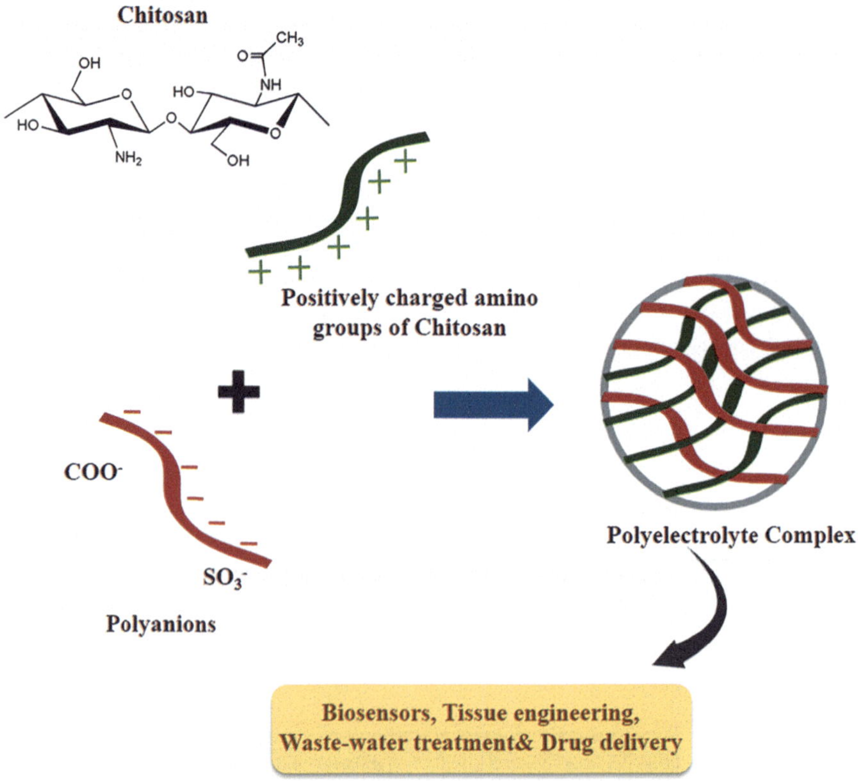

Fig. 12 Schematic representation of polyelectrolyte behavior of chitosan

Complexes formed by combining chitosan with anionic polymers have various applications including biosensors, tissue engineering scaffolds, waste-water treatment, and drug delivery [48]. For instance, the quaternized carboxymethyl chitosan, oxidized hyaluronic acid, and 3,3′-dithiobis-(propionohydrazide) hydrogel have shown excellent drug release properties [49]. Additionally, scaffolds fabricated from chitosan-alginate polyelectrolyte complexes crosslinked with Ca^{2+} are considered as potential candidates for the regeneration of skeleton muscle tissue [50].

5.2 *Biological Properties*

Chitosan and its derivatives display remarkable properties like antimicrobial, antioxidant, anti-inflammatory, and hemostatic potential [3]. Due to this exceptional behavior of chitin, chitosan, chitooligosaccharides, and chitosan derivatives, they hold significant potential in biomedical applications [13].

5.2.1 Antimicrobial

Chitosan and chitosan derivatives exhibit substantial antimicrobial property. The antimicrobial activity of chitosan is primarily due to its cationic behavior and is influenced by factors such as MW, DDA, chelation, and both hydrophilic and hydrophobic behavior [18]. The antibacterial effect of chitosan occurs because it binds to bacterial DNA and interacts with surface molecules, inhibiting the synthesis of mRNA (Fig. 13) [51]. The mechanism of action of chitosan in combating

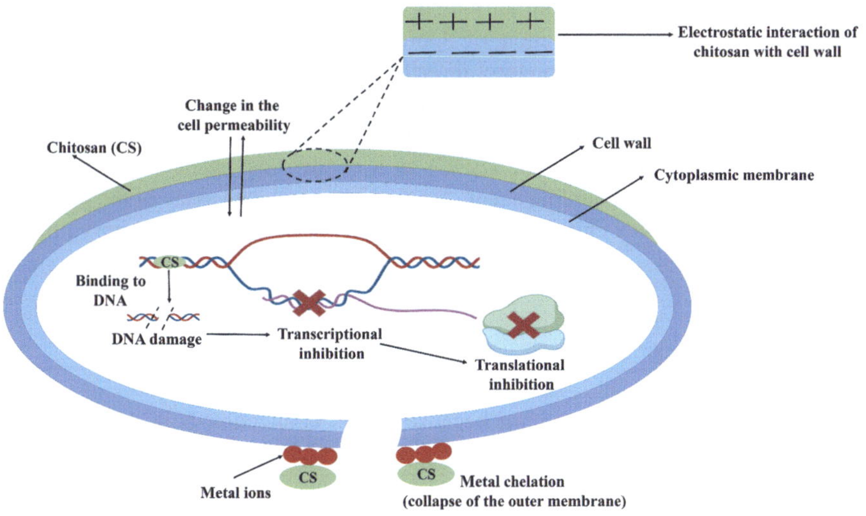

Fig. 13 Illustration of antimicrobial mechanism of chitosan

microbes can be classified as intracellular and extracellular, or a combination of both, depending on the specific location targeted [52]. Since chitosan with higher MW is unable to penetrate the cell wall & the cell membrane, it exerts antimicrobial effect through blocking extracellular nutrient uptake from cells and altering cell permeability. Whereas, the low MW chitosan exhibits both intracellular and extracellular processes, impacting mitochondrial function and thereby disrupting RNA synthesis and protein translation process. Teichoic acid and lipopolysaccharide which are present in gram-positive and gram-negative bacteria respectively play an essential role in binding to the chitosan and leading to disruption of cell membrane function by interfering with the cell walls [51]. Damage to the microbial membrane arises due to electrostatic interactions, causing release of intracellular substances. The concentration of chitosan is highly crucial because, when its amino groups interact with the negatively charged bacterial surface, it results in bacterial agglutination [53]. Chitosan and its derivatives exhibit antimicrobial effect against various microbes [13]. Many studies showed that the quaternization process enhances antimicrobial effectiveness against gram-positive and gram-negative bacteria. Quaternary chitosan derivatives exhibited increased antimicrobial effect against *E. coli* and *S. aureus* when formulated with 2-hydroxypropyltrimethyl ammonium chloride [54]. Sadeghi et al. compared the antimicrobial activities in which trimethyl chitosan exhibited maximum inhibition against *S. aureus* compared to N-diethyl methyl chitosan [55]. In another study, an increase in antibacterial activity was observed when carboxyl groups are added as seen in *O*-carboxymethyl chitosan [56]. Chitosan-based gel sponge when developed using tris (hydroxymethyl) methyl glycine-modified chitosan derivative (CTMG) exhibited notable antibacterial activity [57]. Chrysin-loaded polycaprolactone-chitosan (PCL-CTS) nanofibers prepared by electrospinning method hold promise as a potential wound dressing material with efficient antimicrobial activity [58]. It also exhibits antifungal effects against molds and yeasts. These properties are attributed to its fungistatic rather than fungicidal tendencies, hindering germination of spore and growth. Its mode of action is influenced by the cell wall structure, directly disrupting the growth. Additionally, chitosan is anticipated to act more on fungi than bacteria [59]. In another study benzoic acid grafted chitosan derivative was prepared using 1-(3-dimethyl aminopropyl)-3-ethylcarbodiimide (EDC) as a condensing agent and N-hydroxy succinimide (NHS) as a coupling agent. A mycelial growth rate assay of the prepared derivatives revealed that chitosan with benzoic acid has enhanced inhibitory impact on the fungi [60].

5.2.2 Antioxidant

Antioxidants are gaining attention because of the connection between oxidative stress and various diseases [61, 62]. Oxidative stress induces harm to cellular macromolecules and unintended enzyme activation, causing disorders. Chitosan and its derivatives counteract oxidative damage and scavenge free radicals by disrupting the radical chain reaction involved in oxidation (Fig. 14) [63]. This action

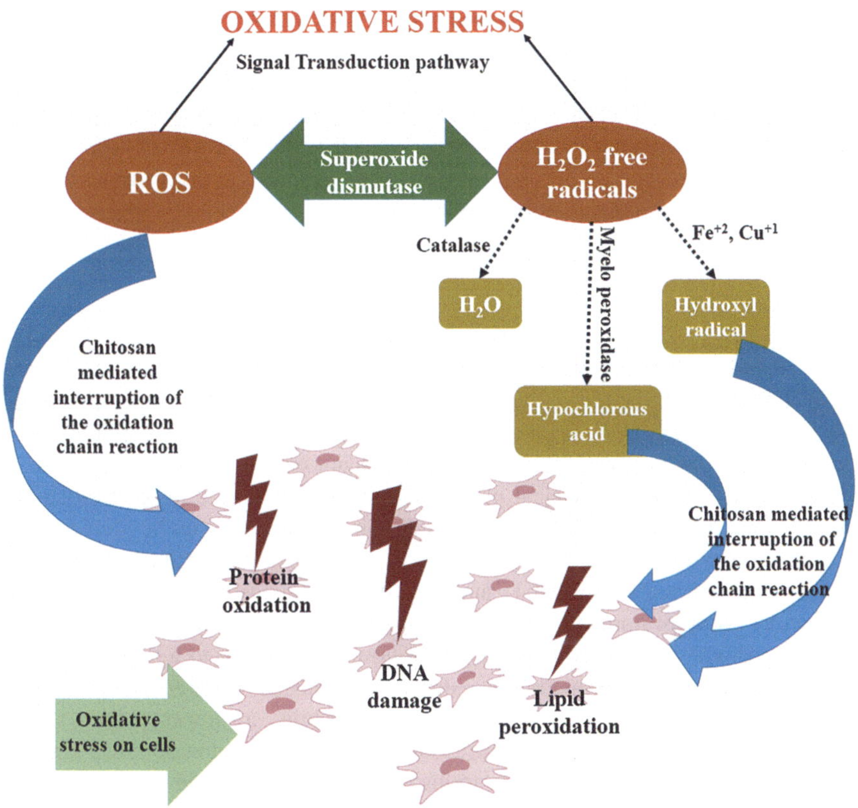

Fig. 14 Antioxidant mechanism of action of chitosan

enhances antioxidant enzyme activity and inhibits lipid peroxidation, thereby increasing the body's antioxidant capacity and potentially delaying the aging process [64]. The antioxidant property of chitosan involves the scavenging of oxygen radicals. This function is affected by DDA and MW of chitosan. Those chitosan oligomers with greater deacetylation levels serve as scavengers for alkyl, hydroxyl, and superoxide radicals [18]. Antioxidant effectiveness is more pronounced in low MW than in higher MW samples. This happens because smaller chains form less intramolecular hydrogen bonds, which increases the availability of reactive groups and enhances radical scavenging ability. Regarding the degree of acetylation, the antioxidant property tends to reduce with an increase in the deacetylation degree [65]. The antioxidant property is due to the functional groups present in the structure. The hydroxyl groups within the structure are available to interact with OH radicals via hydrogen substitution, while OH radicals react with any NH_2 groups to create stable macromolecular free radicals. Those NH_2 groups, when exposed to aqueous solutions, get transformed into NH^+ and then undergo additional reactions with -OH. Thus, the chitosan could exhibit efficient antioxidant properties [66]. Novel chitosan derivatives containing glycine Schiff bases showed a greater antioxidant capacity and free radical scavenging ability with and 100% scavenging capacity for superoxide radical [67]. Chitosan derivatives bearing sulfhydryl group (-SH) have antioxidant activity. The DPPH radical scavenging activity assay and superoxide anion radical scavenging activity assay showed that the derivative has enhanced antioxidant activity, and the derivative with a greater concentration of free sulfhydryl groups showed more potent antioxidant activity [68].

5.2.3 Anti-Inflammatory

Chitosan and its derivatives exhibit anti-inflammatory characteristics. According to the WHO reports, three out of every five individuals experience serious inflammatory conditions such as heart attacks, chronic pulmonary diseases, cardiovascular illness, obesity, and hyperglycemia [69]. It is a major part of the body's response to certain stimuli including physical injury, microbial infection, ultraviolet radiation, and immune responses. However, extended inflammation can be detrimental, playing a role in the development of various diseases, including asthma, rheumatoid arthritis, and multiple sclerosis [64], and the inflammatory responses are mediated by a particular group of immune cells and inflammatory cells within a tissue, aiming to restore structural and operational stability following exposure to a harmful trigger [18]. Anti-inflammatory properties also depend on chitosan's MW. At a higher MW, chitosan exhibits anti-inflammatory activity while a lower MW shows pro-inflammatory activity [70]. These results indicate that chitosan shows encouraging anti-inflammatory effects by various mechanisms, including regulating cytokine expression, scavenging free radicals, and controlling immune responses [71]. Various studies have investigated the anti-inflammatory and pro-inflammatory properties of chitosan and its derivatives [72]. Nguyen et al. synthesized a hydrogel derived from chitin and glucosamine-derived polysaccharide, which synthesizes chitosan hydrogels as a sophisticated drug delivery system with anti-inflammatory properties. The properties of chitosan hydrogel such as pH sensitivity, temperature sensitivity, electric sensitivity, and magnetic strength contributed to improving the anti-inflammatory properties of chitosan [73]. In one of the study, chitosan-alginate NPs exhibited strong antimicrobial activity against *P. acnes* and showed anti-inflammatory properties by suppressing the production of inflammatory cytokine IL-12 production in human monocytes and IL-6 production in human keratinocytes [74].

5.2.4 Hemostatic

Chitosan is well known for its hemostatic applications. Uncontrolled hemorrhaging during surgery, and trauma is crucial and demands immediate action. Chitosan derived from natural sources and possessing distinctive biological properties is frequently used as a local hemostatic agent [75]. Chitosan promotes clotting without

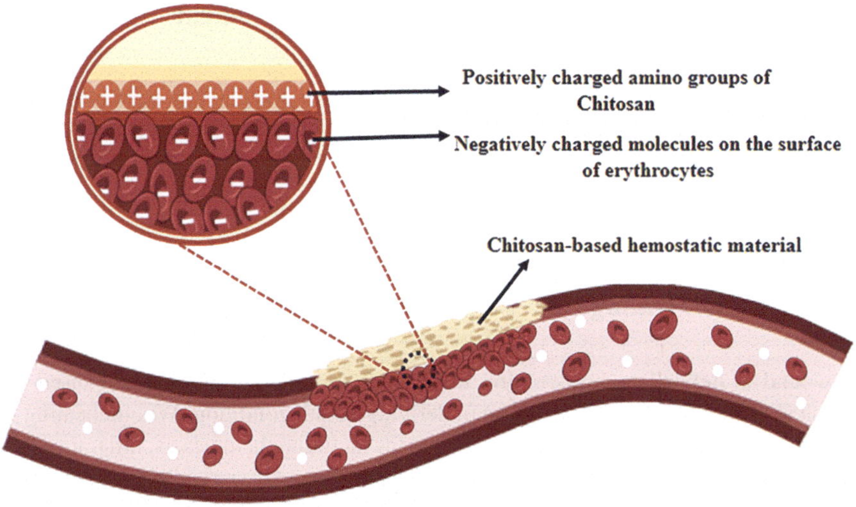

Fig. 15 Hemostatic mechanism of chitosan-based material

triggering the intrinsic pathway, indicating that the hemostatic property of chitosan functions independently. This feature of chitosan holds particular interest since it can help in stopping bleeding in patients with coagulation disorder [76]. The hemostatic property of chitosan is significantly aided by its positive charges, particularly because erythrocyte surfaces are negatively charged [77]. The positive surface charge was able to interact with blood components that carry a negative charge, which helps in stimulating platelet activation and aggregation of blood proteins to initiate the formation of fibrin clots [78]. The interaction between chitosan's positive charges and the negative charges on erythrocytes results in the development of a "mucoadhesive barrier" at the wound area and effectively prevents blood loss (Fig. 15) [77]. The MW and DDA have also a significant effect on the clotting properties of chitosan. Increased deacetylation leads to greater aggregation of erythrocytes and platelets, which is essential for blood clotting [76]. A hemostatic bandage of chitosan nonwoven fabric composed of calcium alginate and adenosine diphosphate showed a higher hemostatic potential compared to commercial products [79]. Yu et al. prepared a hemostatic sponge of chitosan and hydroxypropylmethyl cellulose by lyophilization technique. The developed bandage showed improved rigidity as that of pure chitosan sponges and exhibited excellent hemostatic properties [80].

6 Chitosan-Based Materials and Their Biomedical Applications

Chitosan has unique properties with better biodegradability, biocompatibility, and hemostatic potential which makes it highly suitable for numerous biomedical applications (Fig. 16) [81].

6.1 *Wound Healing*

A wound is defined as a disruption in the skin or mucous membrane caused by physical or thermal damage, or by an underlying health condition. Wound healing is a complex process that involves a series of steps including hemostasis, inflammation, proliferation and remodelling aimed at restoring damaged tissue's integrity. A variety of wound dressings are available to promote the healing procedure [82]. Chitosan has a significant part in every phases of wound healing. It interacts with many cellular processes during the stage. In the early stages of healing, chitosan exhibits hemostatic properties and helps in the infiltration and migration of neutrophils and macrophages thus leading to granulation tissue formation, which allows fibrous tissue formation and re-epithelialization. It also affects the levels of growth factors involved in the wound-healing process. Thus chitosan serves as an excellent agent in promoting wound healing [83].

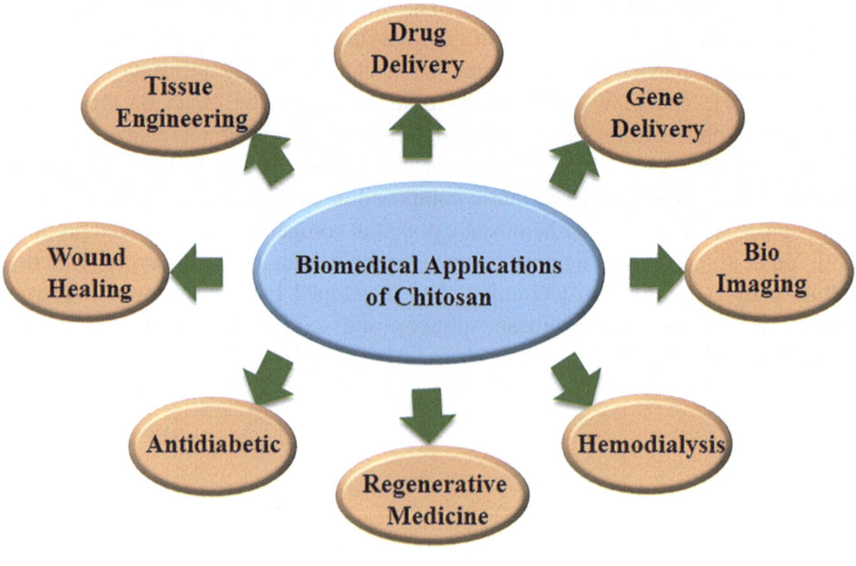

Fig. 16 Biomedical applications of chitosan

Amikacin sulfate and diopside nanoparticles incorporated chitosan dressing scaffold was developed for treating infectious wounds. The prepared chitosan composite has a porous structure which helps in cell infiltration and cell proliferation. The expansion capacity of the scaffold was determined by the swelling studies when applied to the wound area. Drug release from the scaffold demonstrated a sustained release over 7 days. The prepared scaffold exhibited antibacterial efficacy and antibiofilm property. Further studies revealed the cytocompatibility and enhanced attachment of fibroblast cells. Moreover, a scratch assay involving fibroblast cells confirmed the scaffold's ability to facilitate cell migration. Therefore, the developed dressing can act as a promising biomaterial for treating infectious wounds [84]. Skin diseases are among the most widespread diseases in humans and are a major global public health issue. Skin infections are caused by specific pathogens; typical skin conditions include bacterial infections, fungal and yeast infections, and viral infections [85]. For treating skin infections, Pandian et al. developed and characterized a chitosan bandage incorporating octenidine dihydrochloride. The study assessed the anti-biofilm effects of the bandage on *S. aureus* and *C. auris* biofilms, showing a significant reduction in biofilm formation after treatment with the prepared bandage. This research highlights the importance of integrating antimicrobial agents like octenidine dihydrochloride into wound dressings to enhance their therapeutic potential [86]. In another study chitosan bandage incorporated with ciprofloxacin and fluconazole for treating polymicrobial infections was developed. The prepared bandage showed excellent swelling ability and porosity aiding in the absorption of wound exudates effectively and protecting the skin around the wound from excessive moisture. The controlled release of drugs from the dressing helps in attaining an environment devoid of microbes at the wound site for an extended time. Cell culture studies demonstrated that the prepared bandage is biocompatible with human dermal fibroblast cells. Antimicrobial studies exhibited excellent antibacterial property. The in vivo studies using an infected rat wound model revealed a significant reduction in the microbial population at the infected wound site [87].

A multiple-functional wound dressing incorporating lidocaine hydrochloride and mupirocin into chitosan/polycaprolactone using an electrospinning technique was developed and characterized. The nanofibrous structure of the scaffold helps in the interaction between blood cells and demonstrates good blood clotting capacity. The prepared scaffold showed antibacterial activity thereby helping in the enhancement of the wound healing process. Hence, these nanofibrous scaffolds serve as a promising prospect in future clinical aspects [88]. Dual drug-loaded electrospun nanofibers can show a greater therapeutic efficacy on wounds. Poly (vinyl alcohol)/chitosan nanofibers loaded with ciprofloxacin and rutin hydrate were developed and characterized. Dissolution studies were conducted to examine the drug release profile from the nanofibers. Studies showed a greater antibacterial efficacy and enhanced wound healing in full-thickness excisional rabbit wounds. Hence, the fabricated nanofibers have the effective capability to deliver drugs and help in wound healing [89].

Chitosan-poly (hexamethylene) biguanide hydrogel for infectious wound healing was developed and reported. SEM characterization showed a smooth morphology and the FTIR peaks validated antiseptic presence in the hydrogel. The study on water vapor permeability validates the water retention power of the hydrogel and showed greater antibacterial activity. Hence the study showed the developed hydrogel is a potent candidate for treating infectious wounds [90]. Pandian et al. synthesized in-situ silver nanoparticles-incorporated *N,O*-carboxymethyl chitosan hydrogel with ethylenediaminetetraacetic acid-ferric ion complex. The developed hydrogel exhibited injectability and biocompatibility and has antibacterial, anti-biofilm efficacy. Further, SEM and FTIR characterization confirms the presence of silver nanoparticles in the prepared hydrogel. Therefore, the prepared antibacterial and self-healing hydrogel could be utilized for treating infectious wounds [91].

Tigecycline/chitosan-platelet-rich plasma (PRP) composite hydrogel was prepared and characterized. Considering the current treatment strategies the prepared hydrogel has numerous benefits due to its properties such as its adhesiveness in the infection site, less frequent dressing changes, controlled drug release, bacterial growth prevention, and enhanced healing. The ionic cross-linking method was used to prepare tigecycline nanoparticles. The prepared nanoparticles along with activated PRP powder and chitosan hydrogel together form a composite hydrogel which is injectable in nature. The rheology study showed good thermal stability, injectability, and shear-thinning property. The drug release kinetics showed a sustained release of the tigecycline, and the system was found to be cytocompatible. Antibacterial study against *S. aureus* showed enhanced antibacterial activity. In vivo study using drosophila melanogaster model showed enhanced antibacterial activity against *S. aureus.* Hence this prepared hydrogel acts as a suitable medium for the delivery of antibiotics and can be used efficiently to treat skin diseases caused by *S. aureus* [92].

6.2 Drug Delivery

Process of administering pharmaceutical agents into the body to achieve a therapeutic effect is termed "drug delivery" [93]. Chitosan-based materials are extensively studied for oral and topical drug delivery, colon-targeted therapy, gene delivery, and cancer treatment due to their biocompatibility, capacity to interact with bioactive compounds, protect drugs from degradation, improve drug absorption on mucosal tissues, and enable electrostatic binding with DNA and other biomolecules [94]. Chitosan's repeating glycosidic residue contains a 1° amino group that serves as a bonding site, giving chitosan a positive charge. Thus the positive charge provides chitosan pH sensitivity and enables them to create ionic complexes and regulate their release. Chitosan can be modified into several dosage forms because of its adaptability, unique properties, and applications and thus it can act as a suitable candidate for various drug delivery applications [95].

Staphylococcus aureus (S. aureus) is a type of gram-positive cocci commonly existing in our skin and mucous membranes which cause various skin diseases. Another study reported the fabrication of Lysostaphin (LST) incorporated chitosan hydrogel with anti-*staphylococcal* property. Lysostaphin is a bacteriocin which helps in the breaking down of pentaglycine cross bridge within the peptidoglycan of *S. aureus*. From the molecular docking studies, atomic interactions of LST with chitosan were evaluated. The drug release study conducted showed a gradual release of the enzyme over time, with nearly 90–95% of the enzyme being released within 24 h. The hydrogel exhibited a greater antibacterial activity against *S. aureus* by agar well diffusion assay. The study showed inhibition of biofilm formation with a gradual increase in LST concentration using SEM analysis. Thus, the developed LSTC gel shows great promise in drug delivery application for preventing *staphylococcal* infections [96]. Chitosan–gelatin/zinc oxide nanocomposite hydrogel scaffolds (CS–GEL/nZnO) were developed with antibacterial and drug delivery properties. Drug release studies showed that zinc oxide nanoparticles (nZnO) effectively controlled the release of naproxen, sustaining the release for up to 2 days. The prepared hydrogel showed an enhanced antibacterial activity. Quantification of the antibacterial activity showed that greater the nZnO concentration, greater the antibacterial activity. Therefore, naproxen with an antibacterial agent (nZnO) could be very helpful in antibacterial and drug delivery applications [97].

A chitosan-g-Poly (acrylic acid-co-acrylamide) drug carrier for gradual release of curcumin was developed by Kumar et al. The potential of the developed novel drug carrier was studied using pH-regulated kinetic models and in vitro antibacterial assay. Kinetic data showed that there was a greater release observed for curcumin at the range of pH 5.4. The antibacterial studies promote the fact that the grafted chitosan can show significant antibacterial activity than commercial tetracyclines. Thus the developed novel drug carrier can serve as an economical and ecofriendly approach in pharmaceutical field for sustained drug delivery applications [98]. Curcumin-loaded methyl methacrylate (MMA)-modified chitosan (CS) conjugate (CSMMA) was developed using the green method. The developed CSMMA acts as a better gene delivery agent in the transfection efficiency. The results showed that chitosan covalently binds to MMA and forms a heavy porous material. Drug release studies showed a burst release of curcumin initially that is because of the presence of entrapped drug near the surface of nanoparticles. The drug release rate was high at a pH (5.0) which is lower than physiological pH and a maximum of 68% drug entrapment efficiency was found for curcumin-loaded CSMMA nanoparticles. This evidently exhibited that the produced CSMMA nanoparticles can be an efficient target drug delivery system in biomedical field [99].

6.3 Tissue Engineering

In recent years, biomaterial-based tissue engineering has gained significant urge in the healthcare system [100]. Tissue engineering is a multidisciplinary field focused on restoring or regenerating injured tissue or organs by using cell culture [101]. Chitosan can be modified into distinct forms and has a greater capacity to make porous structures for tissue engineering applications. Different compositions were utilized to develop a scaffold with the desired pore size and mechanical properties for various tissue engineering applications [94]. Due to the excellent cell adhesion and proliferation property of chitosan-based scaffolds, it is more used in the bone tissue engineering field. This behavior is related to the scaffold's physical characteristics and its electrostatic interactions, due to the cationic nature of chitosan, with various compounds like cytokines and growth factors [94].

A metal oxide-based biomimetic nanocomposites with chitosan-sodium alginate and TiO_2 nanoparticles was prepared using solvent casting technique. The biocompatibility studies revealed that the biomimetically engineered nanocomposites showed improved suitability for bone tissue engineering applications. Characterizations were done using XRD, PSD, FTIR, and TEM. Conductivity and swelling studies, along with the hydrophilic nature of nanocomposite confirmed using contact angle analysis and they exhibited an enhanced antibacterial activity. Thus, these studies revealed that the developed polymeric nanocomposites possess good physico-chemical and mechanical properties with superior biocompatibility, making them suitable for bone tissue engineering applications [102]. A composite scaffolds of chitosan fortified with mesoporous zinc silicate (mZS) was developed and characterized. Zinc is well known for its antibacterial, osteogenic, and angiogenic potential and silica-based materials are bioactive and help in bone formation and mineralization. Biomineralization studies showed the bioactivity of fabricated scaffolds showed excellent antibacterial activity. Cell attachment within the scaffold network exhibited a greater biocompatibility. The incorporation of mZS showed high porosity, excellent swelling & improved biomineralization, and degradation. Thus the fabricated scaffolds have exceptional properties and have the potential to serve as a template/implant for BTE applications [103].

Cartilage tissue engineering methods have been regarded as a promising approach for repairing articular cartilage [94]. Hydrogel scaffolds for cartilage tissue engineering were prepared by chitosan (CS) and hyaluronic acid (HA) substituted with phenol moieties. The gelation time and water contact angle of the prepared hydrogels were studied using different concentrations of CS and HA and it reduces with increase in the HA content. The addition of HA improves its rheological properties and entanglement density. The prepared hydrogel scaffold exhibited proper cell viability and cell proliferation and demonstrated a higher tendency of MSCs to differentiate into cartilage-like cells. Thus biomimetic hybrid hydrogel constructs are considered highly effective at replicating the cellular environment and could be crucial in advancing cartilage scaffold construction [104]. Jalageri et al. prepared a graphene oxide blended polyvinyl alcohol/chitosan composite hydrogel for articular cartilage tissue engineering application. The engineered hydrogels showed an exceptional porous nature and incorporation of graphene oxide helps in the enhancement of mechanical properties and storage modulus. These studies proved that the composite hydrogel has elastic behavior which resembles the properties of tissues found in mammals. Further, the antibacterial studies showed an enhanced inhibition rate with a higher amount of graphene oxide also the hydrogel showed reduced toxicity and greater cell viability, indicating the suitability of the hydrogel for cartilage regeneration. Hence, the developed composite hydrogel shows significant potential as next-generation biomaterials for applications in articular cartilage [105].

6.4 Hemostasis

Chitosan-based hemostatic agents have become popular for their ability to effectively control bleeding when applied directly to the bleeding site. Chitosan contains protonated NH_2 groups that promote the aggregation of blood cells upon contact with blood, leading to the formation of a clot at the site of bleeding [106]. Chitosan composite hydrogel (CS-nBG) that helps in the enhancement of blood clotting properties of chitosan was prepared by incorporating nanobioglass (nBG) for hemostatic applications. The incorporation of both CS and nBG is intended to facilitate fast blood clotting, which is essential for the effective control of bleeding in crucial situations. nBG contains ions silica, calcium, and phosphate that activate coagulation factors and pathways. The presence of chitosan and nBG was analyzed using SEM and FTIR techniques. Rheological analysis indicated the prepared hydrogel has a greater storage modulus and solid-like behavior. Cytocompatible nature of the prepared composite hydrogel was studied using HUVEC cells. Hemocompatibility study exhibited less than 2% hemolysis rate which falls within the acceptable range for biomaterial with human whole blood. In vitro blood coagulation study took blood clotting time of 212 ± 21 s which is lower than the control Cs and blood alone samples because of the blood cell aggregation by amine group of CS and simultaneously silica, calcium, and phosphate ions of nBG can activate coagulation cascade. In vivo blood clotting studies of the CS-nBG composite hydrogel showed rapid blood clot formation with 54 ± 3 s in a rat model. Thus the developed hydrogel can quickly form blood clots making it a promising candidate for enhancing hemostatic potential [107]. Chitosan, nano whitlockite (nWH) composite hydrogel (CS-nWH) was prepared for hemostatic applications. Nano whitlockite releases Ca^{2+}, Mg^{2+}, and PO_4^{3-} ions that help in the activation of the coagulation cascade. The composite hydrogel showed enhanced elastic modulus and shear-thinning properties. The cytocompatible nature of composite hydrogel (CS-nWH) proved with HUVEC and hemocompatibility study of composite hydrogel exhibited a 5% hemolysis rate which is an allowable limit for biomaterials. In vitro blood clotting results revealed a less blood clotting time of 3.5 ± 1.2 min for composite hydrogel than control CS and whole blood alone sample and is due to the simultaneous action of Ca^{2+}, Mg^{2+}, and PO_4^{3-} ions from nWH and amine groups of CS hydrogel. In vivo experiments were tested on injuries in a rat model and found a reduced time of 62 ± 3 s for achieving hemostasis than control samples. Thus, the hydrogel could be an effective solution for quickly achieving hemostasis during crucial bleeding situations [108]. Sundaram et al. developed a vasoconstrictor and coagulation activator entrapped chitosan containing composite hydrogel with efficient hemostasis properties. The encapsulation of potassium aluminum sulfate (PA) and calcium chloride (Ca) into the chitosan (CS) hydrogel aided in the enhancement of the hemostatic potential of chitosan. The prepared hydrogel demonstrated excellent injectability, shear thinning properties, and cytocompatibility with HUVEC cells. Hemocompatibility studies showed a hemolysis rate of 1 ± 0.5% which is below the acceptable range for biomaterials. The hydrogel exhibited rapid blood clotting time of 112 ± 12 s attributed to the simultaneous interaction of chitosan (CS) with red blood cell (RBC) membranes and platelets, platelet aggregation facilitated by plasminogen activator (PA), and activation of the coagulation cascade by calcium ions (Ca). This process aids in fibrin mesh formation. Moreover, the in vivo studies showed the hydrogel was very effective in achieving hemostasis in a rat model, with a 20 ± 10 s compared to commercially available fibrin sealant. This efficacy is due to the entrapment of PA, which promotes platelet aggregation, and the role of Ca in coagulation cascade activation. The hydrogel is suitable for use in low-pressure bleeding sites and can be beneficial for stopping oozing and bleeding during dental, liver, gastrointestinal, and dermal surgeries. Hence, the developed injectable composite hydrogel could effectively target all stages of hemostasis mechanisms, facilitating rapid blood clotting [109].

7 Conclusion

This chapter gives a concise introduction to chitosan, discussing its synthesis, structure, properties, and the diverse biomedical application of chitosan. Owing to its superior biocompatibility and biodegradability, chitosan has been designated as "Generally Recognized as Safe" by the FDA and is authorized for use in biomedical applications. Chitosan's versatility is attributed to its various exceptional properties such as antibacterial property, hemostatic potential, biodegradability, film-forming capability, biocompatibility, tunable mechanical strength, and muco-adhesiveness. The cationic properties of chitosan, along with its amine and hydroxyl functional groups, make it accessible for various biomedical applications since it is derived from natural sources like crustaceans. Chitosan has become a boon in the healthcare system due to its exceptional properties and versatility, allowing it to be used in multiple forms for diverse biomedical applications. The review encompasses the latest outcome of chitosan and its derivatives across diverse biomedical fields. Thus, chitosan emerges as a promising contender in the field of biomedical applications.

Acknowledgments One of the authors, Arya Lakshmi V. V., is grateful to Amrita Vishwa Vidyapeetham for providing her Research fellowship. The authors are also thankful to Amrita Vishwa Vidyapeetham for providing seed grant fund.

References

1. Morin-Crini N, Lichtfouse E, Torri G, Crini G (2019) Fundamentals and applications of chitosan. In: Crini G, Lichtfouse E (eds) Sustainable agriculture reviews 35: chitin and chitosan: history, fundamentals and innovations, pp 49–123
2. Al-Rooqi MM, Hassan MM, Moussa Z, Obaid RJ, Suman NH, Wagner MH, Natto SS, Ahmed SA (2022) Advancement of chitin and chitosan as promising biomaterials. J Saudi Chem Soc 26:101561

3. Azmana M, Mahmood S, Hilles AR, Rahman A, Arifin MAB, Ahmed S (2021) A review on chitosan and chitosan-based bio nanocomposites: promising material for combatting global issues and its applications. Int J Biol Macromol 185:832–848
4. Petroni S, Tagliaro I, Antonini C, D'Arienzo M, Orsini SF, Mano JF, Brancato V, Borges J, Cipolla L (2023) Chitosan-based biomaterials: insights into chemistry, properties, devices, and their biomedical applications. Mar Drugs 21:147
5. Swain SK, Biswal A (2023) Chitosan: a smart biomaterial. In: Swain SK, Biswal A (eds) Chitosan nanocomposites: bionanomechanical applications, pp 1–25
6. Pellis A, Guebitz GM, Nyanhongo GS (2022) Chitosan: sources, processing and modification techniques. Gels 8:393
7. Ahmed S, Ikram S (2016) Chitosan-based scaffolds and their applications in wound healing. Achieve Life Sci 10:27–37
8. Teixeira-Costa BE, Andrade CT (2021) Chitosan as a valuable biomolecule from seafood industry waste in the design of green food packaging. Biomol Ther 11:1599
9. Zargar V, Asghari M, Dashti A (2015) A review on chitin and chitosan polymers: structure, chemistry, solubility, derivatives, and applications. Chem Rev 2:204–226
10. Rinaudo M (2006) Chitin and chitosan: properties and applications. Prog Polym Sci 31:603–632
11. No HK, Meyers SP (1995) Preparation and characterization of chitin and chitosan – a review. J Aquat Food Prod Technol 4:27–52
12. Domard A, Domard M (2002) Chitosan: structure-properties relationship and biomedical applications. In: Dumitriu S (ed) Polymeric biomaterials, pp 187–212
13. Aranaz I, Alcántara AR, Civera MC, Arias C, Elorza B, Caballero AH, Acosta N (2021) Chitosan: an overview of its properties and applications. Polymers 13:3256
14. Wang W, Meng Q, Li Q, Liu J, Zhou M, Jin Z, Zhao K (2020) Chitosan derivatives and their application in biomedicine. Int J Mol Sci 21:487
15. Jayakumar R, Prabaharan M, Nair SV, Tokura S, Tamura H, Selvamurugan N (2010) Novel carboxymethyl derivatives of chitin and chitosan materials and their biomedical applications. Prog Mater Sci 55:675–709
16. Antony IR, Jayakumar R (2023) Preparation of different types of carboxymethyl chitosan derivatives. In: Jayakumar R (ed) Multifaceted carboxymethyl chitosan derivatives: properties and biomedical applications, pp 19–29
17. Antony IR, Jayakumar R (2023) Introductory aspects of carboxymethyl chitosan derivatives. In: Jayakumar R (ed) Multifaceted carboxymethyl chitosan derivatives: properties and biomedical applications, pp 1–17
18. Harugade A, Sherje AP, Pethe A (2023) Chitosan: a review on properties, biological activities and recent progress in biomedical applications. React Funct Polym 191:105634
19. Kamoun EA (2016) N-succinyl chitosan–dialdehyde starch hybrid hydrogels for biomedical applications. J Adv Res 7:69–77
20. Guo Y, Qiao D, Zhao S, Liu P, Xie F, Zhang B (2024) Biofunctional chitosan–biopolymer composites for biomedical applications. Mater Sci Eng R Rep 159:100775
21. Peng Y, Han B, Liu W, Xu X (2005) Preparation and antimicrobial activity of hydroxypropyl chitosan. Carbohydr Res 340:1846–1851
22. Piraccini BM, Iorizzo M, Lencastre A, Nenoff P, Rigopoulos D (2020) Ciclopirox hydroxypropyl chitosan nail lacquer: a review of its use in onychomycosis. Dermatol Ther 10:917–929
23. Dimassi S, Tabary N, Chai F, Blanchemain N, Martel B (2018) Sulfonated and sulfated chitosan derivatives for biomedical applications: a review. Carbohydr Polym 202:382–396
24. Anushree U, Punj P, Vasumathi BS (2023) Phosphorylated chitosan accelerates dermal wound healing in diabetic wistar rats. Glycoconj J 40:19–31
25. Jayakumar R, Nagahama H, Furuike T, Tamura H (2008) Synthesis of phosphorylated chitosan by novel method and its characterization. Int J Biol Macromol 42:335–339

26. Kou SG, Peters LM, Mucalo MR (2021) Chitosan: a review of sources and preparation methods. Int J Biol Macromol 169:85–94
27. Maddaloni M, Vassalini I, Alessandri I (2020) Green routes for the development of chitin/chitosan sustainable hydrogels. Sustain Chem 1:325–344
28. Souza VG, Pires JR, Rodrigues C, Coelhoso IM, Fernando AL (2020) Chitosan composites in packaging industry-current trends and future challenges. Polymers 12:417
29. Hayes M, Carney B, Slater J, Bruck W (2008) Mining marine shellfish wastes for bioactive molecules: chitin and chitosan ndash; part a: extraction methods. Biotechnol J 3:871–877
30. Reshad RAI, Jishan TA, Chowdhury NN (2021) Chitosan and its broad applications: a brief review. J Clin Exp Investig 12:em00779
31. Bakshi PS, Selvakumar D, Kadirvelu K, Kumar NS (2020) Chitosan as an environment-friendly biomaterial – a review on recent modifications and applications. Int J Biol Macromol 150:1072–1083
32. Pavinatto A, Pavinatto FJ, Delezuk JADM, Nobre TM, Souza AL, Campana-Filho SP, Oliveira Jr ON (2013) Low molecular-weight chitosans are stronger biomembrane model perturbants. Colloids Surf B Biointerfaces 104:48–53
33. Pillai CK, Paul W, Sharma CP (2009) Chitin and chitosan polymers: chemistry, solubility and fiber formation. Prog Polym Sci 34:641–678
34. Habibi Y, Lucia LA (2012) Polysaccharide building blocks: a sustainable approach to the development of renewable biomaterials. Wiley, p 430
35. Yang J, Tian F, Wang Z, Wang Q, Zeng YJ, Chen SQ (2008) Effect of chitosan molecular weight and deacetylation degree on hemostasis. J Biomed Mater Res B Appl Biomater 84:131–137
36. Verma C, Quraishi MA (2021) Chelation capability of chitosan and chitosan derivatives: recent developments in sustainable corrosion inhibition and metal decontamination applications. Curr Res Green Sustain Chem 4:100184
37. Kumirska J, Weinhold MX, Thöming J, Stepnowski P (2011) Biomedical activity of chitin/chitosan-based materials-influence of physicochemical properties apart from molecular weight and degree of N-acetylation. Polymers 3:1875–1901
38. Li Q, Dunn ET, Grandmaison EW, Goosen MF (2020) Applications and properties of chitosan. In: Goosen MF (ed) Applications of chitin and chitosan, pp 3–29
39. Sharkawy A, Barreiro MF, Rodrigues AE (2020) Chitosan-based Pickering emulsions and their applications: a review. Carbohydr Polym 250:116885
40. Hosseinnejad M, Jafari SM (2016) Evaluation of different factors affecting antimicrobial properties of chitosan. Int J Biol Macromol 85:467–475
41. Mujtaba M, Morsi RE, Kerch G, Elsabee MZ, Kaya M, Labidi J, Khawar KM (2019) Current advancements in chitosan-based film production for food technology; a review. Int J Biol Macromol 121:889–904
42. Santos VP, Marques NS, Maia PC, Lima MABD, Franco LDO, Campos-Takaki GMD (2020) Seafood waste as attractive source of chitin and chitosan production and their applications. Int J Biol Macromol 21:4290
43. Rasheed PA, Pandey RP, Jabbar KA, Samara A, Abdullah AM, Mahmoud KA (2020) Chitosan/lignosulfonate nanospheres as "green" biocide for controlling the microbiologically influenced corrosion of carbon steel. Materials 13:2484
44. Ways TMM, Lau WM, Khutoryanskiy VV (2018) Chitosan and its derivatives for application in mucoadhesive drug delivery systems. Polymers 10:267
45. Guerini M, Condro G, Perugini P (2022) Evaluation of the mucoadhesive properties of chitosan-based microstructured lipid carrier. Pharmaceutics 14:170
46. Sandri SRG, Bonferoni MC, Ferrari F, Mori M, Caramella C (2012) The role of chitosan as a mucoadhesive agent in mucosal drug delivery. J Drug Del Sci Tech 22:275–284
47. Wani TU, Pandith AH, Sheikh FA (2021) Polyelectrolytic nature of chitosan: influence on physicochemical properties and synthesis of nanoparticles. J Drug Del Sci Tech 65:102730

48. Hamman JH (2010) Chitosan based polyelectrolyte complexes as potential carrier materials in drug delivery systems. Mar Drugs 8:1305–1322
49. Zhang F, Zhang S, Cui S, Jing X, Feng Y, Coseri S (2024) Rapid self-healing carboxymethyl chitosan/hyaluronic acid hydrogels with injectable ability for drug delivery. Carbohydr Polym 328:121707
50. Ciarlantini C, Francolini I, Silvestro I, Mariano A, d'Abusco AS, Piozzi A (2024) Design of bioactive and biomimetic scaffolds based on chitosan-alginate polyelectrolyte complexes for tissue engineering. Carbohydr Polym 327:121684
51. Beulah P, Jinu U, Ghorbanpour M, Venkatachalam P (2019) Green engineered chitosan nanoparticles and its biomedical applications – an overview. Adv Phytonanotechnol:329–341
52. Ke CL, Deng FS, Chuang CY, Lin CH (2021) Antimicrobial actions and applications of chitosan. Polymers 13:904
53. Kandimalla KK, Borden E, Omtri RS, Boyapati SP, Smith M, Lebby K, Mulpuru M Gadde M (2013) Ability of chitosan gels to disrupt bacterial biofilms and their applications in the treatment of bacterial vaginosis. J Pharm Sci 102:2096–2101
54. Liu P, Meng W, Wang S, Sun Y, Ashraf AM (2015) Quaternary ammonium salt of chitosan: preparation and antimicrobial property for paper. Open Med 10:473–478
55. Sadeghi AMM, Dorkoosh FA, Avadi MR, Saadat P, Rafiee-Tehrani M, Junginger HE (2008) Preparation, characterization and antibacterial activities of chitosan, N-trimethyl chitosan and N-diethylmethyl chitosan nanoparticles loaded with insulin using both the ionotropic gelation and polyelectrolyte complexation methods. Int J Pharm 355:299–306
56. Shariatinia Z (2018) Carboxymethyl chitosan: properties and biomedical applications. Int J Biol Macromol 120:1406–1419
57. Sun L, Zhou J, Lai J, Zheng X, Zhang LM (2024) Multifunctional chitosan-based gel sponge with efficient antibacterial, hemostasis and strong adhesion. Int J Biol Macromol 256:128505
58. Khasteband M, Sharifi Y, Akbari A (2024) Chrysin loaded polycaprolactone-chitosan electrospun nanofibers as potential antimicrobial wound dressing. Int J Biol Macromol 263:130250
59. Abd El-Hack ME, El-Saadony MT, Shafi ME, Zabermawi NM, Arif M, Batiha GE, Khafaga AF, Abd El-Hakim YM, Al-Sagheer AA (2020) Antimicrobial and antioxidant properties of chitosan and its derivatives and their applications: a review. Int J Biol Macromol 164:2726–2744
60. Yang H, Liu Y, Wen F, Yan X, Zhang Y, Zhong Z (2024) Preparation, characterization, antioxidant and antifungal activities of benzoic acid compounds grafted onto chitosan. Int J Biol Macromol 259:129096
61. Singh A, Kukreti R, Saso L, Kukreti S (2019) Oxidative stress: a key modulator in neurodegenerative diseases. Molecules 24:1583
62. Hayes JD, Dinkova-Kostova AT, Tew KD (2020) Oxidative stress in cancer. Cancer Cell 38:167–197
63. Ngo DH, Vo TS, Ngo DN, Kang KH, Je JY, Pham HND, Byun HG, Kim SK (2015) Biological effects of chitosan and its derivatives. Food Hydrocoll 51:200–216
64. Wang W, Xue C, Mao X (2020) Chitosan: structural modification, biological activity and application. Int J Biol Macromol 164:4532–4546
65. Pu S, Li J, Sun L, Zhong L, Ma Q (2019) An in vitro comparison of the antioxidant activities of chitosan and green synthesized gold nanoparticles. Carbohydr Polym 211:161–172
66. Kim KW, Thomas RL (2007) Antioxidative activity of chitosans with varying molecular weights. Food Chem 101:308–313
67. Cui J, Wang Y, Liang X, Zhao J, Ji Y, Tan W, Dong F, Guo Z (2024) Synthesis, antimicrobial activity, antioxidant activity and molecular docking of novel chitosan derivatives containing glycine Schiff bases as potential succinate dehydrogenase inhibitors. Int J Biol Macromol 267:131407

68. Han X, Mi Y, Ji Y, Sun M, Tang H, Dong F, Guo Z (2024) A novel chitosan antioxidant bearing sulfhydryl group: synthesis, characterization and activity assessment. Int J Biol Macromol 261:129816
69. Goyal S, Thirumal D, Rana J, Gupta AK, Kumar A, Babu MA, Kumar P, Sindhu RK (2024) Chitosan based nanocarriers as a promising tool in treatment and management of inflammatory diseases. Carbohydr Polym Technol Appl 7:100442
70. Chang SH, Lin YY, Wu GJ, Huang CH, Tsai GJ (2019) Effect of chitosan molecular weight on anti-inflammatory activity in the RAW 264.7 macrophage model. Int J Biol Macromol 131:167–175
71. Zosangpuii SP (2023) Chitosan a natural anti-inflammatory and wound healing agent: a brief update. Nat Resour Hum Health 4:115–127
72. Kim S (2018) Competitive biological activities of chitosan and its derivatives: antimicrobial, antioxidant, anticancer, and anti-inflammatory activities. Int J Polym Sci 1:1708172
73. Nguyen HT, Do NH, Lac HD, Nguyen PL, Le PK (2023) Synthesis, properties, and applications of chitosan hydrogels as anti-inflammatory drug delivery system. J Porous Mater 30:655–670
74. Friedman AJ, Phan J, Schairer DO, Champer J, Qin M, Pirouz A, Blecher-Paz K, Oren A, Liu PT, Modlin RL, Kim J (2013) Antimicrobial and anti-inflammatory activity of chitosan-alginate nanoparticles: a targeted therapy for cutaneous pathogens. J Invest Dermatol 133:1231–1239
75. Lunkov AP, Zubareva AA, Varlamov VP, Nechaeva AM, Drozd NN (2023) Chemical modification of chitosan for developing of new hemostatic materials: a review. Int J Biol Macromol 23:127608
76. Khan MA, Mujahid M (2019) A review on recent advances in chitosan based composite for hemostatic dressings. Int J Biol Macromol 124:138–147
77. Chen KY, Chen YC, Lin TH, Yang CY, Kuo YW, Lei U (2020) Hemostatic enhancement via chitosan is independent of classical clotting pathways-a quantitative study. Polymers 12:2391
78. Gheorghiţa D, Moldovan H, Robu A, Biţa AI, Grosu E, Antoniac A, Corneschi I, Antoniac I, Bodog AD, Bacila CI (2023) Chitosan-based biomaterials for hemostatic applications: a review of recent advances. Int J Mol Sci 24:10540
79. Yang J, Cai F, Lv Y, Jiang T, Zhao X, Hu X, Zheng Y, Shi X (2024) Chitosan nonwoven fabric composited calcium alginate and adenosine diphosphate as a hemostatic bandage for acute bleeding wounds. Int J Biol Macromol 257:128561
80. Yu C, Lu Y, Pang J (2024) A hemostatic sponge derived from chitosan and hydroxypropylmethylcellulose. J Mech Behav Biomed Mater 150:106240
81. Haider A, Khan S, Iqbal DN, Shrahili M, Haider S, Mohammad K, Mohammad A, Rizwan M, Kanwal Q, Mustafa G (2024) Advances in chitosan-based drug delivery systems: a comprehensive review for therapeutic applications. Eur Polym J 26:112983
82. Borda LJ, Macquhae FE, Kirsner RS (2016) Wound dressings: a comprehensive review. Curr Dermatol Rep 5:287–297
83. Patrulea V, Ostafe V, Borchard G, Jordan O (2015) Chitosan as a starting material for wound healing applications. Eur J Pharm Biopharm 97:417–426
84. Mothilal NP, Pradeep A, Arthi C, Gopal K, Kaliannagounder VK, Park CH, Kumar VA Rangasamy J (2024) Amikacin sulphate loaded chitosan-diopside nanoparticles composite scaffold for infectious wound healing. Int J Biol Macromol 17:130217
85. Upadhyay A, Karumathil DP, Upadhyaya I, Bhattaram V, Venkitanarayanan K (2016) Controlling bacterial antibiotic resistance using plant-derived antimicrobials. In: Versteeg-buschman L (ed) Antibiotic resistance, pp 205–226
86. Pandian M, Kumar VA, Jayakumar R (2021) Antiseptic chitosan bandage for preventing topical skin infections. Int J Biol Macromol 193:1653–1658
87. Raveendran NT, Mohandas A, Menon RR, Menon SA, Biswas R, Jayakumar R (2018) Ciprofloxacin-and fluconazole-containing fibrin-nanoparticle-incorporated chitosan bandages for the treatment of polymicrobial wound infections. ACS Appl Bio Mater 2:243–254

88. Yang S, Li X, Liu P, Zhang M, Wang C, Zhang B (2020) Multifunctional chitosan/polycaprolactone nanofiber scaffolds with varied dual-drug release for wound-healing applications. ACS Biomater Sci Eng 6:4666–4676
89. Alotaibi BS, Khan AK, Kharaba Z, Yasin H, Yasmin R, Ijaz M, Khan M, Murtaza G (2024) Development of poly (vinyl alcohol)-chitosan composite nanofibers for dual drug therapy of wounds. ACS Omega 9:12825–12834
90. Antony IR, Pradeep A, Pillai AV, Menon RR, Kumar VA, Jayakumar R (2023) Antiseptic chitosan-poly (hexamethylene) biguanide hydrogel for the treatment of infectious wounds. J Funct Biomater 14:528
91. Pandian M, Selvaprithviraj V, Pradeep A, Rangasamy J (2021) In-situ silver nanoparticles incorporated N, O-carboxymethyl chitosan based adhesive, self-healing, conductive, antibacterial and anti-biofilm hydrogel. Int J Biol Macromol 188:501–511
92. Nimal TR, Baranwal G, Bavya MC, Biswas R, Jayakumar R (2016) Anti-staphylococcal activity of injectable nano tigecycline/chitosan-PRP composite hydrogel using drosophila melanogaster model for infectious wounds. ACS Appl Mater Interfaces 8:22074–22083
93. Adepu S, Ramakrishna S (2021) Controlled drug delivery systems: current status and future directions. Molecules 26:5905
94. De Sousa VR, da Cunha M, Santos A, Viana de Sousa B, De Araujo NG, De Lima N, Santana L, Menezes RR (2020) A review on Chitosan's uses as biomaterial: tissue engineering, drug delivery systems and cancer treatment. Materials 13:4995
95. Desai N, Rana D, Salave S, Gupta R, Patel P, Karunakaran B, Sharma A, Giri J, Benival D, Kommineni N (2023) Chitosan: a potential biopolymer in drug delivery and biomedical applications. Pharmaceutics 15:1313
96. Nithya S, Nimal TR, Baranwal G, Suresh MK, Anju CP, Kumar VA, Mohan CG, Jayakumar R, Biswas R (2018) Preparation, characterization and efficacy of lysostaphin-chitosan gel against *Staphylococcus aureus*. Int J Biol Macromol 110:157–166
97. Rakhshaei R, Namazi H, Hamishehkar H, Kafil HS, Salehi R (2019) In situ synthesized chitosan–gelatin/ZnO nanocomposite scaffold with drug delivery properties: higher antibacterial and lower cytotoxicity effects. J Appl Polym Sci 136:47590
98. Kumar RV, Verma A, Kumar P, Pandey J (2019) Novel binary grafted chitosan nanocarrier for sustained release of curcumin. Int J Biol Macromol 131:184–191
99. Jaiswal S, Dutta PK, Kumar S, Koh J, Pandey S (2019) Methyl methacrylate modified chitosan: synthesis, characterization and application in drug and gene delivery. Carbohydr Polym 211:109–117
100. Murali VP, Jayakumar R (2023) Natural biopolymers in tissue engineering-role, challenges, and clinical applications. In: Murali VP, Jayakumar R (eds) Natural biopolymers in drug delivery and tissue engineering, pp 409–434
101. Koons GL, Diba M, Mikos AG (2020) Materials design for bone-tissue engineering. Nat Rev Mater 5:584–603
102. Shanmugam BK, Rangaraj S, Subramani K, Srinivasan S, Aicher WK, Venkatachalam R (2020) Biomimetic TiO2-chitosan/sodium alginate blended nanocomposite scaffolds for tissue engineering applications. Mater Sci Eng 110:110710
103. Jindal A, Mondal T, Bhattacharya J (2020) An in vitro evaluation of zinc silicate fortified chitosan scaffolds for bone tissue engineering. Int J Biol Macromol 164:4252–4262
104. Davachi SM, Haramshahi SMA, Akhavirad SA, Bahrami N, Hassanzadeh S, Ezzatpour S, Hassanzadeh N, Kebria MM, Khanmohammadi M, Bagher Z (2022) Development of chitosan/hyaluronic acid hydrogel scaffolds via enzymatic reaction for cartilage tissue engineering. Mater Today Commun 30:103230
105. Jalageri MB, Kumar GM (2024) Graphene oxide reinforced polyvinyl alcohol/chitosan composite hydrogel for cartilage regeneration. Polym Bull 81:1–18
106. Pusateri E, McCarthy SJ, Gregory KW, Harris RA, Cardenas L, McManus AT, Goodwin CW (2003) Effect of a chitosan-based hemostatic dressing on blood loss and survival in a model of severe venous hemorrhage and hepatic injury in swine. J Trauma 54:177–182

107. Sundaram MN, Amirthalingam S, Mony U, Varma PK, Jayakumar R (2019) Injectable chitosan-nano bioglass composite hemostatic hydrogel for effective bleeding control. Int J Biol Macromol 129:936–943
108. Muthiah Pillai NS, Eswar K, Amirthalingam S, Mony U, Varma PK, Jayakumar R (2019) Injectable nano whitlockite incorporated chitosan hydrogel for effective hemostasis. ACS Appl Bio Mater 2:865–873
109. Sundaram MN, Mony U, Varma PK, Rangasamy J (2021) Vasoconstrictor and coagulation activator entrapped chitosan based composite hydrogel for rapid bleeding control. Carbohydr Polym 258:117634

Adv Polym Sci (2025) 295: 35–62
https://doi.org/10.1007/12_2024_182

Published online: 12 November 2024

Chitosan Biomaterials: Applications and Prospects in the Pharmaceutical Sector

Joel Maria Joju and R. Jayakumar

Contents

1 Introduction 36
2 Chitosan and Its Derivatives in Pharmaceutical Applications 39
2.1 As Drug Excipient 39
2.2 Drug Delivery 40
2.3 Gene Delivery 44
2.4 Tissue Engineering 46
2.5 Wound Healing 48
2.6 Hemostasis 50
3 Chitosan as an Adsorbent of Pharmaceutical Contaminants 54
4 Conclusion 55
References 56

Abstract Chitosan is a naturally derived biopolymer obtained by the deacetylation process of chitin. It is a non-toxic, biocompatible, and biodegradable polymer with randomly arranged D-glucosamine and N-acetyl-D-glucosamine units. The free amino groups in the polymer chain can get protonated in acidic pH rendering it cationic and allowing it to interact with other biological macromolecules. The tunable porosity, gelation ability, and structural resemblance of chitosan with glycosaminoglycans (GAGs) make it an ideal candidate for pharmaceutical applications. Chitosan's versatility is further demonstrated by its derivatization and fabrication into different forms such as scaffolds, sponges, beads, nanoparticles, microparticles, hydrogel, cryogel, films, and membranes, expanding its use in the pharmaceutical sector. It also has been used as an excipient in drug formulations along with its multitude of applications in drug delivery, gene delivery, wound

J. M. Joju and R. Jayakumar (✉)
Polymeric Biomaterials Lab, School of Nanosciences and Molecular Medicine, Amrita Vishwa Vidyapeetham, Kochi, India
e-mail: rjayakumar@aims.amrita.edu

healing, and tissue engineering. This chapter focuses on the pharmaceutical aspects and applications of the biopolymer chitosan in the pharmaceutical field.

Keywords Chitosan · Contaminant adsorbent · Drug delivery · Drug excipient · Pharmaceutical applications · Tissue engineering · Wound healing

1 Introduction

Chitosan (CS) is a natural linear polycationic copolymer consisting of irregularly placed N-acetyl-D-glucosamine and D-glucosamine units and is used for variety of pharmaceutical applications in different forms [1]. It is derived from chitin, which is extracted from shrimp and crab shells [2]. The first report on the discovery of chitin was in 1799 by the English Chemist Charles Hatchett during his studies on crustacean marine life. Later in 1811, French scientist Henri Braconnot reported "fongine", now known as "chitin". Down the lane in 1859 Charles Rouget synthesized an acid-soluble form of chitin by modifying it with deacetylation. This new molecule was then named as "chitosan" by Felix Hoppe-Seyler in 1894 [3]. CS is derived by partially deacetylating chitin using NaOH, biological approaches, or by a combination of both processes (Fig. 1) [4].

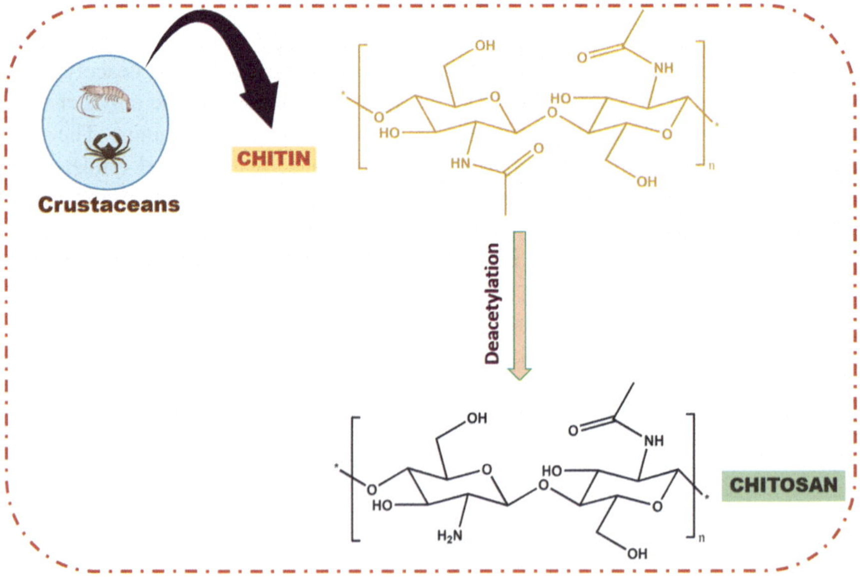

Fig. 1 General synthesis of chitosan from chitin

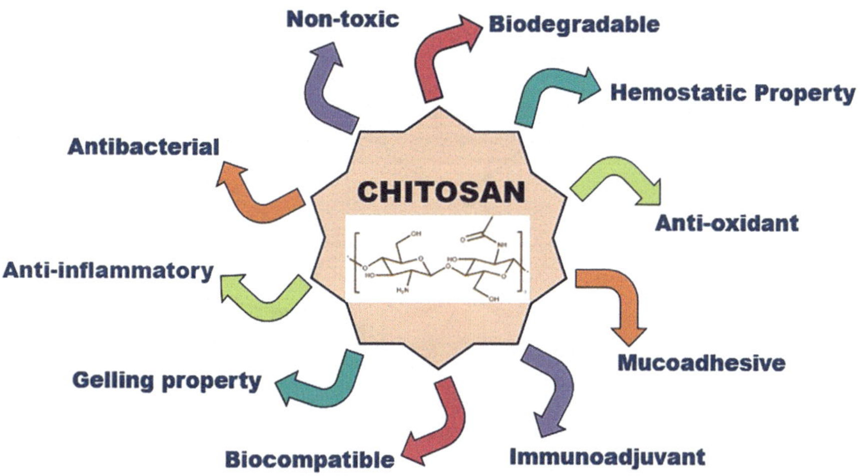

Fig. 2 Properties of chitosan

The extensive number of citations in the Scopus itself provides evidence of the vast studies conducted on CS, the smart material. CS is cationic, while the majority of other biopolymers are either anionic or neutral [5]. This property contributes to the unparalleled applications of CS through its electrostatic complex formation across various medical fields. The presence of reactive amino groups in the structure of CS allows for its vast derivatization. While these chemical modifications would retain the original structural skeleton, they also preserve the native biological properties of CS while adding improved characteristic features to it [6]. The physicochemical properties like molecular weight, viscosity, solubility, and degree of deacetylation greatly impact the suitability of CS in different pharmaceutical applications. CS with a lower deacetylation degree will be appropriate for applications that require cell growth and adhesion whereas applications that need higher flexibility or charge density on the chain require more deacetylated CS [5].

CS is a versatile compound that possesses a vast number of characteristics including antibacterial, anti-inflammatory, anticancer, anti-oxidant [7], hemostatic, gelling, and mucoadhesive properties (Fig. 2) [8] which makes it an ideal candidate across the pharmaceutical, food, agricultural, waste management, textile, and paper industries [9, 10]. The positive amino groups in CS can chelate with calcium and magnesium ions on the cell walls of bacteria and thereby inhibit its growth. CS can also impart antibacterial action by interacting with cell walls or bacterial DNA [11]. It also has anti-inflammatory and anti-oxidative activities via the hydroxyl-amino groups mediated termination of free radicals, which when produced in excess can lead to oxidative stress causing inflammation [7]. These properties enable the use of CS in the pharmaceutical area.

The potential for functional group modification is another factor that contributed to the versatility of CS biopolymer (Fig. 3). The structure of CS contains primary and

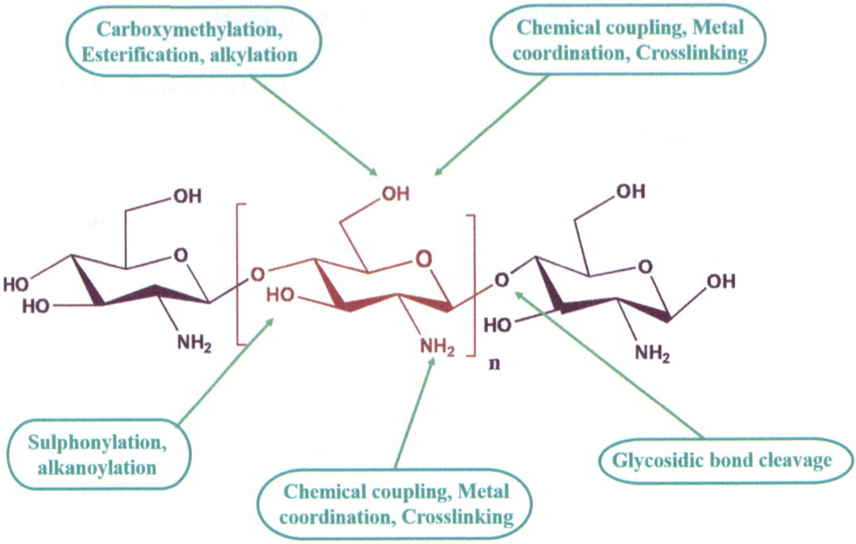

Fig. 3 Possible modifications on chitosan structure

secondary hydroxyl groups (carbon 3 and 6 respectively), primary amino group on carbon 2, acetamide units, and glycosidic linkages, all capable of undergoing chemical reactions leading to the formation of various CS derivatives [11]. Carboxymethyl chitosan, thiolated chitosan, glycol chitosan, succinyl chitosan, phosphorylated chitosan, hydroxypropyl chitosan, and quarternized chitosan [12], are some examples of products of such derivatization. Derivatization can also improve the existing properties of CS including the solubility and biological features. These chemically modified CS derivatives are also employed as pharmaceutical excipients. CS and its derivatives are increasingly used in transdermal drug-delivering systems owing to its skin adhesive properties. Low molecular weight CS finds application as skin permeability enhancer in many of these transdermal drug delivery systems depending on the pH of the site of application. A pH of 7.4 is found to be the optimal pH when CS shows appreciable permeation via the interaction between the protonated amino groups and carboxyl groups of keratins and glycoproteins of cell membranes, whereas in acidic (<6) and basic pH (>8) these interactions are not much effective in enhancing skin permeation [13].

This review chapter focuses on different forms (scaffolds, sponges, nanocomposites, nanoparticles, microparticles, films, hydrogels, and membranes) of CS and its derivatives which have been used in the pharmaceutical field in recent years.

2 Chitosan and Its Derivatives in Pharmaceutical Applications

CS and its derivatives in multiple formulations such as scaffolds, sponges, beads, hydrogels, cryogels, microparticles, nanoparticles, tablets, films, and membranes are being used in numerous pharmaceutical applications (Fig. 4). A detailed review of the applications is discussed in the following sections.

2.1 As Drug Excipient

Pharmaceutical excipients are the non-active components in a drug formulation to support the effectiveness of the drug [14]. The traditional drug excipients are thought to be inert but the latest studies show that the excipients can increase the bioavailability, dissolution, and absorption of the active component by interacting with it [15]. CS as an excipient has gained importance due to its flexibility in use [16], its ability to act as a disintegrant in tablet formulations, or as direct tablet compression [17]. CS can enhance the permeability across epithelia along with its mucoadhesive properties [18]. The mucoadhesive property of CS is attributed to the possible interactions between negatively charged mucin moieties and positively charged amino groups [19]. In a previous study, using the spray drying method, CS, crospovidone and mannitol (Cop-CCM) containing co-processed excipient was

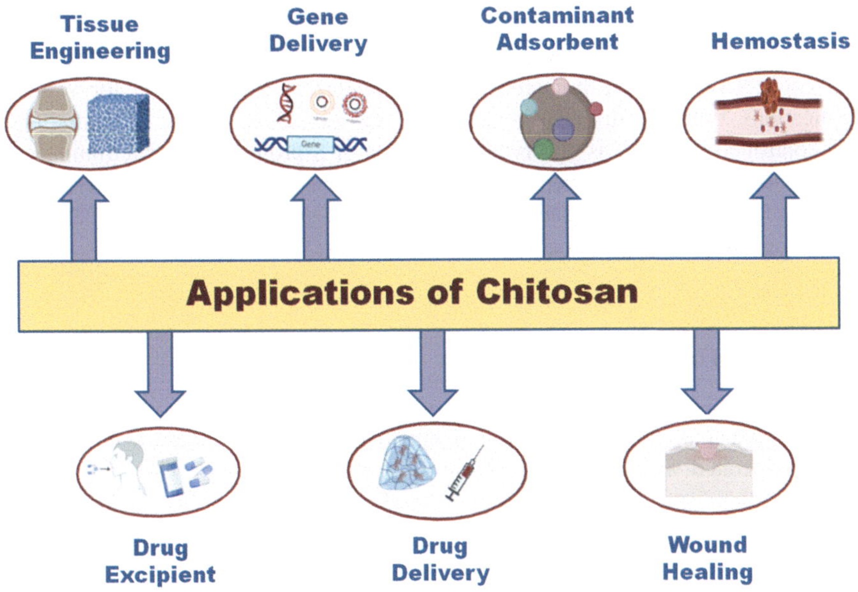

Fig. 4 Applications of chitosan

prepared to overcome the disadvantages of poor flowability and lower density of CS. Cop-CCM containing 50% CS, 43% mannitol, and 6.5% crospovidone was chosen as optimum using Central Composite Design (CCD). Using the SeDeM diagram expert system, Cop-CCM was characterized. The compressibility and flow nature of the prepared Cop-CCM was further verified by the Heckel and Kawakita plot. The drug release study of Cop-CCM was compared with paracetamol and the results suggested that the formulated co-processed excipient is suitable for its use in the pharmaceutical field [20]. Quick-disintegrating tablets that disintegrate faster when placed on the tongue can be consumed without water consumption [21]. To achieve this faster disintegration, disintegrants are included in the drug formulation and they can play this role effectively. Meloxicam tablet with quick integration was formulated using wet granulation technique making use of CS with medium molecular weight. When a compression force of 10.8 and 11.0 kN was applied, meloxicam tablets with good dissolution, mechanical strength, and faster disintegration were successfully prepared [22].

In another study, *N*-cinnamyl substituted *O*-amine functionalized CS (cinnamyl-chitosan), microcrystalline cellulose (MCC), and CS-based blends were formulated with acetaminophen as the active agent. The larger particle sizes of the cinnamyl-chitosan and CS blends resulted in the formulation of tablets with good compaction, powder flowability, and no defects. The blends exhibited superior mechanical as well as drug-delivering properties. CS when modified to cinnamyl chitosan earned antibacterial and antioxidant properties and this makes cinnamyl chitosan-based excipient a potential candidate for drug delivery [23]. CS is modified with NAC-6-MNA, a ligand obtained through the linking between *N*-acetyl-cysteine and 6-mercaptonicotinamide. This *S*-protected CS derivative has improved mucoadhesive properties due to the disulfide bond interaction with the cysteine-rich domains of the mucus layer. Two *S*-protected CS formulations (CS-MNA-160 and CS-MNA-320) were prepared with metronidazole encapsulation for vaginal applications. While the drug was released in a few minutes in the case of unmodified CS, controlled release was observed in the case of S-protected CS. Metronidazole was found to stay for longer periods in the vaginal tissues and CS-MNA-320 showed activity against *E. coli.* Based on these results, *S*-protected CS-MNA-320 is a promising excipient for vaginal applications [19].

2.2 Drug Delivery

An important challenge involved with treating diseases therapeutically is the toxicity issues associated with high doses of the drug. At the same time, the drug should remain in the system for sufficient time to show its effectiveness. In this regard systems that can release the drug in a controlled manner in a specific time frame are of great importance since this could address the toxicity problem. Based on the type of drug, target site, administrating route, disease condition, and existing complications, drug vehicles are designed to attain the needed therapeutical effect [13]. CS exhibits a variety of biological properties such as antibacterial, anti-inflammatory, and anticancer [24]. In drug delivery systems, the positive charge of CS enables it to interact with drug molecules of anionic character to form stable and soluble systems [25]. The pH sensitivity of CS, arising from the presence of amino groups, will aid in the drug release profile [26]. At acidic pH, the protonated amino groups in the skeleton of CS will repel each other and this will lead to swelling and the eventual release of the drug molecule encapsulated within [27]. The limitation of using CS in drug delivery applications due to poor solubility can be overcome by chemically modifying CS to its derivatives of better solubility [6].

CS, dopamine (DA), and inulin aldehyde (IA)-based hydrogel as a drug delivery system was fabricated in a recent study. Release of indomethacin encapsulated in the hydrogel demonstrated that the density of the cross-linking determined the drug release profile. Indomethacin release was slow and sustained resulting from the higher cross-linking. When the concentration of IA was 3 mg/ml, the hydrogel system showed the best sustained release of indomethacin [28]. Sayyar et al. successfully synthesized CS/hyaluronic acid (HA) – magnetic Montmorillonite composite hydrogel (CS/HA-mMMt) as a dual drug delivery system for Curcumin (CUR)/Ciprofloxacin (CIP). This pH-dependent system showed sustained release of CUR and CIP at normal conditions and an upsurge of release when magnetic field was applied. When the amount of mMMt was more, the system exhibited decreased release of the drugs. The release profile can be further controlled by adjusting the magnetic field that can be applied externally. The in vitro scratch assay demonstrated the hydrogel's potential for use in wound healing [29]. Another delivery system was fabricated by incorporating CS nanoparticles loaded with rosuvastatin in sodium alginate (SA)-poly (vinyl alcohol) (PVA) hydrogel. CS nanoparticles with 3 wt% of drug were first prepared by the method of ionic gelation. These nanoparticles were then incorporated with SA-PVA hydrogel in the ratio of 7:3 to obtain the system with the desired mechanical properties. In the first 4 h, 13% of the loaded rosuvastatin drug was released, followed by 67% in the next 4 h, and the remaining 20% was released between 8 and 24 h [30].

CS films incorporated with CS depolymerization products (CDP) with varying molecular weights and degree of deacetylation, demonstrating successful release of CIP were fabricated and studied. CIP was loaded with the CS/CDP films using glutaraldehyde as the crosslinker. Prepared films showed sustained release of CIP when the CDP had low molecular weight, degree of polymerization and were crosslinked for a duration of 48 h. The CIP release from the uncrosslinked films was faster when compared to the crosslinked films. Additionally, the films that were crosslinked for a longer duration exhibited a slower and more sustained release profile than those crosslinked for 24 h, due to decreased porosity. This is because, increased crosslinking results in increase in compactness, thereby reducing the porosity. These formulated films also exhibited pH-sensitive behavior, showing slower release in neutral pH than in the acidic pH [31].

One of the important attributes of CS is its ability to resist bacterial growth. There are numerous factors that can influence the activity of CS against these microorganisms. Type of microorganism, deacetylation degree, molecular weight, the source of

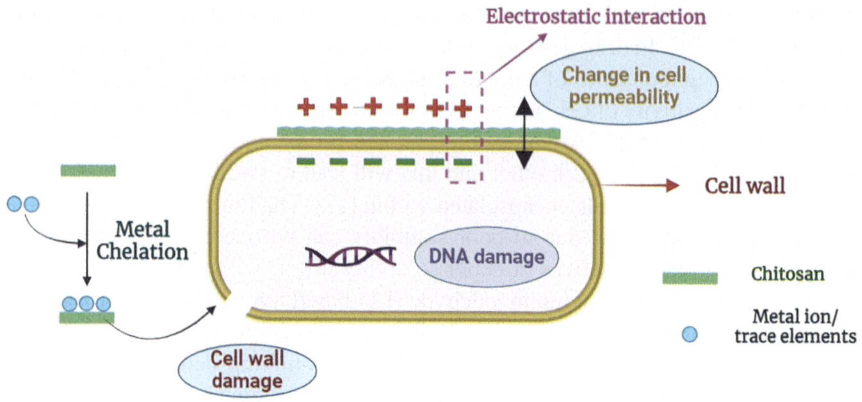

Fig. 5 Antibacterial mechanism of chitosan

CS, type of derivative/composite, pH, and temperature are some such determining factors. The general mode of action of CS against these organisms is through the electrostatic interaction of the positive groups of CS and anionic molecules of the bacterial cell surface. This leads to permeabilization of the cell wall, leakage of intracellular matter, and ultimate cell death (Fig. 5) [32]. The antibacterial effect of CS is enhanced by the derivatization or composite fabrication or by the use of CS matrices as antibiotic drug carriers. For example, for the controlled delivery of serotonin receptor agonist sumatriptan succinate (SS) drug, CS/tripolyphosphate/graphene oxide (CS/TPP/GO) nanocomposite hydrogel was prepared. Compared to CS hydrogel, the incorporation of GO is attributed to slow and sustained drug release. Cytotoxicity studies on HEK 293 cells showed that the composite is biocompatible. Antibacterial studies on *E. coli* and *S. aureus* showed positive results, showing the composite's antibacterial effectiveness. The drug release kinetics of the nanocomposites investigated using the Korsmeyer-Peppas model showed that the k value of hydrogels in the buffer of pH 1.2 was lower when compared with a buffer of pH 7.4. This result indicates that the drug release of CS/TPP/GO hydrogel is pH-sensitive. When studied in buffers of the same pH, when the GO content increased, the k value decreased suggesting that the introduction of GO helped in slow and sustained drug release [33]. With the aim of upgrading the antibacterial effect of CS, CS gel laden with colistimethate sodium was optimized to treat multidrug-resistant Gram-negative bacteria. When the antibacterial activity was examined against *E. coli*, *K. pneumonia,* and *P. aeruginosa,* the hydrogel showed a clear zone of inhibition with diameters 1.7 ± 0.13 cm (*E. coli*), 1.8 ± 0.08 cm (*K. pneumonia*), and 1.6 ± 0.29 cm (*P. aeruginosa*). The hydrogel also demonstrated activity against multidrug-resistant clinical isolates of the above three bacteria. The drug released from the hydrogel was effective enough to inhibit the biofilm formed by the clinical isolates of *E. coli*, *K. pneumonia,* and *P. aeruginosa* indicated by the no to reduced red projections when observed under a fluorescence microscope [34]. In an another study, a cumulative release of 81.23% PHMB was obtained from a hydrogel which also displayed in vitro antibiofilm and in vitro antibacterial activity against a number of ATCC and clinical isolates [35].

To study the anti-staphylococcal activity, Lysostaphin (LST) – CS gels were fabricated. The release assay showed 90–95% release of lysostaphin within 24 h. The in vitro agar diffusion method, anti-biofilm assay, and ex vivo assay on porcine skin model against *S. aureus* showed the potential of LST-CS gel to be used to treat staphylococcal infections [36]. In another study, CS hydrogel containing antibiotic tigecycline (tg) and platelet-rich plasma (PRP) was formulated to achieve anti-staphylococcal activity. The in vitro release of tg from tg-CSNPs-CSPRP gel showed 37 ± 2.3% burst release after 24 h, and a 58% release within 7 days. The sustained release of tg from the gel contributed to the in vitro anti-staphylococcal activity of the gel. Long-term anti-staphylococcal activity of the gel is rooted in the slow and sustained release of the drug from the gel system. Further, the ex vivo porcine skin model assay and in vivo anti-staphylococcal activity on the *Drosophila Melanogaster* model confirmed the potential of the optimized gel to treat *S. aureus*-infected wounds [37].

Octenidine dihydrochloride (Ocd) (0.05% and 0.1%) containing antiseptic CS bandage was developed for wound treating purposes in another study. The initial first-day sudden release of Ocd was replaced by sustained release of the drug in the next 6 days with a cumulative release of 96.2 ± 0.35% for CS-Ocd 0.05% and 99.5 ± 0.2% for CS-Ocd 0.1% bandages. The reduced optical density obtained from in vitro antibacterial study against *S. aureus* and *C. auris* and biofilm inhibition make these bandages a potential Ocd drug carrier in treating infected injuries [38]. All these literatures give insight into CS in formulating devices that could be effectively used in controlling bacterial growth while treating wounds.

CS-based composites and matrices are also being extensively used as anticancer drug vehicles. Lignin/CS (Lig/CS) nanoparticles were synthesized by co-assembling them both electrostatically, which can be used as the carriers of antitumor drugs docetaxel (DTX) and CUR. A release percentage of 51% and 50% were observed for CUR and DTX respectively from the Lig/CS system at pH 5.4, which is higher than the percentage of release of DTX and CUR observed at normal physiological condition (pH 7.4) [39]. In another study, a hydrogel with acrylamide grafted CS, gold nanoparticles, and silver nanoparticles (CS-g-PAAm/Au–Ag-NPs) were synthesized to study the delivery of fluorouracil drug and its efficiency in fighting against liver cancer cells. This pH-sensitive hydrogel exhibited 97% of fluorouracil release within 300 s. Au nanoparticles increased the ability of Ag nanoparticles to fight off liver cancer cells while reducing the cytotoxicity. The prepared hydrogel demonstrated its ability to encapsulate the drug in acidic pH and release it in intestinal pH, which makes it an ideal candidate for oral drug delivery of anticancer drugs [40]. SA-CS nanoparticles (ACNPs) were synthesized to encapsulate and deliver amygdalin to cancer cells. 90% encapsulation efficiency was observed with a sustained release of the drug for over a 10-h duration. The drug release was observed to be a pH-dependent. A sustained and increased pattern of drug release was observed with increased pH. The anticancer effect of ACNPs on H1299 cells further proved its potential in anticancer drug delivery applications [41]. In another study, the chemotherapeutic drug oxaliplatin (OXPt) was entrapped in CS nanoparticles with mucoadhesive properties to evaluate the ex vivo porcine mucosa penetration. Burst release was observed first, followed by slow and controlled release. The nanoparticle further increased the OXPt penetration threefold when used on oral mucosa. Under iontophoresis, the penetration of OXPt was further increased to a level sufficient to cause the death of the cancer cells. These results indicate the synthesized nanoparticle's potential for use in oral cancer treatment [42].

The diverse biological functions exhibited by proteins and peptides make them important therapeutic agents, and their integration into pharmaceutical formulations may be effective in treating chronic conditions [43]. CS and its derivatives are being used in protein and peptide delivery since it can pose as a host capable of protecting the guests incorporated within and also because it can adhere to mucosal surfaces [44]. For instance, an antihypertensive peptide KGYGGVSLPEW-laden CS microparticle integrated with CS oral films was developed for buccal delivery. Enhanced penetration was able to establish with these mucoadhesive films while the CS microparticles contributed towards the protection of the peptide embedded within [45]. CS can also be utilized for the oral administration of insulin, a peptide hormone, as this cationic molecule can interact with the negative mucus layer lining the gastrointestinal tract (GIT). This enhances the contact time with the mucosal layer and ultimately results in the efficient absorption of the encapsulated insulin. CS can open up the tight epithelial junctions and allow the insulin to overcome the epithelial barriers [46]. In another research, with the intention of formulation of nanoparticle delivery systems for protein or peptide delivery, bovine serum albumin (BSA)-infused CS-tripolyphosphate nanoparticles were optimized. These nanoparticles showed high BSA encapsulation and steady release of BSA over a 2-week period [47]. In short, it can be concluded that CS and its derivatives can be utilized in various drug delivery applications as a carrier.

2.3 *Gene Delivery*

Gene delivery refers to the means of delivery of nucleic acids (DNA and RNA) into recipient cells to induce transgene expression to obtain the possible outcomes of recovery or regeneration while treating diseases [48, 49]. Being negatively charged, the cell membrane denies the entry of anionic naked DNA, which will be easily degraded by proteases when left unprotected [50]. Therefore, efficient delivery of genes along with the development of effective vectors is a need of the hour. Gene carriers are mainly classified into viral vectors, which offer superior transfection efficiency but come with higher toxicity issues, and non-viral vectors, which have lower cytotoxicity [50]. CS is included in the non-viral vector category and is successfully applied in many gene delivery applications. The cationic characteristic of chitosan allows it to electrostatically bind with DNA and carry out the gene delivery [50]. pH of the media, degree of deacetylation [51], charge on the CS molecule, and molecular weight are some of the factors that can influence the transfection capability of CS/DNA polyplex. An increase in the charge density on the CS that comes with higher deacetylation degree will favor CS-DNA binding while the binding ability and the molecular weight are related inversely. The chain length of CS and the pH of the media is also related inversely to CS-DNA polyplex formation [48].

Potential gene vehicles of CS nanoparticles (CNPs) were synthesized which are capable of passing through the blood-brain barrier and are effective in treating brain cancer and other brain diseases. The average size of CS and CS-green fluorescent protein (GFP) nanoparticles were 316 and 289 nm respectively, while the zeta potential was 28.8 and 10.6 mVs respectively. The GFP-tagged plasmid's in vitro transfection in brain cancer MG-U87 cell line and HEK293-293 cell line were evaluated to determine the efficiency of the CS-GFP nanoparticles. The fluorescent microscopy results showed that the efficiency of transfection for the synthesized CS-GFP nanoparticles was 27.41% in HEK293 cells and 20.56% of cells in U-87 cells respectively, which is higher than the transfection found for commercial reagent, CTR. Also, 53% efficiency was observed for the nanoparticles on the U-87 cells which is again more than the efficiency showed for CTR when fluorescent activated cell sorting (FACS) analysis was evaluated. Further, the in vivo study on Balb/c mice confirmed the nanoparticles' ability to cross the blood-brain barrier [52]. In another study, nanoparticles of polyelectrolytes fabricated from human SET1 (hSET1) antisense oligonucleotide loaded trimethyl chitosan (TMC)-HA, TMC-SA, and TMC-dextran sulfate were synthesized. The primary screening and D-method were used to select TMC-HA nanoparticles for further studies. The studies on MCF7 cells showed a considerable decrease in viability when evaluated for the oligonucleotide-loaded synthesized nanoparticles. Additionally, real-time PCR results revealed a 44% reduction in hSET1 expression. In vivo studies further showed the nanoparticles' ability to be effective against colon and intestine cancer [53]. Wang et al. fabricated a platform for gene delivery to improve the regeneration of nerves by loading nanoparticles of RCP (Arginine-Glycine-Aspartate (RGD)-functionalized CS-graft) gene vector combined with pJUN on poly-L-lactic acid (PLLA)-silk fibroin (SF) parallel fibrin (PSPF). The in vivo experiments indicated that this RCP/pJUNPSPF@PGA scaffold can deliver the gene vector and boost the NGF, BDNF, and VEGF by transfecting c-JUN in Schwann cells [54].

Polyplexes of TMC and HA were synthesized and studied as a potential intravitreal siRNA delivery platform. The cellular uptake of polyplexes was obtained from confocal microscopy. The quantitative proof of significant transfection was given by a 90% mean fluorescence intensity from flow cytometry profiles in both HUVEC and RPE cell lines. When evaluated using RT-PCR, the polyplexes demonstrated a 70% reduction in VEGFR-2 expression. Furthermore, when injected intravitreally, the polyplexes were capable of permeating through the retinal barriers within 1 h. In vivo experiments also showed a reduction in laser-induced CNV lesions. These results show the polyplexes' potential to be an ocular gene vehicle [55]. CD44 targeting siRNA delivering platform that can interfere with the oncogene Bcl2 was developed using hyaluronic dialdehyde and CS nanoparticles (siRNA@CS-HAD NPs) for efficaciously treating bladder cancer. The siRNA loading efficiency into the nanoparticle was observed to be 95% using UV spectroscopy. Moreover, the transfection rate when quantified via flow cytometry showed more fluorescence intensity for siRNA@CS-HAD NPs. Immunofluorescence results and fluorescence images of the mice suggested that the siRNA@CS-HAD NPs are capable of targeted siRNA delivery to the tumor site in an in vivo model [56].

2.4 Tissue Engineering

Tissue engineering combines biology with engineering principles to develop biomaterials to achieve damaged tissue repair caused by disease or injury. Developing proper designs for the material is crucial in achieving the goals of tissue engineering [57]. CS possesses favorable characteristics such as biocompatibility, low toxicity, and biodegradability, giving it significant importance as a raw material for biomaterial fabrication. Another property that qualifies CS for the fabrication of biomaterials is the ability of CS to promote ECM generation via its biodegraded N-acetyl-D-glucosamine [58]. The development of different formulations of CS, such as beads, films, hydrogels, scaffolds, sponges, and microspheres, are being used in skin, bone, and cartilage tissue engineering. The main features of chitosan that can contribute to bone and cartilage tissue engineering are listed in Fig. 6.

Fabrication of CS-based materials that could mimic the complex bone and cartilage characteristics is of great importance. In a recent study, CS, HA, and hydroxyapatite (HAp) based scaffolds were fabricated with the incorporation of GO, functionalized graphite (f-Gr) and functionalized multi-walled carbon nanotube

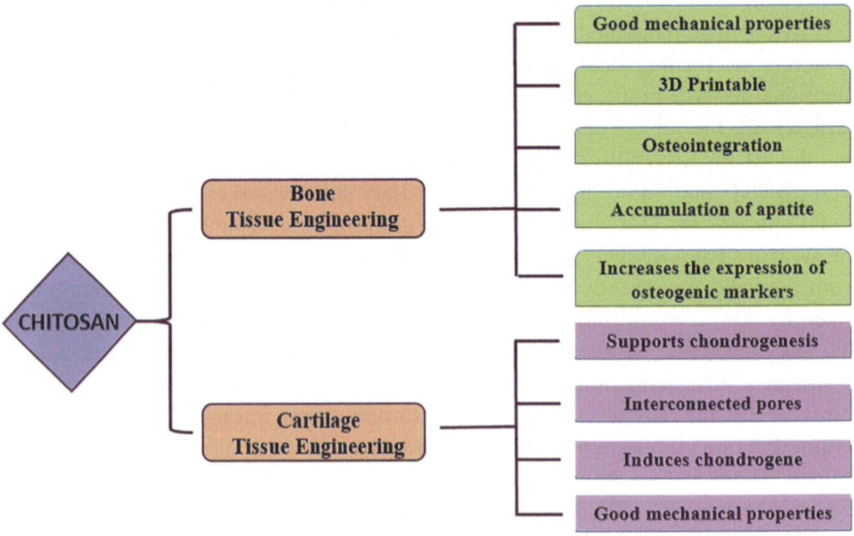

Fig. 6 Properties of chitosan that enable bone and cartilage regeneration

(f-MWCNT). CS/HA/HAp/GO, CS/HA/HAp/f-Gr, and CS/HA/HAp/f-MWCNT scaffolds were studied while keeping the CS/HA/HAp scaffold as a control. Hydroxyapatite, a human bone mineral component, is incorporated to boost the mechanical strength of the scaffold in association with its osteoinductive ability. Along with its thermal, mechanical, and hydrophilic properties, earlier studies show that f-Gr promotes osteoblast adhesion and proliferation. Functionalized multi-walled carbon nanotubes (f-MWCNT) in the formulation gave the scaffold solid protein absorption, cell adhesion, and differentiation properties. The lyophilized scaffolds exhibited structures with varying pore sizes and the GO, f-MWCNT, and f-Gr incorporation enhanced the mechanical strength of the scaffold. Also, when used in conjunction with CS/HA/HAp, f-MWCNT contributed to cell differentiation [59].

In another work, a nanocomposite scaffold was fabricated using CS, poly (2-hydroxyethyl methacrylate) (PHEMA), and modified SiO_2 nanoparticles. CIP drug was added to this nanocomposite scaffold which showed encapsulation efficiency of 20%. The drug release profile showed a burst release of CIP at pH values 4.2 and 7.4 in the first 5 h, and then maintained a sustained release up to 120 h. The maximum value for in vitro drug release was observed at pH 4.2 due to the weakening of the hydrogen bond between CIP and the polymer matrix in the acidic range. The scaffold showed antibacterial, blood-biocompatible as well as protein adsorption properties along with osteoblast cell proliferation [60]. These properties of the thus fabricated scaffold made it eligible for bone tissue engineering. Injectable CS hydrogel with improved osteoinduction and mechanical strength was developed by loading CS-modified halloysite nanotubes (mHNTs) with icariin, a bone inducer, and incorporating it with CS/Glycerophosphate (CS/GP) hydrogel. The hydrogel is formed via sol-gel transition with reduced gelation time. The hydrogel was further encapsulated with human adipose-derived stem cells (hASCs). HNT modification with CS reduced initial burst release and increased the entrapment efficiency and loading capacity. Furthermore, the differentiated hASCs encapsulated in the scaffold demonstrated increased proliferation and mechanical strength [61]. Therefore, the developed injectable hydrogel showed the potential to be used in bone tissue engineering.

Bioengineering materials that can provide support and a proper environment for re-growing the damaged cartilage tissue, as these show little propensity to self-heal are called cartilage tissue engineering. These biomaterials should be biologically active, and should have 3D structural features, mechanical strength, significant porous nature, and degradation rate [62]. CS can be used for this biomaterial fabrication owing to CS's ability to promote chondrogenic activity, ability to interact with GAGs of negative charge, and to induce expression of proteins that are cartilage-specific [63]. 3D bioprinting is a rapidly evolving technique for developing cells, biomaterials, and biologics for fabricating tissue or cell analogues for tissue engineering applications [64]. 3D-printed stable CS inks were formulated by adding EDTA followed by physically crosslinking it with calcium for printing templates for hydrogel which can achieve cartilage engineering via chondrocyte culturing. This method is used to overcome the disadvantages of pristine CS including reduced solubility and low mechanical strength. The ability of chondrocyte cells to maintain its phenotype further proved the bioink's potential to be used in tissue engineering [65]. In another study, gellan gum-glycol CS matrix was fabricated to learn its potential in cartilage tissue engineering. Significant in vitro cell proliferative, cell viability results, its ability to enhance the mRNA expression of genes involved with cartilages, and the ability to synthesize GAGs further pointed out its tissue engineering aspects [66]. In a different study, CS/alginate/fibrin injectable nanocomposite gels were fabricated with the incorporation of CS nanoparticles encapsulated with strontium ranelate (SCNPS) for cartilage tissue engineering. Strontium ranelate is an osteoporotic drug that also contributes to cartilage ECM generation by stimulating chondrocyte anabolism and its encapsulation in CS nanoparticles ensures the sustained release of strontium ranelate. The quantification experiments showed significant increase in GAG synthesis and collagen deposition which indicated chondrogenesis [67]. In a previous research, CS/PVA/GO electrospun nanofibers were fabricated. It was found that the scaffold optimized with increased roughness improved the chondrocyte proliferation [68]. All these CS-based studies further expand the possibility of using CS in various bone and cartilage tissue engineering.

Regeneration of the initial anatomy and restoration of physiological functions of the skin after an injury or disease are the main goals of skin tissue engineering [69]. Skin acts as a protective barrier against the external environment and it is of great importance to protect and keep the integrity of the skin safe in place. Healthy skin is composed of different types of cells like fibroblasts, keratinocytes, endothelial, and melanocytes thus making up the epidermis, dermis, and subcutis [70]. According to the skin defect, different approaches exist to re-establish the native skin integrity. In this regard, pharmaceutical and cosmetic industries have a growing need for skin substitutes to replace damaged human skin [71]. Wound healing is an important field of application where skin tissue engineering is put to great use. This field of study became more significant as an approach to overcome the limitations of pain, infections, and scarring involved with the use of tissue-engineered skin autografts and allografts as wound healing therapies [72].

2.5 *Wound Healing*

Wound healing is the biological process of regaining the structure and functions of the largest organ of the body, the skin, after an injury or illness. It progresses through four overlapped steps named hemostasis, inflammation, proliferation, and remodeling [73]. It is a complex process involving several types of cells (fibroblasts, endothelial cells, keratinocytes, etc.) and mediators (growth factors, cytokines, etc.). The time frame of the healing process will depend on the type of wound, the cause or the reason for the wound formation, and the complications involved like infection [74] and prolonged inflammation [75]. The chronic condition arises as a result of polymicrobial infections which can lead to biofilm formation that might prolong the healing process and therefore should be treated with utmost care and calculated management strategies. This polymicrobial community may consist of *S. aureus, E. coli, P. aeruginosa, Candida* sp., *Proteus* sp., *Klebsiella* sp., etc. [76]. The multitude of features exhibited by CS via its antimicrobial action, swelling, and porous properties make CS an ideal option for fabricating wound healing dressing matrices. In a recent study, scaffolds of ε-poly-L-lysine (ε-PL) and N-acetylcysteine (NAC) incorporated CS bioconjugated with collagen hydrogels were prepared. Collagen – NAC – ε-PL (CNE), Collagen – CS (CCS), Collagen – CS – NAC – ε-PL (CCSNE), CS – NAC – ε-PL (CSNE), and CS and collagen (C) solutions (controls) are the six different formulations prepared for this study. The live-dead assay of the scaffolds was evaluated using calcein staining. The C, CCS, and CS-treated cells showed total confluence while the CNE, CSNE, and CCSNE showed lower confluence, but not dead. The scaffolds containing NAC and ε-PL showed antibacterial activity against *S. aureus* (MRSA), *Acinetobacter baumani*, and *Klebsiella pneumoniae carbapenemase* (KPC). Further, the CSNE formulation showed 70% wound closure in in vivo studies [77]. Using the technique of lyophilization, three scaffolds (SC-1, SC-2, SC-3) of collagen-CS with the incorporation of CIP were fabricated with a porosity of 91 ± 0.56%. Dual crosslinking with glutaraldehyde and EDC/NHC was performed to improvise the mechanical strength of the scaffolds, along with favorable swelling, degradation, excellent hemocompatibility, cell attachment, proliferation, and migration of fibroblast cells on the scaffold. SC-3 showed activity against *E. coli* and *S. aureus* with a diameter of the zone of inhibition of 20 ± 1.8 mm and 18 ± 1.4 mm respectively. Slow and sustained release of the antibiotic was observed for SC-1, while SC-2 and SC-3 showed rapid release that is required for rapid wound healing at the early stages of wound healing. In vivo studies with SC-2 showed full recovery of the wound while the wound closure when treated with the control was only 82.25%. Also, the histological studies showed that the scaffold has great potential to be used in skin tissue engineering applications [78].

A recent study engineered a wound dressing via lyophilization from CS hydrogel entrapped with amikacin sulfate (AK) and diopside nanoparticles (CS-nDE-AK). A 7-day sustained release of 69.6 ± 8.2% amikacin sulfate was obtained which followed the Higuchi model kinetics. The in vitro antibacterial activity, in vitro anti-biofilm activity, ex vivo antibacterial activity against *E. coli* and *S. aureus*, and in vitro scratch assay exhibit the potential aspect of using this dressing in managing and improving the healing of infected wounds [79]. In another research work, nanoparticles of CIP-laden fibrin and fluconazole-laden fibrin were incorporated to CS separately, to fabricate bandages with wound healing characteristics (Fig. 7). The swelling ratio value and porosity results indicated the exudate absorption capacity of the bandage in managing chronic wounds. The in vitro antimicrobial activity showed against *S. aureus*, *E. coli*, and *C. albicans*, and the ex vivo antimicrobial activity against a culture of a mixture of all three above microbes, and reduced in vivo microbial activity on a rat model highlighted the bandage's ability to be used in infected wound healing treatment [80].

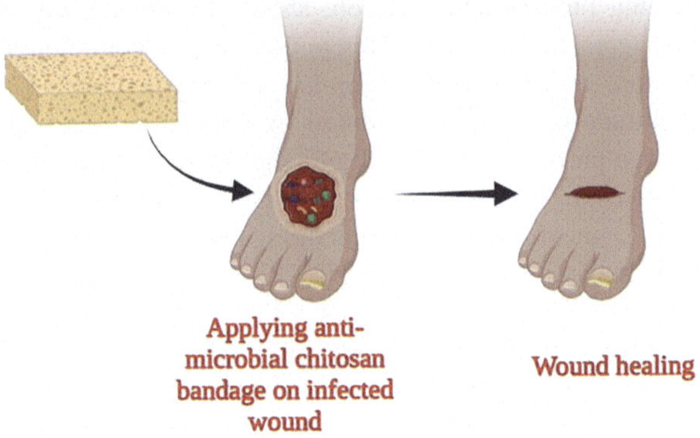

Fig. 7 Antimicrobial chitosan bandage for wound healing

In another study, injectable coacervate CS-HA hydrogels incorporated with pro-angiogenic deferoxamine molecule and deferoxamine nanoparticles were developed to achieve angiogenesis. In vitro drug release assay of free deferoxamine loaded gel showed a burst release of 40% of deferoxamine within 36 h and then sustained release over 5 days, whereas the deferoxamine nanoparticle incorporated gel showed release of deferoxamine over 10 days due to the entrapment of the drug within the hydrogel. In vitro tube formation assay on HUVECs showed positive results for both free deferoxamine and deferoxamine nanoparticles but the release profile of these respective molecules from the gel system was not sufficient to give significant tube formation. Further in vivo experiments on mice showed maximum blood vessel formation for deferoxamine nanoparticle-incorporated hydrogel which indicated the angiogenic potential of the hydrogel which can contribute to the healing of ischemic wounds [81].

2.6 *Hemostasis*

"Heme" and "stasis", two words that stemmed from Greek roots paved the way for the term "Hemostasis" which means "halt of blood" [82]. The body reacts to a traumatic injury via the process called hemostasis which is basically the body's effort to arrest the blood loss by means of forming clot. Out of 7–8% of the total weight of the blood, if the exsanguination exceeds more than 40%, the condition of the patient becomes perilous [83]. Hemostasis broadly follows the stages of constriction of blood conduits, temporary clot formation, and initiation of coagulation cascades to realize the goal of clotting [84]. Bone wax, sutures, and vascular clips are some of the typical methods of controlling blood loss [85]. To further strengthen the process, various hemostatics are being utilized in the medical field. CS is such a naturally sourced biopolymeric material which is used for its hemostatic properties on account of multiple positively charged amino groups present in its backbone chain [86]. The negative part of RBCs and platelets gets attracted towards the cationic part of CS to accomplish the clotting [87, 88]. But this itself is not adequate

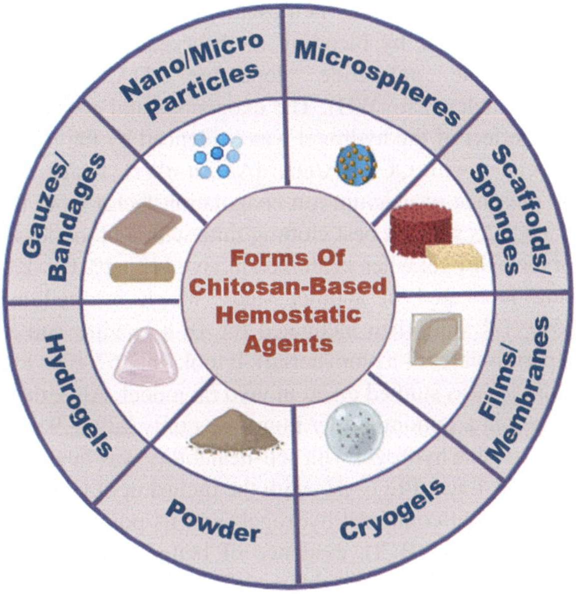

Fig. 8 Different forms of chitosan-based hemostatic agents

to stop a heavily bleeding wound. Therefore, improved hemostasis is achieved by the derivatization of CS and by incorporating other active agents with it [88]. Even though the number of positive amino groups increases the CS-induced hemostatic effect via platelet and erythrocyte aggregation, a higher deacetylation degree can cause intramolecular hydrogen bonding with hydroxyl groups thereby limiting the availability of positive amino groups and this may result in decreased platelet and erythrocyte aggregation [89]. The ability to fabricate different forms of CS-based hemostatics that could be applied directly to induce hemostatic effect broadens the application of CS and its derivatives (Fig. 8) [86]. TraumaStat®, ChitoFlex®, HemCon®, Celox®, and Chitoseal® are some brands of CS-hemostatics available in the market [85, 90].

CS and CS derivatives in a hydrogel formulation have gained much attention, as these hydrogels showed satisfactory adhesion and hemostatic properties along with its exudate absorbing capacity and ability to maintain moist environment required for the faster healing of the wound. Sundaram et al. developed 2%CS-0.25% PA-0.25%Ca hydrogel by integrating potassium aluminum sulfate and calcium chloride in CS hydrogel to impart coagulation properties. In vitro clotting time, prothrombin time (PT), activated partial thromboplastin time (aPTT), and thrombin time (TT) values obtained for this optimized gel were 112 ± 12 s, 20.8 ± 2.1 s, 11.5 ± 0.8 s, and 26.8 ± 0.4 s respectively. When compared with Fibrin sealant and Floseal commercial products, the time took for the blood to form a clot was only 20 ± 10 s in the in vivo rat liver hemorrhage model, while the time taken in a femoral artery injured rat model was 105 ± 31 s [91].

A hydrogel that has hemostatic, anti-infectious, and wound-healing properties was engineered by blending methacrylate anhydride quaternized CS (QCSMA), methacrylate anhydride dopamine (DAMA), and Zn-doped whitlockite nanoparticles (Zn-nWH). The exogenous and endogenous pathway-directed hemostatic effect of the hydrogel was evidenced by the notable aPTT, PT time reduction associated with QCSMA/DAMA3/Zn-nWH, which was believed to be because of the presence of calcium ion coagulation factor from the whitlockite nanoparticles. 129 ± 22 s was the best clotting time with a blood loss of only 27 ± 5 mg obtained for the hydrogel when examined in vivo [92]. 2% CS-4% nano whitlockite hydrogels were fabricated in another study which resulted in a quick clotting time of 3.5 ± 1.2 min when incubated for 48 h in vitro and hemolysis percentage below 5 (permissible for a biomaterial). It took only 62 ± 3 s to realize the clotting when the hydrogel was studied on an in vivo rat model. Also, the time taken for hemostasis to happen in a femoral artery injury was only 229 ± 9 s [93].

Injectable hydrogel with self-healability was fabricated from *N*-succinyl chitosan, HA, and 4-arm-PEG-NH_2 with the inclusion of Ca^{2+} ions. The in vitro experiments with calcium-contained hydrogels and its powder precursor showed a lower blood clotting index (BCI), evidence of better clotting capacity. Further evidences of hemostatic ability were obtained from the in vivo studies which resulted in significantly reduced blood loss for NSC-OHA-based hydrogels (0.08 ± 0.02 to 0.16 ± 0.03 g) [94]. In another study, carboxymethyl chitosan (CMC)-oxidized hyaluronic acid-based biodegradable hydrogel with comparable clot time (38.9 ± 8.7 s) and blood loss property (217.1 ± 28.9 mg) as that of a fibrin sealant was fabricated and studied in an in vivo liver injured SD rat model [95]. *N, O*-CMC-based self-healing hydrogel with the in situ incorporation of silver nanoparticles (AgNPs) was developed via carboxyl-metal-ligand coordinate interaction. The prepared hydrogel showed in vitro antibacterial activity against a number of ATCC strains of *S. aureus, E. coli,* MRSA, *P. aeruginosa, K. pneumonia* and in vitro anti-biofilm activity against *S. aureus, E. coli,* and *P. aeruginosa*. This hemocompatible hydrogel showed hemostatic potential with a clotting time of 4.29 ± 0.59 min for *N, O*-CMC and 4.25 ± 0.85 min for *N,O*-CMC/AgNPs hydrogels, which can be attributed to the coagulation of erythrocytes and thrombocytes. This hydrogel finds application as a tissue sealant for incisional infected wounds [96]. The conjoined results of in vivo and in vitro experiments on a set of shear-thinned hydrogels fabricated from CS and GO (CSGO) provided evidence of potential formulation for improved hemostatic wound-healing application. 0.99%, 0.34%, and 0.67% were the obtained hemolysis ratio for CSGOs with varied GO concentration, which were all under the permitted value. Further, the in vivo studies on a liver injury rat model gave results that favored faster clotting time (80 ± 16.7 s) and reduced blood loss (300 ± 0 mg) [97].

Combining two materials with extremely different chemical, physical, or mechanical features to create a novel material that retains the characteristics of its parent raw materials is called composite material [98]. In a previous study, the wet granulation technique was used in developing porous CS-$CaCO_3$ composite granules with hemostatic properties. The results from aPTT, TT, and PT assays indicated

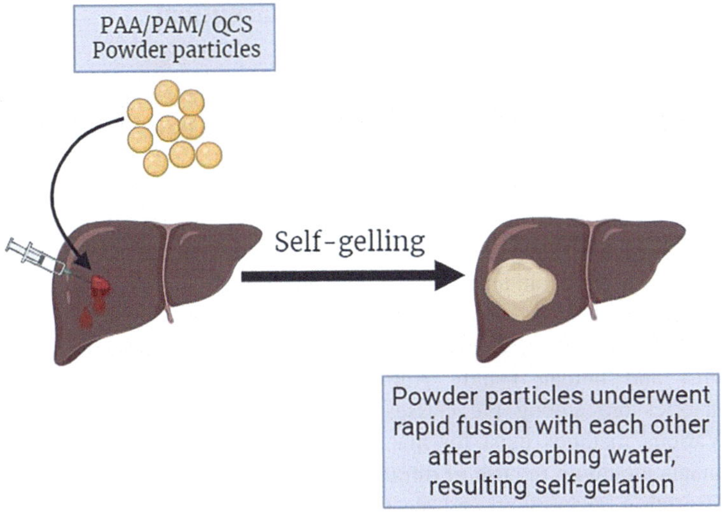

Fig. 9 Illustration showing self-gelling property of PAA/PAM/QCS powder

that the developed composite accelerates intrinsic coagulation. The composite displayed a hemolysis percentage of less than 5, SEM images showed its ability to cluster the platelets on its surface, and in vivo experiments on a rat model further validated its hemostatic and wound healing potential [99]. In another study, silica, calcium, and phosphate-containing CS nano bioglass (2%CS-5%nBG) hydrogel was fabricated. The composite demonstrated a blood clotting time of 212 ± 12 s in vitro, 54 ± 3 s for liver injury in vivo, and 185 ± 9 s for femoral artery injury in vivo, highlighting its effective blood clotting properties [100].

CS in other formulations is also being widely used in hemostatic applications. For example, CS-calcium phosphate nanoflake wound dressing was fabricated, with effective blood clotting characteristics. Hemolysis percentage of 3.4, the ability to induce the aggregation of platelets, and the 15 s blood clotting time all pointed towards the dressing's clotting abilities [101]. Oxidized dialdehyde cellulose and CS crosslinked foam sponges were fabricated with effective blood coagulation characteristics. The time taken for the blood to clot in the tail vein of mice was 163.3 ± 5.8 s and the time taken in a rabbit model was 200.0 ± 34.6 s in vivo [102]. Tao et al. created nanofiber membranes of polyethylene oxide, CS/polyethylene oxide (PEO), and kaolin nanofiber membranes via electrospinning. The time taken for the coagulation of whole blood was 43 ± 1.4 s, while the in vivo rat liver injury model demonstrated a clotting time of 57.3 ± 4.1 s [103]. Polyacrylic acid, polyacrylamide, and quaternate CS containing self-gelling powder (PAA/PAM/QCS) were synthesized in another study (Fig. 9) [104], and CS with silver and tannic acid in a cryogel formulation for hemostatic application was developed and studied in a different study [105]. All these studies displayed that CS can be used as a hemostatic agent in different occasions.

3 Chitosan as an Adsorbent of Pharmaceutical Contaminants

One of the main concerns in today's world is the threat posed by pharmaceutical compounds on water resources thereby indirectly threatening the ecosystem itself. Once they enter the water body, these pharmaceutical compounds can persist due to their lack of biodegradability [106] potentially causing long-term effects. They can pose life-threatening complications for aquatic life and other species, including humans who consume contaminated water. While developing novel pharmaceutical compounds to treat various diseases is crucial, it is equally important to find effective methods to eliminate these contaminants from the environment to prevent harm to ecosystems and human health. Antibiotics, hormones, analgesics, and anti-inflammatories are among the classes of existing pharmaceutical contaminants [106]. Various methods exist for managing and removing these pharmaceutical contaminants including the use of ultrafiltration membranes, biodegradation, ozonation, adsorption, absorption, and electrocoagulation. Among these, adsorption is often preferred due to its low cost and simplicity [107]. The abundant hydroxyl and amino groups in the polymer chain of CS facilitate an efficient adsorption process for these contaminants [108].

In a study, activated carbon-CS beads (ACCB) were synthesized for the removal of anti-inflammatory drug diclofenac (DIF) from the effluents. The fibrous network in the structure of ACCB was considered useful in facilitating adsorption due to its increased surface area. When the pH of the DIF solution was 6, electrostatic interactions between the negatively charged DIF and positively charged ammonium groups of CS, weak π–π interactions between the aromatic groups on both species and hydrogen bonding aided the adsorption of DIF on ACCBs. Also, multilayer adsorption of the adsorbate (DIF) on the adsorbent (ACCB) via chemisorption was confirmed through the pseudo-second-order kinetics and Freundlich isotherm [109]. Mahmoodi et al. synthesized hydrogel systems of GO–CS and amine graphene oxide–CS (AGO–CS) as adsorbents for removing DIF from aqueous sources. A mass ratio of 2:5 for GO or AGO and CS was selected as the optimum while the optimal pH found was 5, at which the hydrogel systems showed 90.42% and 97.06% adsorption for GO–CS and AGO–CS respectively. Langmuir model fitted the most in this case which revealed monolayer adsorption of the drug on the adsorbent. A higher removal efficiency was found at lower temperatures as the adsorption was found to be exothermic. The adsorbents showed significant removal efficiency and recoverability even after 3 cycles of adsorption-desorption and hence has great potential as an adsorbent of DIF [110].

In another study, ibuprofen (IBU), rifampicin (RIF), and streptomycin (STM) adsorption on mud-crab shell extracted CS was studied which followed pseudo-second-order kinetics. For RIF and STM, Langmuir model fitted the most revealing homogenous monolayer adsorption of these on CS adsorbent while for IBU, Freundlich isotherm gave evidence of multilayer adsorption. The relatively smaller values of mean free energy of adsorption (4.226 J mol^{-1} for RIF, 6.202 J mol^{-1} for

STM, and 2.988 J mol^{-1} for IBU) indicated weak interactions between the adsorbent and adsorbate. In this study, RIF showed the most affinity towards CS, then IBU and then STM. The desorption process of these drugs from CS is done by simply washing with water or slightly acidic or basic solutions, which then can be reused for further adsorption-desorption cycles [111]. In another study, a floating zwitter-ionic chitosan-MEDSP composite material was fabricated, demonstrating the capability to remove a mixture of DIF, ketoprofen (KPF), salicylic acid (SAL), IBU, and paracetamol (PAR) from an aqueous system. One of the main factors that affect the adsorption is the pH of the solution and in this case, at the determined optimal pH of 4, DIF showed the highest adsorption of 92.5%, followed by SAL of 82.5%, KPF of 82%, PAR of 67.5%, and IBF of 42%. Also, the optimum pH of desorption was determined to be 10. After 10 cycles of adsorption-desorption, the one that lost its adsorption capacity more was DIF, which showed the highest adsorbing power in the first cycle followed by SAL and KPF. PAR and IBF showed zero loss even after 10 cycles [112].

Adsorbent sponges composed of carboxymethyl cellulose, carboxyalkyl chitosan, and sulfonated graphene oxide (CMc/SGO–GCC) to eliminate sulfonamide antibiotic contaminants were fabricated and reported. The adsorption of sulfamethoxazole (SMX) and sulfapyridine (SPD) on the adsorbent were found to be pH dependent with an optimal pH of 6 and 8 respectively. Even after 5 cycles, the adsorbent was able to maintain 87% adsorption capacity which makes it a potential adsorbent for removing sulfonamide pollutants [113]. A nanocomposite bead composed of nano-ZnO, Fe_3O_4, CS, and SA was fabricated via green synthesis which is capable of degrading 94.77% of CIP and 93.31% of sulfamethoxazole (SMX) when irradiated with UV-C light under determined optimal conditions of pH and temperature. Pseudo-first-order kinetics and the Langmuir model fitted the most for both antibiotics. The SMX and CIP photodegradation involved hydroxyl and superoxide radicals and the degraded products showed less toxicity compared to the parent compound. This novel green CS nanocomposite is a potential remover of pollutants by photodegradation [114]. These studies sum the effective use of CS in different formulations for the removal of pharmaceutical contaminants.

4 Conclusion

This chapter gives a brief summary of the pharmaceutical applications of chitosan and its derivatives. The structural and biological properties inherited by chitosan and chitosan derivatives account for its wide-range applications in this sector. Its properties of antibacterial and anti-inflammatory render it a fitting candidate for tissue engineering, wound healing, and hemostatic applications. Its low-toxic, biocompatible features are being exploited in formulating drug compositions. The possible electrostatic linkage between cationic chitosan and other anionic counterparts leads to designing chitosan and its derivative-composed drug and gene delivery vehicles. Also, the adsorbing-desorbing property qualifies it as an adsorbent of pharmaceutical pollutants that are being generated in massive amounts. In short, chitosan, the smart material, is being explored and can continue to be explored in the future and expand its use in the pharmaceutical sector.

Acknowledgements One of the authors, Joel Maria Joju, is grateful to Amrita Vishwa Vidyapeetham for providing her Research fellowship. The authors are also thankful to Amrita Vishwa Vidyapeetham for providing seed grant fund.

References

1. Ali A, Ahmed S (2018) A review on chitosan and its nanocomposites in drug delivery. Int J Biol Macromol 109:273–286
2. Al-Rooqi MM, Hassan MM, Moussa Z, Obaid RJ, Suman NH, Wagner MH, Natto SS, Ahmed SA (2022) Advancement of chitin and chitosan as promising biomaterials. J Saudi Chem Soc 26:101561
3. Crini G (2019) Historical review on chitin and chitosan biopolymers. Environ Chem Lett 17: 1623–1643
4. Khan A, Alamry KA (2021) Recent advances of emerging green chitosan-based biomaterials with potential biomedical applications: a review. Carbohydr Res 506:108368
5. Biswal A, Swain SK (2023) Chitosan: a smart biomaterial. In: Swain SK, Biswal A (eds) Chitosan nanocomposites: bionanomechanical applications, pp. 1–25
6. Shete A, Chavan A, Potekar P, Yadav G, Shah N (2024) Modification of physicochemical properties of chitosan to improve its pharmaceutical and agrochemical potential applications. Int J Biol Macromol 267:131404
7. Kim S (2018) Competitive biological activities of chitosan and its derivatives: antimicrobial, antioxidant, anticancer, and anti-inflammatory activities. Int J Polym Sci 2018:1708172
8. Hasnain MS, Nayak AK (2022) Chitosan as mucoadhesive polymer in drug delivery. In: Hasnain MS, Beg S, Nayak AK (eds) Chitosan in drug delivery, pp 225–246
9. Bakshi PS, Selvakumar D, Kadirvelu K, Kumar NS (2020) Chitosan as an environment friendly biomaterial – a review on recent modifications and applications. Int J Biol Macromol 150:1072–1083
10. Morin-Crini N, Lichtfouse E, Torri G, Crini G (2019) Applications of chitosan in food, pharmaceuticals, medicine, cosmetics, agriculture, textiles, pulp and paper, biotechnology, and environmental chemistry. Environ Chem Lett 17:1667–1692
11. Khanna N, Chatterji T, Singh S, Khanna N (2023) A medico-biological approach to chitosan bionanocomposites. In: Swain SK, Biswal A (eds) Chitosan nanocomposites: bionanomechanical applications, pp 55–80
12. Guo Y, Qiao D, Zhao S, Liu P, Xie F, Zhang B (2024) Biofunctional chitosan–biopolymer composites for biomedical applications. Mat Sci Eng R 159:100775
13. Panonnummal R, Kumar VS, Jayakumar R, Sabitha M (2021) Application of chitosan and its derivatives in transdermal drug delivery. In: Jayakumar R, Prabaharan M (eds) Chitosan for biomaterials IV: biomedical applications, pp 411–446
14. Soni V, Jain P, Haider T, Pandey V, Tekade RK (2018) Documentation protocol in product development including clinical records. In: Tekade RK (ed) Dosage form design parameters, pp 403–440
15. Kalasz H, Antal I (2006) Drug excipients. Curr Med Chem 13:2535–2563
16. Ilium L (1998) Chitosan and its use as a pharmaceutical excipient. Pharm Res 15:1326–1331
17. Khan MI, An X, Dai L, Li H, Khan A, Ni Y (2019) Chitosan-based polymer matrix for pharmaceutical excipients and drug delivery. Curr Med Chem 26:2502–2513

18. Singh MK, Prajapati SK, Mahor A, Rajput N, Singh R (2011) Chitosan: a novel excipient in pharmaceutical formulation: a review. Int J Pharm Sci 2:2266
19. Lupo N, Fodor B, Muhammad I, Yaqoob M, Matuszczak B, Bernkop-Schnürch A (2017) Entirely S-protected chitosan: a promising mucoadhesive excipient for metronidazole vaginal tablets. Acta Biomater 64:106–115
20. Patil S, Pandit A, Godbole A, Dandekar P, Jain R (2021) Chitosan based co-processed excipient for improved tableting. Carbohydr Polym Technol Appl 2:100071
21. Pandey P, Dahiya M (2016) Oral disintegrating tablets: a review. Int J Pharm Sci Rev Res 5: 50–62
22. Draksiene G, Venclovaite B, Pudziuvelyte L, Ivanauskas L, Marksa M, Bernatoniene J (2021) Natural polymer chitosan as super disintegrant in fast orally disintegrating meloxicam tablets: formulation and evaluation. Pharmaceutics 13:879
23. Ren G, Clancy C, Tamer TM, Schaller B, Walker GM, Collins MN (2019) Cinnamyl O-amine functionalized chitosan as a new excipient in direct compressed tablets with improved drug delivery. Int J Biol Macromol 141:936–946
24. Shokati Eshkiki Z, Mansouri F, Karamzadeh AR, Namazi A, Heydari H, Akhtari J, Tabaeian SP, Akbari A (2024) Chitosan and its derivative-based nanoparticles in gastrointestinal cancers: molecular mechanisms of action and promising anticancer strategies. J Clin Pharm Ther 2024:1239661
25. Mu L, Wu L, Wu S, Ye Q, Zhong Z (2024) Progress in chitin/chitosan and their derivatives for biomedical applications: where we stand. Carbohydr Polym 343:122233
26. Ghauri ZH, Islam A, Qadir MA, Gull N, Haider B, Khan RU, Riaz T (2021) Development and evaluation of pH-sensitive biodegradable ternary blended hydrogel films (chitosan/guar gum/PVP) for drug delivery application. Sci Rep 11:21255
27. Rizwan M, Yahya R, Hassan A, Yar M, Azzahari AD, Selvanathan V, Sonsudin F, Abouloula CN (2017) pH sensitive hydrogels in drug delivery: brief history, properties, swelling, and release mechanism, material selection and applications. Polymers 9:137
28. Rahnama H, Khorasani SN, Aminoroaya A, Molavian MR, Allafchian A, Khalili S (2021) Facile preparation of chitosan-dopamine-inulin aldehyde hydrogel for drug delivery application. Int J Biol Macromol 185:716–724
29. Sayyar Z, Mahdavinia GR, Khataee A (2023) Dual-drug (curcumin/ciprofloxacin) loading and release from chitosan-based hydrogels embedded with magnetic montmorillonite/hyaluronic acid for enhancing wound healing. J Biol Eng 17:66
30. Afshar M, Dini G, Vaezifar S, Mehdikhani M, Movahedi B (2020) Preparation and characterization of sodium alginate/polyvinyl alcohol hydrogel containing drug-loaded chitosan nanoparticles as a drug delivery system. J Drug Deliv Sci Technol 56:101530
31. Affes S, Aranaz I, Acosta N, Heras Á, Nasri M, Maalej H (2021) Chitosan derivatives-based films as pH-sensitive drug delivery systems with enhanced antioxidant and antibacterial properties. Int J Biol Macromol 182:730–742
32. Li J, Zhuang S (2020) Antibacterial activity of chitosan and its derivatives and their interaction mechanism with bacteria: current state and perspectives. Eur Polym J 138:109984
33. Jafari Z, Rad AS, Baharfar R, Asghari S, Esfahani MR (2020) Synthesis and application of chitosan/tripolyphosphate/graphene oxide hydrogel as a new drug delivery system for Sumatriptan succinate. J Mol Liq 315:113835
34. Pradeep A, Ashok N, Priya V, Pillai AV, Menon RR, Kumar VA, Jayakumar R (2022) Colistimethate sodium-chitosan hydrogel for treating gram-negative bacterial wound infections. Int J Biol Macromol 214:610–616
35. Antony IR, Pradeep A, Pillai AV, Menon RR, Kumar VA, Jayakumar R (2023) Antiseptic chitosan-poly (hexamethylene) biguanide hydrogel for the treatment of infectious wounds. J Funct Biomater 14:528
36. Nithya S, Nimal TR, Baranwal G, Suresh MK, Anju CP, Kumar VA, Mohan GC, Jayakumar R, Biswas R (2018) Preparation, characterization and efficacy of lysostaphin-chitosan gel against Staphylococcus aureus. Int J Biol Macromol 110:157–166

37. Nimal TR, Baranwal G, Bavya MC, Biswas R, Jayakumar R (2016) Anti-staphylococcal activity of injectable nano tigecycline/chitosan-PRP composite hydrogel using Drosophila melanogaster model for infectious wounds. ACS Appl Mater Interfaces 8:22074–22083
38. Pandian M, Kumar VA, Jayakumar R (2021) Antiseptic chitosan bandage for preventing topical skin infections. Int J Biol Macromol 193:1653–1658
39. Chai Y, Wang Y, Li B, Qi W, Su R, He Z (2021) Microfluidic synthesis of lignin/chitosan nanoparticles for the pH-responsive delivery of anticancer drugs. Langmuir 37:7219–7226
40. Nasef SM, Khozemy EE, Mahmoud GA (2023) pH-responsive chitosan/acrylamide/gold/ nanocomposite supported with silver nanoparticles for controlled release of anticancer drug. Sci Rep 13:7818
41. Sohail R, Abbas SR (2020) Evaluation of amygdalin-loaded alginate-chitosan nanoparticles as biocompatible drug delivery carriers for anticancerous efficacy. Int J Biol Macromol 153:36–45
42. Matos BN, Pereira MN, Bravo MDO, Cunha-Filho M, Saldanha-Araújo F, Gratieri T, Gelfuso GM (2020) Chitosan nanoparticles loading oxaliplatin as a mucoadhesive topical treatment of oral tumors: iontophoresis further enhances drug delivery ex vivo. Int J Biol Macromol 154: 1265–1275
43. Perry SL, McClements DJ (2020) Recent advances in encapsulation, protection, and oral delivery of bioactive proteins and peptides using colloidal systems. Molecules 25:1161
44. Mohammadi Z, Eini M, Rastegari A, Tehrani MR (2021) Chitosan as a machine for biomolecule delivery: a review. Carbohydr Polym 256:117414
45. Batista P, Castro P, Madureira AR, Sarmento B, Pintado M (2019) Development and characterization of chitosan microparticles-in-films for buccal delivery of bioactive peptides. Pharmaceuticals 12:32
46. Alfatama M, Choukaife H, Alkhatib H, Al Rahal O, Zin NZM (2024) A comprehensive review of oral chitosan drug delivery systems: applications for oral insulin delivery. Nanotechnol Rev 13:20230205
47. Yadav P, Yadav AB (2021) Preparation and characterization of BSA as a model protein loaded chitosan nanoparticles for the development of protein-/peptide-based drug delivery system. Futur J Pharm Sci 7:1–9
48. Saranya N, Moorthi A, Saravanan S, Devi MP, Selvamurugan N (2011) Chitosan and its derivatives for gene delivery. Int J Biol Macromol 48:234–238
49. Malek-Khatabi A, Javar HA, Dashtimoghadam E, Ansari S, Hasani-Sadrabadi MM, Moshaverinia A (2020) In situ bone tissue engineering using gene delivery nanocomplexes. Acta Biomater 108:326–336
50. Jabali MK, Allafchian AR, Jalali SAH, Shakeripour H, Mohammadinezhad R, Rahmani F (2022) Design of a pDNA nanocarrier with ascorbic acid modified chitosan coated on superparamagnetic iron oxide nanoparticles for gene delivery. Colloids Surf A Physicochem Eng Asp 632:127743
51. Aranda-Barradas ME, Trejo-López SE, Del Real A, Álvarez-Almazán S, Méndez-Albores A, García-Tovar CG, González-Díaz FR, Miranda-Castro SP (2022) Effect of molecular weight of chitosan on the physicochemical, morphological, and biological properties of polyplex nanoparticles intended for gene delivery. Carbohydr Polym Technol Appl 4:100228
52. Khan IN, Navaid S, Waqar W, Hussein D, Ullah N, Khan MUA, Hussein Z, Javed A (2024) Chitosan-based polymeric nanoparticles as an efficient gene delivery system to cross blood brain barrier: in vitro and in vivo evaluations. Pharmaceuticals 17:169
53. Baghaei M, Tekie FSM, Khoshayand MR, Varshochian R, Hajiramezanali M, Kachousangi MJ, Dinarvand R, Atyabi F (2021) Optimization of chitosan-based polyelectrolyte nanoparticles for gene delivery, using design of experiment: in vitro and in vivo study. Mater Sci Eng C 118:111036
54. Wang Y, Li Y, Huang Z, Yang B, Mu N, Yang Z, Deng M, Liao X, Yin G, Nie Y, Chen T, Ma H (2022) Gene delivery of chitosan-graft-polyethyleneimine vectors loaded on scaffolds for nerve regeneration. Carbohydr Polym 290:119499

55. Chaharband F, Daftarian N, Kanavi MR, Varshochian R, Hajiramezanali M, Norouzi P, Arefian E, Atyabi F, Dinarvand R (2020) Trimethyl chitosan-hyaluronic acid nano-polyplexes for intravitreal VEGFR-2 siRNA delivery: formulation and in vivo efficacy evaluation. Nanomed Nanotechnol Biol Med 26:102181
56. Liang Y, Wang Y, Wang L, Liang Z, Li D, Xu X, Chen Y, Yang X, Zhang H, Niu H (2021) Self-crosslinkable chitosan-hyaluronic acid dialdehyde nanoparticles for CD44-targeted siRNA delivery to treat bladder cancer. Bioact Mater 6:433–446
57. Sultankulov B, Berillo D, Sultankulova K, Tokay T, Saparov A (2019) Progress in the development of chitosan-based biomaterials for tissue engineering and regenerative medicine. Biomol Ther 9:470
58. Amirthalingam S, Rajendran AK, Mani P, Rangasamy J (2021) Perspectives and challenges of using chitosan in various biological applications. In: Rangasamy J, Prabaharan M (eds) Chitosan for biomaterials III: structure-property relationships, pp 1–22
59. Rahman SF, Ghiffary MM, Tampubolon JY, Yulianti ES, Nadhif MH, Katili PA, Hanafiah S, Pangesty AI, Maras MAJ (2024) Effect of graphite, graphene oxide, and multi-walled carbon nanotubes on the physicochemical characteristics and biocompatibility of chitosan/hyaluronic acid/hydroxyapatite scaffolds for tissue engineering applications. J Sci Adv Mater Devices 9: 100719
60. Zare S, Eskandani M, Vandghanooni S, Hossainpour H, Jaymand M (2024) Ciprofloxacin-loaded chitosan-based nanocomposite hydrogel containing silica nanoparticles as a scaffold for bone tissue engineering application. Carbohydr Polym Technol Appl 7:100493
61. Kazemi-Aghdam F, Jahed V, Dehghan-Niri M, Ganji F, Vasheghani-Farahani E (2021) Injectable chitosan hydrogel embedding modified halloysite nanotubes for bone tissue engineering. Carbohydr Polym 269:118311
62. Zhang R, Chang SJ, Jing Y, Wang L, Chen CJ, Liu JT (2023) Application of chitosan with different molecular weights in cartilage tissue engineering. Carbohydr Polym 314:120890
63. Islam MM, Shahruzzaman M, Biswas S, Sakib MN, Rashid TU (2020) Chitosan based bioactive materials in tissue engineering applications – a review. Bioact Mater 5:164–183
64. Demirtaş TT, Irmak G, Gümüşderelioğlu M (2017) A bioprintable form of chitosan hydrogel for bone tissue engineering. Biofabrication 9:035003
65. He Y, Derakhshanfar S, Zhong W, Li B, Lu F, Xing M, Li X (2020) Characterization and application of carboxymethyl chitosan-based bioink in cartilage tissue engineering. J Nanomater 2020:2057097
66. Lee S, Choi JH, Park A, Rim M, Youn J, Lee W, Song JE, Khang G (2020) Advanced gellan gum-based glycol chitosan hydrogel for cartilage tissue engineering biomaterial. Int J Biol Macromol 158:452–460
67. Deepthi S, Gafoor AAA, Sivashanmugam A, Nair SV, Jayakumar R (2016) Nanostrontium ranelate incorporated injectable hydrogel enhanced matrix production supporting chondrogenesis in vitro. J Mater Chem B 4:4092–4103
68. Shamekhi MA, Mirzadeh H, Mahdavi H, Rabiee A, Mohebbi-Kalhori D, Eslaminejad MB (2019) Graphene oxide containing chitosan scaffolds for cartilage tissue engineering. Int J Biol Macromol 127:396–405
69. Bannasch H, Föhn M, Unterberg T, Bach AD, Weyand B, Stark GB (2003) Skin tissue engineering. Clin Plast Surg 30:573–579
70. Böttcher-Haberzeth S, Biedermann T, Reichmann E (2010) Tissue engineering of skin. Burns 36:450–460
71. Stojic M, López V, Montero A, Quílez C, de Aranda IG, Vojtova L, Jorcano JL, Velasco D (2019) Skin tissue engineering. In: Garcia-Gareta E (ed) Biomaterials for skin repair and regeneration, pp 59–99
72. Priya SG, Jungvid H, Kumar A (2008) Skin tissue engineering for tissue repair and regeneration. Tissue Eng Part B Rev 14:105–118

73. Solanki D, Vinchhi P, Patel MM (2023) Design considerations, formulation approaches, and strategic advances of hydrogel dressings for chronic wound management. ACS Omega 8: 8172–8189
74. Loesche M, Gardner SE, Kalan L, Horwinski J, Zheng Q, Hodkinson BP, Tyldsley AS, Franciscus CL, Hillis SL, Mehta S, Margolis DJ, Grice EA (2017) Temporal stability in chronic wound microbiota is associated with poor healing. J Invest Dermatol 137:237–244
75. Andrews KL, Houdek MT, Kiemele LJ (2015) Wound management of chronic diabetic foot ulcers: from the basics to regenerative medicine. Prosthet Orthot Int 39:29–39
76. Mahalakshmi P, Reshma G, Arthi C, Másson M, Rangasamy J (2023) Biodegradable polymeric scaffolds and hydrogels in the treatment of chronic and infectious wound healing. Eur Polym J 198:112390
77. Rodrigues JP, da Costa Silva JR, Ferreira BA, Veloso LI, Quirino LS, Rosa RR, Barbosa MC, Rodrigues CM, Gaspari PBF, Beletti ME, Goulart LR, Corrêa NCR (2024) Development of collagenous scaffolds for wound healing: characterization and in vivo analysis. J Mater Sci Mater Med 35:12
78. Tripathi S, Singh BN, Singh D, Kumar G, Srivastava P (2021) Optimization and evaluation of ciprofloxacin-loaded collagen/chitosan scaffolds for skin tissue engineering. 3 Biotech 11:160
79. Mothilal NP, Pradeep A, Arthi C, Gopal K, Kaliannagounder VK, Park CH, Kumar VA, Rangasamy J (2024) Amikacin sulphate loaded chitosan-diopside nanoparticles composite scaffold for infectious wound healing. Int J Biol Macromol 263:130217
80. Raveendran NT, Mohandas A, Menon RR, Menon AS, Biswas R, Jayakumar R (2018) Ciprofloxacin-and fluconazole-containing fibrin-nanoparticle-incorporated chitosan bandages for the treatment of polymicrobial wound infections. ACS Appl Bio Mater 2:243–254
81. Vignesh S, Sivashanmugam A, Annapoorna M, Janarthanan R, Subramania I, Jayakumar R (2018) Injectable deferoxamine nanoparticles loaded chitosan-hyaluronic acid coacervate hydrogel for therapeutic angiogenesis. Colloids Surf B Biointerfaces 161:129–138
82. Dorgalaleh A, Daneshi M, Rashidpanah J, Yasaghi ER (2018) An overview of hemostasis. In: Dorgalaleh A (ed) Congenital bleeding disorders: diagnosis and management, pp 3–26
83. Guo Y, Wang M, Liu Q, Liu G, Wang S, Li J (2023) Recent advances in the medical applications of hemostatic materials. Theranostics 13:161
84. Du J, Wang J, Xu T, Yao H, Yu L, Huang D (2023) Hemostasis strategies and recent advances in nanomaterials for hemostasis. Molecules 28:5264
85. Gheorghiță D, Moldovan H, Robu A, Bița AI, Grosu E, Antoniac A, Corneschi L, Antoniac L, Bodog AD, Băcilă CI (2023) Chitosan-based biomaterials for hemostatic applications: a review of recent advances. Int J Mol Sci 24:10540
86. Sundaram MN, Pradeep A, Varma PK, Jayakumar R (2021) Different forms of chitosan and its derivatives as hemostatic agent and tissue sealants. In: Jayakumar R, Prabaharan M (eds) Chitosan for biomaterials IV: biomedical applications, pp 1–28
87. Xia Y, Yang R, Wang H, Li Y, Fu C (2022) Application of chitosan-based materials in surgical or postoperative hemostasis. Front Mater 9:994265
88. Zhang D, Hu Z, Zhang L, Lu S, Liang F, Li S (2020) Chitosan-based thermo-sensitive hydrogel loading oyster peptides for hemostasis application. Materials 13:5038
89. Fan P, Zeng Y, Zaldivar-Silva D, Agüero L, Wang S (2023) Chitosan-based hemostatic hydrogels: the concept, mechanism, application, and prospects. Molecules 28:1473
90. Khan MA, Mujahid M (2019) A review on recent advances in chitosan based composite for hemostatic dressings. Int J Biol Macromol 124:138–147
91. Sundaram MN, Mony U, Varma PK, Rangasamy J (2021) Vasoconstrictor and coagulation activator entrapped chitosan based composite hydrogel for rapid bleeding control. Carbohydr Polym 258:117634
92. Yang E, Hou W, Liu K, Yang H, Wei W, Kang H, Dai H (2022) A multifunctional chitosan hydrogel dressing for liver hemostasis and infected wound healing. Carbohydr Polym 291: 119631

93. Muthiah Pillai NS, Eswar K, Amirthalingam S, Mony U, Kerala Varma P, Jayakumar R (2019) Injectable nano whitlockite incorporated chitosan hydrogel for effective hemostasis. ACS Appl Bio Mater 2:865–873
94. Weng H, Jia W, Li M, Chen Z (2022) New injectable chitosan-hyaluronic acid based hydrogels for hemostasis and wound healing. Carbohydr Polym 294:119767
95. Xia L, Wang S, Jiang Z, Chi J, Yu S, Li H, Zhang Y, Li L, Zhou C, Liu W, Han B (2021) Hemostatic performance of chitosan-based hydrogel and its study on biodistribution and biodegradability in rats. Carbohydr Polym 264:117965
96. Pandian M, Selvaprithviraj V, Pradeep A, Rangasamy J (2021) In-situ silver nanoparticles incorporated N, O-carboxymethyl chitosan based adhesive, self-healing, conductive, antibacterial and anti-biofilm hydrogel. Int J Biol Macromol 188:501–511
97. Feng W, Wang Z (2022) Shear-thinning and self-healing chitosan-graphene oxide hydrogel for hemostasis and wound healing. Carbohydr Polym 294:119824
98. Herliana H, Yusuf HY, Laviana A, Wandawa G, Cahyanto A (2023) Characterization and analysis of chitosan-gelatin composite-based biomaterial effectivity as local hemostatic agent: a systematic review. Polymers 15:575
99. He W, Huang X, Zhang J, Zhu Y, Liu Y, Liu B, Wang Q, Huang X, He D (2021) $CaCO_3$–chitosan composites granules for instant hemostasis and wound healing. Materials 14:3350
100. Sundaram MN, Amirthalingam S, Mony U, Varma PK, Jayakumar R (2019) Injectable chitosan-nano bioglass composite hemostatic hydrogel for effective bleeding control. Int J Biol Macromol 129:936–943
101. Akram AM, Omar RA, Ashfaq M (2023) Chitosan/calcium phosphate-nanoflakes-based biomaterial: a potential hemostatic wound dressing material. Polym Bull 80:5071–5086
102. Wei X, Ding S, Liu S, Yang K, Cai J, Li F, Wang C, Lin S, Tian F (2021) Polysaccharides-modified chitosan as improved and rapid hemostasis foam sponges. Carbohydr Polym 264: 118028
103. Liu T, Zhang Z, Liu J, Dong P, Tian F, Li F, Meng X (2022) Electrospun kaolin-loaded chitosan/PEO nanofibers for rapid hemostasis and accelerated wound healing. Int J Biol Macromol 217:998–1011
104. Tan Z, Li X, Yu C, Yao M, Zhao Z, Guo B, Liang L, Wei Y, Yao F, Zhang H, Li J (2023) A self-gelling powder based on polyacrylic acid/polyacrylamide/quaternate chitosan for rapid hemostasis. Int J Biol Macromol 232:123449
105. Xu G, Xu N, Ren T, Chen C, Li J, Ding L, Chen Y, Chen G, Li Z, Yu Y (2022) Multifunctional chitosan/silver/tannic acid cryogels for hemostasis and wound healing. Int J Biol Macromol 208:760–771
106. Karimi-Maleh H, Ayati A, Davoodi R, Tanhaei B, Karimi F, Malekmohammadi S, Orooji Y, Fu L, Sillanpää M (2021) Recent advances in using of chitosan-based adsorbents for removal of pharmaceutical contaminants: a review. J Clean Prod 291:125880
107. Ahmed MJ, Hameed BH (2018) Removal of emerging pharmaceutical contaminants by adsorption in a fixed-bed column: a review. Ecotoxicol Environ Saf 149:257–266
108. da Silva Alves DC, Healy B, Pinto LADA, Cadaval Jr TRSA, Breslin CB (2021) Recent developments in chitosan-based adsorbents for the removal of pollutants from aqueous environments. Molecules 26:594
109. Kumar A, Patra C, Rajendran HK, Narayanasamy S (2022) Activated carbon-chitosan based adsorbent for the efficient removal of the emerging contaminant diclofenac: synthesis, characterization and phytotoxicity studies. Chemosphere 307:135806
110. Mahmoodi H, Fattahi M, Motevassel M (2021) Graphene oxide–chitosan hydrogel for adsorptive removal of diclofenac from aqueous solution: preparation, characterization, kinetic and thermodynamic modelling. RSC Adv 11:36289–36304
111. Shahrin EWE, Narudin NAH, Shahri NNM, Nur M, Lim JW, Bilad MR, Mahadi AH, Hobley J, Usman A (2023) A comparative study of adsorption behavior of rifampicin, streptomycin, and ibuprofen contaminants from aqueous solutions onto chitosan: dynamic interactions, kinetics, diffusions, and mechanisms. Emerg Contam 9:100199

112. Malesic-Eleftheriadou N, Evgenidou E, Lazaridou M, Bikiaris DN, Yang X, Kyzas GZ, Lambropoulou DA (2021) Simultaneous removal of anti-inflammatory pharmaceutical compounds from an aqueous mixture with adsorption onto chitosan zwitterionic derivative. Colloids Surf A Physicochem Eng Asp 619:126498
113. Liu Y, Nie P, Yu F (2021) Enhanced adsorption of sulfonamides by a novel carboxymethyl cellulose and chitosan-based composite with sulfonated graphene oxide. Bioresour Technol 320:124373
114. Roy N, Kannabiran K, Mukherjee A (2022) Studies on photocatalytic removal of antibiotics, ciprofloxacin and sulfamethoxazole, by Fe_3O_4-ZnO-chitosan/alginate nanocomposite in aqueous systems. Adv Powder Technol 33:103691

Adv Polym Sci (2025) 295: 63–116
https://doi.org/10.1007/12_2024_178

Published online: 10 December 2024

Harnessing Chitosan and Its Derivatives for Innovative Antimicrobial Biomaterials

Sudip Mukherjee, Subhankar Maity, Dipanjana Patra, and Jayanta Haldar

Contents

1 Introduction 64
2 Structural Parameters Modulating Antimicrobial Properties of the Chitosan 66
  2.1 Source of Chitosan 66
  2.2 Molecular Weight (Mw) 67
  2.3 Degree of Deacetylation (DD) 67
  2.4 pH 67
  2.5 Functionalized Chitosan Derivatives 68
3 Chitosan as Anti-Infective Coatings to Prevent Implant Associated Infections 71
  3.1 Chitosan-Based Antimicrobial Coatings on Orthopedic Implants 71
  3.2 Chitosan-Based Antimicrobial Coatings on Catheters 80
4 Chitosan as Anti-Infective Wound-Healing Biomaterials 86
  4.1 Chitosan-Based Antimicrobial Wound-Healing Hydrogels 86
  4.2 Chitosan-Based Antimicrobial Hemostatic Sponges 90

Subhankar Maity and Dipanjana Patra contributed equally to this work.

S. Mukherjee and S. Maity
Antimicrobial Research Laboratory, New Chemistry Unit, Jawaharlal Nehru Centre for Advanced Scientific Research (JNCASR), Bengaluru, Karnataka, India
e-mail: sudipmukherjee@jncasr.ac.in; subhankar@jncasr.ac.in

D. Patra
Chemistry and Physics of Materials Unit, Jawaharlal Nehru Centre for Advanced Scientific Research (JNCASR), Bengaluru, Karnataka, India
e-mail: dipanjana@jncasr.ac.in

J. Haldar (✉)
Antimicrobial Research Laboratory, New Chemistry Unit, Jawaharlal Nehru Centre for Advanced Scientific Research (JNCASR), Bengaluru, Karnataka, India

School of Advanced Materials, Jawaharlal Nehru Centre for Advanced Scientific Research (JNCASR), Bengaluru, Karnataka, India
e-mail: jayanta@jncasr.ac.in

4.3 Chitosan-Based Antimicrobial Hemostatic Cryogels 97
5 Chitosan-Based Nano/Microgels Formulations 99
6 Conclusions and Future Outlook 99
References 101

Abstract The escalating global threat of antimicrobial resistance (AMR) poses a critical public health concern and a significant challenge in modern disease treatment. Chitosan, a naturally occurring polysaccharide, emerges as a promising solution due to its inherent antimicrobial properties, biocompatibility, biodegradability, and environmentally benign nature. Chitosan, due to the presence of cationic amine groups in the molecular backbone, exhibits broad-spectrum antimicrobial activity against Gram-positive and Gram-negative bacteria as well as fungi, gaining significant attention for different biomedical applications. This chapter deals with the recent advancements in the chitosan-based antimicrobial biomaterials as a potential intervention against the recent surge in infectious diseases. It provides a brief overview of the structural parameters dictating the antimicrobial activity of chitosan and detailed chemical modifications performed to enhance its efficacy. Subsequently, the chapter deals with the usage of chitosan and its derivatives to prevent implant-associated infections, such as those in orthopedic implants and catheters, through antimicrobial surface coatings employing both covalent and non-covalent encapsulation of biocides. Additionally, it covers the application of chitosan-based biomaterials in anti-infective wound dressings. Various approaches for incorporating antimicrobial properties into diverse wound healing biomaterials such as hydrogels, sponges, and cryogels through covalent and non-covalent modifications of chitosan and its derivatives are presented. Furthermore, the development of chitosan as therapeutic biomaterials, such as nano and microgels, to cure different systemic infections is discussed. Finally, the chapter concludes with the recent challenges in the clinical application of chitosan-based biomaterials and provides insightful comments to overcome those hurdles.

Keywords Antimicrobial biomaterials · Chitosan as antimicrobial coatings · Chitosan as antimicrobial polymer · Chitosan as antimicrobial wound dressing · Implant-associated infections

1 Introduction

The global rise of antimicrobial resistance (AMR) is a critical public health concern and a significant challenge in modern disease treatment, requiring action at all levels [1, 2]. In 2019, AMR was estimated to cause 4.95 million fatalities, with 1.3 million directly attributed to drug-resistant infections, potentially leading to an estimated 10 million deaths per year worldwide by 2050 [3]. In addition to the different resistance mechanisms, a myriad of infectious microbes can colonize and adhere to all possible biotic and abiotic surfaces, forming microbial biofilm [4, 5]. These biofilms consist of phenotypically diverse populations encased in the self-produced extracellular matrix, posing additional challenges by impeding antibiotic permeation and resulting in chronic and persistent infections [5–7]. This alarming scenario necessitates the development of alternative therapeutic interventions.

In the recent past, microbial membrane targeting antimicrobial polymers has emerged as one of the promising solutions [8–10]. These polymers were designed by emulating the structural features of naturally occurring antimicrobial peptides (AMPs) [11]. Despite the recent advancement in the field of synthetic antimicrobial polymers, they often suffer from poor biocompatibility [8–10, 12]. To address this challenge, employing naturally occurring polysaccharides like chitin, chitosan, dextran, starch, etc. are explored for their high biocompatibility, hydrophilicity, and biodegradability. Among these, chitosan stands out due to its inherent antimicrobial properties attributed to the presence of amine groups [13–15]. These amine groups get protonated at physiological pH and electrostatically interact with the negatively charged microbial membranes, similar to AMPs [14]. Chitosan is a straight-chain polymer of β(1→4)-linked 2-amino-2-deoxy-D-glucopyranose with some residual D-glucosamine units, derived from deacetylation of chitin [14]. Chitosan and its derivatives are effective against bacteria, fungus, and yeasts with good biocompatibility [13–15]. Chitosan exerts its antimicrobial activity majorly by three primary processes: a) destruction of cell walls, b) metal chelation, and c) complexation with DNA [13–16]. Additionally, chitosan can hinder bacterial adherence onto surfaces, because of its hydrophilicity and charge neutralization, making it ideal material for surface coatings on biomedical devices and implants [17].

The versatility of chitosan is enhanced by its cationic nature, allowing it to form multilayer structures or complexes with negatively charged polymers [18]. Further, the presence of nucleophilic amine (at C-2) and hydroxyl groups (at C-3 and C-6) offers versatile modification sites for various applications [13, 19]. Taking advantage of these, chitosan-based different smart antimicrobial biomaterials are developed with different stimuli responsive including pH, temperature, light, etc. [20, 21]. In addition to its antimicrobial properties, chitosan also possesses hemostatic, antioxidant, and wound healing abilities [22–24]. Therefore, different commercial wound dressings, for instance, films, sponges, fibers, and hydrogels are developed using chitosan as their base materials [22, 25]. Due to the presence of amine groups at C-2 position (gets protonated in physiological pH), it can interact with the negatively charged cell membranes of red blood cells and platelets facilitating their adhesion, aggregation, and activation thereby accelerating the blood coagulation [24–26]. Further, chitosan promotes the infiltration of inflammatory cells at injury sites, enhancing the host immune response, and neutralizes negatively charged bacterial antigens, mitigating hyperinflammation [22, 23, 25, 27].

This chapter explores the advancements in the field of biomaterials based on chitosan and its derivatives for tackling AMR. Although chitin-based different biomaterials have also been developed by scientist all over the world [13], in this chapter we have focused only on chitosan and its derivatives as antimicrobial biomaterial. With a brief introduction to different structural parameters in dictating the antimicrobial activity of chitosan, different chemical modifications performed to enhance its activity are delineated in detail. The usage of chitosan and its derivatives to prevent implant-associated infections through antimicrobial surface coatings is discussed. Additionally, the application of chitosan-based biomaterials as anti-infective wound dressings is comprehensively presented. The chapter concludes with an overview of recent challenges in the field and a future outlook on overcoming these hurdles.

2 Structural Parameters Modulating Antimicrobial Properties of the Chitosan

Understanding the antibacterial mechanism of chitosan is challenging because numerous parameters influence its antibacterial activity [19]. The main factors affecting its antibacterial activity are: (a) chitosan source [28], (b) M_w [29–32], (c) degree of deacetylation (DD) of chitosan [33–35], and pH [36, 37]. It also depends on other independent factors like temperature [19], cell growth phase [19], nature of microorganism [18, 19]. In addition, scientists all over the world chemically modified chitosan to enhance its antimicrobial activity [38]. The primary antimicrobial activity of chitosan stems from the permeabilization of bacterial or fungi cells [39, 40]. The electrostatic interaction between the phospholipids of bacterial cells and the positive charges of chitosan leads to release of cytoplasmic content, ultimately resulting in the death of the bacterial cells [16, 41]. However, the comparison in antibacterial activity against Gram-positive and Gram-negative bacteria is still controversial. The amino groups and hydroxyl groups are present in chitosan chelate metal cations (e.g., Ca^{2+} or Mg^{2+}) at higher pH, causing the bacterial cell wall to break down as well as prevent the transport of essential nutrients [42, 43]. Further positively charged chitosan can interact with microbial DNA, inhibiting mRNA and protein production. This mechanism depends on the ability of chitosan to permeate cell membrane with the molecular weight of chitosan playing a crucial role [42, 43]. However, chitosan must break through the bacterial cells' plasma membrane, influenced by various circumstances, to start these unique intracellular reactions. This section will detail how different structural parameters dictate chitosan's antimicrobial activity.

2.1 Source of Chitosan

As it is well known, chitosan is a naturally occurring biocompatible polymer produced by deacetylating chitin, which is found in various animals, including crawfish, shrimp, crab, and squid pens [28]. Recent research has suggested that fungi can manufacture chitosan [44]. *Aspergillus brasiliensis* was used to extract fungal chitosan. However, various microorganisms like *E. coli, Enterobacteriaceae,* and *S. aureus* are comparably less susceptible to get killed by fungal chitosan compared to commercial chitosan [45]. However, according to another study, shiitake chitosan outperformed crab chitosan regarding antibacterial activity [46]. Chitosan derived from ground crab leg shells was compared to chitosan derived from whole crab leg shells, which exhibited higher nitrogen content, DD, and enhanced antibacterial activity against *S. aureus, E. coli, S. enteritidis,* and *L. monocytogenes* [47]. Along with the source, it was also reported that the antibacterial activity of the chitosan also depends proportionally on the concentration [48].

2.2 Molecular Weight (Mw)

The current literature provides contradicting results regarding the impact of M_w on the antimicrobial activity of chitosan, and the findings need to be more consistent. In comparison to chitosan oligomers, chitosan exhibited more antibacterial activity and had superior bactericidal effect on Gram-positive bacteria than Gram-negative bacteria [28]. The antibacterial activity of chitosan against *S. aureus* increased as M_w increased from 300 kDa. However, the antibacterial efficacy against *E. coli* decreased as M_w increased. In some other reports, chitosan with low M_w exhibited more excellent antibacterial activity against *E. coli* compared to chitosan with high M_w. Chitosan can also decrease the growth rates of *C. tropicalis* and *C. parapsilosis* [49]. But, the Mw of chitosan did not have any control over the growth rate of *Candida albicans* [50]. Therefore, further investigations are required to elucidate the relationship between molecular weight and the antimicrobial activity of chitosan.

2.3 Degree of Deacetylation (DD)

The antibacterial capabilities of chitosan are significantly influenced by its DD; the greater the DD of chitosan, the greater the positive charge that results from the amino protonation of chitosan, and the stronger its antibacterial activity [34, 35, 51]. For instance, chitosan's antibacterial activity against *E. coli* enhanced even more as its DD (74% to 96%) increased. A similar effect was also found against other micro-organisms like chitosan with a DD of 81.5% had more antibacterial activity than chitosan with a DD of 62.7% against *Salmonella Enteritidis*, *E. coli*, *L. monocytogenes*, and *S. aureus* [47].

2.4 pH

The pH has a significant impact on chitosan's antibacterial action because it is exclusively soluble in acidic media, and it becomes polycationic in nature at pH values below 6.5 [52]. Electrostatic interaction between the cationic-charged protonated chitosan and anionic phospholipids in microbial cell membrane can cause cell damage which further leads to release of cytoplasmic components [53]. When the pH increases above 7 it loses its antibacterial efficacy as all the amine groups become deprotonated which reduces the interaction and the solubility [54].

2.5 Functionalized Chitosan Derivatives

Even being an amino sugar, chitosan is less soluble in aqueous solution (pH = 7), which restricts its application as well as its activity. To increase its solubility various research groups have done various modifications to the amine and the free hydroxyl group. However, selectively functionalizing amine over hydroxyl or vice versa in one pot remains challenging. So, scientists have used several protecting groups to mask the nucleophilic the 3- and 6-hydroxyl groups or 2-amino group and carried further functionalization. The protecting groups used for the amino group are phthaloyl [55], and acetyl, whereas the hydroxyl groups were protected by triphenylmethyl [56], trimethylsilyl [57], tertiarybutyldimethylsilyl [58], and acetyl. Further to improve its solubility in aqueous medium, several functional groups have been introduced such as trimethyl, guanidine, pyridinium, 2-hydroxy-3-trimethylammoniumyl, trimethyl ammonium, and quaternary alkyl groups, to provide a permanent positive charge to the polymer, thereby enabling bioactivity measurements at pH 7.

2.5.1 Chitosan Containing Quaternary Ammonium Groups

Quaternary chitosan derivatives with permanent positive charge can be synthesized either by direct quaternization at the C-2 amine or by adding quaternary ammonium centers (QACs) at any of the three reactive sites (the two OH groups and the NH_2). The easiest way to quaternize the amine group (at C-2) in chitosan is to treat it with methyl iodide and NaOH to generate *N,N,N*-trimethyl chitosan (TMC). Several research have been published that employed this method to achieve varying degrees of quaternization in TMC [59]. However, when *N*-methylation was carried out in this reaction condition, it also resulted in partial *O*-methylation [60]. Later investigations have discovered methods for obtaining TMC without *O*-methylation in an acidic or heterogeneous media. However, 100% quaternized TMC (DS = 1) can only be achieved by treating the TBDMSO-protected chitosan with methyl iodide (MeI) and Cs_2CO_3 in NMP as the reagents (Fig. 1a) [61]. Quaternary ammonium groups can be added to unmodified chitosan through reactions with epoxides like glycidyl-trimethylammonium chloride (GTMAC) (Fig. 1b) [62], alkyl halides like 3-chloro-2-hydroxypropyltrimethylammonium chloride [63], or reductive alkylation followed by methylation. Similarly, 3,6-*O*-ditertiarybutyldimethyl silyl-protected chitosan was used for obtaining N-selective modifications, such as quaternary *N*- (2-(*N, N,*

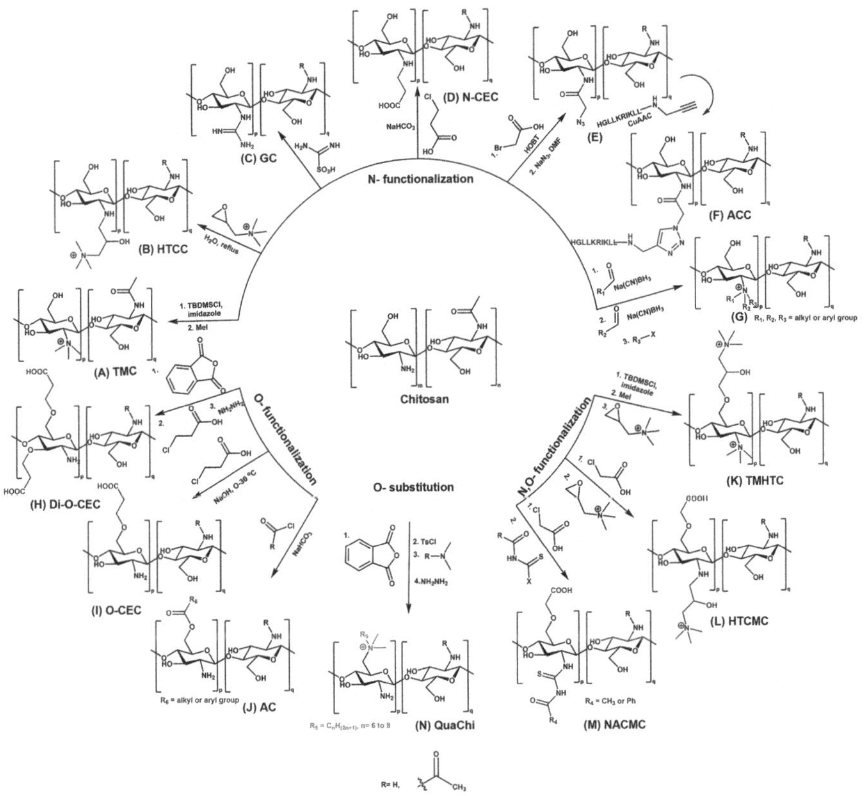

Fig. 1 Synthesis of various chitosan derivatives used for antimicrobial applications. *N*-substituted chitosan derivatives: (**a**) Trimethyl chitosan (TMC); (**b**) *N*-(2-hydroxy) propyl-3-trimethylammonium chitosan chloride (HTCC); (**c**) *N*-guanidiniumyl chitosan chloride (GC); (**d**) *N*-carboxyethyl chitosan (*N*-CEC); (**e**) azido-conjugated chitosan; (**f**) anopolin-conjugated chitosan (ACC); (**g**) different alkyl group-conjugated quaternary chitosan. *O*-substituted chitosan derivatives: (**h**) Di *O*-carboxyethyl chitosan (Di-*O*-CEC); (**i**) *O*-carboxyethyl chitosan (*O*-CEC); (**j**) acetylated chitosan (AC). *N*, *O*-substituted chitosan derivatives: (**k**) *N,N,N*-trimethyl-*O*-[(2-hydroxy-3-trimethylammonium)propyl] chitosan (TMHTC); (**l**) *N*-(2-Hydroxy) propyl-3-trimethylammonium-*O*-[carboxymethyl] chitosan (HTCMC); (**m**) *N*-acylthiourea-*O*-[carboxymethyl] chitosan (NACMC). C-substituted chitosan derivatives: (**n**) C-quaternary ammonium chitosan derivative (QuaChi)

N-trialkyl ammoniumyl and 2-pyridiniumyl)-acetyl) chitosan. Along with the amine group, researchers also have modified the C-6-hydroxyl group to incorporate the permanent positive charge on chitosan. To achieve the same, Haldar and co-workers have protected the amine group with phthalic anhydride followed by tosylation and quaternization with *N,N*-dimethyl alkyl group at the C6 position (Fig. 1o) [64].

2.5.2 Chitosan Modified with Free Amino or Hydroxyl Groups

Chitosan was treated with NaOH to create alkyl halides that had free -NH_2 groups (like aminoethyl) or substituted -NH_2 groups (like dimethyl aminoethyl and diethylaminomethyl). However, the reactivity of terminal amines which is higher compared to other reactive centers in chitosan led to the formation of linear or branched polyethyleneimine derivatives [36]. Derivatives of selective *O*-modified chitosan were synthesized by treating guanidine with aminoiminomethanesulfonic acid (AIMSOA) to conjugate it with chitosan in an aqueous medium (Fig. 1c) [65]. The resulting conjugates had different Mw values (ranging from 13 to 164 kDa) based on the starting chitosan and by varying molar ratios of AIMSOA one can get guanidinylated chitosan derivatives with degrees of substitution (DS) [65]. In other research, cyanamide was used to create guanidinylated chitosan derivatives that were heated to high temperatures in acidic environments [66]. Commonly, chitosan was heated to produce its hydroxyalkyl derivatives by reacting it with different epoxides (Fig. 1b). In certain circumstances, the desired product can be obtained straight from the epoxide production process by nucleophilic replacement with the 2-amino or 6-hydroxyl group of chitosan in an alkaline environment [67]. Maintaining a slightly alkaline pH (pH 8) and using a low NaOH concentration were critical factors to achieve high degrees of substitution (DS) without reagent hydrolysis. Additionally, 3-glycidyl-5,5-dimethylhydantoin and chitosan were reacted in acidic medium at 65 °C to create N-halamine-chitosan derivatives, which had a DS range of 0.35 to 0.45 [68].

2.5.3 Chitosan Modified with Carboxyalkyl Groups

One of the very commonly used chitosan derivatives are carboxymethyl chitosan (CMC) derivatives, which are typically made by dissolving a chitosan suspension in isopropanol, then adding NaOH solution and monochloroacetic acid [69, 70]. Researchers have found that incorporation of the carboxymethyl moiety at the reactive centers (-NH_2 and -OH) can be controlled by modulating reaction conditions. For instance, to prepare *N*, *O*-CMC, chitosan was mixed with isopropanol, and NaOH for 1 h. Then, add the monochloroacetic acid and heat the mixture for three hours at 60°C (Fig. 1d). With the same reagents present, selective O-methyl carboxylation was accomplished by keeping the temperature between 18 and 30°C (Fig. 1i). Similarly, N, O-carboxyethyl chitosan (CEC) synthesized by treating chitosan with NaOH/isopropanol and chloropropionic acid at 60°C [71]. Mono- and di-N-CEC derivatives were formed by reacting chitosan with $NaHCO_3$ and chloropropionic acid in water (Fig. 1h). Further the degrees of substitution can be controlled by varying equivalents of chloropropionic acid [72].

2.5.4 Chitosan Derivatives with Multiple Functionalities

Chitosan derivatives with several functional groups at either or both of the *N* and *O* positions have been synthesized and their antimicrobial activity was investigated. As an example *N,N,N*-trimethyl-*O*-[(2-hydroxy-3-trimethylammonium)propyl] chitosan was synthesized using TMC as a precursor (Fig. 1k) [73]. Variations in reaction durations can also lead to varying degrees of O-carboxymethylation in TMCs with different trimethylation levels [73, 74]. Epoxidation was used to prepare water-soluble chitosan derivatives with dual functional groups (*O*-(2-hydroxy-3-trimethylammonium)propyl (Fig. 1d) and *N*-benzylimine, *N*-benzyl, or *N*-benzyl-*N,N*-diethyl [62]. Inserting various functional groups at the amino position in *O*-CMC derivatives, such as dendrimers, Schiff's bases (Fig. 1g), acylthiourea derivatives (Fig. 1m), quaternized derivatives, terephthaloylthiourea cross-linked derivatives, thiosemicarbazone, and poly(N-vinylimidazole), is considerably easier [63, 75–77].

A wide range of cationic peptides and amino acids have been combined with chitosan. For instance, in order to create chitosan–arginine conjugates, the carboxyl group of arginine had to be activated using NHS/EDC at a certain molar ratio before being coupled with the amino group of chitosan under a range of reaction conditions [78]. The same coupling technique was used to graft acid-functionalized chitosan onto a protected polypeptide. In another study, NCA ring-opening polymerization, which was initiated by the amine groups in the chitosan, was utilized to create copolymers utilizing 6-O-triphenylmethyl chitosan and NCA monomers (lysine and phenylalanine) in DMF. Following deprotection with HBr, the final conjugates were obtained, and their antibacterial activity against fungus, bacteria, and Gram-positive and Gram-negative bacteria was assessed [79]. Using copper-mediated "click chemistry," the short antimicrobial peptide anoplin was attached to azido-functionalized chitosan. Alkyne-functionalized anoplin monomers were grafted onto the biopolymer using both their C- and N-terminals (Fig. 1f). The hemolytic activity of these conjugates toward human red blood cells and their antibacterial activity were assessed against *Escherichia coli* and *Staphylococcus aureus* [80].

3 Chitosan as Anti-Infective Coatings to Prevent Implant Associated Infections

3.1 Chitosan-Based Antimicrobial Coatings on Orthopedic Implants

Implant-associated infection is a serious medical problem that arises from microbial colonization of implanted devices, including prosthetic joints, dental implants, and cardiac devices [4, 81, 82]. Bacteria, fungi, or other pathogens may adhere to the implant surface during surgery or thereafter by hematogenous dispersion, developing biofilms resistant to conventional antibiotic therapy [4]. These complex infections frequently evade the host's immune system and are resistant to conventional antibiotic therapy, necessitating surgical debridement and implant rejection [4, 83]. The problem of infections linked to titanium implants is a major one in the field of medical implants. Titanium, 316L stainless steel, and zirconia which are highly valued for their endurance and biocompatibility in orthopedic, dental, and other medical implants are not immune to microbial colonization despite their inertness [84]. To tackle this problem one obvious and easy solution is to render the implant surface antimicrobial. Scientists have used various strategies, from conventional antibiotics to inorganic antimicrobials to antimicrobial peptides, to coat these implants.

The natural polymer chitosan, which is made from chitin, has become an alternative to coat the surface due to the antimicrobial properties of chitosan and its derivative with excellent biocompatibility. Along with that, chitosan and its derivative have the potential to act as a binding agent to encapsulate various antimicrobials including antibiotics, inorganic antimicrobials, etc. and it can retain the antimicrobial agent on the coated surfaces for long-term antimicrobial activities [17, 85–88]. Along with that due to the cationic character of the chitosan at physiological pH, it attracts the negatively charged protein as well as cell compared to bare Ti. Further, due to the increase of cell adhesion, collagen secretion and ALP activity increases substantially. In this section, we have discussed recent advancements in chitosan-based coating to tackle orthopedic implant-associated infections [89]. On Ti6Al4V implants surface, vancomycin-loaded chitosan/hydroxy apatite (CS/HAp) composites can be electrochemically immobilized.

Vancomycin and CS form hydrogen bonds that fill the HAp coatings' pores, resulting in a smooth, dense sheet. After alteration, polar water molecules broke hydrogen bonds, causing the initial burst release of vancomycin to reduce to 55% and further continue for >30 days. Enhanced antibiotic elution kinetics help sustain steady antibacterial action against *S. aureus* in treating chronic osteomyelitis in animal models. The vancomycin-loaded CS/HAp composite coating enhances osteointegration on Ti6Al4V implants by promoting osteoblast-like cell proliferation, differentiation, and mineralization [90]. Researchers have used electrophoretic deposition (EPD) to create a novel antibacterial implant covering by using antibiotic-loaded gelatin nanospheres embedded in a chitosan matrix (Fig. 2b). They fine-tuned the wettability and surface roughness by varying the weight ratio between chitosan and gelatin nanospheres. Significantly, the coatings showed burst-type release of moxifloxacin and sustained release of vancomycin (Fig. 2d). The coatings adhered firmly to stainless steel, and the gelatin microsphere distributed firmly to the chitosan matrix (Fig. 2c). With strong antibacterial properties and great cytocompatibility, the coating provides a promising method to develop various antibiotic release-based systems to avoid orthopedic implant-associated infections (Fig. 2e–h) [91]. Ti implants' biocompatibility and antimicrobial qualities may also be enhanced by the CS modification. In another report the surface of the Ti substrate was modified with TiO_2 nanotube arrays that can be coated with CS/alginate multilayer sheets after incorporating gentamicin inside it. By preventing bacteria from adhering to the

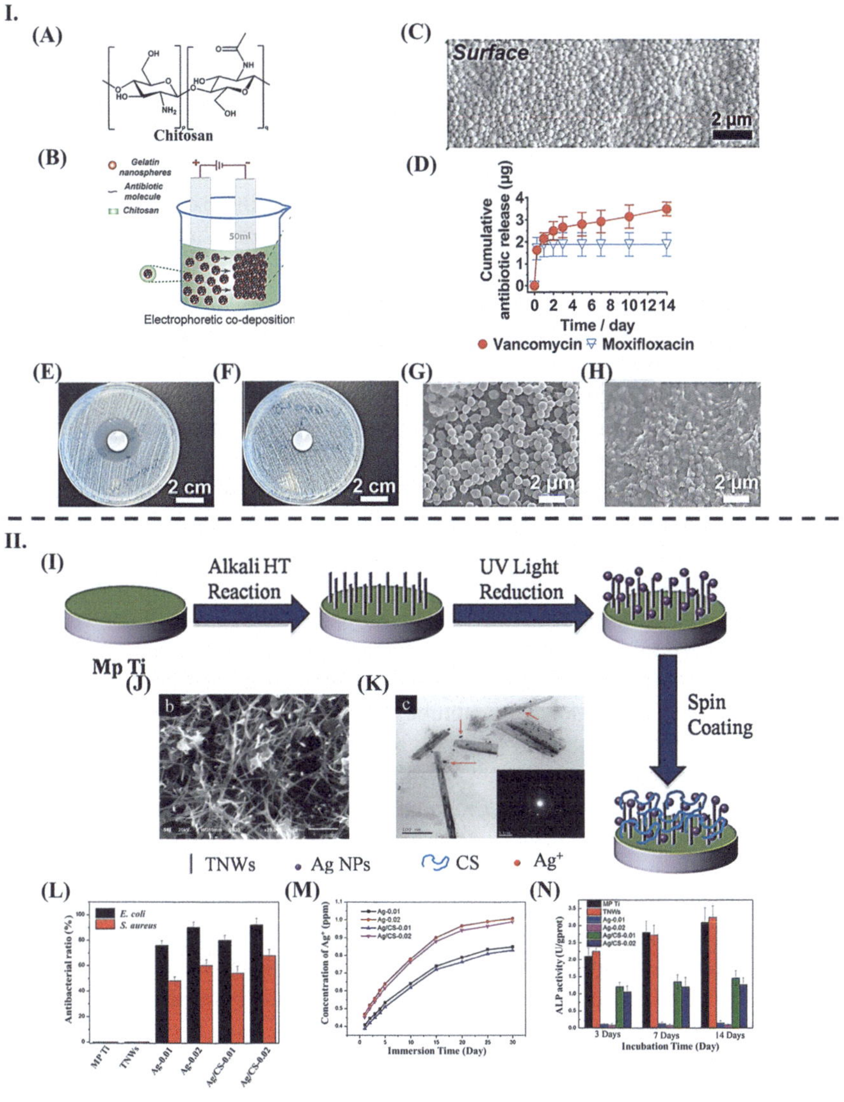

Fig. 2 (I) Antibiotic-loaded chitosan coating on orthopedic implant: (**a**) Chemical structure of chitosan; (**b**) Schematic fabrication of electrophoretic co-deposition of antibiotic-loaded chitosan coated gelatin nanospheres (GN) on 316 L stainless steel; (**c**) SEM image of the coated surface; (**d**) Vancomycin and moxifloxacin release kinetics in PBS; Zone of inhibition of (**e**) uncoated and (**f**) vancomycin coated stainless steel against S. aureus; SEM of S. aureus growth on GNs coating (**g**) without antibiotics and (**h**) with vancomycin; (II) Inorganic antimicrobial loaded chitosan coating on Titanium (Ti); (**i**) Schematic illustration of chitosan loaded Ag-nanoparticle coating on Ti nanowire (TNW); (**j**) SEM and (**k**) TEM of the coated TNW (Inset: SAED pattern of Ag nanoparticles; (**l**) Antimicrobial activity of Ag-np coated TNW at different concentrations; (**m**) Silver release kinetics from Ag-np coated TNW; (**n**) Alkaline phosphatase (ALP) activity of Ag-np coated TNWs. Figures are reprinted with permission from ref [24] and [104]. Copyright 2016 and 2020 American Chemical Society

surface and managing the release of antibacterial drugs from TiO_2 nanotube arrays, the alteration can improve osseointegration, decrease bone infection, and speed osteoblast growth [92]. TiO_2-SiO_2/CS-Lysine-modified Ti alloys also worked well as local delivery systems for gentamicin during the course of treating osteomyelitis. Because of their many nanopores and active OH- and NH_2 groups, which give the implants a great affinity for gentamicin molecules, the composite implants exhibit significant gentamicin loading efficiencies. Additionally, these ceramic/biopolymer-coated implants show notable corrosion resistance in vitro in addition to the initial burst and sustained release of gentamicin [93]. Electrophoretic deposition (EPD) can also be an efficient way to load antimicrobials. EPD immobilizes vancomycin-loaded gelatin/CS coatings on titanium implants without the need for hazardous chemical crosslinking.

The ingrowth of biomineralization and vascularization can be aided by gelatin and CS. These composites have two unique properties: they enhance cellular contact at the bio-interface and inhibit the growth of bacteria as well as biofilms. For this reason, they are quite promising in terms of eliminating osteomyelitis and encouraging bone regrowth [94, 95]. Hyaluronic acid/CS polyelectrolyte multilayers deposited by LBL gave Ti6Al4V alloys exceptional antibacterial properties. The CS-based coating increases hydrophilic behavior, akin to a smooth surface, and decreases the contact angle of the implant surface. It helps avoid implant-related infections by inhibiting bacterial adherence and proliferation. Additionally, triclosan's immobilization aids in lowering the amount of microorganisms on the surface [96]. Gentamicin-loaded nanotube delivery systems can be used to modify TiZr alloys with nanostructured surfaces, which was done by incorporating gentamicin inside the nanotube followed by dip coating the gentamicin-loaded surface with chitosan for several times. Compared to the primary TiZr implant, this composite implant has more molecular polarity and is more hydrophilic. The intentional and focused therapy effects of the implants for osteomyelitis are made possible by the thickness of the CS film, which regulates the drug release with predictable kinetics. The implants with CS modifications have the ability to reduce inflammation in the short term, exhibit enhanced antibacterial properties in the medium term, and facilitate fracture healing in the long run [97]. Researchers created chitosan-based (CHI) hydrogel coatings on Ti6Al4V with crosslinking agents genipin (Ti-CHIGP) and polyethylene glycol (Ti-CHIPEG). These coatings displayed high *in-vivo* biocompatibility, facilitating cell proliferation and differentiation. Additionally, they not only repel but also damage *E. coli* and *S. aureus*. Further, their antibacterial qualities were improved by loading various drugs like cefuroxime, tetracycline, and amoxicillin antibiotics, and non-steroidal anti-inflammatory drug (NSAID) acetylsalicylic acid, which make it a potential solution to reduce implant-related infections [98]. Pishbin *et al.* produced 45S5 bioactive glass (BG particles (size: 1.6–26.7 μm), CS (DA: 85%), and gentamycin (GS)) comprising composite coatings using EPD. The deposition was done in pure CS solution, CS/BG solution, and CS/BG/GS solution for 800, 400, and 400 s, respectively at a single voltage of 10 V. The coatings liberated the GS and improved the HAp production in SBF, demonstrating bioactivity. The releasing kinetics investigation supported the proliferation of the bacteria for up to 10 days while demonstrating the suppression of growth for the first two days. Along with that, the incorporation of the silver further increases the antibacterial efficacy of the coating. However, due to the high loading of the silver, it induces cytotoxicity against the MG63 cell line [99, 100]. Sulfobutyl ether-beta-cyclodextrin and gamma-cyclodextrin were combined to create unique ciprofloxacin-loaded chitosan nanoparticles. The resultant zeta potential values were approximately +30 mV, and the nanoparticle size ranged from 426 to 552 nm. Two nosocomial strains of *Staphylococcus aureus* were able to have their growth inhibited in vitro by this antibacterial coating, with a reduction of around 20 times when compared to controls. After 48 h, 7 days, and 14 days of culture, there was no discernible decrease in the vitality, adhesion, or gene expression of MG63 osteoblast-like cells [101]. Chitosan, bioactive glass nanoparticles, and tetracycline composite were drop casted on hydrothermally etched titanium substrate followed by drop casting of Melittin which completely eradicate MRSA and induce osteogenic properties compared to pristine titanium [102]. Vancomycin-loaded oxidized alginate (oalg) and dexamethasone-loaded chitosan-coated BSA nanoparticle deposited layer by layer on the dopamine-modified Ti surface. Dexamethasone and vancomycin will release slowly and reduce the bacterial burden of *S. epidermidis* significantly. Along with that it is also nontoxic and improves the ALP and AST activity [103].

Apart from the antibiotics, various research groups have used chitosan and its derivative to encapsulate various inorganic antimicrobials onto the implant surfaces. Porous calcium phosphate microsphere coatings with silver nanoparticles have been produced by Jennings and colleagues to give implant coatings osteoconductive and antibacterial properties. Chitosan has been combined with these microspheres, and an alkyloxy silane reaction has been used to attach them to titanium. Microspheres made of calcium phosphate were loaded at a weight percentage of 30% within chitosan coatings, and the concentration of silver was adjusted between 0% and 50%. *P. gingivalis*, *P. denticola*, and *S. aureus* are among the anaerobic and aerobic pathogenic microorganisms whose bacterial viability is reduced by up to 90% when concentrations of silver loaded on calcium phosphate microspheres within the chemically attached coating are increased. This innovative coating may lower the risk of infection in dental and orthopedic implant applications [104]. Electrochemical deposition was used to create a bio-composite coating with chitosan, silver, and hydroxyapatite on an anodized titanium substrate. The coating had a uniform distribution of silver and measured 6.2 ± 0.7 μm in thickness. Because the surface of the produced coatings released silver, they demonstrated outstanding action against *E. coli* and *S. aureus*. Additionally, it was strengthened by the silver and chitosan synergistic impact. It was also discovered that the produced coatings were harmless to MC3T3-E1 cells [105] (Table 1).

Xu and colleagues have created one-dimensional titanate nanowires (TNWs) that incorporate chitosan (CS) and Ag nanoparticles to enhance the biocompatibility and surface antibacterial activity of titanium implants. The Ag nanoparticles were coated with the CS nanofilm using a spin-assisted layer-by-layer assembly technique (Fig. 2i). The findings showed that Ag nanoparticles were uniformly dispersed

Table 1 Chitosan-based antimicrobial coatings on orthopedic implants

Antimicrobial agent	Immobilization technique	Surface	Tested pathogens	Mechanism of antimicrobial activity	Reference
Vancomycin	Electrochemical immobilization	Titanium	*S. aureus*	Release active	[90]
	Electrophoretic deposition	Stainless steel	*S. aureus*	Release active	[94, 95]
	TiO_2–SrHAP as the first layer, followed by vancomycin incorporated				
	Chitosan/Gelatin as the second layer				
Vancomycin and moxifloxacin	Electrophoretic deposition	Titanium	*S. aureus*	Release active	[91]
Gentamicin	Electrodeposition of chitosan and lysin followed by gentamicin	Titanium	*S. aureus, E. coli*	Release active	[93]
	Drop casting of gentamicin on TiO_2 nano tube followed by sodium alginate and chitosan film deposition	Titanium	*S. aureus, E. coli*	Release active	[92]
	Electrophoretic deposition of bio-glass chitosan and gentamicin	Titanium	*S. aureus*	Release active	[99]
	Drop casting of gentamicin followed by dip coating of chitosan	Titanium	*S. aureus, E. coli*	Release active	[97]
Cefuroxime, tetracycline, amoxicillin, and acetylsalicylic acid	Crosslinking of titanium and drug-loaded chitosan by ginipin and PEG-COOH	Titanium	*S. aureus, E. coli*	Release active	[98]
Triclosan	Dip coating in hyaluronic acid and chitosan solution once after anther for multiple time	Titanium	*S. aureus*	Release active	[96]
Ciprofloxacin	Drop-casting of ciprofloxacin loaded chitosan nanoparticles	Titanium	*S. aureus*	Release active	[101]
Ag^+	Electrophoretic deposition of silver, chitosan and hydroxy apatite	Titanium	*S. aureus, E. coli*	Release active	[105]
		Titanium		Release active	[106, 107]

	Chitosan and silver were electrodeposited on poly (acrylic acid)-coated titanium substrates		*S. aureus and P. aeruginosa*		
Ag^{+} and chlorohexidine	Electrospinning the mixture of chitosan, silver nitrate, and poly(ethylene oxide) followed by drop casting of the chlorohexidine	Titanium	MRSA	Release active	[108]
ZnO nanoparticles	Atomic layer deposition of ZnO followed by chitosan coating	Titanium	*S. aureus, E. coli*	Release active	[109]
Ag nanoparticles	CS nanofilm was deposited on the Ag nanoparticles through a spin-assisted layer by layer assembly method	Titanium	*S. aureus, S. epidermidis*	Release active	[106]
	Crosslinking of gallic acid conjugated carboxymethyl chitosan with dopamine-coated titanium by hydrogen peroxide, followed by immersion in $AgNO_3$	Titanium	*S. aureus, E. coli*	Release active	[110]
	Mixtures of Ag nanoparticles, nano catalase, chitosan, and glutaraldehyde instantly drop casted on the hydroxyl-modified titanium surface	Titanium	*S. aureus, E. coli*	Contact active	[111]
	Ag-loaded calcium phosphate microsphere coated with chitosan and crosslinked with titanium by triethoxsilylbutyraldehyde	Titanium	*S. aureus, P. denticola, and P. gingivalis*	Release active	[104]
Hydroxypropyltrimethyl	Layer by layer deposition of HACC	Titanium	*S. epidermidis, S. aureus*	Release active	[112]
Ammonium chloride chitosan (HACC)	Layer-by-layer deposition of HACC	Titanium	*S. epidermidis and S. aureus*	Release active	[113

throughout the surface and effectively transported by TNWs (Fig. 2j, k). Additionally, the Ag nanoparticles were effectively coated with a CS nanofilm. Antibacterial experiments revealed that the modified samples with a greater starting concentration of $AgNO_3$ solution had improved antibacterial activity. Additionally, the antibacterial activity of the TNWs may be further enhanced by a CS nanofilm (Fig. 2l, m). Tests using cell viability and ALP demonstrated that Ag release hampered MC3T3 growth, proliferation, and differentiation. However, as incubation times extended, CS was able to progressively mitigate the harmful effects of Ag (Fig. 2m) [106]. Using atomic layer deposition (ALD) on chitosan (CS)-modified carbon nanotubes (CNTs), one-dimensional (1D) ZnO nanostructures were first placed onto the surfaces of biomedical implants to enhance the antibacterial activity against *S. aureus* and *E. coli* by 39% and 8%, respectively. ZnO and CNT cytocompatibility can be enhanced by CS. One way to adjust Zn concentration is to alter the number of cycles in ALD procedures. By adjusting the amount of ZnO present, this hybrid coating can not only give medical implants strong self-antibacterial efficacy against *E. coli* and *S. aureus* by >73% and 98%, respectively but it can also influence the proliferation and osteogenic differentiation of osteoblast [109]. An antibacterial coating based on chitosan and silver was developed by electrodeposition techniques on poly(acrylic acid)-coated titanium substrates. The silver salt was added during the chitosan deposition step which provides a stable and scalable silver incorporation on the surface. The loaded silver released from the surface for a period of 21 days and effectively neutralizing the surface attached *S. aureus* and *P. aeruginosa* without compromising the biocompatibility against MG63 osteoblast cell line at a 0.3% silver surface atomic percentage [107]. Silver and chlorohexidine-loaded sleeves were developed electrospinning the mixture of chitosan, silver nitrate, and poly(ethylene oxide) followed by drop casting of the chlorohexidine onto the sleeves. These sleeves were directly wrapped on the titanium nail before implanting inside the infected rabbit tibia. The wrapped surface showed around 3 log reduction in the MRSA burden compared to control with minimizing the residual inflammation [108]. Ren et al. developed a hydrogel-based coating onto the polydopamine-modified titanium surface with carboxymethyl chitosan in the presence of hydrogen peroxide. Further, they have in-situ generated silver nanoparticles inside the hydrogel matrix by using $AgNO_3$ as a silver source. The control release of silver nanoparticles reduced the bacterial burden onto the surface and the presence of chitosan provides good biocompatibility towards mammalian cell line [110]. To achieve a functionalized coating on the surface of titanium alloy, Zhang et al. first created glutaraldehyde crosslinked chitosan hydrogels, which were then loaded with silver nanoparticles (nAg) and catalase nano-capsules (n (CAT)). n(CAT) markedly decreased macrophage tumor necrosis factor (TNF-α) expression, elevated osteoblast ALP and osteopontin expression, and improved osteogenesis in chronic inflammatory circumstances. In addition, *S. aureus* and *E. coli* were prevented from growing on the titanium surface by nano-Ag [111]. Wang and co-workers developed zoledronate corrosion resistance coating on AZ31 Mg alloy by combining the fluoride pre-treatment approach with dip coating. In seven days, the chitosan-based coatings demonstrated a 50–60%

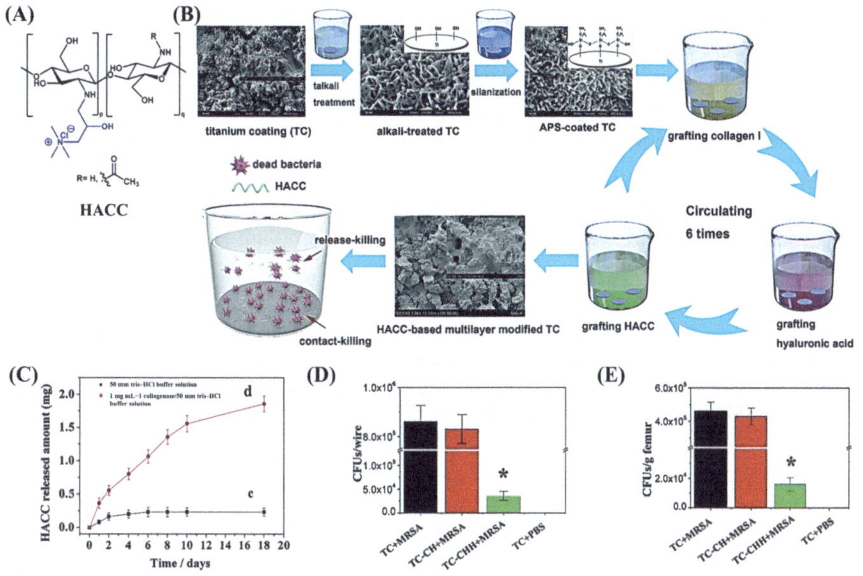

Fig. 3 Surface coating of antimicrobial chitosan derivative, HACC; (**a**) Schematic illustration of layer-by-layer HACC grafting on the titanium with SEM images of the surfaces; (**b**) HACC release kinetics in different buffer; Viable bacterial count in in-vivo femur infection model (**c**) on the wire; (**d**) in rat femur. Figures are reprinted with permission from Ref. [112]. Copyright 2016, The Royal Society of Chemistry

sustained drug release (of the initial 0–0.08 mg/mL burst). Even after 14 days of immersion half of the drugs were retained hydrogen-bonded to chitosan. This made it easier for chitosan coatings to be used effectively in clinical orthopedic applications [114]. Functional group substitution allows for modifications to the physical and chemical characteristics of CS, and its derivatives can positively influence the development of implants' antibacterial and osteogenic qualities. The LBL covalent-immobilized approach is utilized to create the multilayer coating on the porous Ti that is based on hydroxypropyltrimethyl ammonium chloride chitosan (HACC) (Fig. 3a). By inhibiting bacterial colonization and biofilm development, the CS derivative-mediated surface change can stop bone deterioration *in-vivo*. Collagenase-mediated breakdown results in the delayed release of HACC (Fig. 3b), which helps suppress colonized bacteria while also killing off planktonic bacteria. An innovative possible tactic to stop implant-related infections in orthopedics is the CS derivative HACC-assisted surface modification (Fig. 3c, d) [112]. Considering the antimicrobial activity and osseointegration properties, Peng *et al.* had developed quaternary hydroxypropyltrimethyl ammonium chloride chitosan (HACC) by varying the degree of substitution of hydroxypropyltrimethyl ammonium chloride (6%, 18%, 44%). After extensive activity and toxicity study, they have concluded that antibacterial activities of the HACC with 18% or 44% substitution were significantly higher than the others and the 6% or 18% substitution were not cytotoxic and did not interfere with the proliferation and osteogenic differentiation of hMSCs. Based on that they have concluded that HACC with an 18% substitution was a potential pharmaceutical that can inhibit the growth of bacteria and has good biocompatibility with osteogenic cells. Then they coated it on the titanium using APTES as a covalent linker. HACC-coated Ti prevents bacterial colonization as biofilm formation on the surface when it is infected with 10^5 CFU/mL of *S. epidermidis* and *S. aureus.* In addition, it also enhanced the osseointegration properties for improved bone regeneration around the implant [113, 115].

Over the years people have developed various strategies by using chitosan and chitosan derivatives to bring the anti-infective as well as osseointegration properties to the orthopedic implant. However, not a single formulation came to the market due to its purity and potential immunogenicity. In the current scenario, there are still a lot more things to be done to develop a suitable antibacterial titanium implant using chitosan or chitosan derivatives which could address the current limitations.

3.2 Chitosan-Based Antimicrobial Coatings on Catheters

According to statistics, medical devices are responsible for 60–70% of HAIs, with catheter-associated urinary tract infections (CAUTIs) being a leading cause [9, 116, 117]. Healthcare-associated infections (HAIs) can result in higher healthcare expenses, higher morbidity and death rates, and longer hospital stays [118]. Risk factors for HAIs include being over 65, being in the hospital for more than 7 days, having a central venous catheter (CVC), and undergoing urine catheterization [119]. Most urinary tract infections (UTIs) in healthy adults are caused by uropathogenic *Escherichia coli* (UPEC), *Staphylococcus epidermidis*, *Staphylococcus aureus,* and *Candida* spp. [116, 119]. Treating CAUTIs and infections associated with CVCs has become challenging, as higher doses of antibiotics are often required to eradicate the bacteria and clear the infection [120, 121]. These implant-associated infections (IAIs) are now treated with high-dose antimicrobials based on the severity of the infection. If symptoms continue, revision surgery may be necessary [10, 122]. Conventional antimicrobial medicines sometimes fail to treat illnesses due to biofilms and resistant microorganisms. Developing effective ways to reduce device colonization and biofilm formation is critical [123, 124]. The potential of chitosan-based coatings for urinary catheters and CVCs has been explored but more studies are yet to be done (Table 2).

3.2.1 Non-functionalized Chitosan-Based Coatings

Chitosan has been extensively studied as a drug delivery vehicle, recently it has been used in the form of antimicrobial coatings for biomedical implants like intravascular catheters as well as urinary catheters [125]. Selected investigations demonstrated that chitosan antimicrobial coatings were created for use on a wide range of

Table 2 Chitosan-based antimicrobial coatings on catheters

Coating	Material	Remarks	Tested pathogens	Reference
Non-functionalized chitosan	Polyethylene	In-vivo, coated catheters decreased fungal load by 2.5-fold, while in-vitro, they reduced *Candida* biofilms by 95%	*C. parapsilosis, C. albicans*	[130]
	Foley urinary	High and low molecular weight CS-coated catheters decreased bacterial adherence by 70% and 58%, respectively	*E. coli, K. pneumoniae*	[144]
	Silicone	CS-coated catheters suppressed *C. albicans*, *S. epidermis*, and mixed biofilms by 68%, 65%, and 60%, respectively	*C. albicans, S. epidermidis*	[137]
	Foley urinary	After 48 h, catheters coated with CS were able to reduce the formation of *K. pneumoniae* and *E. coli* biofilms by up to 6.3 log	*E. coli, K. pneumoniae*	[134]
	Polyurethane	CS-coated surfaces reduced the biofilm metabolic-like activity of *S. epidermidis* by 80.5% and decreased the biofilm activity of *C. albicans* by 87.5% after 24 h and by 90.0% after 48 h	*A. baumannii, C. albicans, S. epidermidis, S. aureus*	[139]
	Polyurethane	In CS catheters, *S. epidermidis* and *C. albicans* biofilms saw a 57.6% and 43.5% reduction in metabolic biomass activity respectively	*A. baumannii, C. albicans, S. epidermidis, S. aureus*	[138]
Non-functionalized chitosan loaded with amoxicillin/clavul anic acid (CoAM)	Silicone	CS-CoAM-coated silicone surfaces were highly effective (>93%) in preventing biofilm development on CVC tube surfaces	*S. aureus*	[145]

(continued)

Table 2 (continued)

Coating	Material	Remarks	Tested pathogens	Reference
Non-functionalized chitosan loaded with hyaluronic acid (HA)/ b-peptide (coumarin-linker-(ACHC-β^3hVal-β^3hLys)$_3$	Polyethylene	Catheter coated with CS-HA loading with b-peptide inhibited biofilm formation in-vitro and in-vivo	*C. albicans*	[146]
Non-functionalized chitosan loaded with Cu ions	Silicone rubber	Coated surfaces showed 100% killing of both *E. coli* and *S. aureus* and showed no inflammation post-implantation even after 5 days in-vivo condition	*E. coli*, *S. aureus*	[147]
Chitosan functionalized with carboxymethyl	Silicone	Carboxymethyl CS coating inhibited *E. coli* and *P. mirabilis* adherence by over 90% after 4 h. The CS coating prevented biofilm growth under both static and flow conditions	*E. coli*, *P. mirabilis*	[140]
	Silicone	After 90 min, carboxymethyl-coated surfaces failed to attach to over 90% of cells. Coated silicone films reduced *S. epidermidis*, *C. tropicalis*, and mixed biofilm production by 64.1%, 66.6%, and 54.7%, respectively	*C. tropicalis*, *S. epidermidis*	[142]
Chitosan functionalized with fatty acid derivative	Poly(ethylene terephthalate) and butylene dilinoleate (50:50)	*E. coli* colonization on CS/fatty acid derivative coatings decreased by about 80%	*E. coli*	[148]
Chitosan functionalized with catechol loaded with Ag NP	Polyurethane	*E. coli* cells significantly decreased on the CS-catechol/ AgNPs-coated surfaces, resulting in a 96% reduction compared to the uncoated films	*E. coli*, *S. aureus*	[149]

(continued)

Table 2 (continued)

Coating	Material	Remarks	Tested pathogens	Reference
Chitosan functionalized with polyvinylpyrrolidone (PVP) loaded with Ag NP	Silicone and polyethylene	Chitosan-AgNPs/PVP composites had higher antibacterial activity than polyethylene films. In addition, these films reduced *S. aureus* and *E. coli* adhesion by up to 100%	*E. coli*, *S. aureus*	[150]
Chitosan functionalized with hyaluronic acid (HA)	Polyethylene	CS-HA-coated catheters reduced biofilm reduction (80%)	*C. albicans*	[141]
Chitosan functionalized with alginic acid	Silicone	Chitosan-alginic acid-coated surfaces inhibited 100% *S. aureus* of biofilm formation	*S. aureus*	[151]

implanted medical devices (IMDs), including central venous catheters (17.5%) and urinary catheters (12.5%) [126]. Chitosan being biocompatible has shown immense potential as an antimicrobial by altering its molecular weight [127]. Low molecular weight chitosan has the potential to penetrate bacterial cell walls and disrupt their cellular functions by binding with DNA and high molecular weight (M_W) chitosan can change bacterial cell membrane potential and enter into the cells through electrostatic interactions with the negatively charged membranes [126, 128]. Chitosan demonstrates both antifungal and antibiofilm properties against *C. albicans* and *C. parapsilosis* biofilms using a central venous catheter model with a 95% reduction of the biofilm in 30 min [129, 130]. Pu and coworkers found that chitosan could reduce fungal metabolic activity of *C. albicans* significantly, because of the physical stress imposed by chitosan on fungal biofilm structure due to penetration of chitosan and binding to DNA and inhibiting RNA synthesis [131]. Sudarshan and coworkers explained that the positively charged chitosan molecules interacted with the negatively charged cell membranes leading to the leakage of cellular matter which imparted antimicrobial properties to chitosan [54].

Chitosan-based coatings demonstrate broad-spectrum antimicrobial properties against Gram-positive and Gram-negative bacteria. *S. aureus* is one of the common pathogens in IAIs [132]. Bulwan and coworkers developed a chitosan-coated silicone film that could effectively suppress *S. aureus* metabolic activity and reduce biofilm formation by 60% compared to uncoated films [133]. Campana *et al.* described how Foley catheters coated with both high M_W and low M_W chitosan reduced *Klebsiella pneumoniae* and *Escherichia coli* adherence *in-vitro* leaving only 20% viable bacteria on the catheters and an 86% reduction in bacterial biofilm on the catheters [134]. They concluded that chitosan solutions with lower molecular weights and pH levels exhibit increased antibacterial action [19, 134]. Similarly, chitosan-coated polyurethane (PU) films reduced the number of *Pseudomonas aeruginosa* and *S. aureus* adhered cells. It proved that on coating chitosan on surfaces, it improves surface hydrophilicity hence making it difficult for bacteria to adhere onto it [135, 136]. Rubini and coworkers investigated the effectiveness of chitosan-coated silicone catheters against *C. albicans*, *S. epidermidis*, and mixed biofilms. Chitosan-coated urinary catheters significantly reduced biofilm formation by 60%-68%, depending on the stage of development [137].

Even though as previously mentioned, both high M_W and low M_W chitosan show potent antimicrobial activity, many researchers have focused on the low M_W [134]. Cobrado and coworkers have found that polyurethane catheter segments coated with low-MW chitosan reduced the metabolic activity of *S. epidermidis* and *C. albicans* biofilms by 80.5% and 87.5%, respectively, after 24 h [138]. Given these encouraging results, the scientists further investigated the effectiveness of coated catheters to suppress *in-vivo* biofilm formation. Chitosan has been shown to minimize pathogen adherence and inhibit re-growth on medical devices, such as urine and central venous catheters [139].

3.2.2 Functionalized Chitosan-Based Coatings

Chitosan can be functionalized by substituting -OH groups with $-CH_2COOH$ to synthesize carboxymethyl chitosan [140]. Wang and coworkers developed silicone films that were cured previously with polydopamine and were coated with carboxymethyl chitosan which inhibited the growth and adherence of *Proteus mirabilis* and *E. coli*. The coating also inhibited biofilm formation by reducing bacterial adhesion up to 90% in 4 h under both static and dynamic conditions [140]. *Candida* spp. is one of the most notorious pathogens responsible for IAIs. Tan and coworkers examined the efficacy of carboxymethyl chitosan-coated silicone films against *Candida* biofilms. It showed 70% inhibition of biofilms [141, 142]. The coated films also inhibited polymicrobial biofilms because the films inhibited the adherence of pathogens [142]. The effects may be due to interactions between -COOH and $-NH_2$ groups, which increase the quantity of $-NH_3^+$ groups. Positive charges in chitosan molecules may enhance their interaction with negatively charged cell membranes, leading to improved antimicrobial efficacy [143]. Peng and coworkers have added hydroxypropyltrimethyl ammonium chloride (HACC) to improve its water solubility, antibacterial activity, and biocompatibility, depending on the level of quaternary ammonium substitution. Polymethylmethacrylate (PMMA) mixed with HACC inhibited MRSA and *S. epidermidis* biofilms [115].

Lee and coworkers have developed a covalent coating for catheters, by functionalizing chitosan with catechol through EDC-mediated chemistry (Fig. 4a). The coating was immobilized through a three-step process on different catheter surfaces like polyvinyl chloride and polyurethane (Fig. 4b). The wettability of these coated surfaces showed more hydration compared to a bare catheter tubing, due to the hydrophilic nature of the coating (Fig. 4c). The robust-frictionless coating was also impregnated with Ag NPs that enhanced the antibacterial activity against *E. coli* and *S. aureus* (Fig. 4d, e) [149].

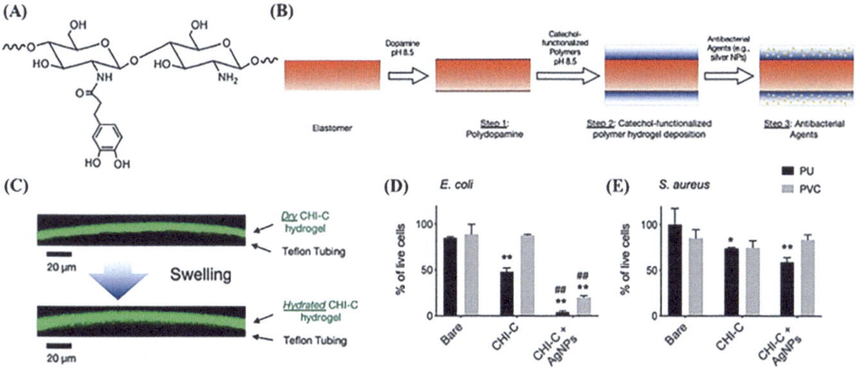

Fig. 4 Catechol-functionalized chitosan-based antimicrobial coating for catheters; (**a**) Chemical structure of catechol-functionalized chitosan (CHI-C); (**b**) Schematic illustration of immobilizing the CHI-C coating along with Ag NPs onto the catheter surface; (**c**) Confocal microscopy picture of FITC-conjugated CHI-C hydrogel coating on PTFE tube, both dry and swelled; Measurement of the adherence of live (**d**) *E. coli* and (**e**) *S. aureus* through Live/Dead BacLight Viability assay. Figures are reprinted with permission from Ref. [144]. Copyright 2019. Frontiers in Materials

Chitosan can encapsulate anionic antibiotics like rifampicin due to electrostatic interactions, which can slowly release [152]. Cao and Sun developed a rifampicin-chitosan coating for long-term medical devices [153]. The technique uses a cationic surfactant to induce antibiotic concentrations on material surfaces. Even after 72 h of incubation, synthetic films remained resistant to *Staphylococcus* spp. adherence [153]. Chitosan and anionic antibiotics create a stable combination and have a regulated diffusion mechanism, leading to high antibacterial durability [154]. Ajdnik and coworkers coated chitosan nanoparticles with amoxicillin/clavulanic acid before depositing them on silicone material. Silicone films coated with amoxicillin/clavulanic acid were highly effective (>93%) in preventing *S. aureus* biofilm development for up to 1 month of exposure [145]. A study by Yang and coworkers introduced a four-step modification technique to apply a chitosan/poly (vinyl alcohol) (PVA) hydrogel layer onto a segmented polyurethane (SPU) catheter [155]. This process involved oxidizing the SPU surface and utilizing the carbodiimide reaction to cross-link the carboxyl groups of SPU with the amine groups of the chitosan/PVA hydrogel through amide bonds. The results demonstrated that the chitosan/PVA-coated SPU exhibited significant antibacterial effects against *S. aureus*, *P. aeruginosa*, and *E. coli*, outperforming a hydrophilic coating approved for various medical devices, including urinary catheters [155]. Kara and coworkers modified synthetic polyurethanes by combining chitosan's antibacterial characteristics and heparin's antiadhesive property. The immobilization procedure required activating polyurethane films with plasma and then contacting them with an acrylamide solution to connect the amine groups [156]. Chitosan was bound by treating samples with glutaraldehyde. Heparin-free chitosan samples showed antibacterial activity against *S. aureus*, *S. epidermidis*, *E. coli*, and *P. aeruginosa* [156]. Niemczyk and coworkers designed a coating procedure for covalently bonding amphiphilic chitosan-fatty acid derivatives (CHFA) to a multiblock thermoplastic elastomer composed of poly(ethylene terephthalate) (PET) hard segments and butylene dilinoleate (DLA) soft segments in a 50:50 weight ratio, a class of polymer with biomedical significance [148]. Plasma treatment was paired with oxygen flushing to stimulate the surface, followed by carbodiimide coupling chemistry to covalently link CHFA to the surface and create effective coatings [148]. Bacterial colonization assays showed a significant reduction (over 80%) in *E. coli* colonization after coating with chitosan−linoleic and chitosan−α-linolenic derivatives, demonstrating its potential. Numerous research suggests that chitosan can be utilized to provide protective coverings for medical devices, such as catheters. However, more *in-vitro* and *in-vivo* investigations are needed [12].

4 Chitosan as Anti-Infective Wound-Healing Biomaterials

Moisture and nutrients present in wounds create favorable conditions for microbial colonization and proliferation, necessitating wound dressings with antimicrobial properties [157–160]. Wound-related infections pose significant risks, as microbes can easily invade and proliferate in tissues, impeding the healing process and potentially leading to severe complications such as sepsis and mortality [160, 161]. Chitosan, renowned for its biocompatibility, biodegradability, hemostatic properties, nontoxicity, and antimicrobial efficacy, finds extensive application in wound dressings [162]. Further, chitosan due to its positive charges, can bind and scavenge negatively charged bacterial antigens, for instance, lipopolysaccharide (LPS) and lipoteichoic acid (LTA) [163]. Therefore, it can mitigate the inflammation at the infected site by inhibiting the activation of NF-κB (nuclear factor kappa-B) and reducing the production of pro-inflammatory cytokines, such as tumor necrosis factor-α (TNF-α) [27, 163, 164]. Although chitosan-based biomaterials are widely used in different wound healing applications from ocular injuries, bone injuries, periodontal injuries, etc., there are several extensive reviews covering this [13, 15, 22]. In this section, we focused on the application of chitosan and its derivative as antimicrobial wound dressing formulations to prevent infections and facilitate wound healing.

4.1 Chitosan-Based Antimicrobial Wound-Healing Hydrogels

Chitosan is a linear polysaccharide comprised of β (1→4) linked residues of *N*-acetyl-2 amino-2-deoxy-D-glucose and 2-amino-2-deoxy-D-glucose. The excellent biocompatibility and the presence of abundant amino groups on the backbone of chitosan facilitates its robust application for the development of numerous biomaterials [21, 162, 165]. Various antibacterial agents such as minocycline, gentamicin/ciprofloxacin, norfloxacin, and sulfadiazine were incorporated into chitosan-based hydrogels to develop novel antimicrobial biomaterials to tackle infections (Table 3)

Table 3 Chitosan-based antimicrobial wound healing hydrogels

Hydrogel matrix	Antimicrobial component	Cross-linking method	Mechanism of antibacterial activity	Tested pathogens	Reference
Chitosan glycol	Colistin	Imine bond cross-linking with aldehyde-modified PEG	Release active	*P. aeruginosa*	[168]
Chitosan	Tobramycin	Imine bond cross-linking with oxidized dextran	Release active	*P. aeruginosa*	[169]
Chitosan	Ciprofloxacin	Imine bond cross-linking with PEG-glyoxylic aldehyde	Release active	*E. coli*	[170]
Chitosan	Tigecycline	Non-covalent interaction with platelet-rich plasma	Release active	*S. aureus*	[171]
Chitosan	ZnO nanoparticle	Imine bond crosslinking with oxidized cellulose	Release active	*S. aureus, E. coli*	[172]
Chitosan	Ag nanoparticle	Non-covalent interaction	Release active	*S. aureus, E. coli*	[180]
Chitosan	Au nanoparticle	Non-covalent interaction	Release active	*S. aureus, P. aeruginosa*	[181]
Chitosan	Antimicrobial peptide, Piscidin-1 with β-glycerophosphate	Non-covalent interaction	Release active	*A. baumannii*	[182]
Chitosan	Hydrogen peroxide and antimicrobial peptide, RKKKKLLRKKC	Non-covalent interaction	Release active	MRSA	[183]

(continued)

Table 3 (continued)

Hydrogel matrix	Antimicrobial component	Cross-linking method	Mechanism of antibacterial activity	Tested pathogens	Reference
Chitosan and γ-poly(glutamic acid) (γ-PGA)	Chitosan	Electrostatic interaction	Contact active	*S. aureus, E. coli*	[184]
N-succinyl chitosan and dextran aldehyde	N-succinyl chitosan	Imine bond cross-linking	Contact active	*S. aureus, P. aeruginosa, S. pyogenes, E. coli, C. perfringens*	[185]
Dimethyldecylammonium chitosan – graft-poly(ethylene glycol) methacrylate (DMDC-Q-g-EM)	Dimethyldecylammonium chitosan	Covalent cross-linking with poly(ethylene glycol) diacrylate	Contact active	*P. aeruginosa, E coli, S. aureus,* and *F. solani*	[186]
Quaternary lipophilic chitosan derivative (QuaChi) and gelatin	QuaChi	Amide coupling	Contact active	MRSA, VRSA, *S. epidermidis, A. baumannii, P. aeruginosa*	[65]
N-(2-hydroxypropyl)-3-trimethylammonium chitosan chloride (HTCC) and polydextran aldehyde (PDA)	HTCC	Imine bond crosslinking	Contact active	MRSA, VRSA	[188]
HTCC and PDA	HTCC and vancomycin	Imine bond crosslinking	Both contact and release active (dual active)	MRSA, VRSA	[189]
HTCC-AgCl nanocomposite and PDA	HTCC and AgCl nanoparticle	Imine bond crosslinking	Dual active	MRSA, VRSA, *E. coli, A. baumannii, P. aeruginosa, K pneumoniae*	[200]

[159, 162]. Despite superior properties of conventional injectable hydrogels, their desirable performances deteriorate because of certain deformation at macro or microscale. This shortens the lifetime and applicability of hydrogel. To address this issue, smart hydrogels with self-healing properties have been introduced that can restore their structures and functionality even after damage which leads to long-lasting materials. In this regard, a self-healable hydrogel encapsulated with colistin was prepared by chitosan glycol and aldehyde-modified PEG to eradicate the bacterial burden in *P. aeruginosa*-infected burn wounds (Table 3) [166]. A self-healing hydrogel was fabricated by crosslinking quaternized chitosan and oxidized dextran through imine bond which showed sustained release of tobramycin was reported for healing of *P. aeruginosa*-infected burn wound (Table 3) [167]. An injectable PEG-chitosan hydrogel loaded with ciprofloxacin antibiotic has been developed for sustained delivery of the drug to prevent infections and promote wound healing (Table 3) [168]. Nimal and coworkers have developed a potent anti-staphylococcal chitosan-platelet-rich plasma (PRP) hydrogel incorporated with tigecycline-loaded chitosan nanoparticles which was reported as a promising therapeutic gel to heal an infected wound (Table 3) [169]. In many cases, certain inorganic metal nanoparticles have been incorporated into these hydrogels [170–174]. Noble metal particles, specifically gold (Au) and silver (Ag) nanoparticles (NPs), have the capability to be integrated into the hydrogel matrix (Table 3) [123, 175, 176]. These composite scaffolds exhibit improved physical and biological properties. The metal ions released from the hydrogel scaffold induced bacterial cell death by binding to microbial DNA to prevent replication, damaging proteins, disrupting cell membranes, and generating oxidative stress [123, 176]. Various chitosan-based hydrogels encapsulated with metal ions and metallic nanoparticles have been developed to mitigate microbial infections (Table 3) [123, 177]. A novel silver nanoparticle containing chitosan hydrogel with ultrahigh mechanical strength has been developed. This novel hydrogel exhibited potent bactericidal activity against *S. aureus* and *E. coli* with notable wound healing properties (Table 3) [178]. A gold nanoparticle incorporated chitosan-based non-toxic hydrogel was reported with bactericidal activity against *S. aureus* and *P. aeruginosa* [179]. Rezaei *et al.* have developed a novel thermo-responsive chitosan hydrogel using β-GP. The incorporation of an antimicrobial peptide (Piscidin-1) in the hydrogel demonstrated antibacterial activity against resistant clinical strains of *A. baumannii* (Table 3) [180]. Another chitosan-based hydrogel was developed for the codelivery of hydrogen peroxide and an antimicrobial peptide RKKKKLLRKKC to cure MRSA biofilm-infected wounds (Table 3) [181].

In addition to biocide-loaded hydrogels, various inherently active chitosan-based hydrogels were also developed (Table 3) [20, 21, 182, 183]. M.B. Chan-Park and coworkers developed a polycationic antimicrobial hydrogel using dimethyldecylammonium chitosan (with high quaternization)-graft-poly(ethylene glycol) methacrylate (DMDC-Q-g-EM) and poly(ethylene glycol) diacrylate (Table 3) [184]. The hydrogel exhibited excellent antibacterial activity against *P. aeruginosa, E. coli, S. aureus,* and *F. solani* through bacterial membrane suctioning "anion sponge" effect caused by the polycationic hydrogel. This hydrogel was also uniformly coated through UV irradiation with high biocompatibility in the rabbit conjunctiva model [184]. Recently, our group also developed a lipophilic quaternary chitosan-based antimicrobial hydrogel by crosslinking with gelatin which displayed broad-spectrum antibacterial activity [64]. Further, we developed N-(2-hydroxypropyl)-3-trimethylammonium chitosan chloride (HTCC), a biocompatible chitosan derivative with superior water solubility as well as antimicrobial activity compared to chitosan [185]. HTCC was further cross-linked with polydextran aldehyde via imine bond to obtain an antimicrobial sealant (Fig. 5a, b). This biocompatible contact active sealant kills bacteria through disruption of their membrane, able to prevent sepsis thereby enhancing mice survivability in a cecum ligation and puncture (CLP) [186]. This hydrogel also accelerated wound healing in a rat model (Fig. 5c). Further our group has designed a dual-active strategy by incorporating vancomycin into imine bond crosslinked HTCC-polydextran aldehyde-based injectable hydrogel (IHV), which released antibiotic for prolonged time (Fig. 5d–f) [187]. Inherent antibacterial nature of the hydrogel matrix enhances antibacterial potency, resulting in excellent revealed ~6 log reduction in bacterial burden inside the gel as well as in the distal tissues in murine MRSA subcutaneous infection model (Fig. 5g). In a recent paper, similar strategy was implemented by incorporating AgCl nanoparticles instead of vancomycin in PVA-HTCC hydrogel (Fig. 6a) [188]. Due to superior antibacterial activity of AgCl nanoparticles, it was able to kill planktonic and stationary phase cells of both Gram-positive and Gram-negative bacteria and also eradicated bacterial biofilms. L-Proline was used in the hydrogel matrix to facilitate and induce the wound healing process through collagen deposition, angiogenesis, and extracellular matrix formation. The hydrogels expedite wound healing process through collagen deposition proved through Masson trichrome and H&E staining (Fig. 6b, c). Moreover, it exhibited excellent in-vivo efficacy in wound infection model in reducing the bacterial burden (Fig. 6d). Importantly, the shear-thinning feature of P-BAC streamlines the administration procedure, making it more practical and usable. Later on, the similar hydrogel, BacSeal (devoid of L-Proline), was also developed as an ocular sealant with rapid and broad-spectrum antibacterial activity (Fig. 7a) [189]. Most importantly, it was able to maintain ocular pressure up to 75 mm-Hg when applied in human eye injury model (Fig. 7b, c). In *ex- vivo* human cornea infection model, the hydrogel exhibited ~99% reduction in bacterial count (Fig. 7d–f). Overall, our findings showed that this readily administrable hydrogel has the potential to be a diverse solution for successful wound healing with strong anti-infective capabilities.

4.2 Chitosan-Based Antimicrobial Hemostatic Sponges

Nowadays, antimicrobial hemostatic sponges are manifested as promising biomaterials to treat wound infections due to their microporous structure, excellent fluid absorption capability, and high cell proliferation ability [23, 25, 190]. Chitosan has been extensively used as a hemostatic material owing to the presence of -NH_3

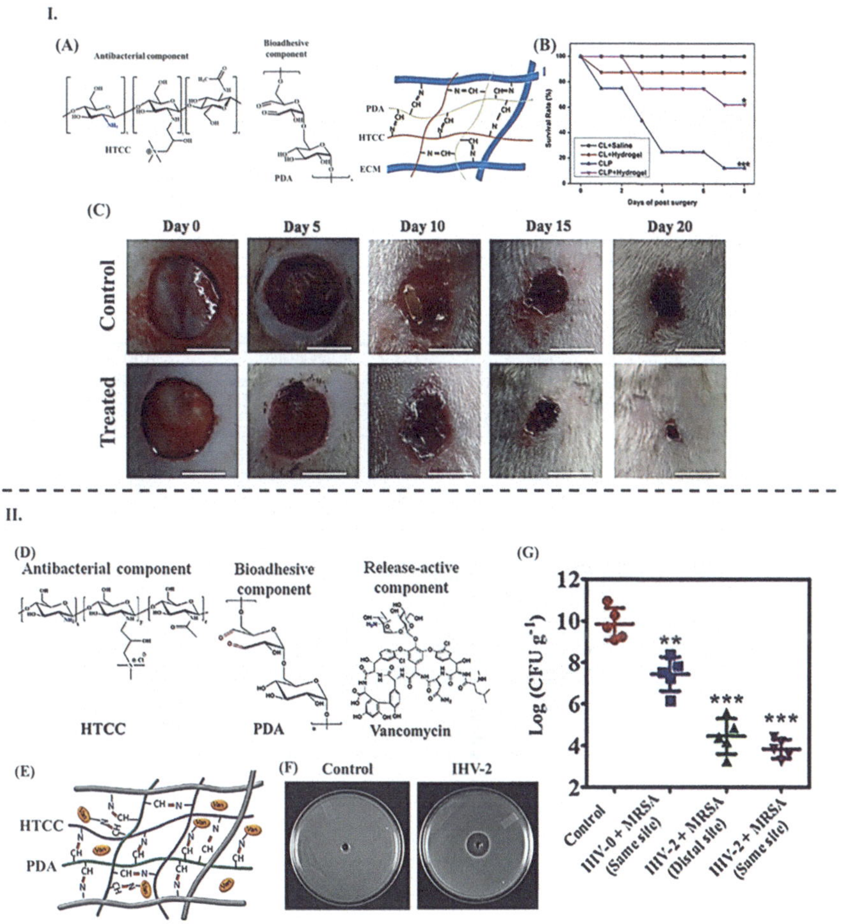

Fig. 5 (I) HTCC and PDA-based contact active antimicrobial hydrogels; (**a**) Chemical structure of HTCC and PDA used to formulate the hydrogel; (**b**) HTCC and PDA were cross-linked through imine bond; (**c**) wound healing study upon hydrogel treatment. (II) Dual-active (both release and contact active) antimicrobial injectable hydrogel (IHV); (**d**) IHV was developed by mixing HTCC, polydextran aldehyde (PDA), and vancomycin; (**e**) the hydrogel was fabricated through imine bond cross-linking of PDA and HTCC; (**f**) release active antibacterial property of IHV tested through zone of inhibition (ZOI) study of IHV against vancomycin; (**g**) in-vivo antimicrobial efficacy of IHV against subcutaneous MRSA infection in mice. Figures are reprinted with permission from Ref. [186, 187]. Copyright 2016 and 2018 American Chemical Society

moieties (resulting from the protonation of amino group at physiological pH) which electrostatically interact with the anionic phospholipid of platelet and red blood cells (RBCs) [25, 26, 191, 192]. This enhances adhesion, aggregation, and activation of both platelet and RBCs accelerating the hemostatic process. Chitosan-based various hemostats are therefore approved and commonly used in market such as Celox, TraumaStat, HemCon, Axiostat, etc. [26, 191]. For instance, the FDA-approved

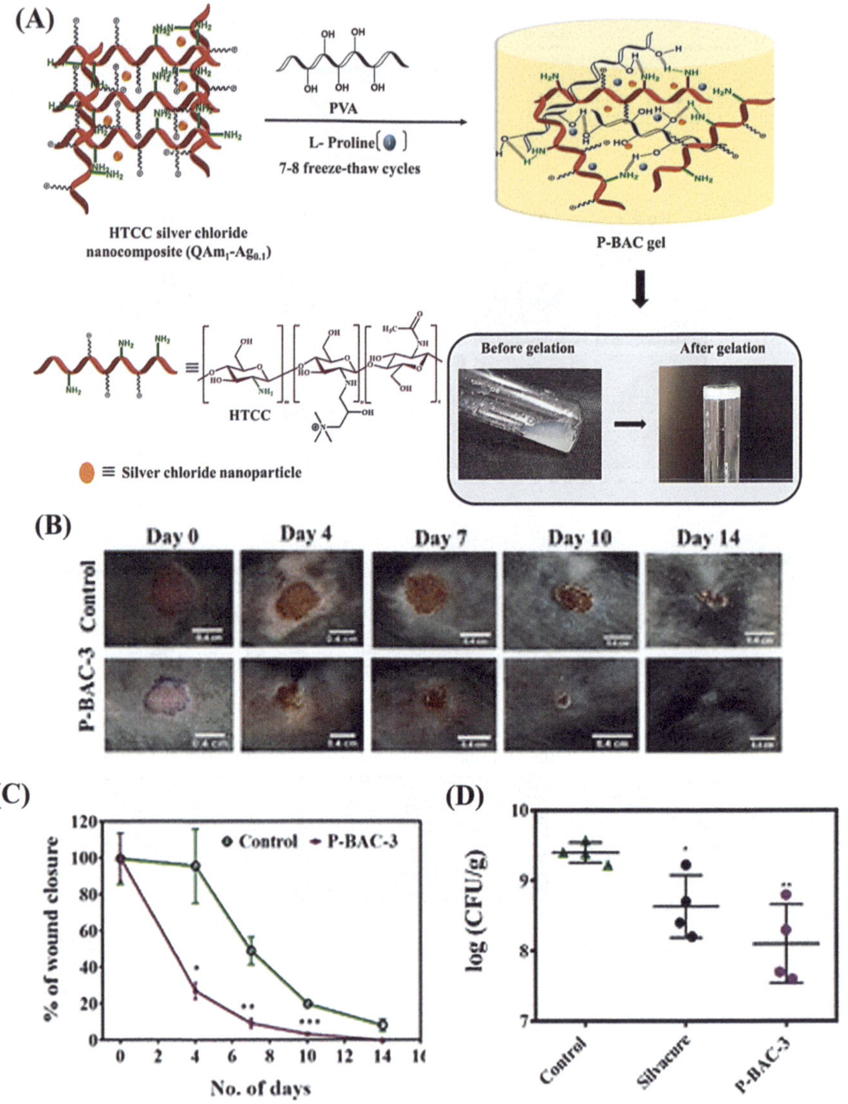

Fig. 6 Dual-active antimicrobial hydrogel composed of HTCC and AgCl nanocomposite. (**a**) Schematic development of dual-active hydrogel (P-Bac) composed of PVA-HTCC-AgCl and L-Proline; Wound healing ability of P-Bac hydrogel: (**b**) digital images of the wound and (**c**) quantification of wound closure; (**d**) In-vivo efficacy of P-Bac hydrogel in wound infection model. Figures are reprinted with permission from Ref. [188]. Copyright 2023 American Chemical Society

XStat™ (RevMedx, Wilsonville, USA) is a miniature cellulose sponge coated with chitosan that expands rapidly to fill wound cavities and apply additional pressure to achieve rapid hemostasis [193]. However, poor to moderate antimicrobial activity of

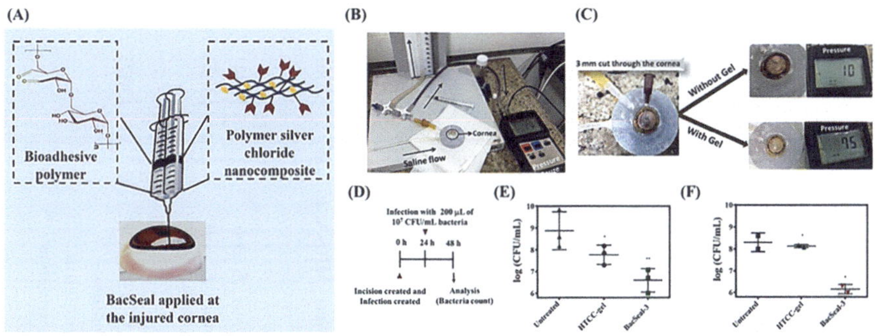

Fig. 7 Dual-active ocular sealant, BacSeal for corneal injury and infection. (**a**) BacSeal was formed through imine bond cross-linking of bioadhesive polymer polydextran aldehyde (PDA) and HTCC-AgCl nanocomposite; Sealant property of BacSeal, (**b**) Experimental set-up and (**c**) measurement of intra-ocular pressure (IOP) upon BacSeal treatment injured cornea in ex-vivo human eye-injury model; (**d**) Experimental scheme and antibacterial activity of BacSeal treatment after 24 h post-infection against (**e**) *P. aeruginosa* ATCC 27853 and (**f**) *P. aeruginosa*-L-2377/1. Reprinted with permission from Ref. [189]. Copyright 2024 Elsevier

these hemostats limits their usage in case of traumatic injuries [24]. Various biocides (antibiotics, silver nanoparticles, and antimicrobial agents) encapsulated or intrinsically active antimicrobial sponges have been developed to cure deep tissue infection [24, 25]. Raveendran *et al.* have developed a released-active chitosan-based antimicrobial sponge which was incorporated with fibrin nanoparticle encapsulated antibiotics mixture (ciprofloxacin and fluconazole) [194]. This antibacterial and antifungal sponge revealed compelling efficacy to treat polymicrobial infected wounds. Like antibiotics, various metal nanoparticle loaded sponges also have been developed [123]. Poly-L-lysine-modified silver nanoparticle grafted graphene was crosslinked with chitosan in the presence of glutaraldehyde, which reduced blood clotting time compared to pristine chitosan dressing [195]. You et al. have reported the silver nanoparticle-impregnated chitosan and collagen crosslinked sponge matrix to tackle wound infections [196]. Silver nanoparticle-embedded chitosan sponges were prepared by in-situ reducing silver nitrate solution using a reducing agent and thiol-modified chitosan [197]. Thiol and amino groups of sponge matrix interact with blood cells and activate them, therefore the sponges rapidly inhibited blood loss in rat liver puncture model and rabbit ear injury model. Later on another chitosan-based anti-infective wound dressing was developed by entrapping Au–Ag combined nanoparticle into the chitosan matrix, exhibiting faster silver release compared to only silver-encapsulated chitosan wound dressing [198]. Ampicillin-immobilized chitosan sponges were developed by conjugating 3-aminopropyl triethoxysilane-modified chitosan with ampicillin through amide bond formation, and further crosslinked with genipin to form the sponge [199]. These sponges demonstrated antimicrobial potency against *S. aureus*, *E. coli*, and *P. aeruginosa* in in-vitro, ex-vivo, and in-vivo models. There are few other reports of silver

Table 4 Chitosan-based antimicrobial hemostatic sponges

Sponge compositions	Antimicrobial agent	Cross-linking method	Mechanism of antibacterial activity	Tested pathogens	Reference
Chitosan	Fibrin nanoparticle encapsulated ciprofloxacin and fluconazole mixture	Non-covalent interaction	Release-active	*S. aureus, E. coli, and C. albicans.* Eradicate polymicrobial biofilm	[196]
Poly-L-lysine-modified silver nanoparticle grafted graphene crosslinked with chitosan	Ag nanoparticle	Imine bond cross-linking using glutaraldehyde	Release-active	*S. aureus, E. coli*	[197]
Chitosan and collagen crosslinked sponge	Ag nanoparticle	Electrostatic and hydrogen bonding	Release-active	*E. coli, P. aeruginosa*	[198]
Thiol-functionalized chitosan sponge	Ag nanoparticle	Non-covalent crosslinking	Release-active	*S. aureus, E. coli, P. aeruginosa*	[199]
Chitosan	Au–Ag combined nanoparticle	Non-covalent crosslinking		*S. aureus, E. coli*	[200]
3-Aminopropyl triethoxysilane-modified chitosan	Ampicillin	Amide bond formation using genipin	Release active	*S. aureus, E. coli, P. aeruginosa*	[201]
Oxidized nanocellulose, quaternized chitosan, and collagen	Quaternized chitosan	Electrostatic and contact-active hydrogen bonding self-assembly		*S. aureus,* MRSA, *E. coli*	[206]
Alkylated chitosan	Alkylated chitosan	Non-covalent interaction	Contact active	*S. aureus, E. coli*	[204]
Tris(hydroxymethyl) CTMG methyl glycine-modified chitosan derivative (CTMG)		Non-covalent interaction	Contact active	*S. aureus, E. coli*	[205]
HTCC-AgCl nanocomposite and nanoparticle gelatin	HTCC and AgCl	Amide coupling	Both contact and release active (dual active)	MRSA, VRSA, *E. coli, A. baumannii, P. aeruginosa, K. pneumoniae*	[207]

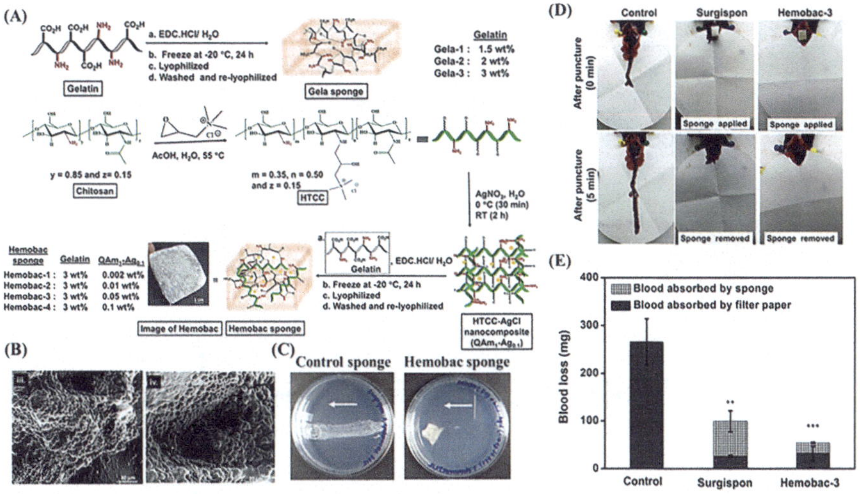

Fig. 8 Dual-active antibacterial hemostatic sponge (Hemobac). (**a**) Schematic development of dual-active Hemobac sponge composed of HTCC-AgCl nanocomposite and Gelatin; (**b**) Hemostatic mechanism exerted by Hemobac sponge through agglomeration of red blood cells; (**c**) Antibacterial activity of control gelatin sponge (Surgispon) and Hemobac sponge through dragging method; (**d**) Digital images of the mice liver puncture model; (**e**) Quantification of the blood absorbed by sponge and the filter paper during the experiment. Figures are reprinted with permission from Ref. [204]. Copyright 2022 American Chemical Society

nanoparticles encapsulated chitosan–collagen sponges to control hemostasis and infection mitigation [200] (Table 4).

In addition to biocides-loaded antimicrobial sponges various inherently active antimicrobial sponges have been developed [190, 201–203]. Yuan *et al.* have recently developed an antimicrobial sponge which was engineered by crosslinking of oxidized nanocellulose, chitosan, and collagen via electrostatic and hydrogen bonding self-assembly [204]. This triple-component sponge showed potent wide-spectrum bactericidal activity with potent hemostatic and *in-vivo* biodegradation properties. Similar to dual-active hydrogels, our group has also engineered dual-active hemostatic sponges (Fig. 8) [205]. The sponges were tailored by amide bond cross-linking of gelatin, and HTCC-AgCl nanocomposite AgCl nanocomposites were further encapsulated to enhance antimicrobial properties (Fig. 8a). The blood clotting property of the sponges originates from the adherence, agglomeration, and activation of blood cells as observed in SEM images (Fig. 9b). Hemobac sponges exhibited broad-spectrum antibacterial activity against both planktonic and stationary phase bacteria (Fig. 8c). These biocompatible Hemobac sponges exhibited excellent and hemostatic properties in liver puncture and femoral vein incision model in mice (Fig. 8d, e).

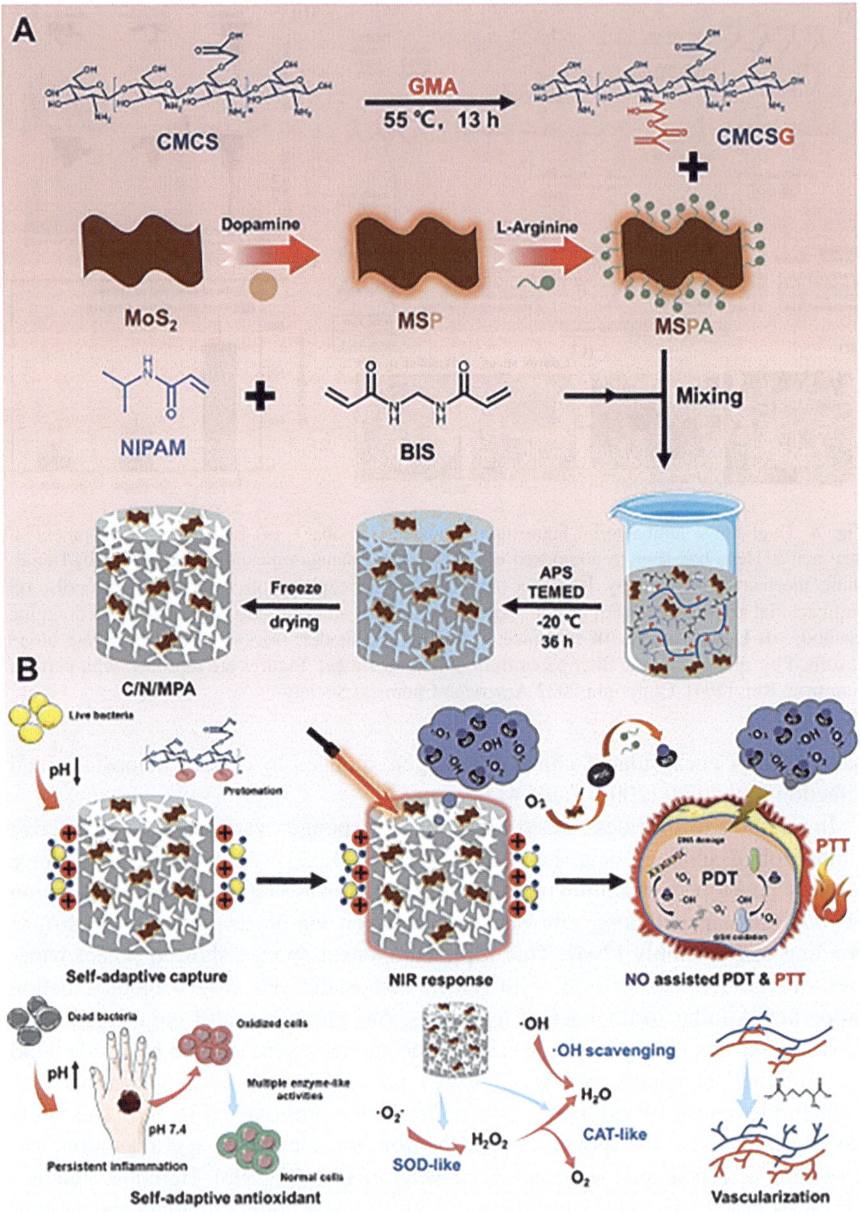

Fig. 9 (**a**) Schematic illustration of the preparation and application of multiple stimuli-responsive nanozyme-based cryogel; (**b**) Mechanism of multiple stimuli-responsive antibacterial and self-adaptive wound management. Figures are reprinted with permission from Ref. [213]. Copyright 2022 John Wiley and Sons

4.3 Chitosan-Based Antimicrobial Hemostatic Cryogels

Recently, cryogels have emerged as an alternative biomaterial in different applications due to their interconnected porous structure and high fluid absorption properties [206–208]. Further, cryogels exhibit robust mechanical and rapid shape recovery properties thereby providing additional pressure when applied in case of non-compressible hemorrhages [209]. Baolin Guo and co-workers have immensely contributed in the field of developing hemostatic cryogels to arrest lethal non-compressible hemorrhages, some of them also possessing antimicrobial activity [210–213]. In one of the recent reports, carboxymethyl chitosan-based multiple stimuli-responsive NO-releasing antibacterial cryogels were prepared (Fig. 9a) [214]. These self-adaptive cryogels in acidic environment of wound infections were able to capture bacteria and effectively reduce the bacterial burden through NO-assisted photothermal and photodynamic therapy (PTT and PDT) (Fig. 9b). The cryogels were prepared through cross-linking of chitosan methacrylate and N-isopropylacrylamide (NIPAM) in cryo-condition using bisacrylamide (BIS) as a cross-linker (Fig. 9b). L-arginine-loaded MoS2 @polydopamine (MSPA) was encapsulated into the system as NO-releasing carrier. Upon near-infrared (NIR) irradiation, NO is released which exerts NO-assisted PDT&PTT antibacterial ability. Further, the release of L-arginine in early wound healing stage induces angiogenesis. Therefore, these cryogels effectively mitigate wound infection, diminish oxidative stress through catalase-like activity, and expedite angiogenesis and collagen deposition to heal the infected wounds. Recently, Feng and co-workers developed another antibacterial hemostatic cryogel using chitosan, oxidized gallic acid, and hemin (Table 5) [215]. Hemin was used as a photothermal agent to effectively eradicate bacterial infection from the wound site. The cryogels exhibited rapid blood-triggered shape recovery and faster hemostasis compared to commercially used gauze and gelatin sponges. Further, the cryogel efficiently accelerated infected wound healing by mitigating inflammation and promoting vascularization. Cao and co-workers prepared shape memory cryogels using quaternized chitosan and mesoporous bioactive glass under −20°C [216]. These sponges composed of quaternized chitosan exert the antibacterial activity against both Gram-positive and Gram-negative bacteria whereas bioactive glass was responsible for rapid hemostasis.

Despite its extensive usage in wound dressing materials, one of the major limitation of chitosan-based hemostats is their slow *in-vivo* degradability [25, 165]. Further, high tissue adhesive nature of these chitosan-based materials increases difficulty and risk of rebleeding during their removal, causing significant patient discomfort. These aspects should be taken into account while developing wound healing materials.

Table 5 Chitosan-based antimicrobial hemostatic cryogels

Cryogel compositions	Antimicrobial agent	Cross-linking method	Mechanism of antibacterial activity	Tested pathogens	Reference
Methacrylated N-(2-hydroxypropyl)-3-trimethylammonium chitosan chloride	Quaternized chitosan	Radical crosslinking at cryo condition	Contact-active	*S. aureus, E. coli, and P. aeruginosa*	[212]
Carboxymethyl chitosan	Nitric oxide (NO), radical crosslinking photothermal and at cryo condition photodynamic therapy		Release active (NO) as well as through photothermal and photo-dynamic therapy	MRSA and *E. coli*	[216]
Chitosan, oxidized gallic acid, and hemin	Hemin-assisted photothermal therapy	Imine bond crosslinking	Hemin-assisted photothermal therapy	*S. aureus*	[217]
Quaternized chitosan and bioac-tive glass	Quaternized chitosan	Glutaraldehyde-mediated imine bond crosslinking	Contact active	*S. aureus, E. coli*	[218]
Quaternized chitosan, and polydopamine	Quaternized chitosan	Imine bond crosslinking between oxidized dopamine and quaternized chitosan	Contact active	*S. aureus, E. coli*	[215]

5 Chitosan-Based Nano/Microgels Formulations

In the last two decades, nano/microgels have gained significant attention as potential drug-delivery vehicles. These are cross-linked hydrogel particles of nano- and micro-sized dimension with high swellability, tissue-permeability, and deformable nature, enabling their application in drug delivery in diverse fields [123, 217–219]. Simultaneously, nanogels and microgels were also incorporated with antibacterial properties by different chemical modifications or encapsulation of antibiotic, metallic nanoparticles, and biocides into the matrix [217, 220–222]. The antibiotic amoxicillin (Amox) was encapsulated in a nanogel formed by self-assembly of maleoyl–chitosan and poly(aspartic acid) through non-covalent interactions [223]. The nanogels were readily formed by the electrostatic interaction between two oppositely charged polymers without the need of any surfactant. In contrast to the conventional chitosan-based nanocarriers, these nanogels did not show any adverse activity by interacting with mammalian cell-membrane proteins. Therefore, the loading of Amox in this nanogel provides an alternative therapy to boost the bioavailability of Amox through a safe cell internalization (endocytosis), without the occurrence of any aggregation. Similarly, another nanogel formulation was prepared by using chitosan hydrochloride and carboxymethyl starch through EDC coupling for sustained delivery of curcumin [224].

Chitosan-based nanogels were also used for antifungal drug delivery. Naturally occurring different essential oils bearing multiple phenylpropanoids possess antifungal properties but suffer from intrinsic hydrophobicity and rapid evaporation. These oils were encapsulated in chitosan-based nanogels to enhance their antifungal activity and stability. The nanogels were formulated by encapsulating C. cyminum essential oils in the chitosan (CS)–caffeic acid (CA) nanogel [225]. Oil encapsulated within the nanogels exhibited sustained release, improved stability, reduced volatility, and superior antimicrobial activity against *Aspergillus flavus*. Although chitosan-based nanogel formulations were extensively used in for drug delivery, nucleic acid, and gene delivery for cancer therapy and rheumatoid arthritis, scientists are exploring it as delivery vehicle to tackle different systemic infections [226–228].

6 Conclusions and Future Outlook

In conclusion, this chapter comprehensively summarized the recent advancements in chitosan-based biomaterials for preventive and therapeutic intervention of antimicrobial resistance. The structural parameters modulating the antimicrobial activity of chitosan and diverse chemical modification performed to enhance antibacterial activity of chitosan were discussed in detail. These modifications have led to the fabrication of different surface coatings on various biomedical surfaces like catheter and titanium implants to prevent the implant-associated infections. Further, the recent development in the field of chitosan-based hemostatic and antimicrobial wound dressings was presented. The chapter also covered advancements in chitosan-based nano/microgels for treating different systemic infections. Although there is a significant progress in chitosan-based anti-infective biomaterials over the last few decades, there exist a few significant challenges that remain for their successful clinical translation.

Even after extensive research, there are contradicting reports on the relationship between molecular weight of chitosan and its antimicrobial activity. Future studies should be conducted with well-standardized parameters, including the source of chitosan, pH, degree of deacetylation, strains of microbes, and temperatures to resolve this issue. Another major challenge in developing antimicrobial coatings for biomedical implants is pro-longing their stability and antimicrobial activity over extended periods to inhibit microbial colonization and proliferation. Very few studies have been performed to assess the chitosan-based coatings in inhibiting or reducing biofilm formation for extended durations [17]. Additionally, prolonged biocompatibility: toxicity as well as foreign body response of these surface coatings were investigated in very few reports, necessitating the need for detailed assessment to apply these coatings in implantable medical devices. Similarly, chitosan-based coatings conjugated with metallic nanoparticles have shown effectiveness in reducing microbial adhesion on medical surfaces; crucial information about their leaching, biocompatibility, and long-term stability is still lacking. Although many studies demonstrated promising in-vitro results, but very few were conducted *in-vivo*, suggesting that most of the anti-infective and anti-adhesive chitosan coatings are yet to reach clinical trials.

One of the major limitations of the chitosan-based hemostatic materials is their slow degradation in-vivo, which should be considered while developing materials for wound healing [25, 26]. Hemostatic materials should ideally "disappear" through removal or degradation [27]. Due to its high mucoadhesive properties, chitosan-based hemostats cause secondary bleeding and pain, while peeling off. Recently, scientists have incorporated additional interfacial layers to reduce tissue adhesion and mitigate the risk of rebleeding during their removal [229]. Additionally, while designing biodegradable materials, the cytocompatibility, thrombogenicity, and immunogenicity of the degradation products should be extensively investigated. In addition, the degradation kinetics of an ideal wound healing agent should match with tissue repair and remodeling rate.

In spite of its excellent biocompatibility, chitosan is yet to secure FDA approval for systemic drug delivery. Concerns regarding its purity, source, and potential immunogenicity contribute to this indecision [15]. Therefore, in-vitro immunocompetence of these biomaterials against different immune cells such as neutrophils and macrophages along with the in-vivo investigation needs detailed assessment with standard and reliable protocols.

The integration of chitosan as an innovative biomaterial into daily life necessitates multidisciplinary collaborations among scientists, biomedical engineers, and industrialists. The collective knowledge accumulated over decades, coupled with ongoing research in chitosan and its derivatives as antimicrobial polymers and biomaterials, holds promise for ushering in a new era in combating antimicrobial resistance.

Acknowledgment Sudip Mukherjee and Dipanjana Patra acknowledge JNCASR for fellowship and Subhankar Maity acknowledges CSIR for fellowship.

References

1. Tacconelli E, Sifakis F, Harbarth S, Schrijver R, van Mourik M, Voss A, Sharland M, Rajendran NB, Rodríguez-Baño J, EPI-Net COMBACTE-MAGNET Group (2018) Surveillance for control of antimicrobial resistance. Lancet Infect Dis 18:e99–e106. https://doi.org/10.1016/S1473-3099(17)30485-1
2. Salam MA, Al-Amin MY, Salam MT, Pawar JS, Akhter N, Rabaan AA, Alqumber MAA (2023) Antimicrobial resistance: a growing serious threat for global public health. Healthcare (Basel) 11. https://doi.org/10.3390/healthcare11131946
3. Piddock LJV (2016) Reflecting on the final report of the O'Neill review on antimicrobial resistance. Lancet Infect Dis 16:767–768. https://doi.org/10.1016/S1473-3099(16)30127-X
4. Arciola CR, Campoccia D, Montanaro L (2018) Implant infections: adhesion, biofilm formation and immune evasion. Nat Rev Microbiol 16:397–409. https://doi.org/10.1038/s41579-018-0019-y
5. Flemming H-C, Wingender J, Szewzyk U, Steinberg P, Rice SA, Kjelleberg S (2016) Biofilms: an emergent form of bacterial life. Nat Rev Microbiol 14:563–575. https://doi.org/10.1038/nrmicro.2016.94
6. Ciofu O, Moser C, Jensen PØ, Høiby N (2022) Tolerance and resistance of microbial biofilms. Nat Rev Microbiol 20:621–635. https://doi.org/10.1038/s41579-022-00682-4
7. Sauer K, Stoodley P, Goeres DM, Hall-Stoodley L, Burmølle M, Stewart PS, Bjarnsholt T (2022) The biofilm life cycle: expanding the conceptual model of biofilm formation. Nat Rev Microbiol 20:608–620. https://doi.org/10.1038/s41579-022-00767-0
8. Si Z, Zheng W, Prananty D, Li J, Koh CH, Kang E-T, Pethe K, Chan-Park MB (2022) Polymers as advanced antibacterial and antibiofilm agents for direct and combination therapies. Chem Sci 13:345–364. https://doi.org/10.1039/d1sc05835e
9. Konai MM, Bhattacharjee B, Ghosh S, Haldar J (2018) Recent progress in polymer research to tackle infections and antimicrobial resistance. Biomacromolecules 19:1888–1917. https://doi.org/10.1021/acs.biomac.8b00458
10. Ghosh S, Mukherjee S, Patra D, Haldar J (2022) Polymeric biomaterials for prevention and therapeutic intervention of microbial infections. Biomacromolecules 23:592–608. https://doi.org/10.1021/acs.biomac.1c01528
11. Magana M, Pushpanathan M, Santos AL, Leanse L, Fernandez M, Ioannidis A, Giulianotti MA, Apidianakis Y, Bradfute S, Ferguson AL, Cherkasov A, Seleem MN, Pinilla C, de la Fuente-Nunez C, Lazaridis T, Dai T, Houghten RA, Hancock REW, Tegos GP (2020) The value of antimicrobial peptides in the age of resistance. Lancet Infect Dis 20:e216–e230. https://doi.org/10.1016/S1473-3099(20)30327-3
12. Jain A, Duvvuri LS, Farah S, Beyth N, Domb AJ, Khan W (2014) Antimicrobial polymers. Adv Healthc Mater 3:1969–1985. https://doi.org/10.1002/adhm.201400418
13. Dhlamini KS, Selepe CT, Ramalapa B, Tshweu L, Ray SS (2024) Reimagining chitosan-based antimicrobial biomaterials to mitigate antibiotic resistance and alleviate antibiotic overuse: a review. Macromol Mater Eng 309. https://doi.org/10.1002/mame.202400018
14. Rabea EI, Badawy ME-T, Stevens CV, Smagghe G, Steurbaut W (2003) Chitosan as antimicrobial agent: applications and mode of action. Biomacromolecules 4:1457–1465. https://doi.org/10.1021/bm034130m
15. Ke C-L, Deng F-S, Chuang C-Y, Lin C-H (2021) Antimicrobial actions and applications of chitosan. Polymers (Basel) 13. https://doi.org/10.3390/polym13060904

16. Raafat D, von Bargen K, Haas A, Sahl H-G (2008) Insights into the mode of action of chitosan as an antibacterial compound. Appl Environ Microbiol 74:3764–3773. https://doi.org/10.1128/AEM.00453-08
17. Teixeira-Santos R, Lima M, Gomes LC, Mergulhão FJ (2021) Antimicrobial coatings based on chitosan to prevent implant-associated infections: a systematic review. iScience 24:103480. https://doi.org/10.1016/j.isci.2021.103480
18. Ma Z, Garrido-Maestu A, Jeong KC (2017) Application, mode of action, and in vivo activity of chitosan and its micro- and nanoparticles as antimicrobial agents: a review. Carbohydr Polym 176:257–265. https://doi.org/10.1016/j.carbpol.2017.08.082
19. Kong M, Chen XG, Xing K, Park HJ (2010) Antimicrobial properties of chitosan and mode of action: a state of the art review. Int J Food Microbiol 144:51–63. https://doi.org/10.1016/j.ijfoodmicro.2010.09.012
20. Hemmingsen LM, Škalko-Basnet N, Jøraholmen MW (2021) The expanded role of chitosan in localized antimicrobial therapy. Mar Drugs. https://doi.org/10.3390/md19120697
21. Taokaew S, Kaewkong W, Kriangkrai W (2023) Recent development of functional chitosan-based hydrogels for pharmaceutical and biomedical applications. Gels 9. https://doi.org/10.3390/gels9040277
22. Wang X, Song R, Johnson M, Sigen A, Shen P, Zhang N, Lara-Sáez I, Xu Q, Wang W (2023) Chitosan-based hydrogels for infected wound treatment. Macromol Biosci 23:e2300094. https://doi.org/10.1002/mabi.202300094
23. Montazerian H, Davoodi E, Baidya A, Baghdasarian S, Sarikhani E, Meyer CE, Haghniaz R, Badv M, Annabi N, Khademhosseini A, Weiss PS (2022) Engineered hemostatic biomaterials for sealing wounds. Chem Rev 122:12864–12903. https://doi.org/10.1021/acs.chemrev.1c01015
24. Hickman DA, Pawlowski CL, Sekhon UDS, Marks J, Gupta AS (2018) Biomaterials and advanced technologies for hemostatic management of bleeding. Adv Mater Weinheim 30. https://doi.org/10.1002/adma.201700859
25. Guo B, Dong R, Liang Y, Li M (2021) Haemostatic materials for wound healing applications. Nat Rev Chem 5:773. https://doi.org/10.1038/s41570-021-00323-z
26. Simpson A, Shukla A, Brown AC (2022) Biomaterials for hemostasis. Annu Rev Biomed Eng 24:111–135. https://doi.org/10.1146/annurev-bioeng-012521-101942
27. Kharaziha M, Baidya A, Annabi N (2021) Rational design of immunomodulatory hydrogels for chronic wound healing. Adv Mater Weinheim 33:e2100176. https://doi.org/10.1002/adma.202100176
28. No HK, Park NY, Lee SH, Meyers SP (2002) Antibacterial activity of chitosans and chitosan oligomers with different molecular weights. Int J Food Microbiol 74:65–72. https://doi.org/10.1016/s0168-1605(01)00717-6
29. Zheng L-Y, Zhu J-F (2003) Study on antimicrobial activity of chitosan with different molecular weights. Carbohydr Polym 54:527–530. https://doi.org/10.1016/j.carbpol.2003.07.009
30. Guerra-Sánchez MG, Vega-Pérez J, Velázquez-del Valle MG, Hernández-Lauzardo AN (2009) Antifungal activity and release of compounds on *Rhizopus stolonifer* (Ehrenb.:Fr.) Vuill. By effect of chitosan with different molecular weights. Pestic Biochem Physiol 93:18–22. https://doi.org/10.1016/j.pestbp.2008.09.001
31. Kingkaew J, Kirdponpattara S, Sanchavanakit N, Pavasant P, Phisalaphong M (2014) Effect of molecular weight of chitosan on antimicrobial properties and tissue compatibility of chitosan-impregnated bacterial cellulose films. Biotechnol Bioprocess Eng 19:534–544. https://doi.org/10.1007/s12257-014-0081-x
32. Li M, Chen X, Liu J, Zhang W, Tang X (2011) Molecular weight-dependent antifungal activity and action mode of chitosan against *Fulvia fulva* (cooke) ciffrri. J Appl Polym Sci 119:3127–3135. https://doi.org/10.1002/app.31831
33. Younes I, Sellimi S, Rinaudo M, Jellouli K, Nasri M (2014) Influence of acetylation degree and molecular weight of homogeneous chitosans on antibacterial and antifungal activities. Int J Food Microbiol 185:57–63. https://doi.org/10.1016/j.ijfoodmicro.2014.04.029

34. Jung EJ, Youn DK, Lee SH, No HK, Ha JG, Prinyawiwatkul W (2010) Antibacterial activity of chitosans with different degrees of deacetylation and viscosities. Int J Food Sci Technol 45: 676–682. https://doi.org/10.1111/j.1365-2621.2010.02186.x
35. Sreekumar S, Wattjes J, Niehues A, Mengoni T, Mendes AC, Morris ER, Goycoolea FM, Moerschbacher BM (2022) Biotechnologically produced chitosans with nonrandom acetylation patterns differ from conventional chitosans in properties and activities. Nat Commun 13: 7125. https://doi.org/10.1038/s41467-022-34483-3
36. Meng X, Xing R, Liu S, Yu H, Li K, Qin Y, Li P (2012) Molecular weight and pH effects of aminoethyl modified chitosan on antibacterial activity in vitro. Int J Biol Macromol 50:918–924. https://doi.org/10.1016/j.ijbiomac.2012.01.018
37. Li J, Wu Y, Zhao L (2016) Antibacterial activity and mechanism of chitosan with ultra high molecular weight. Carbohydr Polym 148:200–205. https://doi.org/10.1016/j.carbpol.2016.04.025
38. Sahariah P, Másson M (2017) Antimicrobial chitosan and chitosan derivatives: a review of the structure-activity relationship. Biomacromolecules 18:3846–3868. https://doi.org/10.1021/acs.biomac.7b01058
39. Costa EM, Silva S, Pina C, Tavaria FK, Pintado MM (2012) Evaluation and insights into chitosan antimicrobial activity against anaerobic oral pathogens. Anaerobe 18:305–309. https://doi.org/10.1016/j.anaerobe.2012.04.009
40. Chung Y-C, Chen C-Y (2008) Antibacterial characteristics and activity of acid-soluble chitosan. Bioresour Technol 99:2806–2814. https://doi.org/10.1016/j.biortech.2007.06.044
41. Liu H, Du Y, Wang X, Sun L (2004) Chitosan kills bacteria through cell membrane damage. Int J Food Microbiol 95:147–155. https://doi.org/10.1016/j.ijfoodmicro.2004.01.022
42. Park S-C, Nah J-W, Park Y (2011) pH-dependent mode of antibacterial actions of low molecular weight water-soluble chitosan (LMWSC) against various pathogens. Macromol Res 19:853–860. https://doi.org/10.1007/s13233-011-0812-1
43. Galván Márquez I, Akuaku J, Cruz I, Cheetham J, Golshani A, Smith ML (2013) Disruption of protein synthesis as antifungal mode of action by chitosan. Int J Food Microbiol 164:108–112. https://doi.org/10.1016/j.ijfoodmicro.2013.03.025
44. Ghormade V, Pathan EK, Deshpande MV (2017) Can fungi compete with marine sources for chitosan production? Int J Biol Macromol. https://doi.org/10.1016/j.ijbiomac.2017.01.112
45. Tayel AA, Ibrahim SIA, Al-Saman MA, Moussa SH (2014) Production of fungal chitosan from date wastes and its application as a biopreservative for minced meat. Int J Biol Macromol 69:471–475. https://doi.org/10.1016/j.ijbiomac.2014.05.072
46. Chien R-C, Yen M-T, Mau J-L (2016) Antimicrobial and antitumor activities of chitosan from shiitake stipes, compared to commercial chitosan from crab shells. Carbohydr Polym 138: 259–264. https://doi.org/10.1016/j.carbpol.2015.11.061
47. Byun SM, No HK, Hong J, Lee SI, Prinyawiwatkul W (2013) Comparison of physicochemical, binding, antioxidant and antibacterial properties of chitosans prepared from ground and entire crab leg shells. Int J Food Sci Technol 48:136–142. https://doi.org/10.1111/j.1365-2621.2012.03169.x
48. Chang AKT, Frias RR, Alvarez LV, Bigol UG, Guzman JPMD (2019) Comparative antibacterial activity of commercial chitosan and chitosan extracted from Auricularia sp. Biocatal Agric Biotechnol 17:189–195. https://doi.org/10.1016/j.bcab.2018.11.016
49. Liu N, Chen X-G, Park H-J, Liu C-G, Liu C-S, Meng X-H, Yu L-J (2006) Effect of MW and concentration of chitosan on antibacterial activity of Escherichia coli. Carbohydr Polym 64: 60–65. https://doi.org/10.1016/j.carbpol.2005.10.028
50. Garcia LGS, de Guedes GMM, da Silva MLQ, Castelo-Branco DSCM, Sidrim JJC, de Cordeiro RA, Rocha MFG, Vieira RS, Brilhante RSN (2018) Effect of the molecular weight of chitosan on its antifungal activity against Candida spp. in planktonic cells and biofilm. Carbohydr Polym 195:662–669. https://doi.org/10.1016/j.carbpol.2018.04.091
51. Kreling V, Falcone FH, Herrmann F, Kemper L, Amiteye D, Cord-Landwehr S, Kehrenberg C, Moerschbacher BM, Hensel A (2024) High molecular/low acetylated chitosans

reduce adhesion of Campylobacter jejuni to host cells by blocking JlpA. Appl Microbiol Biotechnol 108:171. https://doi.org/10.1007/s00253-024-13000-0
52. Lim S-H, Hudson SM (2004) Synthesis and antimicrobial activity of a water-soluble chitosan derivative with a fiber-reactive group. Carbohydr Res 339:313–319. https://doi.org/10.1016/j.carres.2003.10.024
53. Helander IM, Nurmiaho-Lassila EL, Ahvenainen R, Rhoades J, Roller S (2001) Chitosan disrupts the barrier properties of the outer membrane of gram-negative bacteria. Int J Food Microbiol 71:235–244. https://doi.org/10.1016/S0168-1605(01)00609-2
54. Sudarshan NR, Hoover DG, Knorr D (1992) Antibacterial action of chitosan. Food Biotechnol 6:257–272. https://doi.org/10.1080/08905439209549838
55. Kurita K, Ikeda H, Yoshida Y, Shimojoh M, Harata M (2002) Chemoselective protection of the amino groups of chitosan by controlled phthaloylation: facile preparation of a precursor useful for chemical modifications. Biomacromolecules 3:1–4. https://doi.org/10.1021/bm0101163
56. Holappa J, Nevalainen T, Soininen P, Elomaa M, Safin R, Másson M, Järvinen T (2005) N-Chloroacyl-6-O-triphenylmethylchitosans: useful intermediates for synthetic modifications of chitosan. Biomacromolecules 6:858–863. https://doi.org/10.1021/bm049372g
57. Kurita K, Hirakawa M, Kikuchi S, Yamanaka H, Yang J (2004) Trimethylsilylation of chitosan and some properties of the product. Carbohydr Polym 56:333–337. https://doi.org/10.1016/j.carbpol.2004.02.008
58. Rúnarsson OV, Malainer C, Holappa J, Sigurdsson ST, Másson M (2008) Tert-butyldimethylsilyl O-protected chitosan and chitooligosaccharides: useful precursors for N-modifications in common organic solvents. Carbohydr Res 343:2576–2582. https://doi.org/10.1016/j.carres.2008.08.014
59. Rúnarsson ÖV, Holappa J, Nevalainen T, Hjálmarsdóttir M, Järvinen T, Loftsson T, Einarsson JM, Jónsdóttir S, Valdimarsdóttir M, Másson M (2007) Antibacterial activity of methylated chitosan and chitooligomer derivatives: synthesis and structure activity relationships. Eur Polym J 43:2660–2671. https://doi.org/10.1016/j.eurpolymj.2007.03.046
60. Kulkarni AD, Patel HM, Surana SJ, Vanjari YH, Belgamwar VS, Pardeshi CV (2017) N,N,N-trimethyl chitosan: an advanced polymer with myriad of opportunities in nanomedicine. Carbohydr Polym 157:875–902. https://doi.org/10.1016/j.carbpol.2016.10.041
61. Benediktsdóttir BE, Gaware VS, Rúnarsson ÖV, Jónsdóttir S, Jensen KJ, Másson M (2011) Synthesis of N,N,N-trimethyl chitosan homopolymer and highly substituted N-alkyl-N,N-dimethyl chitosan derivatives with the aid of di-tert-butyldimethylsilyl chitosan. Carbohydr Polym 86:1451–1460. https://doi.org/10.1016/j.carbpol.2011.06.007
62. Fu X, Shen Y, Jiang X, Huang D, Yan Y (2011) Chitosan derivatives with dual-antibacterial functional groups for antimicrobial finishing of cotton fabrics. Carbohydr Polym 85:221–227. https://doi.org/10.1016/j.carbpol.2011.02.019
63. Wen Y, Tan Z, Sun F, Sheng L, Zhang X, Yao F (2012) Synthesis and characterization of quaternized carboxymethyl chitosan/poly(amidoamine) dendrimer core-shell nanoparticles. Mater Sci Eng C Mater Biol Appl 32:2026–2036. https://doi.org/10.1016/j.msec.2012.05.019
64. Bhattacharjee B, Ghosh S, Mukherjee R, Haldar J (2021) Quaternary lipophilic chitosan and gelatin cross-linked antibacterial hydrogel effectively kills multidrug-resistant bacteria with minimal toxicity toward mammalian cells. Biomacromolecules 22:557–571. https://doi.org/10.1021/acs.biomac.0c01420
65. Hu Y, Du Y, Yang J, Kennedy J, Wang X, Wang L (2007) Synthesis, characterization and antibacterial activity of guanidinylated chitosan. Carbohydr Polym 67:66–72. https://doi.org/10.1016/j.carbpol.2006.04.015
66. Zhai X, Sun P, Luo Y, Ma C, Xu J, Liu W (2011) Guanidinylation: a simple way to fabricate cell penetrating peptide analogue-modified chitosan vector for enhanced gene delivery. J Appl Polym Sci 121:3569–3578. https://doi.org/10.1002/app.34156

67. Sajomsang W, Gonil P, Tantayanon S (2009) Antibacterial activity of quaternary ammonium chitosan containing mono or disaccharide moieties: preparation and characterization. Int J Biol Macromol 44:419–427. https://doi.org/10.1016/j.ijbiomac.2009.03.003
68. Li R, Hu P, Ren X, Worley SD, Huang TS (2013) Antimicrobial N-halamine modified chitosan films. Carbohydr Polym 92:534–539. https://doi.org/10.1016/j.carbpol.2012.08.115
69. Jin Z, Li W, Cao H, Zhang X, Chen G, Wu H, Guo C, Zhang Y, Kang H, Wang Y, Zhao K (2013) Antimicrobial activity and cytotoxicity of N-2-HACC and characterization of nanoparticles with N-2-HACC and CMC as a vaccine carrier. Chem Eng J 221:331–341. https://doi.org/10.1016/j.cej.2013.02.011
70. Geisberger G, Gyenge EB, Hinger D, Käch A, Maake C, Patzke GR (2013) Chitosan-thioglycolic acid as a versatile antimicrobial agent. Biomacromolecules 14:1010–1017. https://doi.org/10.1021/bm3018593
71. Cai Z, Song Z, Yang C, Shang S, Yin Y (2009) Synthesis, characterization and antibacterial activity of quaternized N,O-(2-carboxyethyl) chitosan. Polym Bull 62:445–456. https://doi.org/10.1007/s00289-008-0004-0
72. Kogan G, Skorik YA, Zitnanová I, Krizková L, Duracková Z, Gomes CAR, Yatluk YG, Krajcovic J (2004) Antioxidant and antimutagenic activity of N-(2-carboxyethyl)chitosan. Toxicol Appl Pharmacol 201:303–310. https://doi.org/10.1016/j.taap.2004.05.009
73. Xu T, Xin M, Li M, Huang H, Zhou S, Liu J (2011) Synthesis, characterization, and antibacterial activity of N,O-quaternary ammonium chitosan. Carbohydr Res 346:2445–2450. https://doi.org/10.1016/j.carres.2011.08.002
74. Xu T, Xin M, Li M, Huang H, Zhou S (2010) Synthesis, characteristic and antibacterial activity of N,N,N-trimethyl chitosan and its carboxymethyl derivatives. Carbohydr Polym 81: 931–936. https://doi.org/10.1016/j.carbpol.2010.04.008
75. Lv J, Yin X, Zeng Q, Dong W, Liu H, Zhu L (2017) Preparation of carboxymethyl chitosan nanofibers through electrospinning the ball-milled nanopowders with poly (lactic acid) and the blood compatibility of the electrospun NCMC/PLA mats. J Polym Res 24:60. https://doi.org/10.1007/s10965-017-1224-5
76. Sun L, Du Y, Fan L, Chen X, Yang J (2006) Preparation, characterization and antimicrobial activity of quaternized carboxymethyl chitosan and application as pulp-cap. Polymer 47: 1796–1804. https://doi.org/10.1016/j.polymer.2006.01.073
77. Mohamed NA, Abd El-Ghany NA (2012) Preparation and antimicrobial activity of some carboxymethyl chitosan acyl thiourea derivatives. Int J Biol Macromol 50:1280–1285. https://doi.org/10.1016/j.ijbiomac.2012.03.011
78. Coma V, Freire CSR, Silvestre AJD (2021) Recent advances on the development of antibacterial polysaccharide-based materials: chitosan, cellulose, and starch. In: Ramawat KG, Mérillon J-M (eds) Polysaccharides: bioactivity and biotechnology. Springer, Cham, pp 1–46
79. Li P, Zhou C, Rayatpisheh S, Ye K, Poon YF, Hammond PT, Duan H, Chan-Park MB (2012) Cationic peptidopolysaccharides show excellent broad-spectrum antimicrobial activities and high selectivity. Adv Mater Weinheim 24:4130–4137. https://doi.org/10.1002/adma.201104186
80. Sahariah P, Sørensen KK, Hjálmarsdóttir MÁ, Sigurjónsson ÓE, Jensen KJ, Másson M, Thygesen MB (2015) Antimicrobial peptide shows enhanced activity and reduced toxicity upon grafting to chitosan polymers. Chem Commun 51:11611–11614. https://doi.org/10.1039/c5cc04010h
81. Zimmerli W (2014) Clinical presentation and treatment of orthopaedic implant-associated infection. J Intern Med 276:111–119. https://doi.org/10.1111/joim.12233
82. Ribeiro M, Monteiro FJ, Ferraz MP (2012) Infection of orthopedic implants with emphasis on bacterial adhesion process and techniques used in studying bacterial-material interactions. Biomatter 2:176–194. https://doi.org/10.4161/biom.22905

83. Masters EA, Ricciardi BF, de Bentley KLM, Moriarty TF, Schwarz EM, Muthukrishnan G (2022) Skeletal infections: microbial pathogenesis, immunity and clinical management. Nat Rev Microbiol 20:385–400. https://doi.org/10.1038/s41579-022-00686-0
84. Cyphert EL, Zhang N, Learn GD, Hernandez CJ, von Recum HA (2021) Recent advances in the evaluation of antimicrobial materials for resolution of orthopedic implant- associated infections in vivo. ACS Infect Dis 7:3125–3160. https://doi.org/10.1021/acsinfecdis.1c00465
85. Tao F, Ma S, Tao H, Jin L, Luo Y, Zheng J, Xiang W, Deng H (2021) Chitosan-based drug delivery systems: from synthesis strategy to osteomyelitis treatment – a review. Carbohydr Polym 251:117063. https://doi.org/10.1016/j.carbpol.2020.117063
86. Bumgardner JD, Wiser R, Gerard PD, Bergin P, Chestnutt B, Marin M, Ramsey V, Elder SH, Gilbert JA (2003) Chitosan: potential use as a bioactive coating for orthopaedic and craniofacial/dental implants. J Biomater Sci Polym Ed 14:423–438. https://doi.org/10.1163/156856203766652048
87. Ordikhani F, Simchi A (2014) Long-term antibiotic delivery by chitosan-based composite coatings with bone regenerative potential. Appl Surf Sci 317:56–66. https://doi.org/10.1016/j.apsusc.2014.07.197
88. Kumari S, Tiyyagura HR, Pottathara YB, Sadasivuni KK, Ponnamma D, Douglas TEL, Skirtach AG, Mohan MK (2020) Surface functionalization of chitosan as a coating material for orthopaedic applications: a comprehensive review. Carbohydr Polym:117487. https://doi.org/10.1016/j.carbpol.2020.117487
89. Lin H-Y, Chen J-H (2013) Osteoblast differentiation and phenotype expressions on chitosan-coated Ti-6Al-4V. Carbohydr Polym 97:618–626. https://doi.org/10.1016/j.carbpol.2013.05.048
90. Yang C-C, Lin C-C, Liao J-W, Yen S-K (2013) Vancomycin-chitosan composite deposited on post porous hydroxyapatite coated Ti6Al4V implant for drug controlled release. Mater Sci Eng C Mater Biol Appl 33:2203–2212. https://doi.org/10.1016/j.msec.2013.01.038
91. Song J, Chen Q, Zhang Y, Diba M, Kolwijck E, Shao J, Jansen JA, Yang F, Boccaccini AR, Leeuwenburgh SCG (2016) Electrophoretic deposition of chitosan coatings modified with gelatin Nanospheres to tune the release of antibiotics. ACS Appl Mater Interfaces 8:13785–13792. https://doi.org/10.1021/acsami.6b03454
92. Liu P, Hao Y, Zhao Y, Yuan Z, Ding Y, Cai K (2017) Surface modification of titanium substrates for enhanced osteogenetic and antibacterial properties. Colloids Surf B Biointerfaces 160:110–116. https://doi.org/10.1016/j.colsurfb.2017.08.044
93. Mohan Raj R, Priya P, Raj V (2018) Gentamicin-loaded ceramic-biopolymer dual layer coatings on the Ti with improved bioactive and corrosion resistance properties for orthopedic applications. J Mech Behav Biomed Mater 82:299–309. https://doi.org/10.1016/j.jmbbm.2017.12.033
94. Nancy D, Rajendran N (2018) Vancomycin incorporated chitosan/gelatin coatings coupled with TiO2-SrHAP surface modified cp-titanium for osteomyelitis treatment. Int J Biol Macromol 110:197–205. https://doi.org/10.1016/j.ijbiomac.2018.01.004
95. David N, Nallaiyan R (2018) Biologically anchored chitosan/gelatin-SrHAP scaffold fabricated on titanium against chronic osteomyelitis infection. Int J Biol Macromol 110:206–214. https://doi.org/10.1016/j.ijbiomac.2017.11.174
96. Valverde A, Pérez-Álvarez L, Ruiz-Rubio L, Pacha Olivenza MA, García Blanco MB, Díaz-Fuentes M, Vilas-Vilela JL (2019) Antibacterial hyaluronic acid/chitosan multilayers onto smooth and micropatterned titanium surfaces. Carbohydr Polym 207:824–833. https://doi.org/10.1016/j.carbpol.2018.12.039
97. Stoian AB, Demetrescu I, Ionita D (2020) Nanotubes and nano pores with chitosan construct on TiZr serving as drug reservoir. Colloids Surf B Biointerfaces 185:110535. https://doi.org/10.1016/j.colsurfb.2019.110535
98. Andrade Del Olmo J, Pérez-Álvarez L, Sáez Martínez V, Benito Cid S, Ruiz-Rubio L, Pérez González R, Vilas-Vilela JL, Alonso JM (2023) Multifunctional antibacterial chitosan-based

hydrogel coatings on Ti6Al4V biomaterial for biomedical implant applications. Int J Biol Macromol 231:123328. https://doi.org/10.1016/j.ijbiomac.2023.123328
99. Pishbin F, Mouriño V, Gilchrist JB, McComb DW, Kreppel S, Salih V, Ryan MP, Boccaccini AR (2013) Single-step electrochemical deposition of antimicrobial orthopaedic coatings based on a bioactive glass/chitosan/nano-silver composite system. Acta Biomater 9:7469–7479. https://doi.org/10.1016/j.actbio.2013.03.006
100. Pishbin F, Mouriño V, Flor S, Kreppel S, Salih V, Ryan MP, Boccaccini AR (2014) Electrophoretic deposition of gentamicin-loaded bioactive glass/chitosan composite coatings for orthopaedic implants. ACS Appl Mater Interfaces 6:8796–8806. https://doi.org/10.1021/am5014166
101. Mattioli-Belmonte M, Cometa S, Ferretti C, Iatta R, Trapani A, Ceci E, Falconi M, De Giglio E (2014) Characterization and cytocompatibility of an antibiotic/chitosan/cyclodextrins nanocoating on titanium implants. Carbohydr Polym 110:173–182. https://doi.org/10.1016/j.carbpol.2014.03.097
102. Zarghami V, Ghorbani M, Bagheri KP, Shokrgozar MA (2022) Improving bactericidal performance of implant composite coatings by synergism between Melittin and tetracycline. J Mater Sci Mater Med 33:46. https://doi.org/10.1007/s10856-022-06666-3
103. Han L, Wang Z, Lu X, Dong L, Xie C, Wang K, Chen X, Ding Y, Weng L (2015) Mussel-inspired adhesive and transferable free-standing films by self-assembling dexamethasone encapsulated BSA nanoparticles and vancomycin immobilized oxidized alginate. Colloids Surf B Biointerfaces 126:452–458. https://doi.org/10.1016/j.colsurfb.2014.12.050
104. Jennings JA, Velasquez Pulgarin DA, Kunwar DL, Babu J, Mishra S, Bumgardner J (2015) Bacterial inhibition by chitosan coatings loaded with silver-decorated calcium phosphate microspheres. Thin Solid Films 596:83–86. https://doi.org/10.1016/j.tsf.2015.08.060
105. Yan Y, Zhang X, Li C, Huang Y, Ding Q, Pang X (2015) Preparation and characterization of chitosan-silver/hydroxyapatite composite coatings onTiO2 nanotube for biomedical applications. Appl Surf Sci 332:62–69. https://doi.org/10.1016/j.apsusc.2015.01.136
106. Xu Z, Li M, Li X, Liu X, Ma F, Wu S, Yeung KWK, Han Y, Chu PK (2016) Antibacterial activity of silver doped titanate nanowires on ti implants. ACS Appl Mater Interfaces 8:16584–16594. https://doi.org/10.1021/acsami.6b04161
107. Cometa S, Bonifacio MA, Baruzzi F, de Candia S, Giangregorio MM, Giannossa LC, Dicarlo M, Mattioli-Belmonte M, Sabbatini L, De Giglio E (2017) Silver-loaded chitosan coating as an integrated approach to face titanium implant-associated infections: analytical characterization and biological activity. Anal Bioanal Chem 409:7211–7221. https://doi.org/10.1007/s00216-017-0685-z
108. Shao J, Wang B, Bartels CJM, Bronkhorst EM, Jansen JA, Walboomers XF, Yang F (2018) Chitosan-based sleeves loaded with silver and chlorhexidine in a percutaneous rabbit tibia model with a repeated bacterial challenge. Acta Biomater 82:102–110. https://doi.org/10.1016/j.actbio.2018.10.021
109. Zhu Y, Liu X, Yeung KWK, Chu PK, Wu S (2017) Biofunctionalization of carbon nanotubes/chitosan hybrids on Ti implants by atom layer deposited ZnO nanostructures. Appl Surf Sci 400:14–23. https://doi.org/10.1016/j.apsusc.2016.12.158
110. Ren Y, Qin X, Barbeck M, Hou Y, Xu H, Liu L, Liu C (2021) Mussel-inspired carboxymethyl chitosan hydrogel coating of titanium alloy with antibacterial and bioactive properties. Materials (Basel). https://doi.org/10.3390/ma14226901
111. Zhang T, Qin X, Gao Y, Kong D, Jiang Y, Cui X, Guo M, Chen J, Chang F, Zhang M, Li J, Yin P (2023) Functional chitosan gel coating enhances antimicrobial properties and osteogenesis of titanium alloy under persistent chronic inflammation. Front Bioeng Biotechnol 11:1118487. https://doi.org/10.3389/fbioe.2023.1118487
112. Ao H, Yang S, Nie B, Fan Q, Zhang Q, Zong J, Guo S, Zheng X, Tang T (2019) Improved antibacterial properties of collagen I/hyaluronic acid/quaternized chitosan multilayer modified titanium coatings with both contact-killing and release-killing functions. J Mater Chem B Mater Biol Med 7:1951–1961. https://doi.org/10.1039/c8tb02425a

113. Peng Z, Ao H, Wang L, Guo S, Tang T (2015) Quaternised chitosan coating on titanium provides a self-protective surface that prevents bacterial colonisation and implant-associated infections. RSC Adv 5:54304–54311. https://doi.org/10.1039/C5RA07540H
114. Wang B, Wang J, Li D, Ren K, Ji J (2012) Chitosan/poly (vinyl pyrollidone) coatings improve the antibacterial properties of poly(ethylene terephthalate). Appl Surf Sci 258:7801–7808. https://doi.org/10.1016/j.apsusc.2012.03.181
115. Peng Z-X, Wang L, Du L, Guo S-R, Wang X-Q, Tang T-T (2010) Adjustment of the antibacterial activity and biocompatibility of hydroxypropyltrimethyl ammonium chloride chitosan by varying the degree of substitution of quaternary ammonium. Carbohydr Polym 81:275–283. https://doi.org/10.1016/j.carbpol.2010.02.008
116. Bryers JD (2008) Medical biofilms. Biotechnol Bioeng 100:1–18. https://doi.org/10.1002/bit.21838
117. Flores-Mireles AL, Walker JN, Caparon M, Hultgren SJ (2015) Urinary tract infections: epidemiology, mechanisms of infection and treatment options. Nat Rev Microbiol 13:269–284. https://doi.org/10.1038/nrmicro3432
118. Stamm WE, Norrby SR (2001) Urinary tract infections: disease panorama and challenges. J Infect Dis 183(Suppl 1):S1–S4. https://doi.org/10.1086/318850
119. Ricardo SIC, Anjos IIL, Monge N, Faustino CMC, Ribeiro IAC (2020) A glance at antimicrobial strategies to prevent catheter-associated medical infections. ACS Infect Dis 6:3109–3130. https://doi.org/10.1021/acsinfecdis.0c00526
120. Patra D, Ghosh S, Mukherjee S, Acharya Y, Mukherjee R, Haldar J (2024) Antimicrobial nanocomposite coatings for rapid intervention against catheter-associated urinary tract infections. Nanoscale 16:11109–11125. https://doi.org/10.1039/d4nr00653d
121. Ghosh S, Mukherjee R, Patra D, Haldar J (2023) Engineering photo-crosslinked antimicrobial coating to tackle catheter-associated infections in vivo. ACS Biomater Sci Eng 9:4404–4414. https://doi.org/10.1021/acsbiomaterials.3c00342
122. Busscher HJ, van der Mei HC, Subbiahdoss G, Jutte PC, van den Dungen JJAM, Zaat SAJ, Schultz MJ, Grainger DW (2012) Biomaterial-associated infection: locating the finish line in the race for the surface. Sci Transl Med 4:153rv10. https://doi.org/10.1126/scitranslmed.3004528
123. Bhattacharjee B, Ghosh S, Patra D, Haldar J (2022) Advancements in release-active antimicrobial biomaterials: a journey from release to relief. Wiley Interdiscip Rev Nanomed Nanobiotechnol 14:e1745. https://doi.org/10.1002/wnan.1745
124. Bakkeren E, Diard M, Hardt W-D (2020) Evolutionary causes and consequences of bacterial antibiotic persistence. Nat Rev Microbiol 18:479–490. https://doi.org/10.1038/s41579-020-0378-z
125. Cheung RCF, Ng TB, Wong JH, Chan WY (2015) Chitosan: an update on potential biomedical and pharmaceutical applications. Mar Drugs 13:5156–5186. https://doi.org/10.3390/md13085156
126. Ramstedt M, Ribeiro IAC, Bujdakova H, Mergulhão FJM, Jordao L, Thomsen P, Alm M, Burmølle M, Vladkova T, Can F, Reches M, Riool M, Barros A, Reis RL, Meaurio E, Kikhney J, Moter A, Zaat SAJ, Sjollema J (2019) Evaluating efficacy of antimicrobial and antifouling materials for urinary tract medical devices: challenges and recommendations. Macromol Biosci 19:e1800384. https://doi.org/10.1002/mabi.201800384
127. Huang K-S, Yang C-H, Huang S-L, Chen C-Y, Lu Y-Y, Lin Y-S (2016) Recent advances in antimicrobial polymers: a mini-review. Int J Mol Sci. https://doi.org/10.3390/ijms17091578
128. Francolini I, Vuotto C, Piozzi A, Donelli G (2017) Antifouling and antimicrobial biomaterials: an overview. APMIS 125:392–417. https://doi.org/10.1111/apm.12675
129. Martinez LR, Mihu MR, Han G, Frases S, Cordero RJB, Casadevall A, Friedman AJ, Friedman JM, Nosanchuk JD (2010) The use of chitosan to damage Cryptococcus neoformans biofilms. Biomaterials 31:669–679. https://doi.org/10.1016/j.biomaterials.2009.09.087
130. Martinez LR, Mihu MR, Tar M, Cordero RJB, Han G, Friedman AJ, Friedman JM, Nosanchuk JD (2010) Demonstration of antibiofilm and antifungal efficacy of chitosan against candidal

biofilms, using an in vivo central venous catheter model. J Infect Dis 201:1436–1440. https://doi.org/10.1086/651558
131. Pu Y, Liu A, Zheng Y, Ye B (2014) In vitro damage of Candida albicans biofilms by chitosan. Exp Ther Med 8:929–934. https://doi.org/10.3892/etm.2014.1839
132. Hall TJ, Villapún VM, Addison O, Webber MA, Lowther M, Louth SET, Mountcastle SE, Brunet MY, Cox SC (2020) A call for action to the biomaterial community to tackle antimicrobial resistance. Biomater Sci 8:4951–4974. https://doi.org/10.1039/d0bm01160f
133. Bulwan M, Wójcik K, Zapotoczny S, Nowakowska M (2012) Chitosan-based ultrathin films as antifouling, anticoagulant and antibacterial protective coatings. J Biomater Sci Polym Ed 23:1963–1980. https://doi.org/10.1163/092050611X601711
134. Campana R, Casettari L, Ciandrini E, Illum L, Baffone W (2017) Chitosans inhibit the growth and the adhesion of Klebsiella pneumoniae and Escherichia coli clinical isolates on urinary catheters. Int J Antimicrob Agents 50:135–141. https://doi.org/10.1016/j.ijantimicag.2017.03.031
135. Kara F, Aksoy EA, Yuksekdag Z, Hasirci N, Aksoy S (2014) Synthesis and surface modification of polyurethanes with chitosan for antibacterial properties. Carbohydr Polym 112:39–47. https://doi.org/10.1016/j.carbpol.2014.05.019
136. Magesh H, Kumar A, Alam A, Priyam SU, Sumantran VN, Vaidyanathan R (2013) Identification of natural compounds which inhibit biofilm formation in clinical isolates of Klebsiella pneumoniae. Indian J Exp Biol 51:764–772
137. Rubini D, Vedha Hari BN, Nithyanand P (2021) Chitosan coated catheters alleviates mixed species biofilms of Staphylococcus epidermidis and Candida albicans. Carbohydr Polym 252: 117192. https://doi.org/10.1016/j.carbpol.2020.117192
138. Cobrado L, Silva-Dias A, Azevedo MM, Pina-Vaz C, Rodrigues AG (2013) In vivo antibiofilm effect of cerium, chitosan and hamamelitannin against usual agents of catheter-related bloodstream infections. J Antimicrob Chemother 68:126–130. https://doi.org/10.1093/jac/dks376
139. Cobrado L, Azevedo MM, Silva-Dias A, Ramos JP, Pina-Vaz C, Rodrigues AG (2012) Cerium, chitosan and hamamelitannin as novel biofilm inhibitors? J Antimicrob Chemother 67:1159–1162. https://doi.org/10.1093/jac/dks007
140. Wang R, Neoh KG, Shi Z, Kang E-T, Tambyah PA, Chiong E (2012) Inhibition of Escherichia coli and Proteus mirabilis adhesion and biofilm formation on medical grade silicone surface. Biotechnol Bioeng 109:336–345. https://doi.org/10.1002/bit.23342
141. Tan Y, Leonhard M, Moser D, Schneider-Stickler B (2016) Antibiofilm activity of carboxymethyl chitosan on the biofilms of non-Candida albicans Candida species. Carbohydr Polym 149:77–82. https://doi.org/10.1016/j.carbpol.2016.04.101
142. Tan Y, Leonhard M, Ma S, Moser D, Schneider-Stickler B (2018) Efficacy of carboxymethyl chitosan against Candida tropicalis and Staphylococcus epidermidis monomicrobial and polymicrobial biofilms. Int J Biol Macromol 110:150–156. https://doi.org/10.1016/j.ijbiomac.2017.08.094
143. Fei Liu X, Lin Guan Y, Zhi Yang D, Li Z, De Yao K (2001) Antibacterial action of chitosan and carboxymethylated chitosan. J Appl Polym Sci 79:1324–1335. https://doi.org/10.1002/1097-4628(20010214)79:7<1324::AID-APP210>3.0.CO;2-L
144. Campana R, Biondo F, Mastrotto F, Baffone W, Casettari L (2018) Chitosans as new tools against biofilms formation on the surface of silicone urinary catheters. Int J Biol Macromol 118:2193–2200. https://doi.org/10.1016/j.ijbiomac.2018.07.088
145. Ajdnik U, Zemljič LF, Bračič M, Maver U, Plohl O, Rebol J (2019) Functionalisation of silicone by drug-embedded chitosan nanoparticles for potential applications in otorhinolaryngology. Materials (Basel). https://doi.org/10.3390/ma12060847
146. Raman N, Marchillo K, Lee M-R, de Rodríguez López AL, Andes DR, Palecek SP, Lynn DM (2016) Intraluminal release of an antifungal β-peptide enhances the antifungal and anti-biofilm activities of multilayer-coated catheters in a rat model of venous catheter infection. ACS Biomater Sci Eng 2:112–121. https://doi.org/10.1021/acsbiomaterials.5b00427

147. Gu G, Erişen DE, Yang K, Zhang B, Shen M, Zou J, Qi X, Chen S, Xu X (2022) Antibacterial and anti-inflammatory activities of chitosan/copper complex coating on medical catheters: in vitro and in vivo. J Biomed Mater Res Part B Appl Biomater 110:1899–1910. https://doi.org/10.1002/jbm.b.35047
148. Niemczyk A, Goszczyńska A, Gołda-Cępa M, Kotarba A, Sobolewski P, El Fray M (2019) Biofunctional catheter coatings based on chitosan-fatty acids derivatives. Carbohydr Polym 225:115263. https://doi.org/10.1016/j.carbpol.2019.115263
149. Yang K, Kim K, Lee EA, Liu SS, Kabli S, Alsudir SA, Albrahim S, Zhou A, Park TG, Lee H, Almalik AM, Karp JM, Alhasan AH, Lee Y (2019) Robust low friction antibiotic coating of urethral catheters using a catechol-functionalized polymeric hydrogel film. Front Mater. https://doi.org/10.3389/fmats.2019.00274
150. Wang B-L, Liu X-S, Ji Y, Ren K-F, Ji J (2012) Fast and long-acting antibacterial properties of chitosan-Ag/polyvinylpyrrolidone nanocomposite films. Carbohydr Polym 90:8–15. https://doi.org/10.1016/j.carbpol.2012.03.080
151. Mendoza G, Regiel-Futyra A, Tamayo A, Monzon M, Irusta S, de Gregorio MA, Kyzioł A, Arruebo M (2018) Chitosan-based coatings in the prevention of intravascular catheter-associated infections. J Biomater Appl 32:725–737. https://doi.org/10.1177/0885328217739199
152. Vieira ACC, Chaves LL, Pinheiro M, Lima SC, Neto PJR, Ferreira D, Sarmento B, Reis S (2021) Lipid nanoparticles coated with chitosan using a one-step association method to target rifampicin to alveolar macrophages. Carbohydr Polym 252:116978. https://doi.org/10.1016/j.carbpol.2020.116978
153. Cao Z, Sun Y (2009) Chitosan-based rechargeable long-term antimicrobial and biofilm-controlling systems. J Biomed Mater Res A 89:960–967. https://doi.org/10.1002/jbm.a.32040
154. Zhou W, Jia Z, Xiong P, Yan J, Li M, Cheng Y, Zheng Y (2018) Novel pH-responsive tobramycin-embedded micelles in nanostructured multilayer-coatings of chitosan/heparin with efficient and sustained antibacterial properties. Mater Sci Eng C Mater Biol Appl 90:693–705. https://doi.org/10.1016/j.msec.2018.04.069
155. Yang S-H, Lee Y-SJ, Lin F-H, Yang J-M, Chen K-S (2007) Chitosan/poly(vinyl alcohol) blending hydrogel coating improves the surface characteristics of segmented polyurethane urethral catheters. J Biomed Mater Res Part B Appl Biomater 83:304–313. https://doi.org/10.1002/jbm.b.30796
156. Kara F, Aksoy EA, Yuksekdag Z, Aksoy S, Hasirci N (2015) Enhancement of antibacterial properties of polyurethanes by chitosan and heparin immobilization. Appl Surf Sci 357:1692–1702. https://doi.org/10.1016/j.apsusc.2015.09.227
157. Sen CK (2021) Human wound and its burden: updated 2020 compendium of estimates. Adv Wound Care (New Rochelle) 10:281–292. https://doi.org/10.1089/wound.2021.0026
158. Sorg H, Tilkorn DJ, Hager S, Hauser J, Mirastschijski U (2017) Skin wound healing: an update on the current knowledge and concepts. Eur Surg Res 58:81–94. https://doi.org/10.1159/000454919
159. Falcone M, De Angelis B, Pea F, Scalise A, Stefani S, Tasinato R, Zanetti O, Dalla Paola L (2021) Challenges in the management of chronic wound infections. J Glob Antimicrob Resist 26:140–147. https://doi.org/10.1016/j.jgar.2021.05.010
160. Kumar S, Kumar Paswan S, Verma P, Akanksha, Sah R, Srivastava S, Venketswara Rao C (2022) Current understanding to accelerate wound healing: mechanism and clinical importance. Recent advances in wound healing, https://doi.org/10.5772/intechopen.101429
161. Das S, Baker AB (2016) Biomaterials and nanotherapeutics for enhancing skin wound healing. Front Bioeng Biotechnol 4:82. https://doi.org/10.3389/fbioe.2016.00082
162. Loo HL, Goh BH, Lee L-H, Chuah LH (2022) Application of chitosan-based nanoparticles in skin wound healing. Asian J Pharm Sci 17:299–332. https://doi.org/10.1016/j.ajps.2022.04.001
163. Xu Z, Han S, Gu Z, Wu J (2020) Advances and impact of antioxidant hydrogel in chronic wound healing. Adv Healthc Mater 9:e1901502. https://doi.org/10.1002/adhm.201901502

164. Shukla SK, Sharma AK, Gupta V, Yashavarddhan MH (2019) Pharmacological control of inflammation in wound healing. J Tissue Viability 28:218–222. https://doi.org/10.1016/j.jtv.2019.09.002
165. Wang L, Hao F, Tian S, Dong H, Nie J, Ma G (2022) Targeting polysaccharides such as chitosan, cellulose, alginate and starch for designing hemostatic dressings. Carbohydr Polym 291:119574. https://doi.org/10.1016/j.carbpol.2022.119574
166. Zhu C, Zhao J, Kempe K, Wilson P, Wang J, Velkov T, Li J, Davis TP, Whittaker MR, Haddleton DM (2017) A hydrogel-based localized release of Colistin for antimicrobial treatment of burn wound infection. Macromol Biosci. https://doi.org/10.1002/mabi.201600320
167. Huang Y, Mu L, Zhao X, Han Y, Guo B (2022) Bacterial growth-induced tobramycin smart release self-healing hydrogel for Pseudomonas aeruginosa-infected burn wound healing. ACS Nano 16:13022–13036. https://doi.org/10.1021/acsnano.2c05557
168. Sharma PK, Halder M, Srivastava U, Singh Y (2019) Antibacterial PEG-chitosan hydrogels for controlled antibiotic/protein delivery. ACS Appl Bio Mater 2:5313–5322. https://doi.org/10.1021/acsabm.9b00570
169. Nimal TR, Baranwal G, Bavya MC, Biswas R, Jayakumar R (2016) Anti-staphylococcal activity of injectable nano tigecycline/chitosan-PRP composite hydrogel using Drosophila melanogaster model for infectious wounds. ACS Appl Mater Interfaces 8:22074–22083. https://doi.org/10.1021/acsami.6b07463
170. Kai J, Xuesong Z (2020) Preparation, characterization, and cytotoxicity evaluation of zinc oxide–bacterial cellulose–chitosan hydrogels for antibacterial dressing. Macromol Chem Phys 221:2000257. https://doi.org/10.1002/macp.202000257
171. Haque S, Kotcherlakota R, Bhamidipati P, Muralidharan K, Sreedhar B, Amanchy R, Patra CR (2023) Smartly engineered casein manganese oxide nanobiomaterials and its potential therapeutic angiogenesis applications for wound healing and limb ischemia. Adv Therap. https://doi.org/10.1002/adtp.202300142
172. Cui H, Liu M, Yu W, Cao Y, Zhou H, Yin J, Liu H, Que S, Wang J, Huang C, Gong C, Zhao G (2021) Copper peroxide-loaded gelatin sponges for wound dressings with antimicrobial and accelerating healing properties. ACS Appl Mater Interfaces 13:26800–26807. https://doi.org/10.1021/acsami.1c07409
173. Wahid F, Zhong C, Wang H-S, Hu X-H, Chu L-Q (2017) Recent advances in antimicrobial hydrogels containing metal ions and metals/metal oxide nanoparticles. Polymers (Basel) 9:636. https://doi.org/10.3390/polym9120636
174. Alavi M, Nokhodchi A (2020) An overview on antimicrobial and wound healing properties of ZnO nanobiofilms, hydrogels, and bionanocomposites based on cellulose, chitosan, and alginate polymers. Carbohydr Polym 227:115349. https://doi.org/10.1016/j.carbpol.2019.115349
175. Huang W, Ying R, Wang W, Guo Y, He Y, Mo X, Xue C, Mao X (2020) A macroporous hydrogel dressing with enhanced antibacterial and anti-inflammatory capabilities for accelerated wound healing. Adv Funct Mater 30:2000644. https://doi.org/10.1002/adfm.202000644
176. Roy A, Basuthakur P, Haque S, Patra CR (2021) Silver-based nanoparticles for antibacterial activity: recent development and mechanistic approaches. In: Krishnaraj RN, Sani RK (eds) Microbial interactions at nanobiotechnology interfaces: molecular mechanisms and applications. Wiley, pp 245–301
177. Yang K, Han Q, Chen B, Zheng Y, Zhang K, Li Q, Wang J (2018) Antimicrobial hydrogels: promising materials for medical application. Int J Nanomedicine 13:2217–2263. https://doi.org/10.2147/IJN.S154748
178. Xie Y, Liao X, Zhang J, Yang F, Fan Z (2018) Novel chitosan hydrogels reinforced by silver nanoparticles with ultrahigh mechanical and high antibacterial properties for accelerating wound healing. Int J Biol Macromol 119:402–412. https://doi.org/10.1016/j.ijbiomac.2018.07.060

179. Regiel-Futyra A, Kus-Liśkiewicz M, Sebastian V, Irusta S, Arruebo M, Stochel G, Kyzioł A (2015) Development of noncytotoxic chitosan-gold nanocomposites as efficient antibacterial materials. ACS Appl Mater Interfaces 7:1087–1099. https://doi.org/10.1021/am508094e
180. Rezaei N, Hamidabadi HG, Khosravimelal S, Zahiri M, Ahovan ZA, Bojnordi MN, Eftekhari BS, Hashemi A, Ganji F, Darabi S, Gholipourmalekabadi M (2020) Antimicrobial peptides-loaded smart chitosan hydrogel: release behavior and antibacterial potential against antibiotic resistant clinical isolates. Int J Biol Macromol 164:855–862. https://doi.org/10.1016/j.ijbiomac.2020.07.011
181. Fasiku VO, Omolo CA, Devnarain N, Ibrahim UH, Rambharose S, Faya M, Mocktar C, Singh SD, Govender T (2021) Chitosan-based hydrogel for the dual delivery of antimicrobial agents against bacterial methicillin-resistant Staphylococcus aureus biofilm-infected wounds. ACS Omega 6:21994–22010. https://doi.org/10.1021/acsomega.1c02547
182. Tsao CT, Chang CH, Lin YY, Wu MF, Wang J-L, Han JL, Hsieh KH (2010) Antibacterial activity and biocompatibility of a chitosan-gamma-poly(glutamic acid) polyelectrolyte complex hydrogel. Carbohydr Res 345:1774–1780. https://doi.org/10.1016/j.carres.2010.06.002
183. Aziz MA, Cabral JD, Brooks HJL, Moratti SC, Hanton LR (2012) Antimicrobial properties of a chitosan dextran-based hydrogel for surgical use. Antimicrob Agents Chemother 56:280–287. https://doi.org/10.1128/AAC.05463-11
184. Li P, Poon YF, Li W, Zhu H-Y, Yeap SH, Cao Y, Qi X, Zhou C, Lamrani M, Beuerman RW, Kang E-T, Mu Y, Li CM, Chang MW, Leong SSJ, Chan-Park MB (2011) A polycationic antimicrobial and biocompatible hydrogel with microbe membrane suctioning ability. Nat Mater 10:149–156. https://doi.org/10.1038/nmat2915
185. Hoque J, Adhikary U, Yadav V, Samaddar S, Konai MM, Prakash RG, Paramanandham K, Shome BR, Sanyal K, Haldar J (2016) Chitosan derivatives active against multidrug-resistant bacteria and pathogenic fungi: in vivo evaluation as topical antimicrobials. Mol Pharm 13:3578–3589. https://doi.org/10.1021/acs.molpharmaceut.6b00764
186. Hoque J, Prakash RG, Paramanandham K, Shome BR, Haldar J (2017) Biocompatible injectable hydrogel with potent wound healing and antibacterial properties. Mol Pharm 14:1218–1230. https://doi.org/10.1021/acs.molpharmaceut.6b01104
187. Hoque J, Bhattacharjee B, Prakash RG, Paramanandham K, Haldar J (2018) Dual function injectable hydrogel for controlled release of antibiotic and local antibacterial therapy. Biomacromolecules 19:267–278. https://doi.org/10.1021/acs.biomac.7b00979
188. Bhattacharjee B, Ghosh S, Haldar J (2023) Versatile and user-friendly anti-infective hydrogel for effective wound healing. ACS Appl Bio Mater 6:4867–4876. https://doi.org/10.1021/acsabm.3c00608
189. Bhattacharjee B, Tabbasum K, Mukherjee R, Garg P, Haldar J (2024) Functionalized chitosan based antibacterial hydrogel sealant for simultaneous infection eradication and tissue closure in ocular injuries. Int J Biol Macromol 273:132838. https://doi.org/10.1016/j.ijbiomac.2024.132838
190. Nepal A, Tran HDN, Nguyen N-T, Ta HT (2023) Advances in haemostatic sponges: characteristics and the underlying mechanisms for rapid haemostasis. Bioact Mater 27:231–256. https://doi.org/10.1016/j.bioactmat.2023.04.008
191. Slaughter BV, Khurshid SS, Fisher OZ, Khademhosseini A, Peppas NA (2009) Hydrogels in regenerative medicine. Adv Mater Weinheim 21:3307–3329. https://doi.org/10.1002/adma.200802106
192. Lee VK, Lee T, Ghosh A, Saha T, Bais MV, Bharani KK, Chag M, Parikh K, Bhatt P, Namgung B, Venkataramanan G, Agrawal A, Sonaje K, Mavely L, Sengupta S, Mashelkar RA, Jang HL (2024) An architecturally rational hemostat for rapid stopping of massive bleeding on anticoagulation therapy. Proc Natl Acad Sci U S A 121:e2316170121. https://doi.org/10.1073/pnas.2316170121
193. Bonanno AM, Graham TL, Wilson LN, Ross JD (2020) Novel use of XSTAT 30 for mitigation of lethal non-compressible torso hemorrhage in swine. PLoS One 15:e0241906. https://doi.org/10.1371/journal.pone.0241906

194. Thattaruparambil Raveendran N, Mohandas A, Ramachandran Menon R, Somasekharan Menon A, Biswas R, Jayakumar R (2019) Ciprofloxacin- and fluconazole-containing fibrin-nanoparticle-incorporated chitosan bandages for the treatment of polymicrobial wound infections. ACS Appl Bio Mater 2:243–254. https://doi.org/10.1021/acsabm.8b00585
195. Choudhary P, Ramalingam B, Das SK (2020) Fabrication of chitosan-reinforced multifunctional graphene nanocomposite as antibacterial scaffolds for hemorrhage control and wound-healing application. ACS Biomater Sci Eng 6:5911–5929. https://doi.org/10.1021/acsbiomaterials.0c00923
196. You C, Li Q, Wang X, Wu P, Ho JK, Jin R, Zhang L, Shao H, Han C (2017) Silver nanoparticle loaded collagen/chitosan scaffolds promote wound healing via regulating fibroblast migration and macrophage activation. Sci Rep 7:10489. https://doi.org/10.1038/s41598-017-10481-0
197. Wu Z, Zhou W, Deng W, Xu C, Cai Y, Wang X (2020) Antibacterial and hemostatic thiol-modified chitosan-immobilized AgNPs composite sponges. ACS Appl Mater Interfaces 12: 20307–20320. https://doi.org/10.1021/acsami.0c05430
198. Li Q, Lu F, Zhou G, Yu K, Lu B, Xiao Y, Dai F, Wu D, Lan G (2017) Silver inlaid with gold nanoparticle/chitosan wound dressing enhances antibacterial activity and porosity, and promotes wound healing. Biomacromolecules 18:3766–3775. https://doi.org/10.1021/acs.biomac.7b01180
199. Wu J, Su C, Jiang L, Ye S, Liu X, Shao W (2018) Green and facile preparation of chitosan sponges as potential wound dressings. ACS Sustain Chem Eng 6:9145–9152. https://doi.org/10.1021/acssuschemeng.8b01468
200. Kalantari K, Mostafavi E, Afifi AM, Izadiyan Z, Jahangirian H, Rafiee-Moghaddam R, Webster TJ (2020) Wound dressings functionalized with silver nanoparticles: promises and pitfalls. Nanoscale 12:2268–2291. https://doi.org/10.1039/c9nr08234d
201. Chen Y, Wu L, Li P, Hao X, Yang X, Xi G, Liu W, Feng Y, He H, Shi C (2020) Polysaccharide based hemostatic strategy for ultrarapid hemostasis. Macromol Biosci 20: e1900370. https://doi.org/10.1002/mabi.201900370
202. Du X, Wu L, Yan H, Jiang Z, Li S, Li W, Bai Y, Wang H, Cheng Z, Kong D, Wang L, Zhu M (2021) Microchannelled alkylated chitosan sponge to treat noncompressible hemorrhages and facilitate wound healing. Nat Commun 12:4733. https://doi.org/10.1038/s41467-021-24972-2
203. Sun L, Zhou J, Lai J, Zheng X, Zhang L-M (2024) Multifunctional chitosan-based gel sponge with efficient antibacterial, hemostasis and strong adhesion. Int J Biol Macromol 256:128505. https://doi.org/10.1016/j.ijbiomac.2023.128505
204. Yuan H, Chen L, Hong FF (2020) A biodegradable antibacterial nanocomposite based on oxidized bacterial nanocellulose for rapid hemostasis and wound healing. ACS Appl Mater Interfaces 12:3382–3392. https://doi.org/10.1021/acsami.9b17732
205. Bhattacharjee B, Mukherjee R, Haldar J (2022) Biocompatible hemostatic sponge exhibiting broad-spectrum antibacterial activity. ACS Biomater Sci Eng 8:3596–3607. https://doi.org/10.1021/acsbiomaterials.2c00410
206. Omidian H, Dey Chowdhury S, Babanejad N (2023) Cryogels: advancing biomaterials for transformative biomedical applications. Pharmaceutics. https://doi.org/10.3390/pharmaceutics15071836
207. Henderson TMA, Ladewig K, Haylock DN, McLean KM, O'Connor AJ (2013) Cryogels for biomedical applications. J Mater Chem B Mater Biol Med 1:2682–2695. https://doi.org/10.1039/c3tb20280a
208. Carriero VC, Di Muzio L, Petralito S, Casadei MA, Paolicelli P (2023) Cryogel scaffolds for tissue-engineering: advances and challenges for effective bone and cartilage regeneration. Gels. https://doi.org/10.3390/gels9120979
209. Hou S, Liu Y, Feng F, Zhou J, Feng X, Fan Y (2020) Polysaccharide-peptide cryogels for multidrug-resistant-bacteria infected wound healing and hemostasis. Adv Healthc Mater 9: e1901041. https://doi.org/10.1002/adhm.201901041

210. Zhao X, Guo B, Wu H, Liang Y, Ma PX (2018) Injectable antibacterial conductive nanocomposite cryogels with rapid shape recovery for noncompressible hemorrhage and wound healing. Nat Commun 9:2784. https://doi.org/10.1038/s41467-018-04998-9
211. Huang Y, Zhao X, Zhang Z, Liang Y, Yin Z, Chen B, Bai L, Han Y, Guo B (2020) Degradable gelatin-based IPN cryogel hemostat for rapidly stopping deep noncompressible hemorrhage and simultaneously improving wound healing. Chem Mater 32:6595–6610. https://doi.org/10.1021/acs.chemmater.0c02030
212. Huang Y, Zhao X, Wang C, Chen J, Liang Y, Li Z, Han Y, Guo B (2022) High-strength antibacterial composite cryogel for lethal noncompressible hemorrhage hemostasis: synergistic physical hemostasis and chemical hemostasis. Chem Eng J 427:131977. https://doi.org/10.1016/j.cej.2021.131977
213. Li M, Zhang Z, Liang Y, He J, Guo B (2020) Multifunctional tissue-adhesive cryogel wound dressing for rapid nonpressing surface hemorrhage and wound repair. ACS Appl Mater Interfaces 12:35856–35872. https://doi.org/10.1021/acsami.0c08285
214. Yang Y, Li M, Pan G, Chen J, Guo B (2023) Multiple stimuli-responsive nanozyme-based cryogels with controlled NO release as self-adaptive wound dressing for infected wound healing. Adv Funct Mater. https://doi.org/10.1002/adfm.202214089
215. Zhang Y, Wang Y, Chen L, Zheng J, Fan X, Xu X, Zhou G, Ullah N, Feng X (2022) An injectable antibacterial chitosan-based cryogel with high absorbency and rapid shape recovery for noncompressible hemorrhage and wound healing. Biomaterials 285:121546. https://doi.org/10.1016/j.biomaterials.2022.121546
216. Yao L, Gao H, Lin Z, Dai Q, Zhu S, Li S, Liu C, Feng Q, Li Q, Wang G, Chen X, Cao X (2022) A shape memory and antibacterial cryogel with rapid hemostasis for noncompressible hemorrhage and wound healing. Chem Eng J 428:131005. https://doi.org/10.1016/j.cej.2021.131005
217. Xiong M-H, Li Y-J, Bao Y, Yang X-Z, Hu B, Wang J (2012) Bacteria-responsive multifunctional nanogel for targeted antibiotic delivery. Adv Mater 24:6175–6180. https://doi.org/10.1002/adma.201202847
218. Qu Y, Chu B, Wei X, Lei M, Hu D, Zha R, Zhong L, Wang M, Wang F, Qian Z (2019) Redox/pH dual-stimuli responsive camptothecin prodrug nanogels for "on-demand" drug delivery. J Control Release 296:93–106. https://doi.org/10.1016/j.jconrel.2019.01.016
219. Keskin D, Zu G, Forson AM, Tromp L, Sjollema J, van Rijn P (2021) Nanogels: a novel approach in antimicrobial delivery systems and antimicrobial coatings. Bioact Mater 6:3634–3657. https://doi.org/10.1016/j.bioactmat.2021.03.004
220. Xiong M-H, Bao Y, Yang X-Z, Wang Y-C, Sun B, Wang J (2012) Lipase-sensitive polymeric triple-layered nanogel for "on-demand" drug delivery. J Am Chem Soc 134:4355–4362. https://doi.org/10.1021/ja211279u
221. Obuobi S, Julin K, Fredheim EGA, Johannessen M, Škalko-Basnet N (2020) Liposomal delivery of antibiotic loaded nucleic acid nanogels with enhanced drug loading and synergistic anti-inflammatory activity against S. aureus intracellular infections. J Control Release 324:620–632. https://doi.org/10.1016/j.jconrel.2020.06.002
222. Zhu J, Li F, Wang X, Yu J, Wu D (2018) Hyaluronic acid and polyethylene glycol hybrid hydrogel encapsulating Nanogel with hemostasis and sustainable antibacterial property for wound healing. ACS Appl Mater Interfaces 10:13304–13316. https://doi.org/10.1021/acsami.7b18927
223. Rusu AG, Chiriac AP, Nita LE, Rosca I, Pinteala M, Mititelu-Tartau L (2020) Chitosan derivatives in macromolecular co-assembly nanogels with potential for biomedical applications. Biomacromolecules 21:4231–4243. https://doi.org/10.1021/acs.biomac.0c01008
224. Li X-M, Wu Z-Z, Zhang B, Pan Y, Meng R, Chen H-Q (2019) Fabrication of chitosan hydrochloride and carboxymethyl starch complex nanogels as potential delivery vehicles for curcumin. Food Chem 293:197–203. https://doi.org/10.1016/j.foodchem.2019.04.096
225. Zhaveh S, Mohsenifar A, Beiki M, Khalili ST, Abdollahi A, Rahmani-Cherati T, Tabatabaei M (2015) Encapsulation of Cuminum cyminum essential oils in chitosan-caffeic acid nanogel

with enhanced antimicrobial activity against Aspergillus flavus. Ind Crop Prod 69:251–256. https://doi.org/10.1016/j.indcrop.2015.02.028
226. Zhang X, Malhotra S, Molina M, Haag R (2015) Micro- and nanogels with labile crosslinks – from synthesis to biomedical applications. Chem Soc Rev 44:1948–1973. https://doi.org/10.1039/c4cs00341a
227. Wu H-Q, Wang C-C (2016) Biodegradable smart nanogels: a new platform for targeting drug delivery and biomedical diagnostics. Langmuir 32:6211–6225. https://doi.org/10.1021/acs.langmuir.6b00842
228. Li Y, Maciel D, Rodrigues J, Shi X, Tomás H (2015) Biodegradable polymer nanogels for drug/nucleic acid delivery. Chem Rev 115:8564–8608. https://doi.org/10.1021/cr500131f
229. Indrakumar S, Ghosh S, Dash TK, Mishra V, Tandon B, Chatterjee K (2023) Silk composite interfacial layer eliminates rebleeding with chitosan-based hemostats. Carbohydr Polym 304: 120479. https://doi.org/10.1016/j.carbpol.2022.120479

Adv Polym Sci (2025) 295: 117–142
https://doi.org/10.1007/12_2024_183

Published online: 12 November 2024

Chitosan Beads for Drug Delivery

Sevda Şenel and Oğuz Şahin

Contents

1 Chitosan and Chitosan Derivatives 118
2 Chitosan Beads 119
3 Bioactive Properties of Chitosan 122
3.1 Wound Healing, Antimicrobial, and Hemostatic Activities 122
3.2 Anti-Inflammatory Activity 123
4 Chitosan Beads for Drug Delivery 123
4.1 Dermal/Transdermal Drug Delivery 124
4.2 Mucosal and Transmucosal Drug Delivery 128
4.3 Oral Drug Delivery 128
5 Microbeads for Transcatheter Arterial Chemoembolization (TACE) 135
6 Perspectives for the Future 136
References 137

Abstract Chitosan is a cationic biopolymer which exerts promising properties such as bioadhesivity, biocompatibility, biodegradability and penetration enhancement as well as bioactive properties including antimicrobial, wound healing, anti-inflammatory, immunostimulatory, and tissue regeneration. With all these properties together, it has become an attracting biomaterial in biomedical field for both therapeutic purposes and delivery of the drugs in different forms such as gels, fibers, films, sponges, micro and nanoparticulate systems, etc. In this chapter, we will focus specifically on chitosan microbeads and their applications in drug delivery. After highlighting the potential of chitosan beads for drug delivery with respect to their physico-chemical and biological properties, their recent applications in the pharmaceutical field will be reviewed. Finally, the possibilities and limitations will be discussed in regard to quality, efficacy, and safety aspects.

S. Şenel (✉) and O. Şahin
Department of Pharmaceutical Technology, Faculty of Pharmacy, Hacettepe University, Ankara, Türkiye
e-mail: ssenel@hacettepe.edu.tr

Keywords Aerogel · Chitosan beads · Drug delivery · Hydrogel · Transcatheter arterial chemoembolization

1 Chitosan and Chitosan Derivatives

Chitosan is obtained by alkaline or enzymatic deacetylation of chitin, which exists in animals, particularly in crustaceans as an important constituent of the exoskeleton, and in certain fungi where it is the principal fibrillar polymer in the cell wall [1, 2]. It is a linear polysaccharide composed of N-acetyl-D-glucosamine (acetylated unit) and β-(1 → 4)-linked D-glucosamine (deacetylated unit) groups, depending on the polymer's degree of deacetylation (DD) (Fig. 1). In general, chitin with a degree of deacetylation above 65–70% is known as chitosan. The process of deacetylation, which involves the removal of acetyl groups from the molecular chain of chitin, results in an amino group (-NH_2). Due to these reactive amino groups, chitosan is positively charged and is soluble in weakly acidic solutions. The charge density is dependent on the pH of the medium and the degree of deacetylation. Besides its solubility, the degree of acetylation and molecular weight of chitosan also affects its biodegradability and bioactive properties.

On the other hand, in order to increase the solubility further, modification of chitosan has been applied through utilization of the free amino and hydroxyl groups of chitosan as reactive functional sites to generate a wide range of chitosan derivatives [3, 4]. The derivatives of chitosan maintain the original physicochemical and bioactive properties of chitosan, and additionally they possess new and/or improved characteristics depending on the nature of modification.

Chitosan has been widely applied in diverse fields such as agriculture, food, cosmetics, biomedical, and environmental [5] (Fig. 2). The versatility of its physicochemical properties as well as its other properties including bioadhesive, penetration enhancing, and bioactive properties (e.g., antimicrobial, anti-inflammatory, immunostimulatory, wound healing, tissue regeneration, hemostatic, etc.) makes chitosan a favorable polymer for biomedical applications [6, 7]. The current review will be only on biomedical applications of chitosan, specifically in bead form.

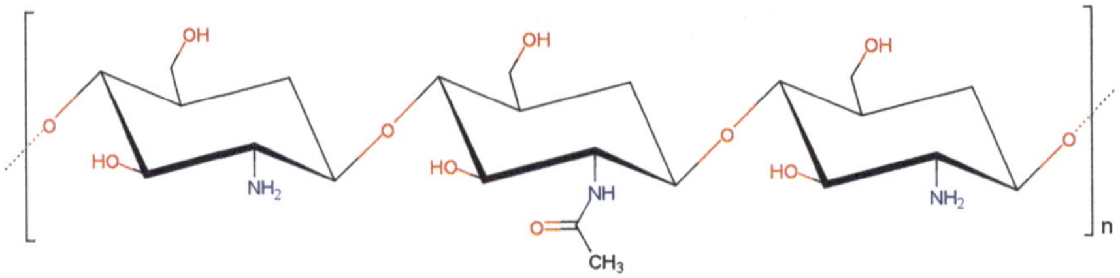

Fig. 1 Molecular structure of chitosan

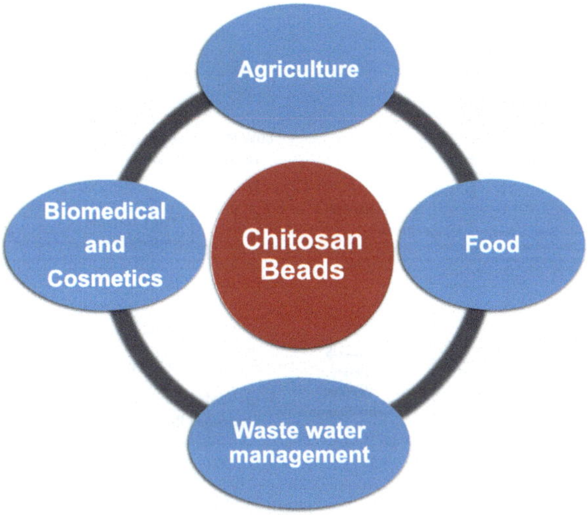

Fig. 2 Application areas of chitosan beads

In the following sections, after a brief introduction to chitosan beads, recent applications of chitosan beads for drug and vaccine delivery as well for its bioactive properties will be reviewed and discussed.

2 Chitosan Beads

Due to its adaptable physicochemical properties, chitosan can be utilized in different forms such as powder, flakes, gels, films, fibers, micro- and nanoparticles, beads, etc., providing the suitable structure for the application aim. Chitosan beads, which we will focus on within the scope of this review, offer numerous advantages, especially in the biomedical area. Besides simplicity in preparation, they provide high drug loading capacity, increased porosity and surface area, and decreased crystallinity, making them favorable for delivery to the desired site in the body [8–12]. There are various structures of chitosan beads and they can be prepared by different methods such as extrusion, emulsion crosslinking, template-sacrifice, ion-imprinting, etc. Chitosan beads are generally in the form of hydrogel or aerogel [13, 14]. Hydrogels are distinctive three-dimensional hydrophilic polymeric network, featuring a loose, interconnected porous structure with significant water retention capabilities. Furthermore, owing to their elongation properties, hydrogels can be molded into different shapes [15]. The water capture depends on the functional groups present on the surface, crosslinking density, pore structure, and solid content. The three-dimensional hydrogel beads are physically or chemically crosslinked networks of chitosan formed in spherical shape and sized in the range of 0.5–1.0 mm of diameter. They can be combined with numerous polymers and

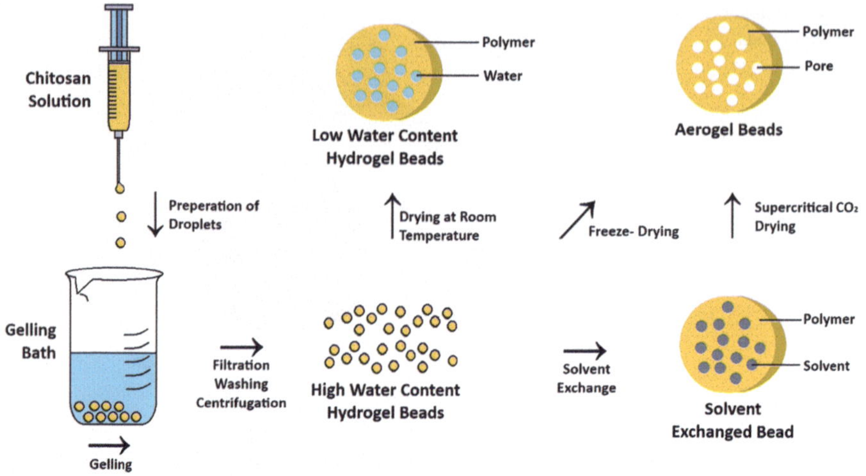

Fig. 3 Preparation of chitosan beads

incorporated with drug molecules [16–18]. In bead formulations, the amount of water in hydrogels can be low or high. Various drying techniques can be applied to reduce the amount of water in the hydrogels. Aerogels, which have high porosity, large surface area, and low density, are traditionally prepared via freeze-drying (freezing and sublimation) or supercritical drying (crossing the critical point of the solvent by increasing the temperature and pressure of a precursor material or gel) [19, 20]. In both methods, the aqueous gel phase is replaced with air. These aerogels can also be rehydrated to form hydrogels with new pore morphologies and physical properties [20–23]. The preparation methods and structure selections of chitosan beads depend on several requirements such as shape, dimension, chemical and thermal stability, release time of the drug, and residuals of the chitosan bead products [21, 24].

The preparation of chitosan beads basically takes place in 3 steps: droplet formation, gelation, and drying (Fig. 3) [21, 25]. In the production of aerogel beads, there may be a further solvent exchange process between the gelation and drying processes. The main purpose of preparation of droplets is to prepare homogeneous droplets by controlling the shape, morphology, and size of the droplets. Firstly, solution of chitosan is prepared. And other polymer or chemical to be included in the formulation can be added to this solution. This solution is then taken into a system that can form droplets in a controlled manner, such as a syringe, and from this system, the solution is added dropwise to the gelation bath [25]. The droplets falling into the gelation bath begin to form bead-shaped gel structures in the presence of gelation medium. The gelation process may occur by different mechanisms: physically triggered by pH, temperature, etc. or chemically triggered by ionic or covalent interactions [21]. The beads obtained are purified via filtration, washing, and centrifugation. Finally, beads with high water content are obtained.

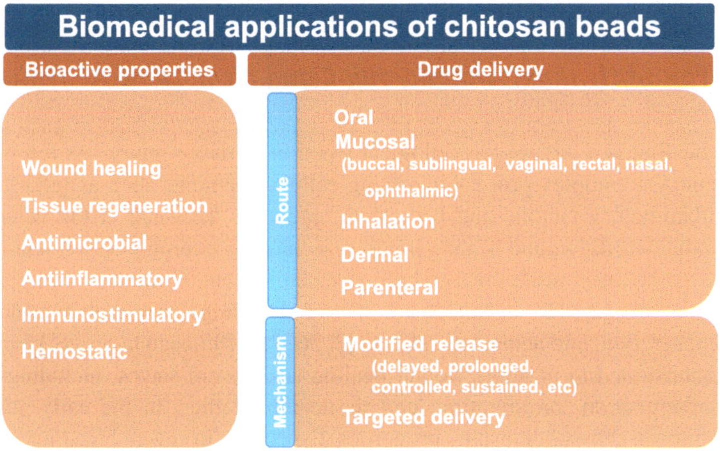

Fig. 4 Biomedical applications of chitosan beads

In order to obtain a hydrogel bead with low water content, hydrogel beads can be kept at room temperature for 1 to 2 days. For aerogels, after the formation of the hydrogel, solvent exchange and solvent removal by a drying method is applied. The preparation steps of the gel determine the formation of the porous framework, where drying is the crucial step to transform the wet gel into a porous solid [20, 23].

Currently, chitosan hydrogel beads form become an attractive material for utilization in various biomedical applications (Fig. 4). Through physical conditioning as beads, chitosan becomes more stable, with superior adsorption properties (e.g., increased surface area and porosity, etc.), with higher drug loading efficiency and with higher penetration across the membranes [7, 26]. Chitosan beads can be prepared also in combination with different polymers such as alginate, cellulose derivatives, gelatin, polyethylene glycol, etc. in order to improve the properties of the system as well as to reinforce it [27–29]. For example, in a study where colonic delivery was aimed, calcium pectinate was incorporated into the chitosan beads to avoid the degradation of the system before reaching the colon [30]. As calcium pectinate is insoluble in the gastric environment and cannot be degraded by gastric enzymes, the beads are expected to reach the colon where calcium pectinate degrades with pectinolytic enzymes and provides drug release at the site. In another study, a bead formulation was developed using carboxymethyl chitosan and calcium alginate combination to improve the gelation of the beads [31]. For further enhancement of the physical strength and stability of the chitosan bead formulations, crosslinkers such as glutaraldehyde, genipin, tripolyphosphate, epichlorohydrin, etc. can also be added [32–34].

3 Bioactive Properties of Chitosan

3.1 Wound Healing, Antimicrobial, and Hemostatic Activities

Wound healing is a natural physiological reaction to tissue injury. As a result of a tissue injury, a complex process involving cellular elements such as inflammatory cells, inflammatory factors, and hemostatic agents is activated at the wound site [35]. In other words, wound healing involves a complex interplay between numerous cell types, cytokines, mediators, and the vascular system. Chitosan hydrogel beads have been investigated for enhancement of wound healing due to their antimicrobial, tissue regeneration, and hemostatic effects [2, 36–38]. Chitosan hydrogel beads have been demonstrated to promote wound healing at different stages, including hemostasis, inflammation, proliferation, and tissue remodeling. In the early stages of injury, chitosan shows hemostatic properties. Especially in severe bleeding, chitosan is actively used in granule or bandage formulation to stop bleeding immediately [39, 40]. Afterward, it promotes the migration of neutrophils and macrophages to the wound, and due to its antimicrobial activity, it ensures that the wound is free of pathogens enhancing the fibrous tissue formation and re-epithelialization process [36, 41]. At this stage, chitosan enhances the collagen production and the expression of some growth factors [42]. Besides their bioactive properties, the chitosan hydrogel beads can be used as delivery systems for antimicrobials, growth factors, stem cells, etc., which would further accelerate wound healing. It is possible to tailor the chitosan hydrogel beads by modification of chitosan or combining with other polymers which allows enhanced wound healing by providing appropriate amount of moisture in the wound area, accelerating tissue regeneration, protecting the wound from secondary infections [37, 38, 43].

The mechanism of antimicrobial activity of chitosan against gram-negative and gram-positive bacteria and fungi is explained by disruption of the cell membrane [44]. Positive charge from amino groups of chitosan interacts with negatively charged components on the microbial membrane, resulting in the alteration of cell permeability and lysis of the cell membrane [45–48]. The antimicrobial property of chitosan has been shown to be dependent on its degree of deacetylation, molecular weight, pH, ionic strength as well as the microorganism type [45]. Antimicrobial activity can be enhanced by chemical modification of functional groups [3].

Bleeding caused by trauma or surgery is a serious health problem, and uncontrollable bleeding can result in death [49, 50]. Therefore, developing safe, effective, and convenient hemostatic materials is important to mitigate uncontrolled bleeding in emergencies [51]. Among the hemostatic materials, chitosan has been very attractive due to both its hemostatic activity and other favorable properties such as bioadhesion, biodegradability, and biocompatibility. In general, chitosan has been used in various forms for wound care applications. Due to its cationic structure, chitosan promotes hemostasis by activating and aggregating platelets which have negative surface charges. With this interaction, the local concentration of platelets and coagulation factors in the wound increases and bleeding stops [52–54]. Chitosan hydrogel beads have also been evaluated for their hemostatic activity alone or in combination with hemostatic drugs [55, 56].

3.2 Anti-Inflammatory Activity

Chitosan has been used alone or in combination with other drugs for its anti-inflammatory activity in different areas such as periodontitis, inflammatory bowel disease, and various injuries [57–59]. It has been shown to mitigate inflammation through various mechanisms, such as modulating cytokine expression, scavenging free radicals, and regulating immune responses [60–64]. Studies have shown that the molecular weight of chitosan significantly influences its anti-inflammatory effect. Decreased anti-inflammatory effect was reported with the decrease in molecular weight [64, 65]. In a recent study, chitosan-agar-based microbeads loaded with celecoxib were shown to enhance the anti-inflammatory effect in the rat paw edema model [66]. In another study, chitosan-gelatin hydrogel beads were prepared by a cross-linking reaction between chitosan and gelatin triggered by photo-induced radicals generated from riboflavin and enhanced antibacterial activity was obtained against *E. coli* [67].

4 Chitosan Beads for Drug Delivery

Drug delivery systems are designed to achieve the desired therapeutic effect without any side effects. It is important that these systems deliver the drug at right dose, at right time, at right site, and for desired period of time. Currently, for this purpose, modified release (delayed, controlled, extended, pulsatile, etc.) dosage forms are used where the rate and/or site of release of the drugs are different from that of the immediate release dosage form administered by the same route [68]. This deliberate modification is achieved by special formulation design and/or manufacturing methods, which can be administered through oral, parenteral, dermal, and mucosal routes. Up to date, considerable advances have been achieved in the development of different modified release systems based on polymers, enabling both hydrophilic and hydrophobic drugs to be delivered over an extended period of time to the site of action. Furthermore, besides enhanced drug efficacy, enhanced safety and patient compliance can be obtained using such polymer-based drug delivery systems. Numerous synthetic (e.g., cellulose derivatives, acrylic acids, etc.) and natural polymers (e.g., alginates, chitosan, pectin, etc.) have been used for drug delivery in different forms such as pellets, micro- and nanoparticles, films, gels, sponges, etc. [16, 69–73]. Major drawback associated with most of these polymeric systems is the use of harsh chemicals as well as elevated temperature during preparation, while with chitosan it is possible to overcome these hurdles [21, 74]. Especially with chitosan beads, which are prepared by ionotropic gelation and emulsion gelation method, utilization of solvents and elevated temperature is avoided making them attractive for drug delivery. Furthermore, due to the chemical versatility of chitosan, it is possible to tailor these beads according to their particle size and distribution, surface charge, loading efficiency, swelling ratio and mucoadhesivity, which would allow to achieve the desired efficacy [6, 75]. With chitosan beads, it is possible to enhance the contact time of the delivery system on the application site while extending the release of the drug, thereby improving the bioavailability of the drug. Due to their size, it is possible to transport the drugs to the site of action at reduced doses which also results in reduced undesired effects. Furthermore, by means of encapsulation, the drug solubility as well as stability can also be enhanced [76].

In the following sections, the application of chitosan in drug delivery will be reviewed according to the administration routes.

4.1 Dermal/Transdermal Drug Delivery

The skin is the largest body organ, comprising 15% of body weight and covering a surface area between 1.5 and 2.0 m^2, which makes it the most accessible organ, offering a promising route for drug delivery [77–79]. Besides ease of application and high patient compliance, opportunity of both local and systemic delivery, avoidance of adverse systemic effects and prevention of hepatic first-pass metabolism and degradation in gastrointestinal tract can be provided by delivery through skin [80, 81]. Local delivery which is also referred as dermal delivery is the transport of the drug applied on the skin to various layers of skin, whereas transdermal delivery refers to penetration of the drug through the skin and absorption to blood, resulting in systemic delivery. Yet, drug delivery through skin is not easy because the protecting function of human skin, mainly the stratum corneum layer imposes physicochemical limitations, acting as the main barrier for dermal and transdermal drug delivery. The lipophilicity and molecular weight of the drug are the main limiting parameters for a drug to be delivered passively via the skin. Various formulation strategies have emerged over years to overcome these hurdles and enhance the drug delivery across skin. For a successful formulation, besides enhanced penetration and efficacy, safety of the system as well as ease of application and cost-effectiveness is also important. Due to its bioadhesivity, biocompatibility, penetration-enhancing effect, and cationic surface charge, chitosan has been widely used for dermal and transdermal drug delivery system [82, 83]. These delivery systems have been prepared in different forms such as gels, films, woven dressings, fibers, micro and nanoparticulate systems, sponges, etc. There is a growing interest in composite chitosan-based systems for dermal and transdermal drug delivery, due to their multiple functions, including hemostasis, antibacterial activities, the stimulation of cell division and differentiation, and the acceleration of tissue healing (Table 1).

Table 1 Recent studies on chitosan beads for topical (dermal and mucosal) drug delivery

Drug	Properties of chitosan	Other ingredients	Aim/delivery route	Preparation method	In vitro studies	In vivo studies	Results	Reference
Zinc Oxide	Chitosan DD: %90 (Tianjin Umbrella Science and Technology-China)	Polyvinyl alcohol PD: 1750	Antibacterial Wound healing (Dermal)	Extrusion	– Antibacterial activity against *E. coli and S aureus* – In vitro cytotoxicity test MG63 osteosarcoma cell line and MCF7 human breast cancer cell line – Hemolysis and erythrocyte aggregation assay	Animal model: 8 week-old male Kunming mice (20 g). A partial thickness wound with a length of 1 cm, was created on the back of the mice – Macroscopic and histopathologic wound examination – Antibacterial activity	– Enhanced antibacterial activity – Increased wound healing – Biocompatible – No hemolysis	[91]
Vancomycin hydrochloride	Chitosan – DD: 90% – Mol Wt: 200–400 kDa (Heppe Medical Chitosan GmbH-Germany)	-	Chronic wound healing (Dermal)	Extrusion Particle size: 1.48–4.49 mm	– Swelling – Collagenase activity – Drug release – Antimicrobial activity against *S. aureus* – Cytotoxicity in BALB/3T3 fibroblasts	-	– High swelling property – Does not inhibition of collagenase activity – Initial burst out of drug following slow release – Antibacterial activity – Acceptable viability	[92]

(continued)

Table 1 (continued)

Drug	Properties of chitosan	Other ingredients	Aim/delivery route	Preparation method	In vitro studies	In vivo studies	Results	Reference
Resveratrol	Chitosan (Zhengzhou Gebei's Food Additive Co., Ltd.-China)	Chitosan beads containing Nano lipid carriers	Antioxidant (Dermal)	Extrusion	– Antioxidant activity using 2,2-diphenyl-1-picrylhydrazyl (DPPH) – Drug release – Permeation across RYJ-6B cells	–	– Enhanced antioxidant activity – Sustained drug release – Enhanced drug permeation	[93]
Resveratrol	– Chitosan (Zhengzhou Gebei's Food Additive Co., Ltd., China)	Chitosan beads containing pickering emulsion (ODO, primrose oil, RH-40, starch modified-OSA	Antioxidant (Dermal)	Extrusion Particle size: 1-1.5 mm	– Antioxidant activity – Drug release – Cytotoxicity study in L929 cells – Drug permeation through excised porcine skin		– Stable formulation – Enhanced antioxidant activity – Enhanced drug permeation and sustained drug release – Biocompatible	[94]
Fluconazole Metronidazole	Chitosan – DD: >%75 – Viscosity 800–2,000 (cps) (Fluka BioChemika)	Vaginal hard gelatin capsules containing chitosan beads	– Antifungal – Antibacterial (Vaginal mucosa)	Extrusion and ionotropic gelation Particle size: 631 – 935 μm	– Drug release – Antifungal activity against *Candida albicans* – Antibacterial activity against *Escherichia coli*	–	– Suitable bioadhesion – Sustained drug release – Enhanced antifungal and antibacterial activity	[48]

					– Ex vivo bioadhesion study in New Zealand adult female albino rabbit vagina			
Tizanidine HCl	Chitosan oligosaccharide lactate – DD: >%90 (Sigma-Aldrich)	Chitosan wafer containing chitosan beads	Treatment of spasticity (Oral mucosa)	Extrusion and ionotropic gelation Particle size: 1.2–2.2 mm	– Drug release	Healthy human volunteers – Mucoadhesion study – Pharmacokinetic study	– Suitable mucoadhesion – Sustained drug release – Cmax: 0.887 + 0.240 ng/ml – Tmax: 6 h – Increased AUC	[90]
Metronidazole	Chitosan – DD: 75–85% – Mol Wt: 50 kDa (Sigma-Aldrich)	Alginate	Antibacterial (Vaginal mucosa)	Ionotropic gelation Particle size: 1.42–1.75 mm	– Drug release – Swelling study – Antibacterial activity against *Bacteroides fragilis* – Ex vivo mucoadhesion using porcine vaginal mucosa	–	– Sustained drug release – Low swelling property – Mucoadhesion – Enhanced antibacterial activity	[89]

Mol Wt: Molecular weight; DD: Deacetylation degree PD: Polymerization degree; ODO: Octyl and decyl glycerate; RH-40: PEG-40 hydrogenated castor oil; OSA: octenyl succinic anhydride

4.2 Mucosal and Transmucosal Drug Delivery

Mucosa is the soft tissue that lines the body's canals and organs in the digestive, respiratory, and reproductive systems. Mucosal delivery systems are designed for delivery of drugs to or across mucosal surfaces such as buccal, ocular, nasal, rectal, and vaginal [84]. Similar to that of dermal delivery, mucosal delivery possesses advantages such as overcoming the first pass effect and gastrointestinal degradation, enhanced bioavailability and higher patient compliance, but also exerts protecting function acting as a microbial, chemical, mechanical, and immune barrier [85]. For certain drugs such as peptides and proteins, which are charged and hydrophilic, and have high molecular weight, it is challenging to cross the mucosal barrier and reach the target site locally in the mucosa or reach blood circulation for systemic effect. In order to overcome the barriers, drug delivery systems have to exhibit various properties such as mucoadhesivity, permeation enhancing property, and extended drug release. Chitosan has been a promising material for mucosal drug delivery due to its cationic charge, mucoadhesive, and penetration-enhancing properties [85, 86]. Due to its bioadhesive property, chitosan allows the delivery system to attach to the mucosal surface for longer period of time, providing longer retention time of the drug at the site of action and prolonged release [87]. Furthermore, besides mucoadhesion, due to its cationic charge it enhances the penetration of the drug across the mucosa. Finally, being biocompatible and exerting occlusion property, it has been safely used for mucosal delivery in different dosage forms [88]. Due to the large particle size of chitosan beads (>1 mm), it is physically unsuitable to apply chitosan beads to some mucosal surfaces such as ocular and pulmonary. However, studies showing mucosal application of chitosan beads have mostly examined their application to sites such as the oral mucosa and vaginal mucosa [48, 89, 90]. In general, these locally applied formulations are aimed to treat the infections at the application site (Table 1).

4.3 Oral Drug Delivery

Oral delivery is the most preferred drug delivery system owing to advantages such as self-administration, cost-effectiveness, being pain-free, high patient compliance, and flexibility in the design of dosage forms. However, there are also challenges in oral drug delivery, which are mainly the poor bioavailability due to the low solubility and low permeability of the drug, and the stability of the drug in the gastrointestinal environment. The gastrointestinal tract is a system with a large surface area and exerts different environmental conditions in its different regions, hence the stability, solubility, and permeability of drugs in the gastrointestinal tract may be different in different regions [88, 95]. The behavior of the drug in the gastrointestinal tract can be dependent on the properties of the drug and physiological conditions such as pH, microbial colonization, and enzymes. Among the strategies to stabilize and solubilize the drugs and control its release, polymer-based delivery systems have attracted the main attention, especially to target the drug to the specific region of the gastrointestinal tract [96–98].

Chitosan has been widely investigated for oral drug delivery. Due to its mucoadhesive property, it can prolong the retention of the drug on intestinal mucosa, thereby increasing drug uptake/absorption. Furthermore, chitosan and its derivatives have been shown to enhance the permeation of the compounds across the intestinal mucosa [99]. Interaction between mucin and chitosan-based systems has been shown to change the attachment of the system to the mucosal surfaces and thereby changing the efficacy of the drug [100]. The concentration, charge, degree of deacetylation of chitosan as well as the pH of the medium is important for this interaction. Although it is not easy to predict the in vivo interaction between chitosan and mucin in the gastrointestinal tract, owing to mucoadhesion resulting from this interaction, it is possible to enhance the bioavailability of the drugs. Furthermore, these systems can also be used for increasing the lubrication of the mucosal surfaces in the treatment of local diseases such as ulcerous colitis, which is shown to be related to colon cancer [101].

In many studies, chitosan beads have been investigated for controlled or targeted release of the drug and also to improve its stability in the gastrointestinal system [95]. Chitosan is used alone or in combination with other polymers such as pectin, guar gum, alginate, and cyclodextrin [102–105]. Besides being the main component of the hydrogel bead, chitosan can be used for coating the beads based on a different polymer, in order to change the surface charge or to provide bioadhesive property [106]. In general, chitosan beads have been for gastric and colon-targeted delivery [30, 107]. For gastric delivery, floating or mucoadhesive chitosan beads were investigated [95]. For floating systems, the density of the delivery system is designed to be lower than that of the gastric fluid so that the delivery system floats in the stomach and provides extended release of the drug [108] For colon-targeted systems, for both local and systemic effect, the drug is released in response to the colonic environment without early drug release in the upper GI tract. Hence, it is essential to consider the physiological properties of the gastrointestinal system and the micro-environment surrounding disease site(s) for the successful development of colon-targeted drug delivery systems [109]. In Table 2, the recent studies on chitosan beads for oral drug delivery are summarized. Chitosan beads alone or in combination with other polymers have been demonstrated to enhance the bioavailability of the drugs. In a study, gastro-retentive chitosan microbeads for delivery of carvedilol were developed with a lower density than that of the gastric fluid, which enabled them to float [110]. In vivo studies were performed in New Zealand rabbits, comparing the effect of floating beads to that of carvedilol suspension. In vitro-in vivo correlation was studied using GastroplusTM software, which utilizes the Advanced Compartmental Absorption and Transit model" (ACAT) model. Bioavailability of floating chitosan beads was found to be higher than that obtained with the suspension. In another study, alginate hydrogel bead containing amoxicillin was coated with chitosan to enhance the mucoadhesive property of the system [106].

Table 2 Recent studies on chitosan beads for oral drug delivery

Drug	Properties of chitosan	Formulation	Aim	Preparation method	In vitro studies	In vivo studies	Results	Reference
Azathioprine	Chitosan – Mol WT: 100,000–300,000 kDa (Acros organics – Belgium)	Chitosan beads filled in acid-resistant	Colon-targeted treatment of inflammatory bowel disease (IBD)	Ionic gelation Particle size: 1.2–2.1 mm	Drug release following incubation in simulated gastric medium (pH 1.2) for 2 h, followed by 4 h in phosphate buffer (pH 6.8), and then for additional 24 h in acetate buffer at pH 4.0 (simulating IBD colon)	Colitis induced New Zealand white rabbits – Pathological and biochemical parameters (clinical activity score, index of tissue edema, mortality rate, macroscopic score, histological features, and biochemical analysis of inflammatory mediators and antioxidants)	– Drug release avoided in gastric conditions and minimized amount lost in the small intestine, complete drug release in colonic conditions, demonstrating IBD-targeted delivery – Improved efficacy	[111]
Anti-A/B toxin immunoglobulin of egg yolk (IgY)	Chitosan *(properties and source not stated)*	Chitosan-calcium pectinate microbeads	Colon-targeted delivery against *Clostridium difficile* toxin A and B	Extrusion Particle size: 1.0–1.4 mm	– Swelling in artificial stomach fluid and intestine fluid – In vitro IgY release performed under a pH gradient (pH 1.2; 6.8; 7.2), pectinolytic	Wistar rats fed 1 ml of 2% pectin every day for 3 days to activate the expression of pectinolytic enzyme in the colon flora – Drug applied by oral gavage – IgY concentration in the	In vitro release – Protection of drug in the upper gastrointestinal tract, thus significantly improving the site-specific release of IgY in the colon – Higher swelling in intestinal fluid than gastric fluid In vivo release	[30]

					enzymes also added	extracts obtained from the stomach, intestine, and colon of rats sacrificed every 2 h – In vivo IgY release efficiency determined by the amount of recovered IgY compared to the amount of IgY administered	– 73% IgY released in the colon	
Doxorubicin hydrochloride	– Chitosan (Zhengzhou Huambo Chemical Products, China)	Pectin/chitosan beads	Colon-targeted anticancer drug delivery	Extrusion and ionotropic gelation	– Drug release in simulated gastric fluid and colon fluid	–	Protection of drug release in gastric fluid, higher drug release in the colon fluid	[102]
Silver nanoparticle	Chitosan – Medium molecular weight – Viscosity: 200–800 cP (Aldrich)	Chitosan/silver Nanocomposite Hydrogel beads	Antibacterial	Extrusion Particle size: about 1 mm	– Antibacterial activity against *Escherichia coli* and *Staphylococcus aureus* – Swelling test – Drug release in simulated colon fluid	–	– Enhanced antibacterial – High swelling property – Sustained drug release	[112]

(continued)

Table 2 (continued)

Drug	Properties of chitosan	Formulation	Aim	Preparation method	In vitro studies	In vivo studies	Results	Reference
Celecoxib	Chitosan – DD: ≥75% – Mol Wt: Low (Sigma-Aldrich Germany)	Chitosan/guar gum-based hydrogel beads	Colon-targeted anti-inflammatory drug delivery	Ionotropic gelation Particle size: 1.65–3.64 mm	– In vitro drug release in gastric (pH 1.2) and colon (pH 7.5) fluid – Ex vivo mucoadhesion using goat intestinal mucosa – Ex vivo permeability across goat intestinal mucosa – Ex vivo swelling study on rabbit stomach and intestine	Anti-inflammatory activity in rat paw edema model – Oedema Edema inducement – CRP levels in blood – IL-6 levels in blood Histological investigation in paw tissues (epidermis, dermis, and adipose) of rats	– 74.0% drug release in the colon fluid, while 24.1% drug release in the stomach fluid – Strong mucoadhesion – Enhanced permeability – High swelling property – Enhanced anti-inflammatory activity	[66]
Diclofenac sodium	Carboxymethyl chitosan – DD: 80% – Mol Wt: 9 kDa (Shanghai Macklin Biochemical-Technology Co. Ltd-China)	Sodium alginate/carboxymethyl chitosan-ZnO hydrogel beads	Delayed release anti-inflammatory drug delivery	Extrusion	– Drug release in simulated gastric fluid (pH 1.2) and simulated intestinal (pH 6.8) fluid – Cytotoxic study in MC3T3-E1 cells	– Drug release and pharmacokinetic in Wistar rats – Swelling study in Wistar rats	– High cell viability (90%) – Low drug in gastric fluid, highest released amount in intestinal fluid – Tmax: 5 h – Cmax: 1.46 + 0.16 μg/ml – AUC: 7.73 + 0.31 μg h/ml – Swelling observed	[105]

							in intestine and stomach	
Carvedilol	Chitosan – DD: >90% (Nitta Gelatin India Limited – India)	Gastro-retentive chitosan beads	Model drug	Extrusion and ionotropic gelation	– Drug release in gastric fluid (pH 1.2)	– Pharmacokinetic study in New Zealand rabbits – In vivo-in vitro correlation	– Burst out release following slow drug release – Tmax: 3.67 + 0.58 h – Cmax: 14.16 + 1.43 ng/ml – AUC: 150.6 + 23.5 ng h/ml – 94% drug absorption in duodenum	[110]
Doxorubicin hydrochloride	Chitosan – DD: 89% – Mol Wt: 10 kDa (Biotech Co. – Korea) O-carboxymethyl chitosan – DD: 81% – Mol Wt: 12 kDa	Multilayer sodium alginate beads with porous and non-porous core containing chitosan-based nanoparticles	Anticancer	Ionotropic gelation Non-porous beads particle size: 3.6 ± 0.2 mm Porous beads particle size: 4.3 ± 0.1 mm	– Swelling in stimulated gastric, duodenum, jejunum, ileum fluids – Drug release in gastric, duodenum, jejunum, ileum fluids	Adult male Sprague–Dawley rat (180-220 g) – Distribution of beads in organs (heart, liver, spleen, lung, and kidney) collected from rats excised at 24 h after oral administration	– Low swelling in gastric fluid, high swelling in intestinal fluid – With non-porous beads, no drug release at gastric pH, increased release at higher pHs (colon) – With porous beads, low release at gastric pH, increasing at pH 7.4 – Higher drug amounts in liver, heart and kidney, and relatively lower in lung and spleen	[113]

(continued)

Table 2 (continued)

Drug	Properties of chitosan	Formulation	Aim	Preparation method	In vitro studies	In vivo studies	Results	Reference
Amoxicillin	Chitosan – DD: 90% – Medium molecular weight (Sisco Research Laboratories Pvt. Ltd-India)	Chitosan-coated graphene oxide incorporated sodium alginate hydrogel beads	Antibacterial	Ionotropic gelation	– Swelling at pH 1.2 and pH 7.4 – Drug release at pH 1.2 – Antimicrobial activity against *S. aureus and E. coli* – Cell viability in mouse fibroblast (L929) cell line		– More swelling in gastric pH than colon pH – Sustained release – Enhanced antimicrobial activity – Biocompatible	[106]

Mol Wt: Molecular weight; DD: Deacetylation degree; MPO: Myeloperoxidase enzyme; SOD: Superoxide dismutase; AUC: Area under curve; CRP: C-reactive proteins; ZnO: Zinc oxide

5 Microbeads for Transcatheter Arterial Chemoembolization (TACE)

Transcatheter arterial chemoembolization (TACE) is a common treatment option for patients with inoperable hepatocellular carcinoma or metastases [114, 115]. The technique involves blockage of the hepatic artery supplying the tumor, with or without local delivery of chemotherapeutic agent. This vascular occlusion increases the tumor's susceptibility to the chemotherapeutic agent. In this treatment, besides use of anticancer drugs, solid embolic materials such as gelatin sponge particles and porous gelatin particles are also used [116, 117]. TACE is mostly used in liver cancer. The interruption of blood flow during chemotherapy is a mechanism that leads to tumor shrinkage and necrosis of cancer cells [118]. The ability of beads and microspheres to encapsulate chemotherapeutic agents gives them superiority over other formulations used for this purpose. These structures are referred to as drug-eluting beads (DEB) [119]. The first drug-eluting bead was a synthetic polyvinyl alcohol-based DC Bead®, which was not biodegradable and biocompatible, and required to be removed after treatment [120]. Hence, drug-eluting beads were further developed using biodegradable and biocompatible polymers such as poly(lactic-co-glycolic acid) and chitosan [121–123]. Due to its biocompatibility and biodegradability, chitosan has been widely investigated for embolization and chemoembolization [124–126]. In these studies, chitosan beads are brought to the appropriate particle size and embolism formation is achieved in the vessel where blood flow is targeted to be interrupted. By controlling the swelling properties and the pH-response properties of the chitosan microbeads, it is possible to obtain a successful embolization and prolonged drug release.

In a study by Arya et al. [124], chitosan beads (160 μm) crosslinked with glutaraldehyde were prepared and α-cyclodextrin was used to coat the surface via electrostatic interactions. It was shown in vitro that the developed beads could provide embolization through clustering on each other. In another study, chitosan microbeads containing doxorubicin were prepared using chitosan with different degree of deacetylation [125]. The swelling rate of chitosan microbeads with higher degree of deacetylation was found to be more affected by pH than that of acetylated chitosan microbeads, and both of them were affected by temperature, swelling increasing with increased temperature. The microbeads were investigated in rabbit ear embolization model and ischemic necrosis on ear was visible due to the vascular occlusion after 15 days. Chitosan beads with lower deacetylation degree were suggested as potential biocompatible and biodegradable embolic agents.

In another study, doxorubicin-containing microbeads based on oxidized carboxymethyl cellulose (Mol Wt: 700000) and chitosan (DD: ≥75%) were prepared at different particle sizes (540 ± 50 μm, 580 ± 60 μm, 830 ± 60 μm) using a modified inverse emulsion-crosslinking method [126]. The loaded microbeads were stable in aqueous dispersions without aggregation for a prolonged period of time but degradable in a lysozyme solution. Furthermore, the microbeads were shown to be pH-sensitive, with fast release of drug in acidic environment due to the protonation of carboxylic groups in the microbeads. Faster drug release was obtained with smaller particle-size beads.

6 Perspectives for the Future

Chitosan bead formulations have been investigated both for their bioactive properties and as a drug delivery system for different application routes. Wound healing, anti-inflammatory, antibacterial, and hemostatic properties of chitosan provide a significant advantage for the use of chitosan beads especially in serious skin injuries. On the other side, as a drug carrier system, active substances with many different indications can be incorporated into chitosan beads. The suitability of chitosan beads for different routes of administration such as oral, mucosal, and dermal offers great benefit. Moreover, its inherited properties such as bioadhesion, penetration enhancement, biodegradation, biocompatibility, and its cationic nature provide advantages over non-chitosan products with enhanced efficacy. The use of chitosan beads in drug delivery systems has been shown to increase bioavailability, prolong release time, and provide targeted drug release (especially in the stomach and colon). There are also studies on the use of chitosan beads for embolization and chemoembolization. Since chitosan is a biodegradable material, the necessity to remove the relevant polymer from the embolized vessel after the treatment seen in current applications (non-biodegradable polymers) makes chitosan beads more advantageous. In addition, especially in hepatocellular carcinoma, these beads, which can be loaded with anticancer drugs, provide both embolization and release of anticancer drugs at the target site.

Considering the results of the studies reported up to date, it can be concluded that chitosan beads are promising systems in pharmaceutical field with wide range of applications and there is no doubt that more studies will be focused on chitosan microbeads in the coming years. It is important to take into consideration that the basis of obtaining a successful microbead system is the selection of chitosan with appropriate properties. Especially for biomedical purposes, safety of chitosan is a crucial requirement besides meeting medical quality requirements. Being a naturally derived polymer, due to different sources, and regional differences, chitosan manufacturers are having difficulties in obtaining the same quality batch every time and depending on the source obtained, the properties show variability which also effects the safety and quality of chitosan. Hence standardization of chitosan becomes apparent. At the same time, purification is crucial as specific proteins coming from shellfish may cause allergy in sensitized individuals. In addition, endotoxin (lipopolysaccharide) contamination may cause undesired inflammatory responses. Finally, more clinical studies are needed to show the safety as well as efficacy of the chitosan beads in humans.

References

1. Roberts GAF (1992) Structure of chitin and chitosan. In: Chitin chemistry. Macmillan Education UK, London, pp 1–53
2. Senel S (2020) Current status and future of chitosan in drug and vaccine delivery. React Funct Polym 147:104452
3. Parmaksız S, Şenel S (2023) Carboxymethyl chitosan for drug and vaccine delivery: an overview. In: Jayakumar R (ed) Advances in polymer science. Multifaceted carboxymethyl chitosan derivatives. Springer, Charm, pp 225–255
4. Wang W, Meng Q, Li Q, Liu J, Zhou M, Jin Z et al (2020) Chitosan derivatives and their application in biomedicine. Int J Mol Sci 21(2):487
5. Riva R, Ragelle H, des Rieux A, Duhem N, Jérôme C, Préat V (2011) Chitosan and chitosan derivatives in drug delivery and tissue engineering. In: Jayakumar RPM, Muzzarelli R (eds) Chitosan for biomaterials II. Advances in polymer science, vol 244. Springer, Berlin, pp 19–44
6. Şenel S, Yuksel S (2020) Chitosan-based particulate systems for drug and vaccine delivery in the treatment and prevention of neglected tropical diseases. Drug Deliv Transl Rev 10(6): 1644–1674
7. Rinaudo M (2006) Chitin and chitosan: properties and applications. Prog Polym Sci 31(7): 603–632
8. Parhi R (2020) Drug delivery applications of chitin and chitosan: a review. Environ Chem Lett 18(3):577–594
9. Mikušová V, Mikuš P (2021) Advances in chitosan-based nanoparticles for drug delivery. Int J Mol Sci 22(17):9652
10. Kravanja G, Primožič M, Knez Ž, Leitgeb M (2019) Chitosan-based (Nano) materials for novel biomedical applications. Molecules 24(10):1960
11. Mura P, Maestrelli F, Cirri M, Mennini N (2022) Multiple roles of chitosan in mucosal drug delivery: an updated review. Mar Drugs 20(5):335
12. Bertolino V, Cavallaro G, Milioto S, Lazzara G (2020) Polysaccharides/halloysite nanotubes for smart bionanocomposite materials. Carbohydr Polym 245:116502
13. Hong F, Qiu P, Wang Y, Ren P, Liu J, Zhao J et al (2023) Chitosan-based hydrogels: from preparation to applications, a review. Food Chem X:101095
14. Wei S, Ching YC, Chuah CH (2020) Synthesis of chitosan aerogels as promising carriers for drug delivery: a review. Carbohydr Polym 231:115744
15. Domalik-Pyzik P, Chłopek J, Pielichowska K (2019) Chitosan-based hydrogels: preparation, properties, and applications. In: Mondal M (ed) Cellulose-based superabsorbent hydrogels. Polymers and polymeric composites: a reference series. Springer, Cham, pp 1665–1693
16. Şenel S (2015) Functionalization of marine materials for drug delivery systems. In: Kim S-K (ed) Functional marine biomaterials. Woodhead Publishing, pp 109–121
17. Amiri M, Khazaeli P, Salehabadi A, Salavati-Niasari M (2021) Hydrogel beads-based nanocomposites in novel drug delivery platforms: recent trends and developments. Adv Colloid Interface Sci 288:10
18. Ghosh S, Ghosh S, Sharma H, Bhaskar R, Han SS, Sinha JK (2024) Harnessing the power of biological macromolecules in hydrogels for controlled drug release in the central nervous system: a review. Int J Biol Macromol 254:18
19. De France KJ, Hoare T, Cranston ED (2017) Review of hydrogels and aerogels containing nanocellulose. Chem Mater 29(11):4609–4631
20. Li C-g, Dang Q, Yang Q, Chen D, Zhu H, Chen J et al (2022) Study of the microstructure of chitosan aerogel beads prepared by supercritical CO2 drying and the effect of long-term storage. RSC Adv 12(33):21041–21049
21. Shi W, Ching YC, Chuah CH (2021) Preparation of aerogel beads and microspheres based on chitosan and cellulose for drug delivery: a review. Int J Biol Macromol 170:751–767
22. Guastaferro M, Reverchon E, Baldino L (2021) Polysaccharide-based aerogel production for biomedical applications: a comparative review. Materials 14(7):1631

23. Smirnova I, Gurikov P (2018) Aerogel production: current status, research directions, and future opportunities. J Supercrit Fluids 134:228–233
24. García-González CA, Alnaief M, Smirnova I (2011) Polysaccharide-based aerogels-promising biodegradable carriers for drug delivery systems. Carbohydr Polym 86(4):1425–1438
25. Qu B, Luo Y (2020) Chitosan-based hydrogel beads: preparations, modifications and applications in food and agriculture sectors – a review. Int J Biol Macromol 152:437–448
26. Saheed IO, Suah FBM (2023) Developing nano-micro size chitosan beads using imidazolium-based ionic liquid: a perspective. Int J Biol Macromol 241:124610
27. Buranachai T, Praphairaksit N, Muangsin N (2010) Chitosan/polyethylene glycol beads crosslinked with tripolyphosphate and glutaraldehyde for gastrointestinal drug delivery. AAPS Pharm Sci Tech 11:1128–1137
28. Prabaharan M, Mano J (2004) Chitosan-based particles as controlled drug delivery systems. Drug Deliv 12(1):41–57
29. Park KH, Choi YW, Ryu H, Lee HJ, Moon J-H, Song H-J et al (2022) Controlled drug release using chitosan-alginate-gentamicin multi-component beads. Materials 15(21):7682
30. Xing PP, Shi YA, Dong CC, Liu H, Cheng Y, Sun JB et al (2017) Colon-targeted delivery of IgY against toxin A and B by encapsulation in chitosan-Ca pectinate microbeads. AAPS Pharm Sci Tech 18(4):1095–1103
31. Tavakol M, Vasheghani-Farahani E, Dolatabadi-Farahani T, Hashemi-Najafabadi S (2009) Sulfasalazine release from alginate-N, O-carboxymethyl chitosan gel beads coated by chitosan. Carbohydr Polym 77(2):326–330
32. Jóźwiak T, Filipkowska U, Szymczyk P, Rodziewicz J, Mielcarek A (2017) Effect of ionic and covalent crosslinking agents on properties of chitosan beads and sorption effectiveness of reactive black 5 dye. React Funct Polym 114:58–74
33. Uddin MS, Khand S, Dong C (2024) Effect of crosslinking agents on chitosan hydrogel carriers for drug loading and release for targeted drug delivery. Gels 10(7):421
34. Joseph AM, George B (2022) Cross-linking biopolymers for biomedical applications. In: Thomas S, AR A, Jose Chirayil C, Thomas B (eds) Handbook of biopolymers. Springer, Singapore, pp 1–38
35. Li A, Ma B, Hua S, Ping R, Ding L, Tian B et al (2024) Chitosan-based injectable hydrogel with multifunction for wound healing: a critical review. Carbohydr Polym:121952
36. Patrulea V, Ostafe V, Borchard G, Jordan O (2015) Chitosan as a starting material for wound healing applications. Eur J Pharm Biopharm 97:417–426
37. Rajkumar DSR, Keerthika K, Vijayaragavan V (2024) Chitosan-based biomaterial in wound healing: a review. Cureus 16(2)
38. Matica MA, Aachmann FL, Tøndervik A, Sletta H, Ostafe V (2019) Chitosan as a wound dressing starting material: antimicrobial properties and mode of action. Int J Mol Sci 20(23):5889
39. Kozen BG, Kircher SJ, Henao J, Godinez FS, Johnson AS (2008) An alternative hemostatic dressing: comparison of CELOX, HemCon, and QuikClot. Acad Emerg Med 15(1):74–81
40. Millner RW, Lockhart AS, Bird H, Alexiou C (2009) A new hemostatic agent: initial life-saving experience with Celox (chitosan) in cardiothoracic surgery. Ann Thorac Surg 87(2):e13–e14
41. Park CJ, Clark SG, Lichtensteiger CA, Jamison RD, Johnson AJW (2009) Accelerated wound closure of pressure ulcers in aged mice by chitosan scaffolds with and without bFGF. Acta Biomater 5(6):1926–1936
42. Baxter RM, Dai T, Kimball J, Wang E, Hamblin MR, Wiesmann WP et al (2013) Chitosan dressing promotes healing in third degree burns in mice: gene expression analysis shows biphasic effects for rapid tissue regeneration and decreased fibrotic signaling. J Biomed Mater Res A 101(2):340–348
43. Liu X, Ma L, Mao Z, Gao C (2011) Chitosan-based biomaterials for tissue repair and regeneration. In: Jayakumar R, Prabaharan M, Muzzarelli R (eds) Chitosan for biomaterials II. Advances in polymer science, vol 244. Springer, Berlin, pp 81–127

44. Akca G, Özdemir A, Öner ZG, Şenel S (2018) Comparison of different types and sources of chitosan for the treatment of infections in the oral cavity. Res Chem Intermed 44:4811–4825
45. Dai T, Tanaka M, Huang Y-Y, Hamblin MR (2011) Chitosan preparations for wounds and burns: antimicrobial and wound-healing effects. Expert Rev Anti Infect Ther 9(7):857–879
46. Rabea EI, Badawy ME-T, Stevens CV, Smagghe G, Steurbaut W (2003) Chitosan as antimicrobial agent: applications and mode of action. Biomacromolecules 4(6):1457–1465
47. Tang H, Zhang P, Kieft TL, Ryan SJ, Baker SM, Wiesmann WP et al (2010) Antibacterial action of a novel functionalized chitosan-arginine against gram-negative bacteria. Acta Biomater 6(7):2562–2571
48. Abass HA, Kamel R (2014) Formulation and evaluation of in situ forming polymeric drug delivery systems for mixed vaginal infection. J Pharm Res Int 4(19):2281–2295
49. Du X, Wu L, Yan H, Jiang Z, Li S, Li W et al (2021) Microchannelled alkylated chitosan sponge to treat noncompressible hemorrhages and facilitate wound healing. Nat Commun 12(1):4733
50. Hickman DA, Pawlowski CL, Sekhon UDS, Marks J, Gupta AS (2018) Biomaterials and advanced technologies for hemostatic management of bleeding. Adv Mater 30(4):1700859
51. Guo Y, Wang M, Liu Q, Liu G, Wang S, Li J (2023) Recent advances in the medical applications of hemostatic materials. Theranostics 13(1):161
52. Fan P, Zeng Y, Zaldivar-Silva D, Agüero L, Wang S (2023) Chitosan-based hemostatic hydrogels: the concept, mechanism, application, and prospects. Molecules 28(3):1473
53. Gheorghiță D, Moldovan H, Robu A, Bița A-I, Grosu E, Antoniac A et al (2023) Chitosan-based biomaterials for hemostatic applications: a review of recent advances. Int J Mol Sci 24(13):10540
54. Zhang S, Lei X, Lv Y, Wang L, Wang L-N (2023) Recent advances of chitosan as a hemostatic material: hemostatic mechanism, material design and prospective application. Carbohydr Polym:121673
55. Wang Y, Fu Y, Li J, Mu Y, Zhang X, Zhang K et al (2018) Multifunctional chitosan/dopamine/diatom-biosilica composite beads for rapid blood coagulation. Carbohydr Polym 200:6–14
56. Fathi P, Sikorski M, Christodoulides K, Langan K, Choi YS, Titcomb M et al (2018) Zeolite-loaded alginate-chitosan hydrogel beads as a topical hemostat. J Biomed Mater Res B 106(5): 1662–1671
57. Özdoğan AI, İlarslan YD, Kösemehmetoğlu K, Akca G, Kutlu HB, Comerdov E et al (2018) In vivo evaluation of chitosan based local delivery systems for atorvastatin in treatment of periodontitis. Int J Pharm 550(1–2):470–476
58. Chung MJ, Park JK, Park YI (2012) Anti-inflammatory effects of low-molecular weight chitosan oligosaccharides in IgE–antigen complex-stimulated RBL-2H3 cells and asthma model mice. Int Immunopharmacol 12(2):453–459
59. Ouyang Y, Zhao J, Wang S (2023) Multifunctional hydrogels based on chitosan, hyaluronic acid and other biological macromolecules for the treatment of inflammatory bowel disease: a review. Int J Biol Macromol 227:505–523
60. Friedman AJ, Phan J, Schairer DO, Champer J, Qin M, Pirouz A et al (2013) Antimicrobial and anti-inflammatory activity of chitosan–alginate nanoparticles: a targeted therapy for cutaneous pathogens. J Invest Dermatol 133(5):1231–1239
61. Maita KC, Avila FR, Torres-Guzman RA, Garcia JP, Eldaly AS, Palmieri L et al (2022) Local anti-inflammatory effect and immunomodulatory activity of chitosan-based dressing in skin wound healing: a systematic review. J Clin Transl Res 8(6):488–498
62. Pangestuti R, Bak S-S, Kim S-K (2011) Attenuation of pro-inflammatory mediators in LPS-stimulated BV2 microglia by chitooligosaccharides via the MAPK signaling pathway. Int J Biol Macromol 49(4):599–606
63. Azuma K, Osaki T, Minami S, Okamoto Y (2015) Anticancer and anti-inflammatory properties of chitin and chitosan oligosaccharides. J Funct Biomater 6(1):33–49

64. Chang S-H, Lin Y-Y, Wu G-J, Huang C-H, Tsai GJ (2019) Effect of chitosan molecular weight on anti-inflammatory activity in the RAW 264.7 macrophage model. Int J Biol Macromol 131:167–175
65. Niu W, Dong Y, Fu Z, Lv J, Wang L, Zhang Z et al (2021) Effects of molecular weight of chitosan on anti-inflammatory activity and modulation of intestinal microflora in an ulcerative colitis model. Int J Biol Macromol 193:1927–1936
66. Batool R, Mudassir J, Khan MA, Zafar S, Rana SJ, Abbas N et al (2023) Fabrication and characterization of celecoxib-loaded chitosan/guar gum-based hydrogel beads. Pharmaceuticals (Basel) 16(4)
67. Fu S, Zhou L, Zeng P, Fu S (2022) Antibacterial chitosan-gelatin hydrogel beads cross-linked by riboflavin under ultraviolet A irradiation. Fibers Polym 23(2):315–320
68. EDQM EC (2023) European Pharmacopoeia (Ph. Eur.) 11th edn
69. Liechty WB, Kryscio DR, Slaughter BV, Peppas NA (2010) Polymers for drug delivery systems. Annu Rev Chem Biomol Eng 1:149–173
70. Thang NH, Chien TB, Cuong DX (2023) Polymer-based hydrogels applied in drug delivery: an overview. Gels 9(7):523
71. George A, Shah PA, Shrivastav PS (2019) Natural biodegradable polymers based nano-formulations for drug delivery: a review. Int J Pharm 561:244–264
72. Sung YK, Kim SW (2020) Recent advances in polymeric drug delivery systems. Biomater Res 24(1):12
73. Borandeh S, van Bochove B, Teotia A, Seppälä J (2021) Polymeric drug delivery systems by additive manufacturing. Adv Drug Deliv Rev 173:349–373
74. Nussinovitch A (2010) Beads as drug carriers. In: Polymer macro-and micro-gel beads: fundamentals and applications, pp 191–230
75. Panonnummal R, Kumar VS, Jayakumar R, Sabitha M (2021) Application of chitosan and its derivatives in transdermal drug delivery. Adv Polym Sci 288:411–446
76. Fuchs K, Duran R, Denys A, Bize PE, Borchard G, Jordan O (2017) Drug-eluting embolic microspheres for local drug delivery–state of the art. J Control Release 262:127–138
77. Jeong WY, Kwon M, Choi HE, Kim KS (2021) Recent advances in transdermal drug delivery systems: a review. Biomater Res 25(1):24
78. Verma S, Utreja P (2022) Exploring therapeutic potential of invasomes, transfersomes, transethosomes, oleic acid vesicles, and cubosomes adopting topical/transdermal route. Micro Nanosystems 14(1):3–20
79. Prausnitz MR, Langer R (2008) Transdermal drug delivery. Nat Biotechnol 26(11):1261–1268
80. Ueno H, Mori T, Fujinaga T (2001) Topical formulations and wound healing applications of chitosan. Adv Drug Deliv Rev 52(2):105–115
81. Hoelgaard A, Møllgaard B (1985) Dermal drug delivery – improvement by choice of vehicle or drug derivative. J Control Release 2:111–120
82. Mohite P, Rahayu P, Munde S, Ade N, Chidrawar VR, Singh S et al (2023) Chitosan-based hydrogel in the management of dermal infections: a review. Gels 9(7):594
83. Panonnummal R, Kumar VS, Jayakumar R, Sabitha M (2021) Application of chitosan and its derivatives in transdermal drug delivery. In: Jayakumar R, Prabaharan M (eds) Chitosan for biomaterials IV: biomedical applications, vol 288. Springer, Cham, pp 411–446
84. Song Y, Wang Y, Thakur R, Meidan VM, Michniak B (2004) Mucosal drug delivery: membranes, methodologies, and applications. Crit Rev Ther Drug Carrier Syst 21(3):195–256
85. Şenel S (2021) An overview of physical, microbiological and immune barriers of oral mucosa. Int J Mol Sci 22(15):7821
86. Rathbone MJ, Pather I, Şenel S (2015) Overview of oral mucosal delivery. In: Rathbone MJ, Senel S, Pather I (eds) Oral mucosal drug delivery and therapy. Springer US, Boston, pp 17–29
87. Şenel S (2010) Potential applications of chitosan in oral mucosal delivery. J Drug Deliv Sci Technol 20(1):23–32
88. Zhang X, Chen G, Zhang H, Shang L, Zhao Y (2023) Bioinspired oral delivery devices. Nat Rev Bioeng 1(3):208–225

89. Cirri M, Maestrelli F, Scuota S, Bazzucchi V, Mura P (2021) Development and microbiological evaluation of chitosan and chitosan-alginate microspheres for vaginal administration of metronidazole. Int J Pharm 598:120375
90. El-Mahrouk GM, El-Gazayerly ON, Aboelwafa AA, Taha MS (2014) Chitosan lactate wafer as a platform for the buccal delivery of tizanidine HCl: in vitro and in vivo performance. Int J Pharm 467(1–2):100–112
91. Yuvaraja G, Pathak JL, Zhang WJ, Zhang YP, Xu J (2017) Antibacterial and wound healing properties of chitosan/poly(vinyl alcohol)/zinc oxide beads (CS/PVA/ZnO). Int J Biol Macromol 103:234–241
92. López-Iglesias C, Barros J, Ardao I, Monteiro FJ, Alvarez-Lorenzo C, Gómez-Amoza JL et al (2019) Vancomycin-loaded chitosan aerogel particles for chronic wound applications. Carbohydr Polym 204:223–231
93. Wu B, Li Y, Li YY, Shi ZH, Bian XH, Xia Q (2022) Nanostructured-lipid carriers-chitosan hydrogel beads carrier system for loading of resveratrol: a new method of topical application. J Biomater Appl 36(8):1444–1457
94. Wu B, Li Y, Li Y, Li H, Ji S, Xia Q (2021) Pickering emulsions-chitosan hydrogel beads carrier system for loading of resveratrol: formulation approach and characterization studies. React Funct Polym 169:105074
95. Baradaran Eftekhari R, Maghsoudnia N, Samimi S, Abedin Dorkoosh F (2019) Application of chitosan in oral drug delivery. In: Jana S, Jana S (eds) Functional chitosan: drug delivery and biomedical applications, pp 43–73
96. Peers S, Montembault A, Ladavière C (2020) Chitosan hydrogels for sustained drug delivery. J Control Release 326:150–163
97. Waqar MA, Mubarak N, Khan AM, Khan R, Shaheen F, Shabbir A (2024) Advanced polymers and recent advancements on gastroretentive drug delivery system; a comprehensive review. J Drug Target 6:655–671
98. Pahwa R, Saini N, Kumar V, Kohli K (2012) Chitosan-based gastroretentive floating drug delivery technology: an updated review. Expert Opin Drug Deliv 9(5):525–539
99. Sangnim T, Dheer D, Jangra N, Huanbutta K, Puri V, Sharma A (2023) Chitosan in oral drug delivery formulations: a review. Pharmaceutics 15(9):2361
100. Collado-González M, González Espinosa Y, Goycoolea FM (2019) Interaction between chitosan and mucin: fundamentals and applications. Biomimetics (Basel) 4(2):32
101. Niu W, Chen X, Xu R, Dong H, Yang F, Wang Y et al (2021) Polysaccharides from natural resources exhibit great potential in the treatment of ulcerative colitis: a review. Carbohydr Polym 254:117189
102. Zhu J, Zhong L, Chen W, Song Y, Qian Z, Cao X et al (2019) Preparation and characterization of pectin/chitosan beads containing porous starch embedded with doxorubicin hydrochloride: a novel and simple colon targeted drug delivery system. Food Hydrocoll 95:562–570
103. Batool R, Mudassir J, Khan MA, Zafar S, Rana SJ, Abbas N et al (2023) Fabrication and characterization of celecoxib-loaded chitosan/guar gum-based hydrogel beads. Pharmaceuticals 16(4):554
104. Malik NS, Ahmad M, Alqahtani MS, Mahmood A, Barkat K, Khan MT et al (2021) β-Cyclodextrin chitosan-based hydrogels with tunable pH-responsive properties for controlled release of acyclovir: design, characterization, safety, and pharmacokinetic evaluation. Drug Deliv 28(1):1093–1108
105. Niu B, Jia J, Wang H, Chen S, Cao W, Yan J et al (2019) In vitro and in vivo release of diclofenac sodium-loaded sodium alginate/carboxymethyl chitosan-ZnO hydrogel beads. Int J Biol Macromol 141:1191–1198
106. Das PN, Raj KG (2024) Chitosan coated graphene oxide incorporated sodium alginate hydrogel beads for the controlled release of amoxicillin. Int J Biol Macromol 254:127837
107. Stavarache CE, Ghebaur A, Dinescu S, Samoilă I, Vasile E, Vlasceanu GM et al (2021) 5-Aminosalicylic acid loaded chitosan-carrageenan hydrogel beads with potential application for the treatment of inflammatory bowel disease. Polymers 13(15):2463

108. Namdev A, Jain D (2019) Floating drug delivery systems: an emerging trend for the treatment of peptic ulcer. Curr Drug Deliv 16(10):874–886
109. Lee SH, Bajracharya R, Min JY, Han J-W, Park BJ, Han H-K (2020) Strategic approaches for colon targeted drug delivery: an overview of recent advancements. Pharmaceutics 12(1):68
110. Praveen R, Verma PRP, Venkatesan J, Yoon DH, Kim SK, Singh SK (2017) In vitro and in vivo evaluation of gastro-retentive carvedilol loaded chitosan beads using Gastroplus™. Int J Biol Macromol 102:642–650
111. Helmy AM, Elsabahy M, Soliman GM, Mahmoud MA, Ibrahim EA (2017) Development and evaluation of chitosan beads for the colonic delivery of azathioprine for treatment of inflammatory bowel disease. Eur J Pharm Sci 109:269–279
112. Yadollahi M, Farhoudian S, Namazi H (2015) One-pot synthesis of antibacterial chitosan/silver bio-nanocomposite hydrogel beads as drug delivery systems. Int J Biol Macromol 79:37–43
113. Li J, Jiang CQ, Lang XQ, Kong M, Cheng XJ, Liu Y et al (2016) Multilayer sodium alginate beads with porous core containing chitosan based nanoparticles for oral delivery of anticancer drug. Int J Biol Macromol 85:1–8
114. Lencioni R, de Baere T, Soulen MC, Rilling WS, Geschwind JFH (2016) Lipiodol transarterial chemoembolization for hepatocellular carcinoma: a systematic review of efficacy and safety data. Hepatology 64(1):106–116
115. Dutta R, Mahato RI (2017) Recent advances in hepatocellular carcinoma therapy. Pharmacol Ther 173:106–117
116. Luo Z, Jiang L, Ding C, Hu B, Loh XJ, Li Z et al (2018) Surfactant free delivery of docetaxel by poly [(R)-3-hydroxybutyrate-(R)-3-hydroxyhexanoate]-based polymeric micelles for effective melanoma treatments. Adv Healthc Mater 7(23):1801221
117. Liu X, Li Z, Loh XJ, Chen K, Li Z, Wu YL (2019) Targeted and sustained Corelease of chemotherapeutics and gene by injectable supramolecular hydrogel for drug-resistant cancer therapy. Macromol Rapid Commun 40(5):1800117
118. Bannerman D, Wan W (2016) Multifunctional microbeads for drug delivery in TACE. Expert Opin Drug Deliv 13(9):1289–1300
119. Namur J, Citron SJ, Sellers MT, Dupuis MH, Wassef M, Manfait M et al (2011) Embolization of hepatocellular carcinoma with drug-eluting beads: doxorubicin tissue concentration and distribution in patient liver explants. J Hepatol 55(6):1332–1338
120. Burrel M, Reig M, Forner A, Barrufet M, de Lope CR, Tremosini S et al (2012) Survival of patients with hepatocellular carcinoma treated by transarterial chemoembolisation (TACE) using drug eluting beads. Implications for clinical practice and trial design. J Hepatol 56(6):1330–1335
121. Lewis AL, Holden RR (2011) DC bead embolic drug-eluting bead: clinical application in the locoregional treatment of tumours. Expert Opin Drug Deliv 8(2):153–169
122. He C, Zeng W, Su Y, Sun R, Xiao Y, Zhang B et al (2021) Microfluidic-based fabrication and characterization of drug-loaded PLGA magnetic microspheres with tunable shell thickness. Drug Deliv 28(1):692–699
123. Kim J, Kwak B, Shim H, Lee Y, Baik H, Lee M et al (2007) Preparation of doxorubicin-containing chitosan microspheres for transcatheter arterial chemoembolization of hepatocellular carcinoma. J Microencapsul 24(5):408–419
124. Arya C, Saez Cabesas CA, Huang H, Raghavan SR (2017) Clustering of cyclodextrin-functionalized microbeads by an amphiphilic biopolymer: real-time observation of structures resembling blood clots. ACS Appl Mater Interfaces 9(42):37238–37245
125. Zhou X, Kong M, Cheng XJ, Feng C, Li J, Li JJ et al (2014) In vitro and in vivo evaluation of chitosan microspheres with different deacetylation degree as potential embolic agent. Carbohydr Polym 113:304–313
126. Weng L, Le HC, Lin J, Golzarian J (2011) Doxorubicin loading and eluting characteristics of bioresorbable hydrogel microspheres: in vitro study. Int J Pharm 409(1–2):185–193

Adv Polym Sci (2025) 295: 143–160
https://doi.org/10.1007/12_2024_180

Published online: 13 November 2024

Mucoadhesive Chitosan in Drug Delivery

Flavia Laffleur

Contents

1 Introduction 144
2 Mucoadhesion 145
  2.1 Concept of Mucoadhesion 145
  2.2 Theories of Mucoadhesion 146
  2.3 Mucoadhesive Assays 148
3 Chitosan 148
  3.1 Origin of Chitosan 148
  3.2 Modification of Chitosan 150
4 Mucoadhesive Drug Delivery 153
  4.1 Chitosan in Ocular Delivery 153
  4.2 Chitosan in Wound Management 154
  4.3 Chitosan in Vaginal Delivery 155
5 Summary 157
References 157

Abstract Mucoadhesive drug delivery systems are of great interest due to the fate of prolonging residence time at the site of absorption/action (I), providing sustained drug release (II), and minimizing drug degradation in the body (III). Various natural, semi-synthetic, and synthetic polymers have been used in the development of mucoadhesive drug delivery systems. Chitosan – a naturel cationic polysaccharide – exhibiting mucoadhesive features, is widely used in the development of mucoadhesive drug delivery systems. Nevertheless, chitosan per se shows some drawbacks as its limitation in mucoadhesive strength, limited water solubility, and pH burden. Therefore, modification of chitosans has been investigated to overcome these limitations. In this chapter, we focus on the concept of mucoadhesion with

F. Laffleur (✉)
Department of Pharmaceutical Technology, Institute of Pharmacy, Leopold Franzens University of Innsbruck, Innsbruck, Austria
e-mail: Flavia.Laffleur@uibk.ac.at

theories showing the beneficial modifications of chitosan for different body sites such as ocular and vaginal mucosae as well as in wound management.

Keywords Chitosan · Drug delivery systems · Mucoadhesion · Vaginal dosage forms

1 Introduction

Mucoadhesive drug delivery is increasingly recognized for its distinct benefits over conventional delivery. The close contact of the formulation with the tissue ensures a targeted drug release at the designated site of action. This release mechanism not only enhances the bioavailability of the drug but also amplifies its local or systemic effects. Functionally modified polymers have unique chemical properties that enable effective delivery to specific mucosal areas. Certain mucoadhesive polymers not only facilitate local and targeted drug delivery but are also able to open tight junctions, thus influencing epithelial transport. Due to their chemical properties and their ability to bypass the first-pass effect through transepithelial transport, mucoadhesive polymers have enhanced drug absorption and a higher therapeutic efficacy by simultaneously lowering frequent dosing. Frequently used mucoadhesive polymers contain polar functional groups such as amide, carboxyl, hydroxyl, sulfate, and thiol, which are able to interact with mucin glycoproteins. Due to the cationic characteristics of chitosan and their ability of electrostatic interactions between the amino groups and the mucin, it has a greater mucoadhesion than methylcellulose, which is a non-ionic polymer [1, 2]. Chitosan is derived from chitin by chitin deacetylase. It is widely been used in a variety of biomedical and drug delivery research due to its low toxicity, biocompatibility, antimicrobial activity, permeation-enhancing effect, and mucoadhesive properties [3]. Furthermore, the use of chitosan with its mucoadhesive properties for drug delivery systems has many advantages: most importantly, the enhanced drug absorption and bioavailability. Moreover, using mucoadhesive chitosan allows the reach of specific organs or tissues in order to minimize adverse effects [4]. The mucoadhesiveness of native chitosan is limited therefore researchers started to launch modified chitosan conjugates such as thiolated, preactivated [5], quaternized, and maleic acid functionalized chitosan [6] to foster its mucoadhesive potential. This book chapter outlines the concept of mucoadhesion based on the various theories. Furthermore, it gives an insight into the origin and modification of mucoadhesive chitosan. The proof of concept is given by the different dosage forms on a variety of mucosae rendering this polymer a unique candidate for adhesive delivery.

2 Mucoadhesion

2.1 Concept of Mucoadhesion

Mucoadhesion designates the manner of pharmaceutical form interacting with chemical structures in the mucus resulting in adhesion [7]. Decisive for mucoadhesion are often mucins, hydrophilic basic glycoproteins included in the mucus, since they are able to interact via van der Waals forces, electrostatic interaction as well as hydrogen bridges and therefore form highly cross-linked porous three-dimensional structures [8, 9]. Advantageous is this opportunity in drug delivery because of the prolonged residence time and small distance, resulting in high permeability [10]. The mucoadhesive process consists of contact, interpenetration, and consolidation phase as shown in Fig. 1.

The contact phase introduces mucoadhesion by the contact between film-forming polymer and mucosa, after which the film hydrates itself by absorption of mucus and gels [10]. The second phase is characterized by physical entanglement between film-forming polymer and mucus. After that, in the consolidation phase, physical adhesion takes place and the chemical bonding between mucosa, film, and mucus components maintains the attachment.

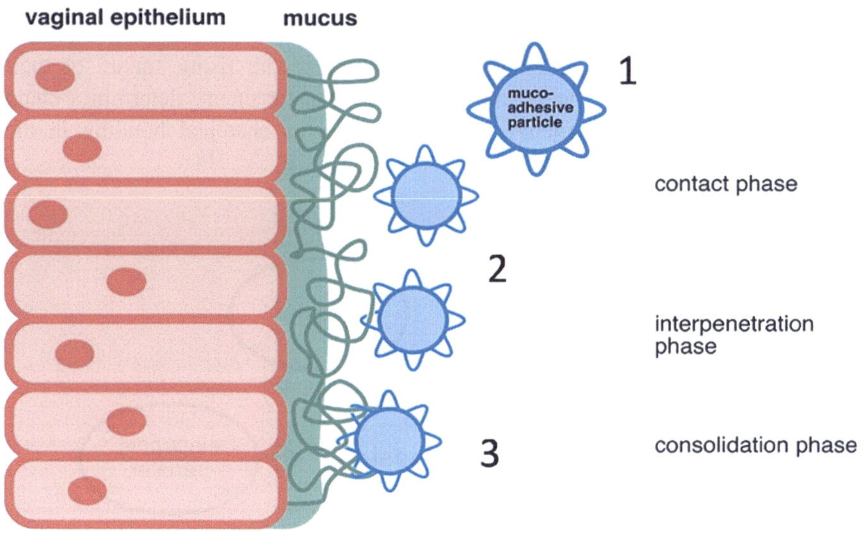

Fig. 1 Mucoadhesion phases: visualization of the process of mucoadhesion of mucoadhesive particles to the mucus layer on vaginal epithelium sectioned in contact (1), interpenetration (2), and consolidation (3) phase adapted from Cazorla-Luna et al. [11]

2.2 *Theories of Mucoadhesion*

Six key theories of adhesion namely wetting theory, electronic theory, adsorption theory, diffusion theory, mechanical theory, and fracture theory are used to investigate mucoadhesion as shown in Fig. 2.

2.2.1 Absorption Theory

Polymer and mucus interact via chemical bonding of covalent or non-covalent character and therefore different strength based on the type of film-forming polymer plus "chemisorption", strong covalent interactions along the contact surface. Depending on the type and strength of the chemical interaction the adherence is either permanent or in the better case reversible. Generally, adhesion is often only due to a composite of different or same interactions, like multiple weak van der Waals forces, enabling total strong mucoadhesion. Absorption theory is particularly important in interpenetration and consolidation phases.

2.2.2 Electronic Transfer Theory

Polymer and mucus interact via electronic transfer caused by their differing electrical potential resulting in an electrical double layer forming on the contact surface. The adhesion is directly related to particle size and Coulomb force, and therefore increases with rising electric charges and decreases with rising square of their distance. Like absorption theory, also electronic transfer theory determines chain entanglement in the interpenetration stage as well as chemical bonding in the consolidation stage.

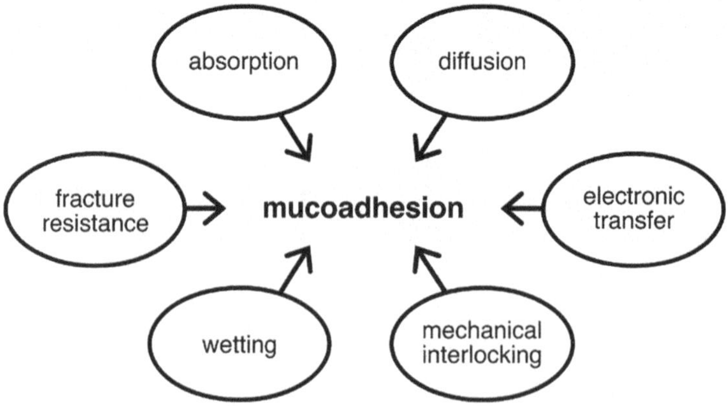

Fig. 2 Summary of the theories describing mucoadhesion adapted from Jawadi et al. [10]

Electrostatic interaction is also possible between two differently charged polymers via the formation of polyelectrolyte complexes. This effect is favored by low difference between environmental pH and polymer pKa. Since intravaginal environment is acidic, in this case film-forming polymers with pKas around 4 are suitable [12], like cationic chitosan, anionic alginate, pectin, xanthan gum, gum arabic, carrageenan, carboxymethylcellulose, and methacrylate. Based on this the film properties can be adapted. For example, swelling can be decreased since the interactions with aqueous medium are inhibited due to the polymer-polymer interaction, which results in stronger mucoadhesion and more sustained drug delivery. Martin-Illana et al. used this effect to formulate trilayers containing chitosan plus xanthan gum, pectin, or karaya gum additional to ethyl cellulose. Addition of chitosan as a third layer showed the influence of interpolymer electrostatic interaction on the film properties, as chitosan decreased the high swelling potential of anionic xanthan gum and karaya gum via electrostatic interaction and increased swelling in pectin films by stabilizing their structure [12].

2.2.3 Fracture Theory

According to the fracture theory, detachment and insufficient adhesion can be explained by the fraction of preferably mucus surface near rigid structures that show weak interaction and no interpenetration. Since adhesion strength of polymers increases with their chain length, fracture of polymer chains results in decreased mucoadhesion and improved fracture resistance in higher mucoadhesive potential.

2.2.4 Wetting Theory

Intermolecular interactions and surface tension based on the molecular contact between a liquid or semi-solid, in this case the gelled film, and solid mucosa cause adhesion according to the gels spreadability. Swelling in the contact phase is attributed to this concept.

2.2.5 Mechanical Inter-Locking Theory

According to this theory, the formulation and an irregular, porous surface adhere via mechanical interlocking. As a result, a more regular, stable, due to the irregularities larger interaction surface is formed. Mechanical inter-locking positively influences the interpenetration stage.

2.2.6 Diffusion Theory

Due to mutual 0.2 to 0.5 μm deep interpenetration and entanglement between polymer and mucin chain system, polymer and mucus attach to each other semi-permanent depending on penetration strength and rate as well as diffusion coefficients, substance characteristics (molecular weight, flexibility, chain structure, and linking), expansion efficiency, and miscibility of the compounds. Diffusion theory is involved in the consolidation phase additional in absorption and electronic transfer.

2.3 Mucoadhesive Assays

Depending on the dosage form and the type of mucosa, different mechanisms are discussed as being appropriate. A general definition of mucoadhesion is hardly feasible due to the multiple adhesion mechanisms and dosage forms such as gels or tablets. These facts have to be taken into account when developing an applicable assay to study mucoadhesion. The comparability of the results is very difficult due to the different parameters for the measurements. To date, there is no standardized method for the study of mucoadhesion. Table 1 gives an overview of published methods for the characterization of mucoadhesive polymers and solid and semi-solid preparations.

3 Chitosan

3.1 Origin of Chitosan

Natural polymers are of plant, animal, or microbial origin and tendentially hydrophilic, except of zein [22]. Also considering their usually high biodegradability and chemical modifiability and low toxicity potential, natural polymers are advantageous [23]. Chitosan is produced by the deacetylation of chitin as shown in Fig. 3, in which some N-acetylglucosamine moieties are converted to glucosamine units.

The presence of large amounts of protonated -NH_2 groups on the chitosan structure is responsible for its solubility in acidic aqueous media, as its pKa is approximately 6.5 [24]. When approximately 50% of all amino groups are protonated, chitosan becomes soluble. Due to its unique properties compared to other biomaterials, such as biocompatibility, biodegradability, hydrophilicity, non-toxicity as well as antifungal, antibacterial, and wound-healing properties, the biopolymer chitosan plays a leading role in drug delivery.

Table 1 Overview of characterization methods for mucoadhesive polymeric dosage forms

Type of test for solid preparation	Description of the test	Reference
Disintegration test for solid dosage forms	Vaginal tablets must disintegrate within 30 min in a water bath at 36°C to 37°C, unless they are intended for prolonged action; the test is performed with three tablets and a special apparatus. Compliance is confirmed if all three tablets tested disintegrate within the specified time	[13]
Resistance to crushing of tablets	The test measures the force required to crush tablets under defined conditions using an apparatus with two jaws, one moving toward the other. The force applied to disrupt each of the 10 tablets by crushing is measured, with results expressed in newtons, indicating the mean, minimum, and maximum values	[14]
Friability of uncoated tablets	The tablet friability test involves tumbling a sample of tablets (6.5 g for those $\leq$650 mg each or 10 tablets for those >650 mg) in a drum at 24–26 r/min for 100 rotations, then weighing before and after to measure weight loss; a mass loss of $\leq$1.0% is deemed acceptable for most products	[15]
Mucoadhesion time measurement assessment	The core principle of a detachment examination involves first attaching the genital tablet to the genital cell membranes, and then immersing them in a solvent. The sample is kept at an average temperature of 37 ± 1° C in either an incubator or a water bath, with or without agitation, depending on the individual study needs. The duration of mucoadhesion retention time is assessed by optically charting it	[16]
Swelling capacity	This test characterizes bioadhesion by measuring a tablet's ability to absorb liquid, thereby increasing its volume and weight, using a modified "Petri dish" method where the tablet's weight change in a lactic acid solution (pH 4.5) indicates its swelling capacity, calculated as a percentage and expressed as mean ± standard deviation	[17]
Type of test for semi-solid preparations	**Description of the test**	**Reference**
Evaluation of mucoadhesive properties	The ex vivo tensile strength test is submerging excised mucosal tissues (such as buccal mucosa, intestinal mucosa, and vaginal mucosa) in a concentrated mucin dispersion with a defined concentration. Afterward, the mucin-coated mucosal tissues are exposed to the wet polymeric substance in physiological solutions for a particular period of time. Next, the maximal force required to separate the mucosal tissues and polymer surfaces following the mucoadhesive bonding is measured	[18]
Spreadability	This can be assessed by using an extensometer. A fixed quantity of the semi-solid formulation is placed on a plate and weighted down by a glass plate under defined	[19]

(continued)

Table 1 (continued)

	conditions. The comparison of the sample areas before and after a defined load provides a measure of the spreadability of the formulation being analyzed. As this is not a pharmacopoeial method, care must be taken to ensure that the methodology provides valid results (e.g., parallel-plate method)	
Swelling rate of hydrogels	In order to calculate the rate at which a hydrogel sample swells, it is common practice to measure the expanding capacity of the sample at different time intervals using free-absorbency potential evaluations. The swelling values that have been measured are then included in a Voigt model. Usually, the evaluation of free-absorbency capacity is performed using the tea-bag, sieve, or filtration techniques	[20]
Viscosity-rotating viscometer method	This method involves recording the torque on a rotor at a constant angular velocity within the gel using absolute viscometers, which provide well-defined flow conditions and absolute viscosity values, or relative viscometers, which offer comparative values under specific conditions without well-defined flow. For non-Newtonian hydrogels, the procedure adjusts for the fluid's shear rate or angular velocity to accurately determine the apparent viscosity, ensuring the measurement's relevance to the gel's application conditions	[19]
Sterility	Sterility testing encompasses evaluating substances, preparations, or articles for the absence of contaminating microorganisms under aseptic conditions to guarantee compliance with required sterility specifications. This procedure includes careful observation of the culture media for microbial growth after incubation, with a satisfactory outcome indicating no detection of contamination within the test parameters. Precautions against microbial contamination and regular monitoring of test conditions are fundamental to maintaining the integrity of the sterility test and its results	[21]

3.2 Modification of Chitosan

When the acetyl groups of chitin, which is highly availabe in shellfish, are fully or in parts removed, chitosan as economical polymer is produced [25]. Chitosan contains D-glucosamine and N-acetylglucosamine, which are connected in positions 1 and 4, forming an in acid aqueous medium good, but in neutral or basic aqueous and organic media poorly soluble, well biocompatible polymer. Additional to environmental conditions like pH and temperature also the film-forming polymer's crystallinity, the ratio between the positions with acetyl and those without acetyl group as well of their molecular weight of the polymer influence its solubility [26]. Deacetylation grade furthermore determines chitosan's conformation and flexibility. Since chitosan includes amino and hydroxyl groups, which are protonated in

Fig. 3 The synthesis of the polymer chitosan from chitin by the enzyme chitin deacetylase

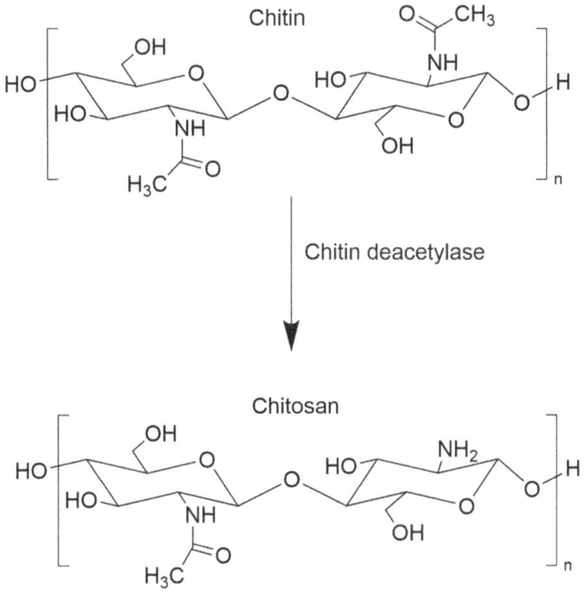

physiological environment, it is able to interact ionically and act as hydrogen bridge donor and acceptor [27]. As high deacetylated chitosan derivates include more free amino groups, they are more favorable in forming hydrogen bonds, which explains the higher solubility. This enhances the mucoadhesion since chitosan as a cationic film-forming polymer is able to form electrostatic interaction with anionic functionalities like SO_4 and sialic of mucin, a compound included in vaginal mucus [9]. Therefore, chitosan is highly mucoadhesive, based on different interaction types: electrostatic interaction, prerequisite pH is lower than 6, hydrogen bonds, as well as hydrophobic interaction presumably by incorporation of other chemical structures. Also because of its antiallergic, anti-inflammatory, immuno-enhancing, anticoagulant, antioxidant, analgesic, anticancer, antimicrobial, and hypercholesteremic activity, chitosan is a commonly used film-forming polymer. Moreover, chitosan has been shown to effectively deliver genetic material. Chitosan can also be used to adapt swelling and drug release since it forms polyelectrolytic complexes with anionic polymers like alginates, sodium carboxymethylcellulose, xanthan gum, karaya gum, or pectin in acidic medium. Nevertheless, plasticizers or other film-forming polymers usually need to be added in chitosan-based formulations to enable mechanical stability.

Native chitosan can be modified, for example, quaternated, thiolated, or polymerized to improve different properties as shown in Fig. 4.

Quaternized chitosan might improve mucoadhesion as well as drug resorption by loosening the tight junctions [28]. Chitosan's solubility can be increased through N-carboxymethyl, as well as O-, N-hydroxyl alkyl modification. Another possibility to modify chitosan is by conjugating it with maleic anhydride, which enhances mucoadhesion as shown by Sanchez Armengol et al. Maleic acid in this case is

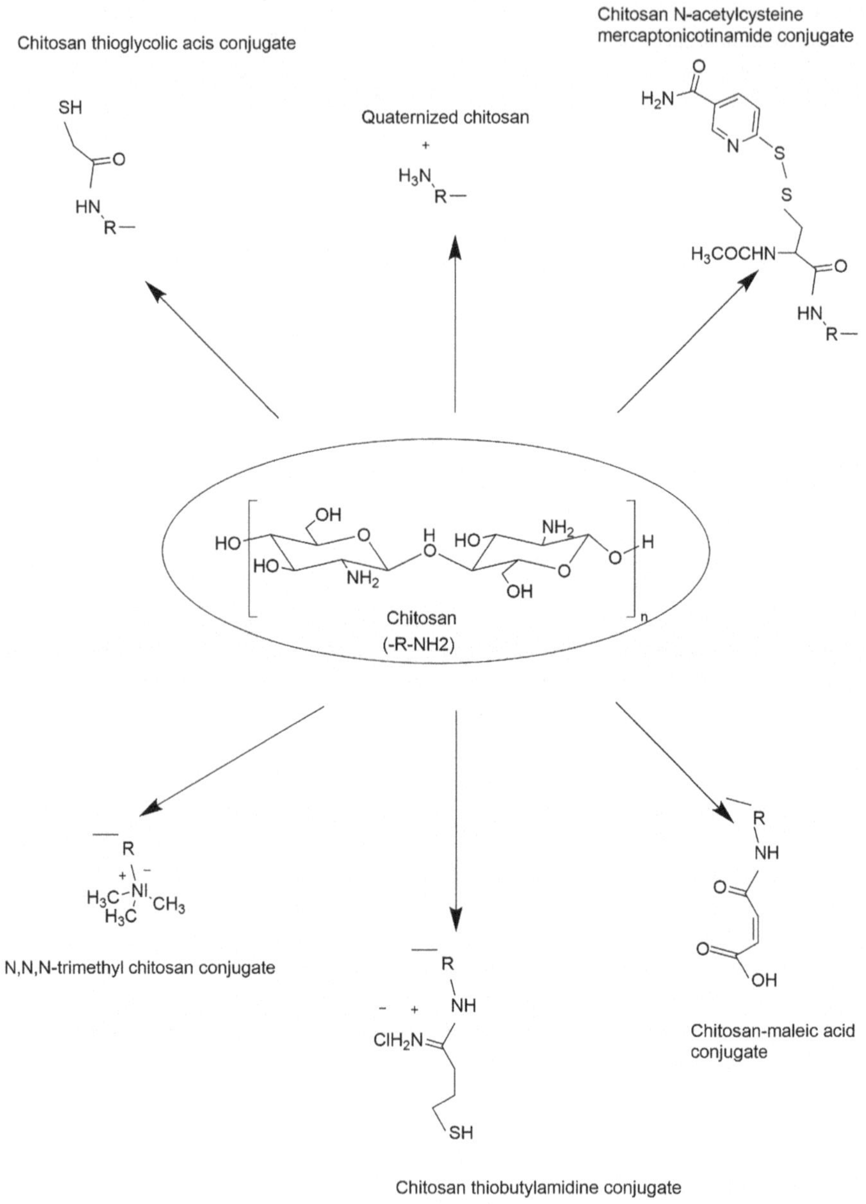

Fig. 4 Overview of possible modifications of chitosan

particularly suitable since it is relatively facile and efficient to incorporate. This is enabled by the included unsaturated imid which can be substituted with different molecules, like polymers, and functional groups. The course of this study showed stronger mucoadhesion compared to native chitosan which may be especially due to covalent "click" interactions with thiol groups. Furthermore, fluid absorption is increased by the modification with maleic acid. Although maleic acid in concentrations higher or equal to 0.1% has a negative impact on cell viability in 24 h toxicity tests, conjugation of chitosan with this compound preserves chitosan's biocompatibility. Therefore, chitosan-maleic acid is a non-toxic, mucoadhesive polymer for drug delivery [29].

4 Mucoadhesive Drug Delivery

4.1 Chitosan in Ocular Delivery

The mainly used and approved polymers in ocular delivery are hyaluronic acid (HA), chitosan, poly(lactic-co-glycolic acid) (PLGA), and poly(ethylene glycol) (PEG). PEG is also used in nanosuspensions in combination with other polymers (e.g., polylactide, chitosan), to enhance precorneal retention and improve drug penetration across the epithelial membrane, thus leading to greater concentrations at the target site [30]. Polymers are increasingly recognized for their efficacy in pharmaceutical formulations, as polymers enhance the administration of specific active ingredients. By using polymers as ocular drug delivery systems the bioavailability can be improved by overcoming ophthalmic barriers, thus leading to minor side effects and enhanced treating outcome [31]. Chitosan, a cationic polysaccharide obtained from chitin, is used as a biocompatible matrix for ocular drug delivery systems and shows antimicrobial, antifungal, and wound-healing properties. Chitosan such as HA can easily be modified without changing their physical characteristics, thus leading to a great diversity. It is utilized in nano-based formulations for the treatment of glaucoma, conjunctivitis, and immune-related degenerative eye diseases [30, 32, 33]. Su et al. developed recently nano-based ocular drug delivery systems namely liposome for prolonged residence time at the site of action. Liposomes are sphere-shaped vehicles consisting of lipid bilayers with an inner aqueous core, which can enclose both hydrophilic and lipophilic drugs. While hydrophilic drugs are placed in the aqueous core, lipophilic drugs accumulate in the lipid bilayer. The main advantage of liposomes is the prolonged residence time and therefore improved absorption of the active ingredient, which also leads to an improved drug penetration into the posterior segment. Liposomes are also used as entrapment compounds to protect the active ingredient from environmental influences. In a recent study, the potential of a new drug delivery system using triamcinolone acetonide-loaded liposomes was designed and assessed. These liposomes, consisting of soybean phosphatidylcholine and cholesterol, were coated with chitosan due to their unique bioadhesive characteristics to the cornea. Chitosan modifies the liposome surface and therefore improves the drug penetration. The modified cationic liposomes exhibited strong binding to the corneal surface and interacted with the negatively charged surface of the eye to prolong the residence time and increase local drug concentration [34].

4.2 Chitosan in Wound Management

Fan et al. created chitosan/gelatin/poly (vinyl alcohol) (PVA) hydrogels using the gamma irradiation method [35]. Chitosan is a common natural amino polysaccharide obtained by deacetylation of chitin [36]. Chitosan is a linear polymer consisting of two repeating linear monomers linked β-1,4-glycosidically: D-glucosamine and N-acetyl-D-glucosamine. Their sequences and contents are different [37]. Modifications are possible due to the reactive hydroxyl groups in its chemical structure. This means that gels with a wide range of solubility patterns and properties can be produced. In an acidic environment, the protonation of the cationic groups produces a chitosan gel, which is suitable for wound healing as it stimulates the broblasts and inhibits the growth of inflammatory cells. Chitosan is non-toxic, and shows great hemostatic potential and bioadhesiveness and bacteriostatic effect against a diverse range of microorganisms. This makes it suitable for use in the pharmaceutical sector [35]. The hydrogels investigated by Fan et al. were irradiated with 40 kGy. When tested for blood coagulation activity, the weight ratio of chitosan/gelatin 1:1 showed an ideal hemostatic effect. Furthermore, these hydrogels showed satisfactory pH sensitivity, water evaporation rate, and swelling capability, so it would be very suitable as a wound dressing. Bayindir Bilgic et al. synthesized a hydrogel material from carboxy methyl cellulose and chitosan polymers to which alpha-tocopherol was added to stimulate cell proliferation. The hydrogel was synthesized using a chemical reaction with ammonium persulfate as the initiator catalyst and N,N′--methylene bisacrylamide as the crosslinker. Alpha-tocopherol effectively increased the activity of the hydrogels produced. The cost-effective, easy-to-synthesize, and natural production of this biomaterial makes it suitable for medical use and allows new strategies for wound healing to be developed [38].

Due to the poor mechanical stability of chitosan biomaterials, it is reinforced or cross-linked with suitable compounds. This is done by incorporating carbohydrates such as cellulose and starch, synthetic polymers, plasticizers, and some proteins. In the pharmaceutical field, chia seeds (*Salvia hispanica* L.) have attracted attention due to their functional properties. When chia seeds come into contact with water, a gelatinous shell forms or a transparent gum is secreted, which is bound to the seed. This gum is an anionic heteropolysaccharide: xyloglucan. It is composed of glucose, xylose, and methylglucuronic acid and has a molecular weight of 800–2000 kDa. Martínez-Ibarra et al. investigated several porous hydrogels made from different mixtures of chitosan and xyloglucan. A mixture of xyloglucan from chia seeds and chitosan was prepared with the addition of glycerol by various stirring processes and controlled order of polymer addition. Hydrogel dressings were produced by combining both polymers and a subsequent lyophilization process. Martínez-Ibarra et al. showed in their study that the incorporation of xyloglucan has a positive effect on the resistance of chitosan and at the same time does not impair its antimicrobial properties. Therefore, the addition of a plant-based polymer would be another way to achieve suitable mechanical strength while maintaining a high capacity to absorb physiological fluids [37].

Hydrogels have the property of absorbing exudates perfectly and creating a moist environment, which accelerates the healing process of the wound and relieves the sensation of pain. The statin pitavastatin can positively influence this process when integrated into the matrix of the hydrogel. Statins have been shown to have an anti-inflammatory and wound-healing effect. Inflammatory reactions in the body can slow wound healing by delaying re-epithelialization and decreasing collagen formation. Pitavastatin has a tissue regenerative property and provides protection against bacterial infections through its antimicrobial action. A study by Eldeeb et al. investigates the fabrication of a bioactive wound dressing loaded with a pitavastatin nanovesicle system and composed of a three-dimensional nanocomposite alginate hydrogel. They use the thin film hydration technique to produce the nanoparticles, using sodium deoxycholate as a surfactant and various lipids and later processed into a calcium alginate hydrogel to form a nanocomposite hydrogel. A drug release of up to 3 days and the highest possible entrapment efficiency was shown by the formulation containing 0.1% sodium deoxycholate and 400 mg Lipoid S45. The studies showed that this biocompatible, bioactive, and biodegradable wound dressing is a safe and promising solution for wound healing [39].

It is not simple to produce a single hydrogel with all the desired functions. Liu et al. investigated a two-layer model for hydrogel wound dressings. The inner layer was a hybrid hydrogel prepared from chitosan-N-2-hydroxypropyl trimethylammonium chloride and N-isopropylacrylamide, and the outer layer was prepared by acrylamide and PVA. By photoinduced electron/energy transfer reversible addition-fragmentation chain transfer (PET-RAFT) polymerization, the two layers of the model could be covalently bonded and showed best interfacial strength [40].

4.3 Chitosan in Vaginal Delivery

The application of solid vaginal delivery forms poses some difficulties. Problematic application may occur due to the self-cleaning effect of the vagina and increased irritation, which may consequently lead to lower compliance. Due to the disadvantages just mentioned, it is all the more important to improve existing formulations or to develop new ones. This phenomenon of mucoadhesion is now the basis for the emergence of modern vaginal tablets. After many years of intensive research, minitablets, tablets, pessaries, and other dosage forms have been developed. These are characterized by prolonged residence time in the vagina, continuous release of active ingredients, and efficacy and convenience for patients. An overview of vaginal delivery systems based on chitosan-containing formulations such as liposomes, tablets, gels, films, and microparticles in the treatment of vaginal bacterial infections caused by *Chlamydia trachomatis* and aerobic vaginitis as well as fungal infections caused by *Candida albicans* is listed in Table 2.

All of this is achieved through the use of natural and synthetic polymers with distinct mucoadhesive properties [8]. Furthermore, chitosan is considered in the

Table 2 Chitosan-based vaginal formulations for the treatment of bacterial and fungal vaginal infections

Pathogen	API(s)	Polymer(s)	Formulation	References
	For bacterial infections			
Aerobic vaginitis	Azithromycin	Chitosan	Liposomes in hydrogel	Ana Čačić et al. [41]
Chlamydia trachomatis	Azithromycin	Chitosan	Liposomes in hydrogel	May Wenche Jøraholmen et al. [42]
	For yeast infections			
Candida albicans	Clotrimazole (CL)	Chitosan	Tablets	Szymanska et al. [43]
	Amphotericin, fluconazole	Chitosan, poloxamer	Liquid crystal precursor mucoadhesive system	Salmazi et al. [44]
	Econazole and miconazole nitrate	Chitosan	Gel	Senyigit et al. [45]
	Econazole nitrate	Chitosan, Na-CMC, poloxamers	Microparticles	Albertini et al. [46]
	Fluconazole	chitosan: pectin (75:25)	Film	Mishra et al. [47]

literature to be highly suitable for the preparation of mucoadhesive vaginal tablets. Szyman'ska et al. studied the properties of clotrimazole tablets prepared under three different conditions with different concentrations of chitosan. Porcine vaginal mucosa, gelatin discs, and mucin gel were used. The findings validated the adhesive characteristics of chitosan. Formulations with 40% and 25% of chitosan were identified as the most favorable options for additional research, as they exhibited extended adhesion to vaginal tissue and sustained release of clotrimazole [43]. Chitosan may exhibit specific alterations in its behavior. Baloglu et al. proposed the use of thiolated polymers to develop preparations containing miconazole nitrates and econazole in combination with thiolated poly (acrylic acid)-cysteine [48]. The tablets incorporating the produced polymer demonstrated a favorable water absorption rate and displayed mucoadhesive qualities. However, it is important to note that thiol groups may be susceptible to oxidation. Therefore, shielded S-chitosan was created through thiolation of the polymer and safeguarding the thiol groups from oxidation with the assistance of an aromatic ligand. In this scenario, NAC-6-MNA was linked to chitosan using a carbodiimide-mediated amide linkage. Protected S-chitosan displayed increased mucoadhesive characteristics, extended the release of metronidazole, which is an antibacterial and antiprotozoal agent, from vaginal tablets, and improved its antimicrobial effectiveness [49]. Given the importance of vaginal pH, we must also mention acid-buffering bioadhesive tablets for the treatment of vaginal infections. An elevation in pH levels can indicate or contribute to mixed vaginal infections. Tablets that incorporate sodium monocitrate as a buffering agent maintain a pH level of 4.4, a characteristic associated with a healthy vaginal environment. In addition to sodium monocitrate, these tablets also contain effective active ingredients such as metronidazole, clotrimazole, and Lactobacillus acidophilus spores to treat infections of the genitourinary tract. L. acidophilus is normally present in the healthy vagina and contributes to acidic pH. In another experiment carried out by Szymanska et al. in 2019, the different thicknesses of the microparticle encapsulations were examined. Microparticles (MP) were prepared by spray drying and differ in mass ratio of chitosan glutamate to zidovudine. For the first formulation, the encapsulation efficiency was $87.2 \pm 0.8\%$, the moisture content was $5.2 \pm 0.1\%$, and a viscosity mPas at 25°C was 170 mPas (MP1). The second formulation had an encapsulation efficiency of $79.0 \pm 1.2\%$, a moisture content of $5.0 \pm 0.2\%$ and a viscosity of 2,420 mPas (MP2). This experiment examined how strongly the different microparticles can adhere to vaginal mucosa. Influence of body movement on adhesive properties and strong mucosal retention is important for penetration into epithelial layer. MP1 was less adhesive MP1 = 280 μJ; MP2 = 450 μJ; placebo = 410 μJ (Wad = work of mucoadhesion expressed in microjoules) than expected but was due to the presence of drugs that interfered with interactions between mucoadhesive polymer and vaginal epithelium. Although placebo formulation was only included as a control, this formulation had the highest mucoadhesion strength, while the MP2 showed highest mucoadhesion work. This shows that encapsulated microparticles can control drug release to some extent and improve adhesion properties [50]

5 Summary

Chitosan is a unique naturally occurring polymer with outstanding properties per se such as low toxicity, biocompatibility, permeation enhancement, and mucoadhesive strength. The potential of its mucoadhesive platform is not fully reached so far. In this chapter, the last recent modification of chitosan was described as thiolation, preactivation, and maleic conjugation. Chitosan-based drug delivery systems such as nano-based microparticles or liposome; solid dosage forms such as tablets; semi-solids, e.g., hydrogels were chosen to show the power of chitosan in mucosal delivery.

References

1. Surendranath M, Rekha MR, Parameswaran R (2022) Recent advances in functionally modified polymers for mucoadhesive drug delivery. J Mater Chem B 10(31):5913–5924. https://doi.org/10.1039/d2tb00856d

2. Naik JB, Pardeshi SR, Patil RP et al (2020) Mucoadhesive micro-/Nano carriers in ophthalmic drug delivery: an overview. BioNanoScience 10(3):564–582. https://doi.org/10.1007/s12668-020-00752-y
3. Ways TMM, Lau WM, Khutoryanskiy VV (2018) Chitosan and its derivatives for application in Mucoadhesive drug delivery systems. Polymers 10(3). https://doi.org/10.3390/polym10030267
4. Armengol ES, Laffleur F (2021) The progress on sulfhydryl modified polymers with regard to synthesis, characterization and mucoadhesion. Int J Pharm 5:592. https://doi.org/10.1016/j.ijpharm.2020.120016
5. Netsomboon K, Jalil A, Laffleur F et al (2020) Thiolated chitosans: are Cys-Cys ligands key to the next generation? Carbohydr Polym 15:242. https://doi.org/10.1016/j.carbpol.2020.116395
6. Armengol ES, Grassiri B, Piras AM et al (2024) Ocular antibacterial chitosan-maleic acid hydrogels: and studies for a promising approach with enhanced mucoadhesion. Int J Biol Macromol 254. https://doi.org/10.1016/j.ijbiomac.2023.127939
7. de Carvalho ACW, Paiva NF, Demonari IK et al (2023) The potential of films as Transmucosal drug delivery systems. Pharmaceutics 15(11). https://doi.org/10.3390/pharmaceutics15112583
8. Osmalek T, Froelich A, Jadach B et al (2021) Recent advances in polymer-based vaginal drug delivery systems. Pharmaceutics 13(6). https://doi.org/10.3390/pharmaceutics13060884
9. Valamla B, Thakor P, Phuse R et al (2022) Engineering drug delivery systems to overcome the vaginal mucosal barrier: current understanding and research agenda of mucoadhesive formulations of vaginal delivery. J Drug Deliv Sci Tec. https://doi.org/10.1016/j.jddst.2022.103162
10. Jawadi Z, Yang CSE, Haidar ZS et al (2022) Bio-inspired Muco-adhesive polymers for drug delivery applications. Polymers 14(24). https://doi.org/10.3390/polym14245459
11. Cazorla-Luna R, Martín-Illana A, Notario-Pérez F et al (2021) Naturally occurring polyelectrolytes and their use for the development of complex-based Mucoadhesive drug delivery systems: an overview. Polymers 13(14). https://doi.org/10.3390/polym13142241
12. Martín-Illana A, Chinarro E, Cazorla-Luna R et al (2022) Optimized hydration dynamics in mucoadhesive xanthan-based trilayer vaginal films for the controlled release of tenofovir. Carbohydr Polym:278. https://doi.org/10.1016/j.carbpol.2021.118958
13. Markl D, Zeitler JA (2017) A review of disintegration mechanisms and measurement techniques. Pharm Res 34(5):890–917. https://doi.org/10.1007/s11095-017-2129-z
14. Juban A, Nouguier-Lehon C, Briancon S et al (2015) Predictive model for tensile strength of pharmaceutical tablets based on local hardness measurements. Int J Pharm 490(1–2):438–445. https://doi.org/10.1016/j.ijpharm.2015.05.078
15. Torisu T, Maruno T, Yoneda S et al (2017) Friability testing as a new stress-stability assay for biopharmaceuticals. J Pharm Sci 106(10):2966–2978. https://doi.org/10.1016/j.xphs.2017.05.035
16. Abidin IZ, Murphy EJ, Fehrenbach GW et al (2023) A systematic review of mucoadhesive vaginal tablet testing. Drug Target Insight 17:5–30. https://doi.org/10.33393/dti.2023.2477
17. Bartoníková K, Spaglová M, Papadakos M et al (2024) Mucoadhesive vaginal tablets containing metronidazole: screening of optimal natural polymer in the composition. Sci Pharm 92(1). https://doi.org/10.3390/scipharm92010010
18. Jussen D, Sharma S, Carson JK et al (2020) Preparation and tensile properties of guar gum hydrogel films. Polym Polym Compos 28(3):180–186. https://doi.org/10.1177/0967391119867560
19. Szulc-Musiol B, Siemiradzka W, Dolinska B (2023) Formulation and evaluation of hydrogels based on sodium alginate and cellulose derivatives with quercetin for topical application. Appl Sci 13(13). https://doi.org/10.3390/app13137826
20. Zhang K, Feng WX, Jin CR (2020) Protocol efficiently measuring the swelling rate of hydrogels. Method:7. https://doi.org/10.1016/j.mex.2019.100779
21. Seyfarth H (1998) Test for sterility according to the regulations of the European Pharmacopoeia/Supplement 1998. Pharm Ind 60(12):1073–1083

22. Notario-Pérez F, Cazorla-Luna R, Martín-Illana A et al (2020) Design, fabrication and characterisation of drug-loaded vaginal films: state-of-the-art. J Control Release 327:477–499. https://doi.org/10.1016/j.jconrel.2020.08.032
23. Dolci LS, Albertini B, Di Filippo MF et al (2020) Development and evaluation of mucoadhesive gelatin films for the vaginal delivery of econazole. Int J Pharm:591. https://doi.org/10.1016/j.ijpharm.2020.119979
24. Aranaz I, Alcántara AR, Civera MC et al (2021) Chitosan: an overview of its properties and applications. Polymers 13(19). https://doi.org/10.3390/polym13193256
25. Mura P, Maestrelli F, Cirri M et al (2022) Multiple roles of chitosan in mucosal drug delivery: an updated review. Mar Drugs 20(5). https://doi.org/10.3390/md20050335
26. Desai N, Rana D, Salave S et al (2023) Chitosan: a potential biopolymer in drug delivery and biomedical applications. Pharmaceutics 15(4). https://doi.org/10.3390/pharmaceutics15041313
27. Riccio BVF, Silvestre ALP, Meneguin AB et al (2022) Exploiting polymeric films as a multipurpose drug delivery system: a review. AAPS PharmSciTech 23(7). https://doi.org/10.1208/s12249-022-02414-6
28. Gao X, Dong D, Zhang C et al (2024) Chitosan-functionalized poly(beta-amino ester) hybrid system for gene delivery in vaginal mucosal epithelial cells. Pharmaceutics 16(1). https://doi.org/10.3390/pharmaceutics16010154
29. Armengol ES, Knoll P, Hupfauf A et al (2023) Chitosan-maleic acid conjugate as potential excipient candidate for oral drug delivery? Future Med Chem 15(3):241–253. https://doi.org/10.4155/fmc-2022-0258
30. Lynch C, Kondiah PPD, Choonara YE et al (2019) Advances in biodegradable nano-sized polymer-based ocular drug delivery. Polymers 11(8). https://doi.org/10.3390/polym11081371
31. Rathod S (2023) Interpenetrating polymeric network (IPNs) in ophthalmic drug delivery. Int Ophthalmol 43(3):1063–1074. https://doi.org/10.1007/s10792-022-02482-4
32. Allyn MM, Luo RH, Hellwarth EB et al (2022) Considerations for polymers used in ocular drug delivery. Front Med (Lausanne):8. https://doi.org/10.3389/fmed.2021.787644
33. Arribada RG, Behar-Cohen F, De Barros ALB et al (2022) The use of polymer blends in the treatment of ocular diseases. Pharmaceutics 14(7). https://doi.org/10.3390/pharmaceutics14071431
34. Su Y, Fan XQ, Pang Y (2023) Nano-based ocular drug delivery systems: an insight into the preclinical/clinical studies and their potential in the treatment of posterior ocular diseases. Biomater Sci 11(13):4490–4507. https://doi.org/10.1039/d3bm00505d
35. Fan LH, Yang H, Yang J et al (2016) Preparation and characterization of chitosan/gelatin/PVA hydrogel for wound dressings. Carbohydr Polym 146:427–434. https://doi.org/10.1016/j.carbpol.2016.03.002
36. Ferreira MOC, de Lima IS, Morais AIS et al (2019) Chitosan associated with chlorhexidine in gel form: synthesis, characterization and healing wounds applications. J Drug Deliv Sci Technol 49:375–382. https://doi.org/10.1016/j.jddst.2018.12.003
37. Martínez-Ibarra DM, Sánchez-Machado DI, López-Cervantes J et al (2019) Hydrogel wound dressings based on chitosan and xyloglucan: development and characterization. J Appl Polym Sci 136(12). https://doi.org/10.1002/app.47342
38. Bilgic MB, Lacin NT, Berber H et al (2019) In vitro evaluation of alpha-tocopherol loaded carboxymethylcellulose chitosan copolymers as wound dressing materials. Mater Technol 34(7):386–393. https://doi.org/10.1080/10667857.2019.1573944
39. Eldeeb AE, Salah S, Amer MS et al (2022) 3D nanocomposite alginate hydrogel loaded with pitavastatin nanovesicles as a functional wound dressing with controlled drug release; preparation, and evaluation. J Drug Deliv Sci Tecnol. https://doi.org/10.1016/j.jddst.2022.103292
40. Liu JZ, Miao JK, Zhao L et al (2022) Versatile bilayer hydrogel for wound dressing through PET-RAFT polymerization. Biomacromolecules 23(3):1112–1123. https://doi.org/10.1021/acs.biomac.1c01428

41. Cacic A, Klaric DA, Keser S et al (2023) A novel approach for the treatment of aerobic vaginitis: azithromycin liposomes-in-chitosan hydrogel. Pharmaceutics 15(5). https://doi.org/10.3390/pharmaceutics15051356
42. Joraholmen MW, Johannessen M, Gravningen K et al (2020) Liposomes-in-hydrogel delivery system enhances the potential of resveratrol in combating vaginal chlamydia infection. Pharmaceutics 12(12). https://doi.org/10.3390/pharmaceutics12121203
43. Szymanska E, Winnicka K, Amelian A et al (2014) Vaginal chitosan tablets with clotrimazole-design and evaluation of mucoadhesive properties using porcine vaginal mucosa, mucin and gelatine. Chem Pharm Bull 62(2):160–167. https://doi.org/10.1248/cpb.c13-00689
44. Salmazi R, Calixto G, Bernegossi J et al (2015) A curcumin-loaded liquid crystal precursor mucoadhesive system for the treatment of vaginal candidiasis. Int J Nanomedicine 10:4815–4824. https://doi.org/10.2147/Ijn.S82385
45. Senyigit ZA, Karavana SY, Eraç B et al (2014) Evaluation of chitosan based vaginal bioadhesive gel formulations for antifungal drugs. Acta Pharm 64(2):139–156. https://doi.org/10.2478/acph-2014-0013
46. Albertini B, Passerini N, Di Sabatino M et al (2009) Polymer-lipid based mucoadhesive microspheres prepared by spray-congealing for the vaginal delivery of econazole nitrate. Eur J Pharm Sci 36(4–5):591–601. https://doi.org/10.1016/j.ejps.2008.12.009
47. Mishra R, Soni K, Mehta T (2017) Mucoadhesive vaginal film of fluconazole using cross-linked chitosan and pectin in vitro and in vivo study. J Therm Anal Calorim 130(3):1683–1695. https://doi.org/10.1007/s10973-017-6402-5
48. Baloglu E, Senyigit ZA, Karavana SY et al (2011) Evaluation of Mucoadhesive vaginal tablets of antifungal drugs prepared with Thiolated polymer and development of a new dissolution technique for vaginal formulations. Chem Pharm Bull 59(8):952–958. https://doi.org/10.1248/cpb.59.952
49. Lupo N, Fodor B, Muhammad I et al (2017) Entirely S-protected chitosan: a promising mucoadhesive excipient for metronidazole vaginal tablets. Acta Biomater 64:106–115. https://doi.org/10.1016/j.actbio.2017.10.014
50. Szymanska E, Krzyzowska M, Cal K et al (2021) Potential of mucoadhesive chitosan glutamate microparticles as microbicide carriers – antiherpes activity and penetration behavior across the human vaginal epithelium. Drug Deliv 28(1):2278–2288. https://doi.org/10.1080/10717544.2021.1992037

Adv Polym Sci (2025) 295: 161–182
https://doi.org/10.1007/12_2024_184

Published online: 12 December 2024

Chitosan Films and Membranes in Drug Delivery

V. V. Arya Lakshmi, Madhumitha Shanmuganathan, G. Shreeman Raj, and R. Jayakumar

Contents

1 Introduction 162
2 Preparation of Chitosan Films and Membranes 163
2.1 Electrospinning 163
2.2 Solvent Casting 164
2.3 Phase Inversion 165
2.4 Thermo-compression Molding 165
2.5 Extrusion 166
2.6 Layer-by-Layer Assembly 166
3 Applications of Chitosan Films/Membranes in Drug Delivery 167
3.1 Antibacterial 167
3.2 Anticancer 169
3.3 Antifungal 171
3.4 Anti-Inflammatory 173
3.5 Protein and Peptide 174
3.6 Nucleic Acid 176
4 Conclusion 178
References 178

Abstract Chitosan, a biocompatible and biodegradable polymer obtained from chitin, has therapeutic properties that attract it more to drug supplies. Chitosan can be fabricated into various forms such as nanoparticles, microparticles, hydrogels, beads, nanofibers, films, membranes, and scaffolds. Chitosan thin films and membranes could be produced by multiple methods, like solvent casting and electrospinning; this will increase their suitability for medical use. Chitosan films

V. V. Arya Lakshmi, M. Shanmuganathan, G. Shreeman Raj, and R. Jayakumar (✉)
Polymeric Biomaterials Lab, School of Nanosciences and Molecular Medicine, Amrita Vishwa Vidyapeetham, Kochi, India
e-mail: rjayakumar@aims.amrita.edu

and membranes could deliver various therapeutic drugs like antibiotics, anti-inflammatory drugs, antifungal agents, anticancer compounds, proteins, peptides, and nucleic acids. Its antibacterial and anti-inflammatory properties enhance therapeutic effects while minimizing side effects. This chapter shows various applications of chitosan films and membranes in drug delivery highlighting their potential to improve their performance.

Keywords Antibacterial · Anticancer · Anti-inflammatory · Chitosan · Drug delivery · Films · Membranes · Protein and peptide delivery

1 Introduction

Drug delivery systems (DDS) are used to deliver drugs to certain tissues, organs, cells, and subcellular compartments, facilitating drug release and absorption through various drug carriers [1]. The concept of this controlled release of medications within the human body becomes a main objective in fabricating advanced drug delivery systems. The major requirement for making these delivery vehicles is to ensure pharmaceutical drugs are delivered in a pharmacokinetically controlled manner [2] and also to overcome issues like limited solubility, drug aggregation, low bioavailability, poor biodistribution, lack of selectivity, or to reduce side effects comes with therapeutic drugs [1]. It also helps to reduce the drug dose and frequency, lowering plasma level fluctuations, and thus lowering side effects, enhancing the therapeutic index and patient adherence [3]. Polymers are employed as drug delivery systems for encapsulating and releasing drugs [4]. Even though a range of synthetic polymers can address the aforementioned conditions, its toxicity and high cost have ultimately obstructed their role [2]. As a result, natural polymers are often preferred for designing drug delivery systems due to their biocompatibility, biodegradability, non-toxicity and the presence of biologically recognizable elements that support cellular processes, making them a more attractive option compared to synthetic polymers [4]. One example of a natural polymer obtained from crustaceans is chitosan (Fig. 1) [2]. Chitosan has attracted attention because of its non-toxic, biocompatible, biodegradable, stable, and sterilizable nature. It also offers several

Fig. 1 Chemical structure of chitosan

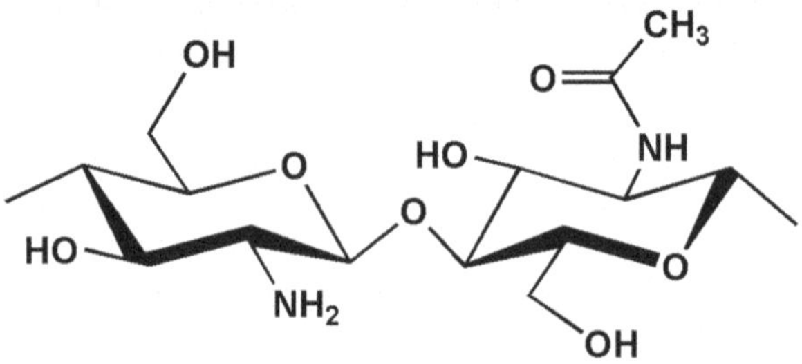

characteristics such as controlled drug release rate, cross-linking ability with other polymers, antimicrobial activity, gel-forming property, adhesion, stimulation of the immune system, activation of macrophages, and gas permeability [4]. Chitosan can be derived from chitin through a partial N-deacetylation process [5]. Chitosan contains units of N-acetyl-D-glucosamine and 2-amino-2-deoxy-β-D-glucopyranose units [6]. Chitosan's properties and applications are controlled by the degree of deacetylation and molecular weight [7]. Chitosan's functionality is affected by its amino group, which has a positive charge, and its hydroxyl group, which carries a negative charge. The amino group is rapidly protonated and enhances chitosan's solubility in acidic aqueous solutions, while the primary hydroxyl group allows for the formation of branched or copolymers, affecting its mechanical properties, and the secondary hydroxyl group can be used to enhance the solubility of chitosan by substitution [8]. Due to their exceptional qualities, chitin and chitosan are widely utilized in various biomedical applications. This is possible because chitosan may be processed into a variety of shapes, including capsules, hydrogels, microsponges, particles, films, and membranes [9]. Drug release from chitosan-based forms is influenced by the polymer's physicochemical properties, such as its bioadhesion to mucin or skin, swelling property, gel-forming properties in different body fluids with varying pH and ion concentrations, and also the presence of copolymers and excipients [5]. Chitosan nanofilms have a large surface area to volume ratio, allowing them for the effective release of drugs. Chitosan membranes, which are thin and flexible, can be made to specific sizes and thus can be used to deliver drugs directly into biological settings [10]. Thus, this chapter concentrates on the preparations, properties, and various drug delivery applications of chitosan films and membranes.

2 Preparation of Chitosan Films and Membranes

Chitosan films and membranes can be processed using different techniques (Fig. 2). These strategies allow the modification of physical and chemical properties of chitosan films and membranes [11].

2.1 Electrospinning

The electrospinning process requires a strong electric potential for two electrodes with opposing charges [12]. In the process, a homogeneous solution is driven by a fine needle with a higher voltage, which results in the development of a Taylor cone and the fabrication of nonwoven fiber films on the plate. Optimization of electrospinning parameters is an essential parameter for the development of chitosan films and membranes. These variables include the voltage applied, the flow rate of solution, the type of solvent used, and the needle–collector gap. The high surface

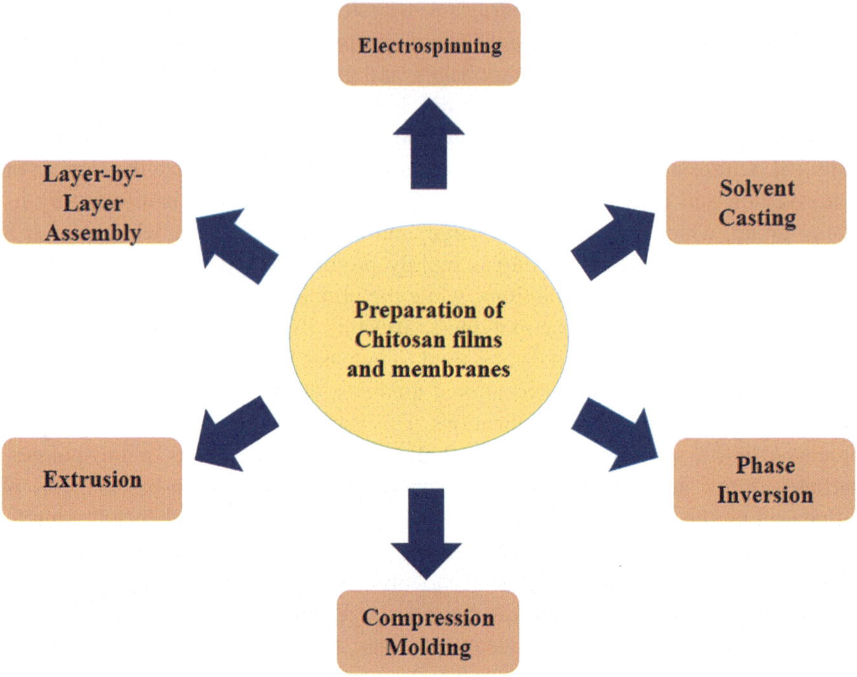

Fig. 2 Preparation methods for chitosan films and membranes

area-to-volume ratio of chitosan nanofilm mats makes them particularly attractive for drug delivery systems, as it significantly enhances their capacity for drug loading [13]. Nanofiber membranes based on pectin and chitosan, incorporated with dihydromyricetin for drug delivery application in wound healing, were developed. The electrospun biomimetic nanofiber membranes exhibited antioxidant and anti-inflammatory effects from the dihydromyricetin delivery and showed enhanced wound healing [14]. Poly (vinyl alcohol) (PVA)/chitosan nanofibrous films incorporated with ciprofloxacin and rutin hydrate were fabricated using an electrospinning method for treating infectious wounds. The study emphasized the successful fabrication of dual drug-loaded nanofibers via the electrospinning technique and highlighted the optimal conditions needed for effective electrospinning [15].

2.2 *Solvent Casting*

The solvent casting method is a commonly utilized technique on account of its simplicity and reasonably affordable. This procedure involves dissolving a polymer and a plasticizer in an evaporative solvent. If a drug is present, it can be suspended or dissolved in the solution, which is then poured into a mold and dried [16]. The solvent casting method allows for the modification of the film's properties by changing the production conditions. This adaptability makes it a reliable and popular choice among many processes for the preparation of polymer films and membranes [17]. Supernak et al. developed a chitosan membrane with different molecular weights as an antibacterial membrane with gentamicin using the solvent casting method. The prepared membranes showed antibacterial activity and did not show any cytotoxic effects on osteoblast or fibroblast cells, indicating their potential for use in various medical applications, including as drug carriers to aid in drug delivery [18].

2.3 *Phase Inversion*

Phase inversion is a highly versatile method for creating polymeric membranes with various morphologies for different applications. The process involves converting a polymer from a solution to a solid state through methods such as controlled evaporation, thermal precipitation, or primarily, immersion precipitation, where a polymer solution is cast on a support and then immersed in a nonsolvent bath, leading to membrane formation through solvent-nonsolvent exchange and phase separation [19]. PVA/CS asymmetrical membranes are excellent wound dressing material for skin regeneration. In another study, PVA/CS membranes were developed using the phase inversion method. This technique enables precise adjustment of processing conditions, leading to the development of highly porous materials with enhanced morphology, mechanical properties, and hydrophilicity [20].

2.4 *Thermo-compression Molding*

An alternative strategy for preparing more sustainable chitosan films/membranes involves using thermo-compression molding [21]. Chitosan films were prepared via thermo-compression molding using chitosan powder, acetic acid, and polyols as plasticizers [22]. This process allowed for faster production by limiting the use of organic solvents, as the biopolymer and acid were mixed in their solid condition. Chitosan films were developed using a new method that avoids organic solvents. This study uses thermomechanical procedures and solid-state mixing of chitosan and citric acid to develop cross-linked films via compression molding. This approach uses low-melting combinations, which eliminates the need for solvents like acetic acid and demonstrates a greater advancement in film preparation [21].

2.5 Extrusion

The extrusion process avoids the need for huge amounts of potentially hazardous solvents and produces improved results than casting. The extrusion process consists of several steps: (i) preparing the formulation and mixing raw materials in specific proportions, (ii) blending the constituents used in film production, (iii) reducing the size of the extrudates with a pelletizer, (iv) evaporation of the pellets, (v) transferring the pellets into another extruder to form sheets, and (vi) using a blown film extruder to combine the layers into a single film. Among the various extrusion techniques, melt extrusion is particularly effective, producing films with excellent mechanical properties, good thermal stability, high transparency, and strong antimicrobial activity. This method reduced production times and eliminated the use of organic solvents by combining the biopolymer and acid in their solid states [11]. Chitosan and poly (3-hydroxypropyl ethyleneimine) elastic films were developed for delivering haloperidol using the extrusion technique. The films were prepared to enhance their mechanical properties for efficient buccal delivery of haloperidol [23].

2.6 Layer-by-Layer Assembly

Layer-by-layer (LbL) assembly is a method used to develop composite films [24]. It involves layering materials with opposite charges on a substrate in stages via electrostatic and other interactions. This technique helps in the management of film parameters such as porosity, weight, and thickness [25]. Dipping, spraying, spinning, and brushing are some of the production techniques for layer-by-layer (LbL). Dipping is a method in which the film preparation method involves immersion of an ionically charged species in a polyelectrolyte solution. After the application of each layer, the substrate is rinsed to get rid of excess polyelectrolytes. This process is repeated to achieve the appropriate number of layers [26]. The spray LbL method involves spraying polyelectrolyte solutions onto a vertical substrate. It is a more efficient and versatile technique than dipping since it can coat small areas. Films produced by this method are generally smoother and thinner [27, 28]. The spin LbL method utilizes a rotating substrate to apply and spread polyelectrolyte solutions, forming multilayers through repeated addition, drying, and washing. It provides advantages such as rapid production, uniform and organized layers, and precise control over film thickness. This method also affects film morphology more significantly with small molecules due to their greater mobility [29, 30]. In the layer-by-layer brushing method, a substrate is attached to a supporting wall and coated using a polyelectrolyte solution, which is then rinsed to develop layers. Such an approach speeds up the production of films, minimizes wastage of raw materials, and is more flexible compared to static dipping methods [25].

3 Applications of Chitosan Films/Membranes in Drug Delivery

3.1 Antibacterial

Chitosan exhibits antibacterial activity, making it suitable for biomedical applications. Chitosan films/membranes can be used for localized drug delivery. Films derived from chitosan have antibacterial properties and are applied to wounds to prevent infections and enhance healing. Chitosan films can be used as drug delivery systems, enabling the controlled release of therapeutic compounds such as antibiotics [31]. Chitosan's cationic nature at acidic pH allows it to interact with anionic surfaces, such as those located in the gastrointestinal tract, enhancing its utility in drug delivery systems. Chitosan works by binding to the negatively charged bacterial membrane of bacteria and changing its permeability leading to disruption of cell membrane, leakage of cell components, and eventually cell death. Chitosan also binds to bacterial DNA which inhibits transcription thus inhibiting bacterial growth. Chitosan with a higher molecular weight can develop a polymer layer on the cell membrane, restricting the entry of nutrients while chitosan with a lower molecular weight can penetrate the cell through permeation. When chitosan adsorbs electronegative material and then enters the cell, it flocculates, disrupting the physiological processes and ultimately killing the bacteria. Additionally, at low concentrations, chitosan adheres to the surface of the cell and disrupts it by allowing intracellular components to leak out. At high concentrations, protonated chitosan coats the cell surface and stops intracellular components from leaking [32].

Ciprofloxacin-loaded chitosan (CS) and chitosan-depolymerization-products (CS/CDP)-based films were prepared using casting technique. Antibacterial activity of ciprofloxacin-loaded chitosan films and chitosan-depolymerization-products (CDP) is studied using the agar diffusion method. Culture suspension of gram-negative pathogenic bacterial strains and gram-positive pathogenic strains were spread over Muller-Hinton agar medium plates. The gram-negative bacteria especially *S. enterica* and *P. aeruginosa* were more inhibited than the gram-positive strains. The CDP addition notably improved the antibacterial properties of CS/CDP films as compared to chitosan films. The drug release profile of CS films and CS/CDP films were studied in PBS solution. More sustained release of ciprofloxacin observed in CS/CDP-based films than CS-based film is because the higher thickness of the films can act as a diffusion barrier in which CFX has to pass a greater distance to cross the matrix of the polymer. Based on the drug release profiles CS/CDP film matrices showed potential as effective drug delivery systems, providing a sustained release of CFX and holding promise for biomedical applications (Fig. 3) [33].

Chitosan has antibacterial and anti-inflammatory properties that may work in combination with the drugs that are being administered [31]. Multilayer membranes made of chitosan, alginate, and silk fibroin incorporated with diclofenac sodium (DS) were developed using a layer-by-layer assembly method. The drug release profile showed a 70% of the DS was released into the simulated exudate fluid media

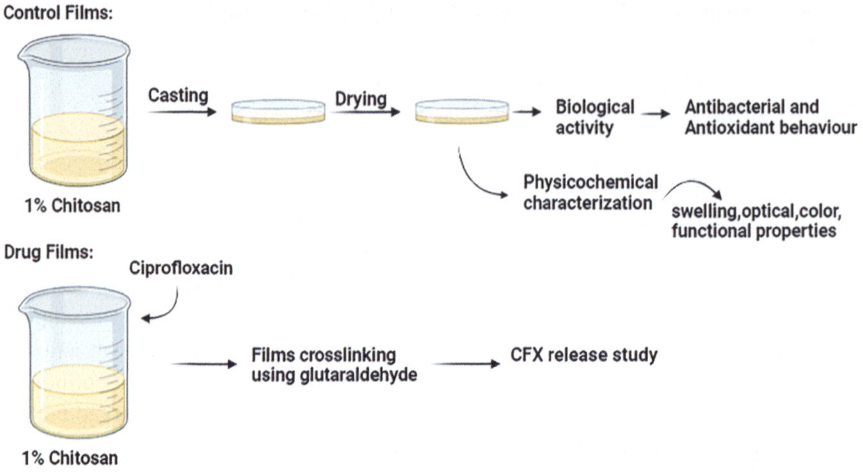

Fig. 3 Preparation and properties of CS-CFX film and CFX-loaded CS/CDP-based film

in a controlled manner. The results suggested that the multilayer membrane provided increased resistance to diffusion, resulting in slower and more controlled release of DS. The microbial permeation study indicated that the positive control exhibited turbidity, indicating that the medium could promote microbial growth while the developed membranes did not show turbidity. This suggests that the membranes hold significant potential as wound dressings by preventing microbial infections [32]. Nanofibrous membranes containing a combination of chitosan, poly (vinyl alcohol) (PVA), and graphene oxide (GO) are prepared and antibiotics such as ciprofloxacin (Cip) and ciprofloxacin hydrochloride (Cip.HCL) are loaded into the nanofibers using electrospinning technique. The GO addition reduced antibacterial activity because the amino groups on the chitosan chains reacted with the epoxy and carboxyl groups on the GO nanosheets but after administering the antibiotic drugs, the nanofibrous membranes demonstrated noticeably increased antibacterial activity. The GO nanosheets with a combination of chitosan regulated the drug release profile, preventing large release initially and the addition of GO enhanced the drug release rate. Thus results suggest that the prepared nanofibrous membranes can act as a potential candidate for wound dressing [34]. Chitosan-based films used in wound dressings can stop infection by preventing bacteria from growing on the surface of the wound. This lowers the risk of sepsis and other issues and promotes healing. Chitosan-based films and membranes can also be made to specifically target and destroy bacterial biofilms, which are frequently resistant to conventional antibiotics [35]. Moxifloxacin-incorporated chitosan-pullulan-silver films were prepared for antibacterial drug delivery and characterized. The antibiotic moxifloxacin was effectively encapsulated in chitosan-pullulan films that had been added with silver nanoparticles. The films were created using a casting method, allowing for uniform distribution of the drug. The presence of incorporated silver contributed to the antibacterial activity of the film. Characterization studies indicated that the films had good tensile strength and swelling ratios. According to the drug release studies, moxifloxacin was released from the films under controlled conditions, and a sustained release pattern was seen over time. Thus, the developed nanocomposite film exhibits substantial potential as a biomaterial for drug delivery applications with enhanced antibacterial activity [36].

3.2 *Anticancer*

Cancer is still a major worldwide health concern, requiring novel strategies for efficient drug delivery. Chitosan can adhere to mucosal surfaces and improve the absorption of drugs. Its cationic nature facilitates interactions with negatively charged cell membranes. Chitosan has the capacity to directly suppress the proliferation of tumor cells through the process of apoptosis. By lowering glycolysis, chitosan inhibits the growth of tumors by lowering the amount of ATP and glucose taken in by the tumor cells. It has recently been discovered that chitosan causes liver cancer cells to break down during necrosis by reducing mitochondrial membrane potential, inducing lipid peroxidation, and neutralizing cell surface charge. The mechanism of action of chitosan membranes in anticancer drug delivery is shown in Fig. 4 [37].

Fluorouracil (5-FU) is a cancer antimetabolite drug often employed in chemotherapy procedures. 5-FU deals with vital biosynthetic activity and causes cytotoxicity in tumor cells by blocking thymidylate synthase (TS) and incorporating its

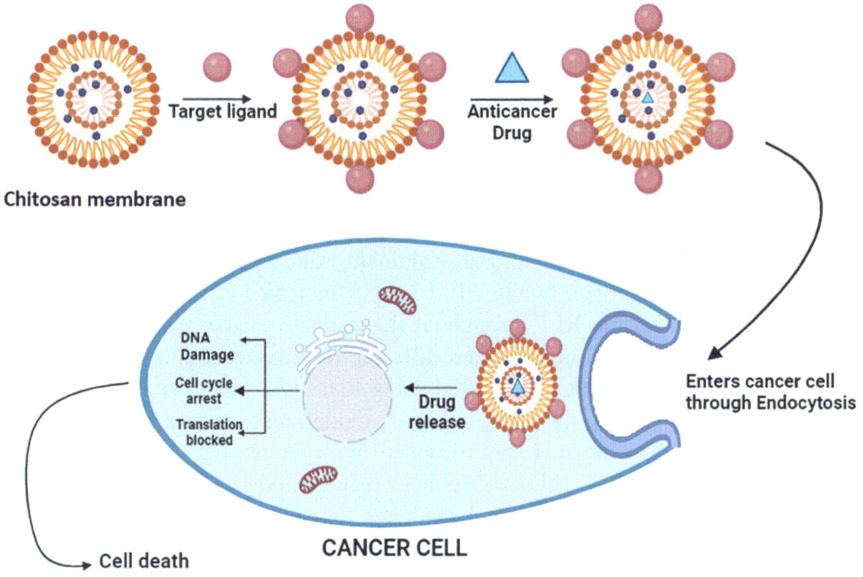

Fig. 4 The mechanistic action of chitosan membranes in anticancer drug delivery

byproducts into RNA or DNA. A chitosan (CS)/montmorillonite (MMT) nanocomposite film containing the anticancer drug 5-fluorouracil was produced using the solvent-evaporation technique. It has been reported that montmorillonite enhances film stability and efficiently enhances the activity of anticancer drugs. The chitosan/montmorillonite film can be applied topically to skin lesions that allows the controlled release of drugs at the tumor site. It has been shown that the chitosan/montmorillonite nanocomposite film has a consistent, porous structure that is advantageous for drug loading and release. 5-FU in the film and the incorporation of montmorillonite into the chitosan matrix are characterized using FTIR and XRD. TGA analysis showed that the nanocomposite film has greater heat stability, which is crucial for its use in drug delivery applications. The developed nanocomposite exhibited a controlled release of 5-FU in the drug release studies. Thus the developed CS nanocomposite film can act as a suitable alternative biomaterial for treating skin cancer [38]. For cancer treatments, chitosan membranes incorporated with pentasodium tripolyphosphate (TPP) and 1, 4-naphthoquinone (NQ14) were developed. The drug release rate of NQ14 depends on the concentration of TPP. A higher TPP concentration resulted in a reduction in NQ14 release from crosslinked chitosan membranes. Hence, varying the TPP concentration could transform crosslinked chitosan membranes into a major delivery system for managing cancer treatment [39]. Traditional drug delivery systems have lots of limitations like uncontrolled drug release, biotoxicity, etc. To overcome these limitations, a drug carrier that contains waterborne polyurethane (WPU) anddoxorubicin-loaded chitosan composite membrane were prepared. In this study, a three-dimensional tumor-targeting drug delivery system called Fe_3O_4@UIO-66-NH_2/graphdiyne (FUGY) that is made up of two components, i.e., graphdiyne (GDY) and a metal-organic framework (MOF) structure called Fe_3O_4@UIO-66-NH_2 (FU). This FUGY acts as a carrier for anticancer drugs. FUGY contains high drug-loading content (around 43.8%). By using fluorescence imaging, it is depicted that FUGY showed greater and more controlled drug delivery than the traditional drug delivery systems. FUGY with chitosan has high biocompatibility that minimizes side effects and treats tumor [40]. Biodegradable nanocomposite films were prepared using chitosan-alginate loaded with the anticancer drug Ginsenoside Rb1. Ginsenoside Rb1 is derived from the herbaceous plant Panax notoginseng and is a member of the dammaran saponin group. FT-IR analysis demonstrated that chitosan, alginate, and Ginsenoside Rb1 were interconnected via hydrogen bonds. SEM images indicated effective dispersion of chitosan and Ginsenoside Rb1 throughout the alginate matrix. Drug release was studied using various pH solutions and differing drug concentrations. The results showed an initial burst release (analyzed using first-order kinetics) and a constant release later (studied using the Higuchi kinetic model) also the rate of drug release increased with increasing pH and dropped with increasing concentration of Ginsenoside Rb1. Thus, chitosan/alginate nanocomposite films loaded with ginsenoside Rb1 can act as a potential candidate for cancer treatment [41].

3.3 Antifungal

Chitosan films and membranes have shown promising results in delivering antifungal agents. The antifungal activity of chitosan can be attributed to its ability to cause permeability and cell death by disrupting the fungal cell membranes. Chitosan films have demonstrated innate antifungal activities, which are influenced by parameters such as molecular weight and formulation. Chitosan's positive charge increases the effectiveness of its antifungal agent by facilitating electrostatic interactions with the negatively charged components of fungal cells [10]. By using the film casting method, chitosan films loaded with fluconazole (FZ)-loaded solid lipid nanoparticles (FZ-SLNs) were prepared (Fig. 5). They were characterized using ATR-FTIR which indicated the presence of the FZ-SLNs in the chitosan matrix and FESEM analysis revealed that presence of FZ-SLNs embedded within the chitosan structure. The antifungal properties of the C-FZ films were tested using *Candida albicans* and exhibited zone of inhibition. Permeation studies on porcine buccal mucosa revealed that FZ from the C-FZ films had a higher permeability, and the FZ-SLNs films demonstrated improved mechanical properties along with enhanced drug release. This study proved that FZ-SLN-containing chitosan films are a potentially effective drug delivery system for treating candidiasis [42].

Trimethyl-functionalized chitosan with polyethersulfone (PES) was fabricated using the phase inversion method. The antifungal property of the membranes was studied against *Aspergillus niger*, where it showed 72% antifungal activity against *A. niger*. The reason behind this is the presence of quaternary ammonium groups that help in the breakdown of fungal cell membranes [43]. Buccal films containing clotrimazole were effectively developed by the mucoadhesive excipients chitosan and alginate, along with the incorporation of graphene oxide (GO) through the solvent casting technique. The incorporation of GO into the films modified the electrostatic interactions between chitosan and alginate, leading to a rapid release of clotrimazole from the films. In agar diffusion assays, the buccal films containing

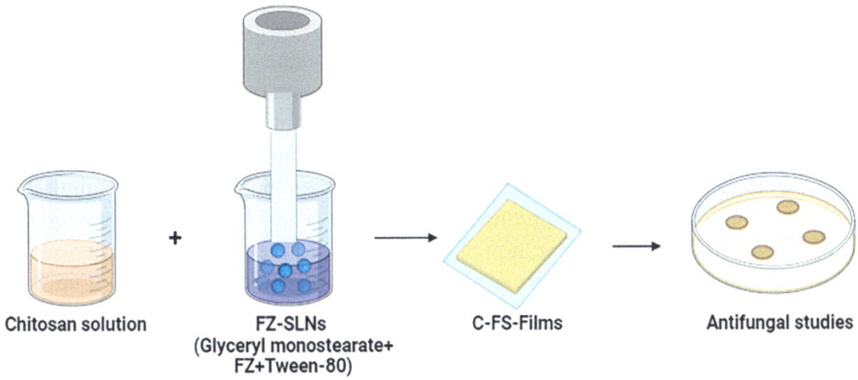

Fig. 5 Synthesis of FZ-SLNs chitosan films

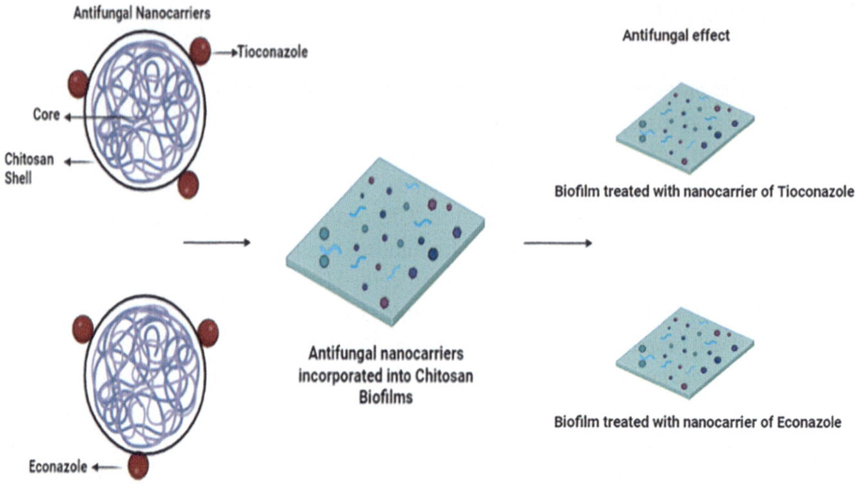

Fig. 6 Image showing solvent casting method to synthesize chitosan films

GO demonstrated increased antifungal activity against *C. albicans*. The films showed notable inhibitory zones, suggesting that the use of graphene oxide improved clotrimazole's ability to fight fungal infections but also facilitated its release [44]. A novel nanocapsule carrier coated with chitosan films was prepared to deliver Tioconazole and Econazole (Fig. 6) and characterized for their physicochemical properties. SEM is used for analyzing morphology, and UV-Vis spectrophotometry measures drug loading capacity, demonstrating a sustained release of the drug. This developed film could offer the controlled drug release benefits of nanocapsules while retaining the properties of a film. Films containing these antifungal nanocapsules might therefore be loaded as a pharmaceutical dosage form and used to treat vaginal candidiasis. Zeta potential studies showed that the membranes are positively charged which indicates enhanced mucoadhesion properties. The antifungal activity was done against *C. albicans* and *A. niger*. A larger zone of inhibition is seen in chitosan-based membrane than in free drugs [45].

Nanofibers made from chitosan (CS), poly (ethylene oxide) (PEO), cellulose nanocrystals (CNC), and acacia plant extract were prepared using electrospinning technique for wound healing applications. Electrospinning chitosan/nanocellulose fibers involve the process in which chitosan and nanocellulose are dissolved to make a homogenous solution. An electric field is applied when the homogenous solution is subjected to electrospinning, producing fine fibers. The fibers were characterized using SEM and FT-IR. Incorporation of CNC enhanced the physical integrity, water vapor transmission rate, and thermal properties of the nanofibers. To determine the efficacy of the fibers, procedures such as disk diffusion assay and minimum inhibitory concentration (MIC) were done against *C. albicans*. As a result, the MIC values were low indicating that chitosan/nanocellulose fibers showed enhanced antifungal activity than pure chitosan. The acacia addition to the optimal CS/PEO/CNC formulation significantly enhanced antifungal activity, emphasizing its effectiveness in preventing or treating infections. The acacia extract showed a sustained release over 24 h, reflecting its controlled action, which supports the wound healing process. Thus, the developed nanofibers system is highly promising as a localized drug delivery system for wound care [46].

3.4 Anti-Inflammatory

Anti-inflammatory drug delivery systems are gaining importance in recent years as there is a dire need for treating various inflammatory diseases. Chitosan has become a viable biomaterial for developing films and membranes for delivering anti-inflammatory medications. The inherent antimicrobial, anti-inflammatory, and antioxidative properties of chitosan enhance the effect of the anti-inflammatory drug incorporated in it. The chitosan films can be optimized to obtain controlled and sustained release by altering the thickness and composition [47].

Effective release of ibuprofen (IBU), an anti-inflammatory drug from chitosan/polyvinyl alcohol blend films, was developed by solvent casting method using Genipin (GEN) as the crosslinker (Fig. 7). FTIR and SEM results were analyzed to comprehend the miscibility of the blend. 0.25 mM concentration of GEN was determined to be the optimum concentration to obtain an improved release profile and reduce burst release. The release studies were conducted by immersing the films in PBS and shaken at 130 rpm. After specified intervals, 3 mL of solution was collected and replaced with same amount of PBS. The collected release solutions were analyzed with UV-VIS spectrophotometer at 265 nm. The cumulative release percentage of films with GEN showed 75.4% after 4 h depicting a more controlled release of IBU from the film compared to the films without GEN. WST-1 assay study shows 96% cell viability suggesting the potential use of these films as a pain-suppressing wound dressing material [47].

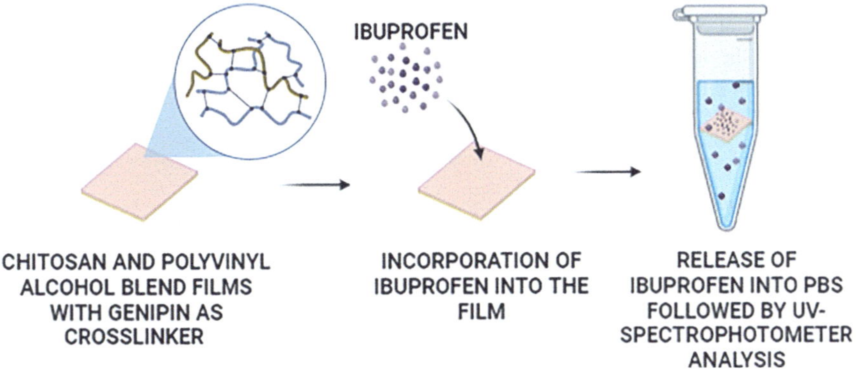

Fig. 7 Preparation of chitosan-poly (vinyl alcohol) blend films incorporated with ibuprofen

To enhance wound healing and skin regeneration chitosan-aloe film embedded with curcumin (CUR) loaded PLGA microparticles was developed. The uniformly embedded films were prepared by high-power ultrasonic emulsification and tape casting methods and were characterized using SEM, ATR-FTIR, and XRD. The drug release rate was measured by adding films in 10 mL of PBS with 10% ethanol and incubating at 37°C for 360 h with agitation. The concentration of CUR was measured at a wavelength of 426 nm. Through this, the encapsulation efficiency and the loading efficiency were determined to be 77.56% and 13.85% respectively. The cumulative release over 360 h was about 89.2%, suggesting a controlled release profile. In vivo studies with full-thickness skin model in Wistar rats exhibited that in comparison to the control films, films embedded with CUR accelerated wound healing and promoted skin regeneration. The released CUR is proven to show anti-inflammatory effects by inhibiting inflammatory mediators like inducible nitric oxide synthase and cyclooxygenase-2 [48]. A study for the efficient delivery of an anti-inflammatory drug, tenoxicam (TNX) with the help of mucoadhesive buccal chitosan films to treat chronic periodontitis was developed. The drug works by inhibiting the activity of cyclooxygenase which has a crucial role in inflammation. TNX-loaded chitosan films were prepared through the method of film casting incorporating polyvinyl pyrrolidone (PVP). PVP acts as a film-forming agent and affects the rate of TNX release from the film. The buccal films were placed in simulated saliva with conditions similar to bodily conditions to determine the release of the loaded TNX. After predetermined intervals, a little supernatant was removed and absorbance was measured using a UV spectrophotometer at 368 nm. The film showed a controlled TNX release over a period of 6 h for effective treatment in periodontitis [49]. Another study aimed at developing a wound dressing material made chitosan and xanthan membrane incorporated with indomethacin (IDM) an anti-inflammatory drug. Incorporation of IDM was done at three different concentrations to obtain the optimum concentration. Chitosan, xanthan gum polyelectrolyte complexes with the addition of different concentrations of IDM were developed using the casting method. All the formulations exhibited an incorporation efficiency of about 90% and above with minimal drug loss. The release studies were investigated by immersing them in PBS and analyzing the aliquots using a UV spectrophotometer at 319 nm. During the initial 5 h, the release reached a plateau form and after 24 h the loaded IDM was completely released from the membrane providing release rates closer to the desired therapeutic dosage for tropical applications [50].

3.5 Protein and Peptide

Chitosan films and membranes can significantly enhance the delivery of proteins and peptides by functioning as effective carriers. These biocompatible materials facilitate a controlled release system, allowing for the gradual release of proteins and peptides directly at the target site.. Chitosan being positively charged showed interacting properties with the negatively charged proteins. The hydroxyl functional groups of

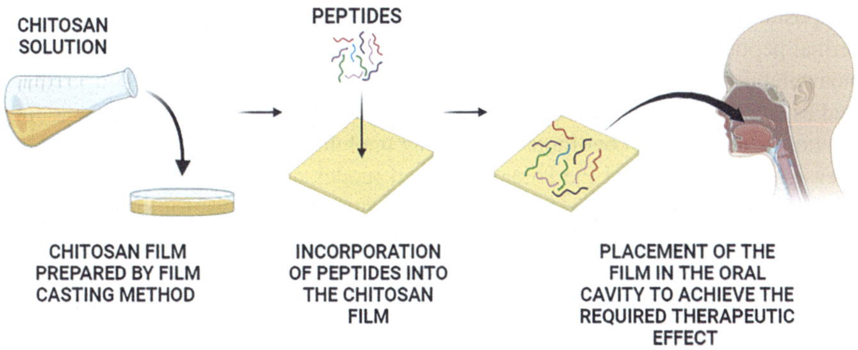

Fig. 8 Peptides loaded chitosan film for oral delivery

the chitosan develop hydrogen bonds with the functional groups of the peptide. Additionally, there is a presence of hydrophobic interactions observed between the non-polar regions of these entities. These interactions lead to the formation of complexes of chitosan with the peptides for their sustained delivery and protection against enzymatic degradation [51].

A study was conducted to develop a pH-responsive nanofibrous membrane of chitosan coated with pectin for antimicrobial peptide (AMP) delivery in the oral cavity (Fig. 8). The membrane showed dual adhesive nature made of chitosan with pectin coating showing adhesion to soft tissues and just the chitosan membrane showing adhesion to hard tissues making it useful in the biomedical field. Since the pH can vary between 4.5 and 6.7 during oral infections, developing a pH-responsive delivery system could be a better alternative to the current therapies. Cationic amphipathic AMPs like D-GL13K and IDR-1018 were incorporated into the nanofibrous membrane through co-electrospinning or surface absorption showing antimicrobial properties against *Streptococci* and providing protection against oral infections. The peptide release studies were carried out by tagging GL13K peptide with TAMRA (fluorescent carboxytetramethylrhodamine analog of D-GL13K) which produces a red fluorescence across different pH. During the preliminary hour, a burst release was noted, followed by a gradual decrease in peptide release. With decreasing pH, the peptide solubility and chitosan degradation increase facilitating the release of peptides from the membrane [52].

Similarly, a nanofibrous membrane fabricated from chitosan and poly (ethylene oxide) (PEO) embedded with AMP NP-10 showed activity against *E. coli* and *S. aureus*. The diameter of the fibers of the membrane increased with an increased concentration of NP-10. The release kinetics was observed for 15 days with the average cumulative release reaching around 88.4% after 15 days. The chitosan membrane absorbs water making it swell and eventually degrade enhancing the release of NP-10 from the membrane. In the first 24 h burst release was obtained due to the immediate interaction of the membrane with the dissolution medium leading to the larger release followed by a sustained release due to the resonant release of the AMP hence making the membrane advantageous in the wound healing field [53]. A composite film consisting of chitosan and tamarind seed polysaccharide for transdermal peptide delivery was developed. Transdermal delivery systems increase the bioavailability of the peptides by avoiding the first-pass metabolism. Drug release studies were conducted by arranging the films into eggshell membranes and adding IP phosphate buffer at conditions resembling the human body. It was stated that the release kinetics followed the Hixson-Crowell model suggesting that the release of the AMP occurs due to the degradation of the film. The release for the initial days ranged between 13.75% and 22.54% and after 5 days was observed to be 78.33% [54]. Chitosan films are embedded with antihypertensive peptide (KGYGGVSLPEW)-loaded chitosan microparticles for oral delivery and were developed using the ionic gelation method. The mean diameter of the microparticle was optimized to be 25 μm with a peptide association efficiency of 76%. The film's adherence potential to the buccal mucosa is enhanced due to the mucoadhesive properties of chitosan which facilitates controlled release of the peptide. The quick release of the peptides occurs due to the disintegration of the film on exposure to saliva within 30 s [55]. Chitosan-polyvinyl alcohol nanofibrous mats through electrospinning encapsulated with fish-derived antioxidant peptides were prepared. The electrospun mats contained a perfect blend of chitosan and polyvinyl alcohol with diameters ranging from 158 to 196 nm. Encapsulation of the antioxidant peptides was attained with over 94% efficiency and its presence was determined by FT-IR and XRD. The peptide release was studied by immersing the mats in PBS at 37°C with agitation (shaking incubator). The release solution was collected at designated time intervals and analyzed using UV spectroscopy at an absorbance of 220 nm. Surface disintegration and diffusion through the matrix attributed to an initial burst release within the first 10 h after a slow phase sustained release for 70 h. MTT assay analysis was performed to check the cytotoxicity of the films for their potential use in the food packaging field [56].

3.6 Nucleic Acid

Nucleic acid delivery, a revolutionary approach in medicine, offers methods to treat genetic disorders by introducing exogenous genetic material into a cell. The therapy is carried out by either substituting a faulty gene with a functional one or by silencing the faulty gene by RNA interference. The gene is delivered by delivering it with a vector. Vector can be viral (modified virus) or non-viral (synthetic particles), which facilitate the entry of the gene into the target cell and offer protection for the gene to be delivered. Though viral vectors are regarded as the most efficient vectors to carry genes into a cell, there is a high risk of causing immune responses in the patient. Non-viral vectors offer a safer alternative and hence are getting increased attention [57]. Chitosan displaying high biocompatibility, biodegradability, and non-toxicity can be employed as non-viral vehicles or for nucleic acid delivery. D-glucosamine monomers of chitosan consist of free amino groups. These positively charged entities facilitate the formation of complexes with the nucleic acid aiding in

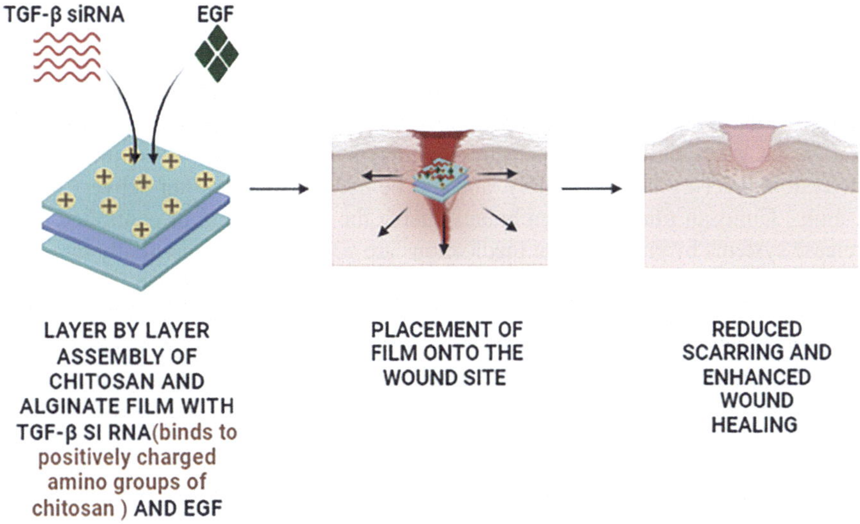

Fig. 9 Wound healing by TGF-β si RNA and EGF incorporated chitosan-alginate films

protecting it from degradation. The transfection efficiency of the carriers from chitosan can be modified by altering the molecular weight hence making chitosan a safer and efficient option for nucleic acid delivery [58].

Chitosan-alginate films loaded with TGF-β si-RNA and EGF showed enhanced wound healing and also impaired collagen deposition at the wound site (Fig. 9). The film also exhibited an antibacterial effect by inhibiting the formation of colonies by *E. coli* and *S. aureus*. The film showed enhanced keratinocytes adhesion onto the surface leading to greater re-epithelialization. The release of si-RNA and EGF was determined by soaking it in PBS (phosphate-buffered saline) at 37°C and analyzed by measuring the absorbance and observed a steady release of si-RNA. In vivo studies of this system showed accelerated wound healing and reduced deposition of extracellular matrix proteins [59].

DNA with lipid carriers (lipoplexes) loaded into a film of alternating layers of glycol chitosan and hyaluronic acid aims to deliver genetic material at localized sites in the body. The film showed good transfection efficiency for at least 7 days in two different cell lines (NIH3T3 fibroblast and HEK293 kidney cells). The film was placed onto a 24-well tissue culture plate and incubated with PBS solution at 37°C with 5% CO_2 with repeated replacement of the PBS daily. DNA quantification from the supernatant was done with a Pico-Green assay before and after the deposition of the film. Additionally, heparin sulfate was added to separate DNA from the lipoplexes. The films showed a 5% release of the embedded DNA for over a week. The film with a larger amount of DNA embedded showed a reduced release rate while the films containing a reduced amount of DNA showed a greater effect on drug release rate [60].

4 Conclusion

Chitosan films and membranes have attained a lot of consideration in the drug delivery field due to their outstanding properties. They can be prepared utilizing a variety of processes, such as electrospinning, solvent casting techniques, phase inversion, layer-by-layer assembly, and can be used in various drug delivery applications. Chitosan films and membranes enable the development of controlled drug release systems by ensuring that medications are administered consistently throughout time, thus improving therapeutic efficiency and patient compliance. This review explains numerous applications of chitosan films and membranes, emphasizing the delivery of antibacterial, antifungal, and anticancer medicines, as well as peptides, proteins, nucleic acids, and anti-inflammatory compounds. It also includes the preparation strategies for chitosan films and membranes, showing their efficiency in various drug-related areas.

Acknowledgments One of the authors, Arya Lakshmi.V. V, is grateful to Amrita Vishwa Vidyapeetham for providing her Research fellowship. The authors are also thankful to Amrita Vishwa Vidyapeetham for providing seed grant fund.

References

1. Li C, Wang J, Wang Y, Gao H, Wei G, Huang Y, Yu H, Gan Y, Mei L, Chen H (2019) Recent progress in drug delivery. Acta Pharm Sin B 9:1145–1162
2. Radha D, Lal JS, Devaky KS (2022) Chitosan-based films in drug delivery applications. Starch-Starke 74:2100237
3. Cirillo G, Hampel S, Spizzirri UG, Parisi OI, Picci N, Iemma F (2014) Carbon nanotubes hybrid hydrogels in drug delivery: a perspective review. Biomed Res Int 1:825017
4. Parhi R (2020) Drug delivery applications of chitin and chitosan: a review. Environ Chem Lett 18:577–594
5. Muhamad II, Selvakumaran S, Lazim NAM (2014) Designing polymeric nanoparticles for targeted drug delivery system. In: Seifalian A, Mel A, Kalaskar DM (eds) Nanomedicine, pp 287–313
6. Amirthalingam S, Rajendran AK, Prabaharan M, Jayakumar R (2021) Perspectives and challenges of using chitosan in various biological applications. In: Prabaharan M, Jayakumar R (eds) Chitosan for biomaterials III: structure-property relationships, pp 1–22
7. Jayakumar R, Prabaharan M, Nair SV, Tamura H (2010) Novel chitin and chitosan nanofibers in biomedical applications. Biotechnol Adv 28:142–150
8. Jayakumar R, Reis RL, Mano JF (2007) Synthesis and characterization of pH-sensitive thiol-containing chitosan beads for controlled drug delivery applications. Drug Deliv 14:9–17
9. Kato Y, Onishi H, Machida Y (2003) Application of chitin and chitosan derivatives in the pharmaceutical field. Curr Pharm Biotechnol 4:303–309
10. Desai N, Rana D, Salave S, Gupta R, Patel P, Karunakaran B, Sharma A, Giri J, Benival D, Kommineni N (2023) Chitosan: a potential biopolymer in drug delivery and biomedical applications. Pharmaceutics 15:1313
11. Barik M, BhagyaRaj GVS, Dash KK, Shams R (2024) A thorough evaluation of chitosan-based packaging film and coating for food product shelf-life extension. J Agric Food Res 15:101164

12. Cui C, Sun S, Shaohua W, Chen S, Ma J, Zhou F (2021) Electrospun chitosan nanofibers for wound healing application. Eng Regen 2:82–90
13. Dhurai B, Saraswathy N, Maheswaran R, Sethupathi P, Vanitha P, Vigneshwaran S, Rameshbabu V (2013) Electrospinning of curcumin loaded chitosan/poly (lactic acid) nanofilm and evaluation of its medicinal characteristics. Front Mater Sci 7:350–361
14. Guo H, Ran W, Jin X, Huang Y, Long F, Xiao Y, Gan RY, Wu Y, Gao H (2024) Development of pectin/chitosan-based electrospun biomimetic nanofiber membranes loaded with dihydromyricetin inclusion complexes for wound healing application. Int J Biol Macromol 278:134526
15. Alotaibi BS, Khan AK, Kharaba Z, Yasin H, Yasmin R, Ijaz M, Khan M, Murtaza G (2024) Development of poly (vinyl alcohol)–chitosan composite nanofibers for dual drug therapy of wounds. ACS Omega 9:12825–12834
16. Borbolla-Jimenez FV, Pena-Corona SI, Farah SJ, Jimenez-Valdes MT, Pineda-Perez E, Romero-Montero A, Del Prado-Audelo ML, Bernal-Chavez SA, Magana JJ, Leyva-Gomez G (2023) Films for wound healing fabricated using a solvent casting technique. Pharmaceutics 15: 1914
17. Anbukarasu P, Sauvageau D, Elias A (2015) Tuning the properties of polyhydroxybutyrate films using acetic acid via solvent casting. Sci Rep 5:17884
18. Supernak M, Makurat-Kasprolewicz B, Kaczmarek-Szczepanska B, Pałubicka A, Sakowicz-Burkiewicz M, Ronowska A, Wekwejt M (2023) Chitosan-based membranes as gentamicin carriers for biomedical applications – influence of chitosan molecular weight. Membranes 13: 542
19. Purkait MK, Sinha MK, Mondal P, Singh R (2018) Introduction to membranes. In: Purkait MK, Sinha MK, Mondal P, Singh R (eds) Interface science and technology, pp 1–37
20. Morgado PI, Lisboa PF, Ribeiro MP, Miguel SP, Simoes PC, Correia IJ, Aguiar-Ricardo A (2014) Poly (vinyl alcohol)/chitosan asymmetrical membranes: highly controlled morphology toward the ideal wound dressing. J Membr Sci 469:262–271
21. Guerrero P, Muxika A, Zarandona I, De La Caba K (2019) Crosslinking of chitosan films processed by compression molding. Carbohydr Polym 206:820–826
22. Matet M, Heuzey MC, Pollet E, Ajji A, Averous L (2013) Innovative thermoplastic chitosan obtained by thermo-mechanical mixing with polyol plasticizers. Carbohydr Polym 95:241–251
23. Kaur M, Sharma A, Puri V, Aggarwal G, Maman P, Huanbutta K, Nagpal M, Sangnim T (2023) Chitosan-based polymer blends for drug delivery systems. Polymers 15:2028
24. Kumar S, Mukherjee A, Dutta J (2020) Chitosan based nanocomposite films and coatings: emerging antimicrobial food packaging alternatives. Trends Food Sci 97:196–209
25. dos Santos VL, Araujo RC, Lisboa ES, Lopes AM, de Albuquerque-Junior RL, Cardoso JC, Blanco-Llamero C, Deshpande TA, Anderson HO, Priefer R, Souto EB (2023) Layer-by-layer assembly: a versatile approach for tailored biomedical films and drug delivery. J Drug Deliv Sci Technol 7:105243
26. Sung C, Hearn K, Reid DK, Vidyasagar A, Lutkenhaus JL (2013) A comparison of thermal transitions in dip-and spray-assisted layer-by-layer assemblies. Langmuir 29:8907–8913
27. Aoki PH, Alessio P, Volpati D, Paulovich FV, Riul Jr A, Oliveira Jr ON, Constantino CJ (2014) On the distinct molecular architectures of dipping-and spray-LbL films containing lipid vesicles. Mater Sci Eng C 41:363–371
28. Salomaki M, Peltonen T, Kankare J (2012) Multilayer films by spraying on spinning surface – best of both worlds. Thin Solid Films 520:5550–5556
29. Guan W, Zhou W, Lu J, Lu C (2015) Luminescent films for chemo-and biosensing. Chem Soc Rev 44:6981–7009
30. Zhuk I, Jariwala F, Attygalle AB, Wu Y, Libera MR, Sukhishvili SA (2014) Self-defensive layer-by-layer films with bacteria-triggered antibiotic release. ACS Nano 8:7733–7745
31. Ma J, Wang Y, Lu R (2022) Mechanism and application of chitosan and its derivatives in promoting permeation in transdermal drug delivery systems: a review. Pharmaceuticals 15:459

32. Pacheco MS, Kano GE, de Almeida PL, Lopes PS, de Moraes MA (2020) Silk fibroin/chitosan/alginate multilayer membranes as a system for controlled drug release in wound healing. Int J Biol Macromol 152:803–811
33. Affes S, Aranaz I, Acosta N, Heras A, Nasri M, Maalej H (2021) Chitosan derivatives-based films as pH-sensitive drug delivery systems with enhanced antioxidant and antibacterial properties. Int J Biol Macromol 182:730–742
34. Yang S, Zhang X, Zhang D (2019) Electrospun chitosan/poly (vinyl alcohol)/graphene oxide nanofibrous membrane with ciprofloxacin antibiotic drug for potential wound dressing application. Int J Biol Macromol 20:4395
35. Kedir WM, Abdi GF, Goro MM, Tolesa LD (2022) Pharmaceutical and drug delivery applications of chitosan biopolymer and its modified nanocomposite: a review. Heliyon 1:8
36. Shah A, Ashames AA, Buabeid MA, Murtaza G (2020) Synthesis, in vitro characterization and antibacterial efficacy of moxifloxacin-loaded chitosan-pullulan-silver-nanocomposite films. J Drug Deliv Sci Technol 55:101366
37. Alhodieb FS, Barkat MA, Barkat HA, Ab Hadi H, Khan MI, Ashfaq F, Rahman MA, Hassan MZ, Alanezi AA (2022) Chitosan-modified nanocarriers as carriers for anticancer drug delivery: promises and hurdles. Int J Biol Macromol 217:457–469
38. Cardoso HP, Rodrigues JFB, da Silva HN, Galdino TP, Luna CBB, Fook MVL, Montazerian M, Baino F, de Lima Silva SM (2024) Chitosan/montmorillonite nanocomposite film as anticancer drug carrier: a promising biomaterial to treat skin cancers. Ceram Int 50: 18528–18539
39. Tomaz FA, de Carvalho SSM, Barbosa CR, Silva LSM, Gutierrez SMA, de Lima BAG, Fook LMV (2018) Ionically crosslinked chitosan membranes used as drug carriers for cancer therapy application. Materials 11:2051
40. Feng Z, Zheng Y, Zhao L, Zhang Z, Sun Y, Qiao K, Xie Y, Wang Y, He W (2019) An ultrasound-controllable release system based on waterborne polyurethane/chitosan membrane for implantable enhanced anticancer therapy. Mater Sci Eng C 104:109944
41. Hoang T, Ramadass K, Loc TT, Mai TT, Giang LD, Thang VV, Tuan TM, Chinh NT (2019) Novel drug delivery system based on ginsenoside Rb1 loaded to chitosan/alginate nanocomposite films. Int J Nanosci Nanotechnol 19:3293–3300
42. Kraisit P, Yonemochi E, Furuishi T, Mahadlek J, Limmatvapirat S (2022) Chitosan film containing antifungal agent-loaded SLNs for the treatment of candidiasis using a Box-Behnken design. Carbohydr Polym 283:119178
43. Tabriz A, Alvi MAUR, Niazi MBK, Batool M, Bhatti MF, Khan AL, Khan AU, Jamil T, Ahmad NM (2019) Quaternized trimethyl functionalized chitosan based antifungal membranes for drinking water treatment. Carbohydr Polym 207:17–25
44. Huang J, Jacobsen J, Larsen SW, Genina N, Van de Weert M, Mullertz A, Nielsen HM, Mu H (2020) Graphene oxide as a functional excipient in buccal films for delivery of clotrimazole: effect of molecular interactions on drug release and antifungal activity in vitro. Int J Pharm 589: 119811
45. Calvo NL, Sreekumar S, Svetaz LA, Lamas MC, Moerschbacher BM, Leonardi D (2019) Design and characterization of chitosan nanoformulations for the delivery of antifungal agents. Int J Mol Sci 20:3686
46. Ribeiro AS, Costa SM, Ferreira DP, Calhelha RC, Barros L, Stojkovic D, Sokovic M, Ferreira IC, Fangueiro R (2021) Chitosan/nanocellulose electrospun fibers with enhanced antibacterial and antifungal activity for wound dressing applications. React Funct Polym 159:104808
47. Aycan D, Yayla NA, Aydin YA (2020) Chitosan polyvinyl alcohol blend films for ibuprofen encapsulation: fabrication, characterization and kinetics. Polym Degrad Stab 181:109346
48. Liu X, You L, Tarafder S, Zou L, Fang Z, Chen J, Lee CH, Zhang Q (2019) Curcumin-releasing chitosan/aloe membrane for skin regeneration. Chem Eng J 359:1111–1119
49. Ashri LY, El Sayeh AF, Ibrahim MA, Alshora DH (2020) Optimization and evaluation of chitosan buccal films containing tenoxicam for treating chronic periodontitis: in vitro and in vivo studies. J Drug Deliv Sci Technol 57:101720

50. Lopes SA, Veiga IG, Bierhalz ACK, Pires ALR, Moraes AM (2019) Physicochemical properties and release behavior of indomethacin-loaded polysaccharide membranes. Int J Polym Mater Polym Biomater 68:956–964
51. Khubiev OM, Egorov AR, Kirichuk AA, Khrustalev VN, Tskhovrebov AG, Kritchenkov AS (2023) Chitosan-based antibacterial films for biomedical and food applications. Int J Mol Sci 24:10738
52. Boda SK, Fischer NG, Ye Z, Aparicio C (2020) Dual oral tissue adhesive nanofiber membranes for pH-responsive delivery of antimicrobial peptides. Biomacromolecules 21:4945–4961
53. Yu L, Dou S, Ma J, Gong Q, Zhang M, Zhang X, Li M, Zhang W (2021) An antimicrobial peptide-loaded chitosan/polyethylene oxide nanofibrous membrane fabricated by electrospinning technology. Front Mater 8:650223
54. Malviya R, Tyagi A, Fuloria S, Subramaniyan V, Sathasivam K, Sundram S, Karupiah S, Chakravarthi S, Meenakshi DU, Gupta N, Sekar M (2021) Fabrication and characterization of chitosan-tamarind seed polysaccharide composite film for transdermal delivery of protein/peptide. Polymers 13:1531
55. Batista P, Castro P, Madureira AR, Sarmento B, Pintado M (2019) Development and characterization of chitosan microparticles-in-films for buccal delivery of bioactive peptides. Pharmaceuticals 12:32
56. Hosseini SF, Nahvi Z, Zandi M (2019) Antioxidant peptide-loaded electrospun chitosan/poly (vinyl alcohol) nanofibrous mat intended for food biopackaging purposes. Food Hydrocoll 89: 637–648
57. Raftery R, O'Brien FJ, Cryan SA. Chitosan for gene delivery and orthopedic tissue engineering applications. Molecules 18:5611–5647
58. Karayianni M, Sentoukas T, Skandalis A, Pippa N, Pispas S (2023) Chitosan-based nanoparticles for nucleic acid delivery: technological aspects, applications, and future perspectives. Pharmaceutics 15:1849
59. Mandapalli PK, Labala S, Jose A, Bhatnagar S, Janupally R, Sriram D, Venuganti VVK (2017) Layer-by-layer thin films for co-delivery of TGF-β siRNA and epidermal growth factor to improve excisional wound healing. AAPS Pharm Sci Tech 18:809–820
60. Holmes CA, Tabrizian M (2013) Substrate-mediated gene delivery from glycol-chitosan/hyaluronic acid polyelectrolyte multilayer films. ACS Appl Mater Interfaces 5:524–531

Adv Polym Sci (2025) 295: 183–212
https://doi.org/10.1007/12_2025_191

Published online: 19 February 2025

Chitosan Scaffolds for Drug Delivery

Meghna Mohandas, Anushikaa Ramprasad, Tanushree Suresh, Sri Brindaa Varadarajan, Sri Manjari Ravishankar, and R. Jayakumar

Contents

1 Introduction 184
2 Fabrication Strategies of Chitosan Scaffolds 185
2.1 Freeze-Drying 187
2.2 Solvent Casting Method 187
2.3 Particulate Leaching 188
2.4 3D Printing/3D Bioprinting 188
2.5 Electrospinning 189
3 Antimicrobial Agent Delivery Using Chitosan-Based Scaffolds 190
3.1 Antibiotics 191
3.2 Metals/Metallic Nanocomposites 192
3.3 Metallic Ion-Doped Bioceramics 192
3.4 Carbon Nanomaterials 193
3.5 Antifungal 194
4 Anti-inflammatory Drug Delivery Using Chitosan-Based Scaffolds 195
5 Anticancer Drug Delivery Using Chitosan-Based Scaffolds 198
6 Biomolecule Delivery Using Chitosan-Based Scaffolds 199
6.1 Growth Factors 200
6.2 Peptide and Proteins 200
6.3 Nucleic Acids 202
6.4 Lipids 203
7 Conclusion 204
References 204

Meghna Mohandas and Anushikaa Ramprasad contributed equally with all other contributors.

M. Mohandas, A. Ramprasad, T. Suresh, S. B. Varadarajan, S. M. Ravishankar, and R. Jayakumar (✉)
Polymeric Biomaterials Lab, School of Nanosciences and Molecular Medicine, Amrita Vishwa Vidyapeetham, Kochi, India
e-mail: rjayakumar@aims.amrita.edu

Abstract Chitosan has shown remarkable progress in the drug delivery field, exhibiting impressive biodegradability, biocompatibility, antibacterial ability, and ease of creating versatile scaffolding systems. Chitosan scaffolds are widely used for delivery of drugs, antimicrobial agents, and biomolecules. This review explores the extensive potentiality of chitosan scaffolds for localized or targeted drug delivery for regenerative medicine and tissue regeneration. This review highlights the fabrication methods such as freeze-drying, solvent casting, particulate leaching, electrospinning, and 3D printing that support the advancement of chitosan-based scaffolds with structural and mechanical properties. This review concentrates on the antimicrobial activity of chitosan scaffolds with antibiotics, metals/metal-organic frameworks, and bioceramics that are used combined or alone to impart antibacterial and antifungal properties. In addition, chitosan scaffolds demonstrate the efficacy in anti-inflammatory and anticancer activities and as a promising method for the delivery of biomolecules including growth factors, proteins and peptides, nucleic acids, and lipids for regenerative medicine and tissue regeneration.

Keywords Anticancer · Anti-inflammatory · Antimicrobial · Biomolecule delivery · Chitosan scaffold · Drug delivery · Fabrication · Regenerative medicine · Tissue engineering

1 Introduction

Scaffold-based drug delivery systems are extensively used and studied in the field of tissue engineering and regenerative medicine. Scaffolds exist as three-dimensional structures that provide an apt microenvironment for the growth of the cells and facilitate in guiding them to adhere, migrate, differentiate, and proliferate into a certain phenotype to restore the lost tissue. Scaffolds are recommended as a promising approach for targeted and non-targeted drug delivery systems (DDSs) owing to the fact that they can deliver various biomolecules and drugs in a regulated manner at the specific target site for multiple applications. Numerous parameters are to be satisfied for selecting a suitable biomaterial for developing a scaffold such as bioresorbability, biocompatibility, stability, swelling, mechanical strength, availability, and source. Generally, biomaterials are classified into two types based on their origin, namely, natural and synthetic [1–3]. Among these, the biomaterials that are procured from natural sources are much preferred in biomedical applications due to their striking features, namely, high biodegradability, biocompatibility, minimal toxicity, availability, ECM mimicking ability, etc., which are further illustrated in Fig. 1 [4].

Amid natural polymeric scaffolds, chitosan-based scaffolding systems have gained considerable attention among researchers and have been utilized for numerous applications. Since chitosan-based scaffolds are highly porous and provide structural support for neo-tissue formation, they also contain amino groups that

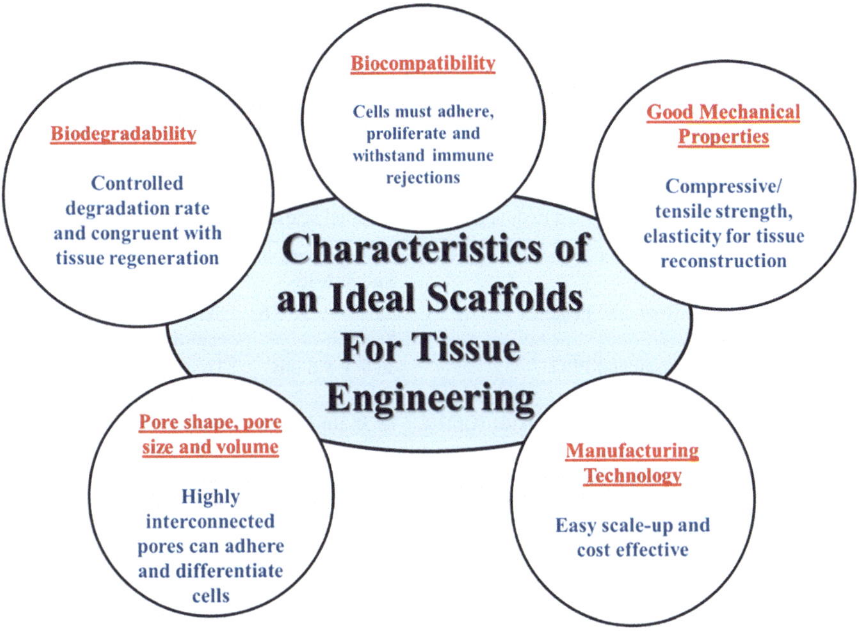

Fig. 1 Characteristics of an ideal scaffold used for tissue engineering

could facilitate mucoadhesion and in situ gelation at physiological conditions. It can also serve as a smart delivery vehicle that delivers drugs in a controlled and sustained manner based on external stimuli [5]. Henceforth, this chapter primarily describes the chitosan-based scaffolds, different fabrication methodologies, and different fields in which they are employed as DDS for the delivery of biomolecules and drugs.

2 Fabrication Strategies of Chitosan Scaffolds

In recent years, many fabrication strategies have emerged for the construction of three-dimensional scaffolds, including freeze-drying, solvent casting, 3D printing, and electrospinning, which are common techniques for chitosan-based scaffold fabrication. The freeze-drying technique generates a porous scaffold with interconnected pores through freezing chitosan solution. Solvent casting/particulate leaching involves casting chitosan solution containing pyrogens, generating cavities within the scaffold, and subsequently eliminating them via leaching with a suitable solvent. 3D printing, an additive manufacturing technique, utilizes layer-by-layer deposition for the creation of three-dimensional construct [6]. Electrospinning is used to generate nanofibrous scaffolds by applying a high-voltage electric field. Electrospun scaffolds encourage cell adhesion, proliferation, and tissue regeneration.

These scaffolds can mimic the natural ECM at the nanoscale, offering an initial environment for new tissue regeneration [6]. Several methods have been reported for scaffold fabrication that utilize chitosan as a polymer, which are discussed in Table 1. A few of the approaches are outlined in the following subheadings.

Table 1 Fabrication methods used to develop chitosan-based scaffolds for bone tissue engineering

Technique	Composition of chitosan-based scaffolds	Average pore size and porosity	Application	References
Freeze-drying	Chitosan and gelatin	227.67 ± 72.0.6 57–60%	Skin tissue engineering	[7]
Freeze-drying	Chitosan and PEG	46.9 ± 8.6 μm	Tissue engineering	[8]
Freeze-drying	Chitosan, graphene oxide (GO)	78.38 μm	Bone tissue engineering	[9]
Freeze-drying	Chitosan, zein, nanohydroxyapatite	220 ± 3 μm 64 ± 0.49	Bone tissue engineering	[10]
Solvent casting	Chitosan, dibasic sodium phosphate, γ-glycidoxypropyltrimethoxysilane	–	Peripheral nerve regeneration	[11]
Solvent leaching	Chitosan, *n*-SiO_2	198–269 μm 70.99–73.23%	Bone tissue engineering	[12]
Particulate leaching	Acidic chitosan, sodium acetate	7–30 μm 60–90%	Tissue engineering	[13]
3D printing	Chitosan, alginate, hydroxyapatite	–	Cartilage regeneration	[14]
3D printing	Chitosan, poly(lactic acid) (PLA)	–	Tissue engineering	[15]
3D printing	Chitosan, hydroxyapatite	70 μm >70%	Tissue engineering	[16]
3D printing	Chitosan, collagen	60–200 μm 83.5%	Therapeutic approach for spinal cord injury	[17]
Electrospinning	Chitosan, poly(ethylene oxide) (PEO)	0.3 mm	Bone tissue engineering	[18]
Electrospinning	Chitosan, gelatin	60–90%	Skin tissue engineering	[19]
Electrospinning	Gold nanoparticle doped poly (caprolactone), (PCL)/chitosan	–	Peripheral nerve tissue engineering	[20]
Electrospinning	Chitosan, polyurethane urea	–	Soft tissue engineering	[21]
Electrospinning	Chitosan/alginate-chitin nano whiskers (bilayer scaffold)	<2 μm	Tissue engineering	[22]
Electrospinning	Chitosan, polypyrene, collagen	–	Tissue engineering	[23]

2.1 *Freeze-Drying*

Freeze-drying or lyophilization is the best conventional method used for the fabrication of porous scaffolds of higher interconnectivity. The method is based on the sublimation to eliminate the water content and result in the fabrication of an anhydrous 3D scaffold. The process involves a three-step procedure; the events initiate freezing of the solution at a lower temperature (−20 to −80°C) followed by sublimation under a vacuum called primary drying and a final step called secondary drying which is accelerated sublimation due to the application of heat [24, 25]. This method is widely applied for manufacturing porous scaffolds of size 20–200 μm and can obtain porosity of around 90% [26]. The porosity can be varied by altering the conditions such as chitosan concentration, volume of the solvent/water, pH of the medium, freeze rate, and temperature. Porous chitosan scaffolds were prepared using freeze-drying, with pre-freezing conducted at −20 and −80°C for different periods to develop two distinct scaffolds having pore sizes of 34 ± 9 and 153 ± 32 μm. The results highlight the significance of higher porosity in facilitating greater diffusion rates [27]. The method also tackles the drawbacks of prolonged processing time, high energy requirement, high cost, harmful effects of solvents, etc.

2.2 *Solvent Casting Method*

Solvent casting is another technique to develop chitosan scaffolds, primarily consisting of two methods. The primary method involves scaffold production by immersing the mold in a chitosan solution. The secondary method involves pouring the polymer solution which was dissolved in volatile solvent into a mold, and upon its evaporation, the polymers solidify to form scaffold. Figure 2 further describes the stage involved in solvent casting. The drawback of this method includes solvent retention within the scaffold, which causes toxicity [28].

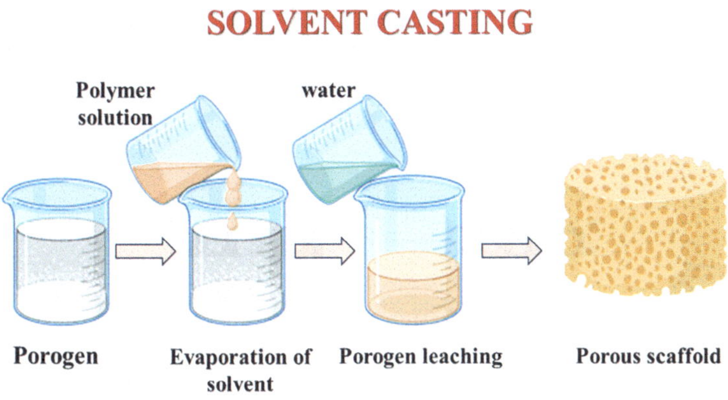

Fig. 2 Schematic depiction of solvent casting method

2.3 Particulate Leaching

This is another process used for fabricating porous chitosan-based scaffolds. It involves the mixing of water-soluble salts or other substances such as (sodium chloride, sodium citrate, sugar), waxes, and porogen into a chitosan solution, which is then cast into a mold of the desired shape and allowed to lyophilize to eliminate the solvent. The resultant salt particles or porogen are leached out by immersion in water or an appropriate solvent to yield a well-porous scaffold. Additionally, Fig. 3 illustrates the different steps that are included in the method. The limitation includes the limited pore shape with respect to the shape of the salt and difficulty in the removal of soluble salt particles from the polymer matrix [29].

2.4 3D Printing/3D Bioprinting

The most innovative method for constructing chitosan scaffolds is 3D printing, which aims to overcome all the obstacles of the traditional methods employed for chitosan scaffold development. The method is also known as additive manufacturing and was first introduced in 1986. The capability of fabricating a scaffold having uniform pore size, shape, and porosity is the major advantage of 3D printing. This fast-emerging technology created a 3D scaffold by adopting an automated approach to assemble a layer-by-layer arrangement, which can be visualized in Fig. 4. This method enables highly precise designing via a CAD file addressed with minimal error despite their multifaceted geometry. Fabrication of chitosan 3D scaffolds alone or in combination with bioactive ceramics and natural or synthetic polymers, along with cells used for 3D bioprinted constructs, is used for tissue repair [30, 31]. Yousefiasl et al. fabricated a 3D printed bioactive chitosan scaffold for craniofacial bone regeneration and showed an increased pore size of 285.5 μm

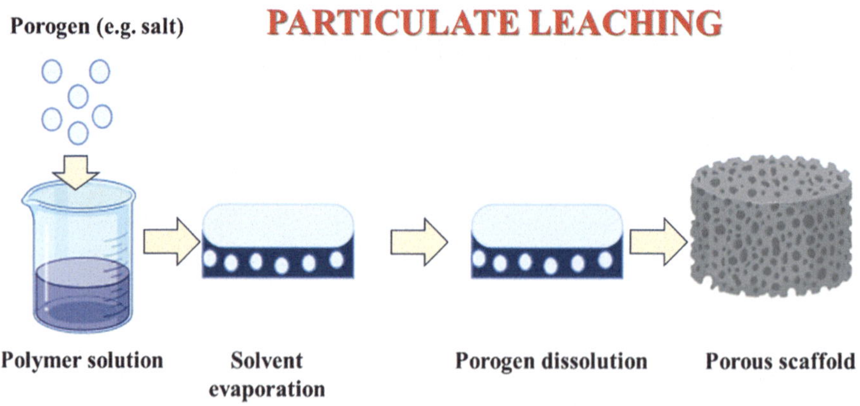

Fig. 3 Schematic depiction of particulate leaching technique

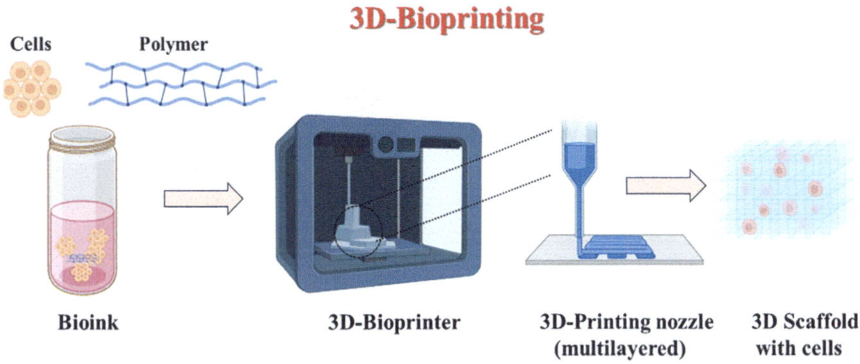

Fig. 4 Schematic depiction of 3D bioprinting technique

demonstrating ideal porosity for supporting new tissue formation [32]. In another study, a high-strength chitosan hydrogel scaffold was prepared via 3D printing without the addition of any organic solvents, which showed excellent mechanical strength of 2.31 Pa with good biocompatibility also accounting for the feasibility of 3D printed chitosan scaffolds for tissue regeneration [33]. However, this method has a few drawbacks like low printing quality, inability to print heat-sensitive materials, and requirement of preformed uniform fibers.

2.5 *Electrospinning*

Electrospinning is a versatile approach used for the development of ultrafine 1D nanofibers, 2D nonwoven membranes, and 3D scaffolds [34]. The fundamental concept of electrospinning comprises an electrodynamic procedure, where a liquid droplet gets elongated as well as stretched into a continuous fiber due to the effect of applied high electric voltage. The components of this technique include a syringe pump, needle, collector, and high-voltage power supply. The electrostatic repulsion generated due to the surface charge of accumulated liquid overcomes the surface tension because the high voltage applied between the tip of the needle and collector forces the droplet into a Taylor cone shape. Once it forms, a charged jet is ejected and extended into a fine fiber-like thing that gets solidified and deposited onto the collector which is depicted in Fig. 5. A continuous and ultrathin polymeric fiber in various hierarchical patterns can be obtained from electrospinning [35]. The human ECM is a three-dimensional structure of 1–500 nm diameter, thus making the technique ideal for cellular spreading, motility, and lineage commitment. Due to the cationic charge of chitosan, high surface tension prevailed in the feed solution making the incorporation of chitosan with auxiliary polymers such as PEG, PVA, PEO, PCL, and PLA [36]. Chitosan-based electrospun fibers can mimic the anisotropic and complex structure of natural bone tissue and have shown to improve

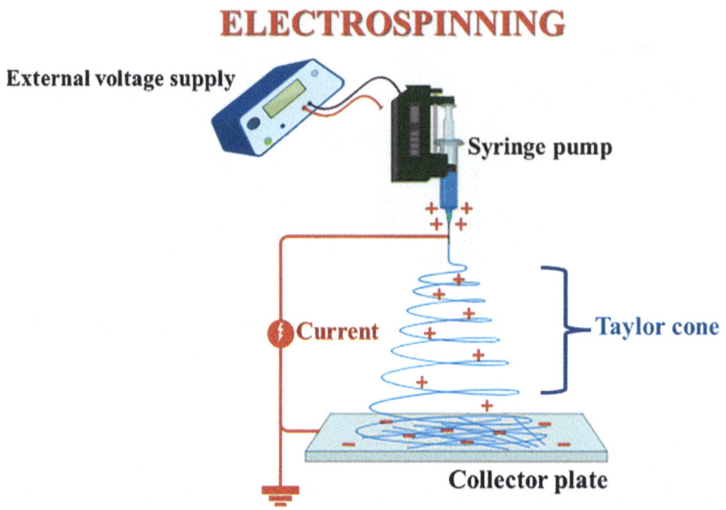

Fig. 5 Illustration of electrospinning setup

osteoblast adhesion, proliferation, and differentiation. Upcoming sections would further discuss about different applications in which chitosan-based scaffolds are employed.

3 Antimicrobial Agent Delivery Using Chitosan-Based Scaffolds

Antimicrobial resistance is a global alarm that diminishes the ability of standard antibiotics to treat widespread bacterial infections. The dramatic increment of multidrug-resistant infections, without a suitable prevention strategy, might compromise cell survival and increase the likelihood of implant failure. The recent reports emphasized the concern about the prevalence of resistance among dominant bacterial infections [37]. Chitosan is an exceptional choice for antimicrobial drug delivery due to its intrinsic antimicrobial properties as well as to deliver extrinsic antimicrobial agents to the infected site for effective wound healing. Understanding the mechanistic action of chitosan in antimicrobial drug delivery is of utmost significance. By understanding the basic mechanisms, researchers can gain a better understanding of how chitosan interacts with microorganisms and develop strategies to enhance the efficacy. The antimicrobial properties of chitosan are due to the cationic charge present in the chitosan backbone [38]. Fabrication of antimicrobial chitosan scaffolds against the bacteria, fungi, and virus using a metal ion, quaternary ammonium compounds, carbon nanomaterials, or combined strategies gave a lot of consideration to the scientists for the effective treatment of antimicrobial infections.

This inherent positive charge of chitosan can attach itself to the negative residue of the bacterial cell wall, which alters the membrane permeability and results in the cell wall rupture. The chitosan can form a polymeric membrane outside the bacterial cell wall and acts as a barrier by the inhibition of oxygen supply for bacterial growth [39]. The positively charged chitosan binds with bacterial DNA, inhibits microbial RNA production, and causes cellular destruction [40]. Li et al. compared the efficiency of non-cross-linked chitosan scaffold and chitosan-β-glycerophosphate cross-linked scaffold in the treatment of oral pathogens. The non-cross-linked chitosan scaffolds showed significant antibacterial activity as chemically cross-linked chitosan scaffold may lose its ability to hold a positive charge. It is confirmed that the antibacterial effect is induced not only by the electrostatic interaction between positively charged chitosan and negatively charged bacterial cells but also by the ability of bacterial cluster development [41].

3.1 Antibiotics

Since Alexander Fleming discovered penicillin in 1928, thousands of antibacterial drugs have been introduced into clinical research. However, due to extensive application, the rise of antimicrobial resistance limits the therapeutic effect of conventional treatments. Prolonged drug release is one of the major investigations for reducing the induction of antibiotic resistance. Owing to the short half-life values, a sustained-release formulation can enhance the desired therapeutic index of antibiotics. The incorporation of antibiotics within chitosan scaffolds shows a time-dependent release and longer duration [42]. A hybrid chitosan-ECM scaffold augmented with rifampicin and minocycline for hemostatic and antimicrobial dressing was prepared, and antibacterial studies showed some resistance to the rifampicin chitosan-ECM scaffold, but not to the minocycline-loaded chitosan-ECM scaffold. The efficacy continued for 2 h and also addressed that the prepared formulation could be effective for 24 h upon further scaffold modifications. The scaffolds augmented with rifampicin and minocycline demonstrated an enhanced antibacterial effect and easy release of drugs from the scaffolds [43]. A colistin/meropenem-loaded nanofibrous chitosan/PVA mat was fabricated against extensively drug-resistant *Acinetobacter baumannii*-skin wound infections. The optimized wound dressing containing 128/256 μg/mL of colistin/meropenem showed exceptional antibacterial activity, and cell viability assays showed good biocompatibility on L929 fibroblast cells. In vivo studies on the subcutaneous mouse model showed no adverse tissue reactions, severe inflammatory responses, or rejection. The engineered colistin/meropenem-loaded chitosan/PVA scaffold showed a synergistic effect of colistin in affecting the bacterial cells and allowed the easy entry of meropenem to the periplasmic space to destroy bacteria [44].

3.2 Metals/Metallic Nanocomposites

The incorporation of metallic nanoparticles like silver and zinc oxide (ZnO) into chitosan scaffolds/bandages gained much attention in clinical scenarios for bone-related infections. The released silver ions from nanoparticles can enhance the bactericidal activity [45]. Chitosan-zinc oxide nanocomposite scaffold enhanced the penetration and accumulation to the deeper tissues in the acidic environment [46]. ZnO nanoparticles incorporated in chitosan scaffold can exhibit high antimicrobial activity against broad-spectrum antibacterial action demonstrating better antibacterial action [47]. Zhang et al. prepared carboxymethyl chitosan/alginate hydrogels with copper, which is integrated with 3D printed polyhydroxyalkanoates/β-tricalcium phosphate scaffolds for MRSA-related infections, thus accelerating bone repair of cranial defects. The synthesized scaffold exhibited outstanding antibacterial activity against gram-positive and gram-negative bacteria. The antibacterial effect of Cu^{2+} disrupts bacterial cell membranes and intracellular protein and DNA. The fluorescence images showed a green fluorescence signal in MRSA cocultured with the developed scaffold, demonstrated high level of ROS generation induced by Cu, and revealed antibacterial potentiality of the prepared scaffold towards MRSA-related bone infections [48]. Karakecili et al. developed fosfomycin, a bactericidal antibiotic carrying an anti-infective platform of 3D chitosan scaffold incorporated with zirconium-based metal-organic framework (MOF)-UiO-66 nanocrystals. A strong antibacterial action against *S. aureus* and a triggered osteogenic differentiation of MC3T3-E1 pre-osteoblasts were observed in the developed scaffold [49]. In another study, chitosan-coated copper metal-organic framework (Cu-MOF) nanoparticles loaded into polydimethylsiloxane (PDMS) nanocomposite sponges were developed for enhanced antibacterial effect. The fabricated chitosan-coated Cu-MOF@PDMS sponges showed a synergistic antibacterial efficacy due to the presence of chitosan and Cu-MOF [50]. The potentiality of chitosan scaffolds incorporated with metals/metallic nanocomposites will aid in osteogenesis with diminished bacterial adhesion to the implanted scaffold.

3.3 Metallic Ion-Doped Bioceramics

The application of bioceramic-based scaffolds can enhance the bactericidal and inhibition effect for enhanced bone-related infections. The degradation characteristics of bioceramics infused with antibacterial ions allow them to work efficiently against bacteria. The ions released due to the degradation of the bioceramic-based scaffold elevate the pH of the surrounding microenvironment and result in the detrimental growth of acidophilic bacteria. The functional elements like Ag^{+}, Zn^{2+}, and Cu^{2+} are doped or coated into bioceramics in the form of oxides, ions, or micro or nanoparticles [51]. Then this functional element is incorporated into

Table 2 List of antibacterial chitosan-based bioceramic/bioglass or combined scaffolds

Scaffold components	Bacterial species	Application	References
Chitosan, hematite-doped bioglass	*Enterococcus faecalis*	Endodontic treatment	[53]
Chitosan, titanium dioxide, bredigite	*Escherichia coli*, *Staphylococcus aureus*	Bone tissue engineering	[54]
Chitosan, gelatin, PLA, *n*-hydroxyapatite	*Escherichia coli*, *Staphylococcus aureus*	Spongy bone regeneration	[55]
Chitosan, hydroxyapatite, α-tricalcium phosphate, titanium and copper ions	*Escherichia coli*, *Staphylococcus aureus*	Orthopedic applications	[56]
Quaternized chitosan, polylactide-*co*-glycolide, hydroxyapatite	*Staphylococcus aureus*, MRSA (ATCC 43300) strains	Bone tissue engineering	[57]
Chitosan, PLA, coated with cerium, copper, silver-doped bioactive glass	*Escherichia coli* (ATCC 25922 strains)	Bone tissue engineering	[58]
Chitosan, PCL, 58 s bioactive glass	*Staphylococcus aureus*	Bone tissue engineering	[59]
Chitosan, alginate, β-cyclodextrin-modified polydopamine-bioactive glass nanoparticles	*Escherichia coli*, *Staphylococcus aureus*	Wound healing and skin regeneration	[60]
Chitosan, collagen, bioglass	*Escherichia coli*, *Staphylococcus aureus*	Periodontal tissue regeneration	[61]
Chitosan, bioactive glass	–	Diabetic wound healing	[62]

chitosan scaffolds to enhance bacteriostatic activity, osteoinductivity, and osteoconductivity to treat osteomyelitis. The clinical studies showed a significant increment in bacterial infections followed by implantation, resulting in the creation of surgical trauma to the patients. The primary obstacle faced in the treatment of orthopedic infections includes bacterial colonization along with biofilm formation on the implant causing foreign body reactions and thus resulting in implant rejections. In this regard, utilizing combined chitosan-bioceramic scaffolds offers enhanced strength to combat infections related to bone implants [52]. Table 2 further discusses the different chitosan-based scaffolds conjugated with bioceramic that are applied for different applications.

3.4 Carbon Nanomaterials

Carbon-based nanomaterials (CNMs) such as carbon nanotubes (CNTs), graphene and GO, fullerenes, and carbon dots (CDs) are regarded as physical analogs of ECM components like collagen fibers, owing to their comparable dimensions. Their outstanding mechanical properties play a major role in reinforcing the organic as well as inorganic artificial scaffolds [63]. Chitosan-reinforced multifunctional graphene-silver-polycationic peptide nanocomposite scaffolds in film and sponge form were fabricated for antibacterial hemostatic and wound dressing [64]. The presence of graphene in the high surface area accounts for its unique mechanical strength and poor degradation properties.

The antibacterial ability of graphene-embedded chitosan-polyvinyl alcohol nanofibers was evaluated against *Escherichia coli*, *Agrobacterium*, and yeast species. The result indicated a remarkable activity towards bacterial species and demonstrated the selective efficiency of graphene in killing prokaryotic cells but not eukaryotic cells. The easy entry of electrons from graphene nanosheet towards prokaryotic cells, due to the absence of nuclear membrane, accounts for the antibacterial ability of graphene. In vivo studies conducted on mouse and rabbit models exhibited effective wound-healing ability for 15 days of treatment [65] Shrestha et al. developed a chitosan-zein scaffold incorporated with multiwalled carbon nanotubes for bone cell regeneration. The antibacterial studies demonstrated a notable increment in bacterial reduction among the groups with multiwalled carbon nanotube scaffolds, indicating the positive influence of the fabricated composite scaffold in antibacterial efficacy along with better growth and differentiation of MC3T3-E1 cells in the absence of osteogenic growth factors [66]. The family of carbon nanomaterial-based scaffolds shows remarkable ability in cell adhesion, osteogenic proliferation, and differentiation.

3.5 Antifungal

The rising fungal resistance poses a global public concern. The currently available antifungals including fungicides limit its usage due to drug resistance. Therefore, the urgency for novel antifungals with good biocompatibility, non-toxicity, and easy abundance has gained wide acceptance. Chitosan has demonstrated strong antimicrobial activity and can prevent a variety of fungi, yeasts, etc. In general, chitosan has been reported to be highly successful in preventing radial growth, spore germination, and germ tube elongation. The antifungal action of chitosan is reported to be fungistatic, not fungicidal [67]. The antifungal properties are due to the electrostatic action of protonated amino groups of chitosan on the fungal cell surface and destroy the cell wall. Chitosan is considered a supreme antifungal candidate due to the alteration of fungal cell surface and triggers the nonspecific suppression of pathogens. The pH-responsive ability of chitosan is due to the weak polyamine class with a pK_a value of 6.5, and thus chitosan displays antifungal activity exclusively in acidic environments [68]. Ciftci et al. developed chitosan- and curcumin-loaded PCL composite fibers and evaluated the antifungal and antibacterial activity. Antifungal and antibacterial activity conducted on *Aspergillus niger* and *Penicillium digitatum* demonstrated a high broad-spectrum of antimicrobial activity at the lowest dose and revealed the potential efficacy of the developed composite fiber for antifungal and antibacterial properties [69]. Nimal et al. fabricated chitosan bandages of ciprofloxacin- and fluconazole-encapsulated fibrin nanoparticles for the treatment of polymicrobial wound infections. The prepared chitosan bandages showed sustained release for 14 days and demonstrated the absorption of surplus exudate from the wound site. The in vitro antimicrobial studies exhibited substantial bacterial and fungal inhibition. In vivo studies conducted on infected polymicrobial rat model revealed significant reduction of both bacterial and fungal growth, thus indicating its potentiality in treating antimicrobial infections [70].

4 Anti-inflammatory Drug Delivery Using Chitosan-Based Scaffolds

Inflammation is the key immune response initiated by the body against both exogenous and endogenous factors like pathogens, cell mutilation, and untreated injuries, respectively. Based on the timeframe, if the inflammation subsides within several hours to days, it is termed acute inflammation. In contrast, if the inflammation persists for more than several days, then it turns out to be chronic inflammation. During acute inflammation, fluid accumulation occurs due to increased blood vessel permeability allowing white blood cells to drift towards the site of tissue damage from main circulation [71]. During acute inflammation, early-phase mediators and late-phase mediators are released, which include vasodilators (NO, PG12, PGE2, PGD2), vasoconstrictors (endothelin), chemotactic factors, and cytokines. Neutrophils are the first responders to vasoactive mediators such as fibrins, kinins, complement factors, histamine, and serotonin [72]. Neutrophils release chemokines that attract other immune cells, especially macrophages. These cells exhibit a dynamic process called NETosis, where networks of extracellular fibers composed of neutrophil DNA sequester pathogens, thereby reducing the microbial load and eventually reducing the inflammation.

In response to chemotaxis, monocytes migrate to inflamed sites, where they differentiate into macrophages as shown in Fig. 6. Acidosis is a common feature at inflamed sites due to excessive oxygen usage by the immune cells. At hypoxic conditions, oxygen-independent glycolysis occurs leading to increased production of lactic acid. ROS-mediated oxidative damage distorts proteins and lipids [72]. Macrophages get polarized into M1 macrophages and M2 macrophages releasing pro-inflammatory and anti-inflammatory cytokines, respectively. Pro-inflammatory cytokines include tumor necrosis factor-α (TNF-α) and interleukin-1 (IL-1), and anti-inflammatory cytokines include IL-10, IL-6, and TGF-β [72]. Polarization of the macrophages occurs in response to the presence of pathogens by pathogen-associated molecular patterns (PAMPs). Pro-inflammatory cytokine level decreases by the production of anti-inflammatory cytokines produced by the M2 macrophages in response to damage-associated molecular patterns (DAMPs) [73]. The balance between M1 and M2 macrophages is crucial for wound healing. At the inflamed site, neutrophils release collagenases, lysozyme, myeloperoxidase (MPO), collagenase,

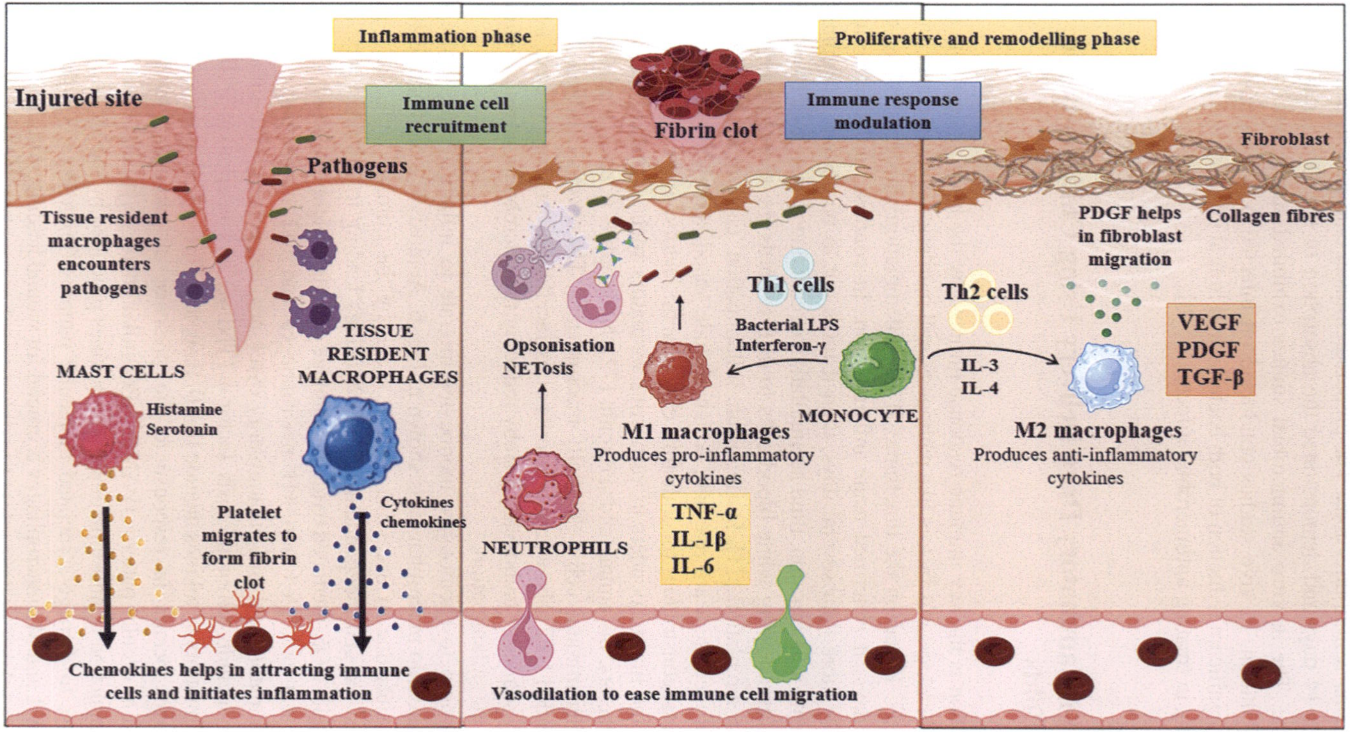

Fig. 6 Interplay of immune cells at the injured site

elastase, and gelatinase. Collagenase and elastase are involved in wound debridement. Lysozyme degrades bacterial cells by hydrolyzing the outer peptidoglycan layer. Myeloperoxidase converts hypochlorous acid produced by neutrophils into hydrogen peroxide and chloride ions; these free radicals produced help in eradicating the infection at the wound site [71]. A group of gelatinases called matrix metalloproteinases (MMP-2 and MMP-9) are tightly regulated by tissue inhibitors of metalloproteinases (TIMPs) [74]. In response to pro-inflammatory cytokines, MMPs gets upregulated and TIMPs are downregulated. The interplay of both MMPs and TIMPs is important to enhance the healing process.

From the above information, it can be concluded that the inflamed area possesses high acidosis and increased levels of free radicals lead to the overexpression of inflammation-associated enzymes, thereby intensifying the inflammatory response. The prolonged release of inflammatory enzymes modulates the expression of drug-metabolizing enzymes (cytochrome P450) leading to reduced drug metabolism [75]. This leads to increased drug concentrations in the systemic circulation, and prolonged exposure to the drug causes potential toxicity to the liver and kidney, which are primarily involved in drug metabolism and excretion, respectively. Further, nonselective delivery of drugs and high doses of drugs cause toxicity to normal cells. To overcome this limitation of traditional therapy, chitosan-based scaffolds can be loaded with specific anti-inflammatory drugs to treat the inflammation locally. Based on the microenvironment of the inflamed area, different delivery systems are used to treat inflammation by considering the sensitivity and specificity displayed at the inflamed site.

1. pH-responsive DDS: The materials respond to the lower pH of the inflamed site to release the therapeutic agents through hydrolysis of acid-labile bonds and protonation of ionizable groups. Acid-cleavable linkages like ortho-esters and acetal or ketal linkages exhibit accelerated hydrolysis in an acidic environment, thereby releasing the drug at the targeted site. At lower pH, amine moieties in polymers like chitosan are protonated, affecting their solubility and facilitating drug release [73].
2. ROS-responsive DDS: Chitosan can be chemically modified with thionic groups to form carboxymethyl chitosan, to assist as an antioxidant. This property helps in protecting cells from injury due to oxidative stress. This system responds to an oxidative environment by releasing therapeutic agents to higher levels of ROS.
3. Enzyme-responsive DDS: MMPs produced in response to higher levels of reactive species cleave chitosan at specific sites to facilitate the release of therapeutic agents [73].

Hence, an introduction of immunomodulatory biomaterials signifies a new tactic in regenerative medicine. These materials are designed to influence the immune system and form a supportive environment for tissue repair rather than simply being passive structures. Chitosan polymer with its adjustable physical and chemical properties helps in altering immune interaction [76]. Cationic polymers are used as cell-free nucleic acid scavengers to prevent the activation of innate immune pathways by inhibiting the Toll-like receptors (TLRs). Cationic polymers are more readily taken by the antigen-presenting cells, leading to more effective antigen presentation to T-lymphocytes and a strong adaptive immune response [77]. Hydrophobic polymers interact more efficiently with the DAMPs. The molecular weight of polymers can affect the anti-inflammatory regulatory T-cells. Lower-molecular-weight chitosan enhances the binding of regulatory T-cell generation via dectin-1-dependent PRRs [78]. Nanomechanics-based experiments provide an idea about the physical interactions at the molecular level, where interaction forces between chitosan and various pathogen recognition receptors (PRRs) were studied. Immunological assays focus on varying responses of molecular weight of chitosan and the ability to enhance Treg generation via dectin-1-dependent PRRs [79]. Fabrication of icariin into polydopamine-poly(L-lactide co-caprolactone)-citric acid-doped chitosan (PDA-PLCL/CC) scaffold highlights the effectiveness of the scaffolds in reducing pro-inflammatory cytokines and promotes chondrogenesis [80]. Similarly, Soriente et al. provide insights into how chitosan-hydroxyapatite scaffolds modulate cytokine production by increasing anti-inflammatory cytokines IL-4 and IL-10, by reducing TGF-β. This promotes a conducive environment for bone healing and regeneration [81]. Alternatively, another composite scaffold made of hydroxyapatite, alginate, chitosan, and fucoidan (HACF) modulates nitrite level, intracellular ROS, and mRNA levels of NF-κB, iNOS, COX-2, TNF-α, IL-1β, and IL-6. HACF has a strong antioxidant capacity to evade the ROS species to reduce oxidative stress. In vitro anti-inflammatory potential of this composite scaffold on LPS stimulated RAW 264.7 macrophages and showed its ability to suppress the inflammation by preventing the translocation of NF-κB [82]. In another study, genipin-cross-linked chitosan scaffolds effectively release pentoxifylline, an anti-inflammatory drug. The final composite scaffold suppresses the production of pro-inflammatory cytokines [83]. Therefore, the versatility of chitosan scaffolds benefits in developing customized scaffolds according to the pathogenesis dominant at the injury site.

5 Anticancer Drug Delivery Using Chitosan-Based Scaffolds

Cancer is a life-threatening disease that causes death due to aberrant call proliferation and spreads to many parts of the body. According to the statistics of 2020, 19.3 million people were at risk of cancer and around 10 million people lost their lives due to cancer. Current treatment modalities for anticancer treatment include chemotherapy, radiotherapy, immunotherapy, and surgical incision of tumors. Though these treatments increase the survival rate of patients, they cause major side effects in patients, including neuropathic issues, organ damage, etc. [84, 85]. Likewise, the oral or intravenous administration can affect the normal healthy cells while targeting the cancerous cells [86]. So, to overcome these limitations, the application of chitosan-based scaffolds loaded with anticancer drugs paves the way for the treatment. The use of therapeutic agents by conjugating them with chitosan shows high anticancer potential and minimal side effects in contrast to the original drugs due to their targeted drug delivery and sustained release of drugs [84]. Different forms of chitosan are used for release of cancer drugs [84, 87].

Chitosan-based materials can be degraded rapidly and are involved in enzymatic and nonenzymatic mechanisms that enhance hydrophobic drug solubility and improve therapeutic biodistribution in cancer cells [88]. Encapsulating drugs as nanocomposites in the form of hydrogels, films, and scaffolds along with chitosan acts as a nanovehicle or nanocarrier for drug delivery to the targeted site and has the potential to revoke the immunosuppressive niche of tumors and minimize the side effects [86, 89]. In an in vivo study, chitosan undergoes enzymatic degradation and decomposes into water and carbon dioxide ensuring no toxic end-product secretions [86]. In the tumor microenvironment, chitosan exhibits improved solubility and sustains better in acidic environments; this property of chitosan supports the development of the pH-sensitive DDS [86]. A scaffold-based drug carrier system has unique properties with specific pore sizes and structural designs that encapsulate the drug molecule within it and respond to biological signals. Different types of scaffolds, like polymeric scaffolds, use synthetic polymers like PEG and PLGA, natural scaffolds (collagen, silk), and nanocarriers [86, 90]. Chitosan-based scaffold tends to possess high porosity in nature that is interconnected with the drug molecules and has the capability of controlled release in the DDS [91]. These chitosan-based scaffolds show anti-tumor activities and are used to inhibit metastasis and proliferation as well as to induce apoptosis [92]. For instance, Luka Dornjak et al. demonstrated that in the case of osteosarcoma cancer, the doxorubicin-loaded chitosan functionalized with boric acid scaffold acts as a DDS of therapeutic agents and anticancer drugs. The results demonstrated that rapid delivery of boron and doxorubicin provides multifunctionality of the material. Those scaffolds demonstrated no cytotoxicity effects when incubated with the tumor cells, and these ingredients showed potential to target the patient-specific tumor cells [93]. Similarly, a group of researchers demonstrated that in ovarian cancer treatment, the bacterial cellulose-incorporated chitosan scaffold showed more mucoadhesive properties to cancerous cells and response to high tumorigenic activity towards cancer cells. The formation of reduced or very small aggregates was observed in the Cs-BC scaffold when compared to pure BC [94].

6 Biomolecule Delivery Using Chitosan-Based Scaffolds

From a tissue engineering perspective, biomolecules are signaling moieties that determine cellular behavior. These molecules exist in all living beings as small molecules and large molecules like protein, growth factor (GF), lipids, polysaccharides, primary and secondary metabolites, and other naturally derived compounds. To regulate cellular fate, it is a prerequisite to control the delivery of such biomolecules at the site of action using scaffolds. On that account, biologically concocted materials are required to regulate the release and presentation of biomolecules in the cellular microenvironment for tissue remodeling at the target site [95, 96]. Upcoming sections will discuss the different biomolecules, challenges in delivery, and the way chitosan-based scaffolds mitigate them.

6.1 Growth Factors

GFs are biomolecules that play a major role in tissue remodeling by controlling the cellular development process such as cell differentiation, proliferation, and migration. These soluble proteins attach to the transmembrane receptors of the specific target cell and alter their cell fate by activating signaling cascades. To gain precise control over tissue development at the wound site, manipulating the secretions and regulating the release kinetic profiles of GF can greatly help [97, 98]. In spite of having many attractive features, GF often exhibits several inadequacies such as (1) brief half-lives, (2) insufficient stability owing to extensive proteolysis in physiological conditions, (3) enzymatic degradation, (4) administrating at higher doses for maintaining efficacy, (5) adverse reaction at higher concentration, and (6) exorbitant.

Hence, there comes the necessity for a three-dimensional polymeric framework that can effectively deliver or release GF in a sustained manner without degrading the protein itself. This method can facilitate extending the therapeutic efficacy of GF for a longer time at the site of action [99, 100]. Henceforth, several study groups worldwide investigate synthetic, natural, or combinatorial polymeric systems to deliver GF effectively. Among natural polymers, chitosan-based scaffolds are widely studied as a bioactive DDS for GF delivery owing to their fascinating features including biocompatible, polycationic, bacteriostatic, mucoadhesive, and biodegradable properties [101, 102]. For instance, a group of scientists fabricated a chitosan/polycyclodextrin (CHT/PCD) composite scaffold for vascular endothelial growth factor (VEGF) delivery to improve vascularization in bone. The outcome exhibited that the composite scaffold could release VEGF in a regulated manner in such a way that it could induce angiogenesis upon testing with endothelial cells; hence, the data suggest that the scaffold could be a promising DDS for VEGF [103]. Further emphasis on different GFs and their respective chitosan-based delivery systems is shown in Table 3.

6.2 Peptide and Proteins

Peptides are a class of compounds that contain a series of precisely organized amino acids of molecular weight ranging from 0.5 to 5 kDa that could be utilized for therapeutic purposes. Proteins and peptides can be formulated in such a way that they can bind to a vast array of biomolecules and are highly specific, considering the fact that they could make connections at various sites of the target molecule. This

Table 3 Different GFs and their chitosan-based delivery systems

Growth factors	Function	Chitosan-based scaffold for delivery	Outcomes	References
VEGF	Angiogenesis	VEGF-laden fibrin nanoparticle incorporated within chitosan-hyaluronic acid composite scaffold	• The scaffold enables effective controlled-release kinetic for inducing effective angiogenesis • The composite scaffold was able to induce angiogenesis	[104]
BMP	Bone formation	BMP 2 and HVP peptide-grafted chitosan scaffold	• The overall scaffold promoted cellular adhesion, proliferation, and osteoblast differentiation in vitro	[105]
TGF-β	Chondrocyte growth	Silk fibroin-chitosan scaffold loaded with TGF-β microspheres	• The overall scaffold enabled controlled release for 16 days • The results demonstrated that the novel system enhanced chondrogenesis in vivo	[106]
FGF	Nerve regeneration	Chitosan scaffold loaded with bFGF	• The scaffold enabled regulated release kinetics of bFGF • The composite scaffold had mechanical strength that benefits neuronal regeneration • The composite scaffold enabled the effective restoration of the 20 mm sciatic nerve in vivo	[107]
IGF	Wound healing	Chitosan sponge/PCL nanofiber composite scaffold loaded with IGF 2	• The scaffold enabled regulated release kinetics of IGF 2 • The scaffold demonstrated good biocompatibility, wettability, and vapor transmission, making it a better candidate for wound dressing in vitro	[108]

property of proteins and peptides enables researchers to utilize them in various applications such as anticancer, immunotherapy, regulating hormones, and communicable and noncommunicable diseases. Despite having many appealing features, they demonstrate few drawbacks, namely, reduced permeation through biological membranes and hence low bioavailability via oral and transdermal administration. The mode of administration still prevails to be intravenous, which is invasive and requires multiple doses that ultimately lead to patient incompliance. Consequently, many attempts are made to fabricate a delivery system that enables stability of the loaded peptide or protein and that ensures targeted and sustained release through various modes of administration. On that account, chitosan has been extensively utilized as an effective DDS for peptides and proteins owing to the fact that chitosan enables protection of the payload against physiological conditions, can penetrate biological barriers owing to their mucoadhesive property, and can access tight junctions among the epithelial cells [109–111]. On this basis, Khosravimelal et al. fabricated an electrospun nanofiber scaffold made up of chitosan/silk fibroin loaded with a cationic antimicrobial peptide derived from cecropin-melittin. The objective of the study was to subject it to a multidrug-resistant bacterial infected wound and assess the healing capability. In this study, chitosan was employed for its bacteriostatic, biodegradable, nontoxic, and feasible properties. These results suggested that the scaffold could effectively induce an antibacterial effect without causing cytotoxicity in vitro, and hence, it acts as an apt delivery vehicle for peptide delivery [112]. Similarly, a group of scientists formulated chitosan fiber hydrogel embedded with a combination of bioactive peptides like RGI and KLT by the wet electrospinning and mechanical stretching method. It is proposed to serve as a filler for nerve conduit injury made in the sciatic nerve in rats and then assess the nerve regeneration capability of the formulation in vivo. They utilized chitosan for its features, namely biocompatibility, degradability, and ease of fabrication into a nanofiber. At the end of the study, the scaffold was able to aid in effectively restoring sciatic nerve defects and is promising for clinical translation [113].

6.3 *Nucleic Acids*

Nucleic acid (DNA and RNA) delivery has certain benefits in protein antigens by vaccinations. To create multi-stimuli response-based scaffolds and their membranes as DDS, chitosan with its derivatives was extensively utilized [114, 115]. Conversely, they were typically used as DDS/wound-healing devices [116]. Electrolytic procedures, lyophilization, precipitation, and electrospinning of physicochemical cross-linked materials were among the approaches implemented for producing chitosan-based scaffolds. They functioned as either three-dimensional final shapes or in vivo gelling-produced stimuli-sensitive hydrogels. Depending on the intended purpose, they were charged with proteins, growth factors, nucleic acids, antibiotics, anti-inflammatory or analgesic drugs, or cells in addition to their exterior matrix function in cell proliferation [117].

The previous studies focused on the mRNA vaccines distributed under the surface of the skin even though most vaccines are typically given into the muscle for response and fewer side effects, at the injection site. For nucleic acid vaccines, specifically injecting them under the skin might trigger a robust immune reaction compared to other methods. However, a specific method is needed to deliver these vaccines through the layer of fat, beneath the skin. However, a delivery channel is required to inject into the subcutaneous fat layer. Based on their findings, the chitosan-liable scaffold is a viable choice for a subcutaneous distribution method of RNA vaccines [118]. The scaffolds have been created and extensively utilized for cell modification, medication, and gene administration but are also used to examine the loading of nucleic acid and its distribution [119, 120]. They found that the amount of protein antigen was noticeably lower than expected in their positive results. The mRNA released from the scaffold triggered a T-cell response and humoral reaction. To enhance the response further, it is suggested to consider administering both protein-based and mRNA vaccines, with the assistance of a chitosan scaffold. After 6–12 months, the long-term immune responses were assessed, particularly IFN-γ and memory B-cells. In the origin of an immune response to foreign antigens, memory B cells proliferate and enhance antigen-specific immunoglobulin levels to remove the antigen. In this context, we considered both the biomaterial and the mRNA they prepared that can boost the system response. The chitosan scaffold can activate cells and trigger the release of dendritic cell costimulatory markers. While using mRNA on its own as a vaccine booster, it is possible that its ability to trigger a response could be lessened by enclosing it in lipoplexes. To the promising effect in this study, the mRNA vaccine-mediated chitosan scaffold delivery is suggested [121].

6.4 Lipids

Lipid-based drug delivery serves as a feasible method to release medications in a defined and site-specific approach. Lipids are in a variety of molecular weights, making them more biocompatible. They are particularly pertinent for drugs not very soluble in water [122]. Hydrophilic and hydrophobic components that compose lipids can self-assemble to create an assortment of structures [123]. Genes and medications can be delivered using positively charged lipids [124]. A cationic head group and its coupler in the head with a water-repellent chain are composed of cationic lipids which are employed to carry genes [125]. As they mesh with the anionic plasma membrane and enhance cellular absorption, cationic lipids are preferred [126]. Nevertheless, cationic lipids have the potential to combine with serum protein to create aggregates [127]. The reticuloendothelial system (RES) also quickly eliminates cationic liposomes [128].

For instance, liposomes composed of neutral or anionic lipids in blood circulation may survive up to five times better [129] than those composed of cationic lipids [130]. Anionic lipid-based liposomes are more stable and facilitate a drug's pH responsiveness-based release in an acidic niche. The liposomes are highly stable at alkaline pH and unleash drugs at acidic pH [131]. Liposomes are double-layered vesicles made of phospholipids along with other amphiphilic compounds. They have acquired prominence as drug carriers [132]. Liposomes can encapsulate both hydrophilic and lipophilic drugs because they are nontoxic, biodegradable, and biocompatible. Liposomal encapsulation protects the drugs and permits the prospective release of a drug. The water-soluble medications may be enclosed in the liposome's intrinsic cavity, whereas hydrophobic drugs may be embedded into the double layer of phospholipids [133, 134]. Furthermore, liposomal drug delivery is examined from many perspectives such as spinal cord injury, neuron destruction, coagulation, and skin wounds [135]. A novel liposome/chitosan-based scaffold or human fibrin gel system has been created as a stable DD for less-molecular-weight water-soluble drugs. Tirofiban is a drug of antithrombosis that was utilized as a prototype medicine. For the production of the composite gel, human fibrin gel-enclosed liposomes based on tirofiban were embedded in chitosan-based scaffolds. The in vitro tirofiban release characteristics from composite gel were investigated by defining the contents of composite gel. The effects reveal that chitosan scaffolds enhance tirofiban release from LCSHFG [136].

7 Conclusion

In this chapter, we reviewed different delivery systems of chitosan-based scaffolds and methods by which they were fabricated. Chitosan-based scaffolds enabled effective delivery of antimicrobial agents and substantially inhibited the growth of both bacteria and fungi. Likewise, it was able to interact with immune cells and deliver anti-inflammatory drugs by which chitosan-based scaffolds effectively healed chronic wounds. Further, chitosan-based scaffolds helped in anticancer drug delivery with sustained release in a regulated manner, which substantially reduced tumor growth and were stable in acidic cancer microenvironments. Additionally, chitosan-based scaffolds facilitated the delivery of various biomolecules, such as GF, lipids, proteins, and lipids, and significantly improved their stability. Hence, chitosan-based scaffolds are one of the best DDSs for a broad range of applications including tissue engineering and regenerative medicine making it a versatile biomaterial in clinical application.

References

1. Echeverria Molina MI, Malollari KG, Komvopoulos K (2021) Design challenges in polymeric scaffolds for tissue engineering. Front Bioeng Biotechnol 9:617141
2. Sultana N (2018) Mechanical and biological properties of scaffold materials. In: Functional 3D tissue engineering scaffolds. Woodhead Publishing, pp 1–21
3. Sachdeva S, Mani A, Vora H, Saluja H, Manka N, Dehane V (2022) Scaffolds in tissue engineering. J Cell Biotechnol 8:95–103
4. Eldeeb AE, Salah S, Elkasabgy NA (2022) Biomaterials for tissue engineering applications and current updates in the field: a comprehensive review. AAPS PharmSciTech 23:267
5. Bharathi R, Shree Ganesh S, Harini G, Vatsala K, Anushikaa R, Aravind S, Abinaya S, Selvamurugan N (2022) Chitosan-based scaffolds as drug delivery systems in bone tissue engineering. Int J Biol Macromol 222:132–153
6. Lu T, Li Y, Chen T (2013) Techniques for fabrication and construction of three-dimensional scaffolds for tissue engineering. Int J Nanomedicine 8:337–350

7. de Souza MF, da Silva HN, Rodrigues JFB, Macêdo MDM, de Sousa WJB, Barbosa RC et al (2023) Chitosan/gelatin scaffolds loaded with Jatropha mollissima extract as potential skin tissue engineering materials. Polymers 15:603
8. Tsai WB, Ahmed IN (2023) The impact of polyethylene glycol-modified chitosan scaffolds on the proliferation and differentiation of osteoblasts. Int J Biomater 2023:4864492
9. Valencia C, Valencia CH, Zuluaga F, Valencia ME, Mina JH, Grande-Tovar CD et al (2018) Synthesis and application of scaffolds of chitosan-graphene oxide by the freeze-drying method for tissue regeneration. Molecules 23:2651
10. Shahbazarab Z, Teimouri A, Chermahini AN, Azadi M (2018) Fabrication and characterization of nanobiocomposite scaffold of zein/chitosan/nanohydroxyapatite prepared by freeze-drying method for bone tissue engineering. Int J Biol Macromol 108:1017–1027
11. Fregnan F, Ciglieri E, Tos P, Crosio A, Ciardelli G, Ruini F et al (2016) Chitosan crosslinked flat scaffolds for peripheral nerve regeneration. Biomed Mater 11:045010
12. Nga NK, Huyen TTT, Dung TN (2023) Solvent casting-particulate leaching synthesis of a nano-SiO2/chitosan composite scaffold for potential use in bone tissue engineering. Vietnam J Chem 61:605–611
13. Lim JI, Lee YK, Shin JS, Lim KJ (2011) Preparation of interconnected porous chitosan scaffolds by sodium acetate particulate leaching. J Biomater Sci Polym Ed 22:1319–1329
14. Sadeghianmaryan A, Naghieh S, Yazdanpanah Z, Sardroud HA, Sharma NK, Wilson LD et al (2022) Fabrication of chitosan/alginate/hydroxyapatite hybrid scaffolds using 3D printing and impregnating techniques for potential cartilage regeneration. Int J Biol Macromol 204:62–75
15. Hikmawati D, Zulaihah S, Aminatun A (2020) The design of 3D printed polylactic acid (PLA) morphology with coating of chitosan. AIP Conf Proc 2314:050005
16. Chavanne P, Stevanovic S, Wüthrich A, Braissant O, Pieles U, Gruner P et al (2013) 3D printed chitosan/hydroxyapatite scaffolds for potential use in regenerative medicine. Biomed Eng/Biomed Tech 58:000010151520134069
17. Shi XY (2020) Three-dimensional printed collagen/chitosan scaffold improves neurological recovery after spinal cord injury in rats. Chin J Tissue Eng Res 53:3474–3479
18. Vondran JL, Sun W, Schauer CL (2008) Crosslinked, electrospun chitosan–poly (ethylene oxide) nanofiber mats. J Appl Polym Sci 109:968–975
19. Pezeshki-Modaress M, Zandi M, Rajabi S (2018) Tailoring the gelatin/chitosan electrospun scaffold for application in skin tissue engineering: an in vitro study. Prog Biomater 7:207–218
20. Saderi N, Rajabi M, Akbari B, Firouzi M, Hassannejad Z (2018) Fabrication and characterization of gold nanoparticle-doped electrospun PCL/chitosan nanofibrous scaffolds for nerve tissue engineering. J Mater Sci Mater Med 29:1–10
21. Vieira T, Silva JC, Do Rego AB, Borges JP, Henriques C (2019) Electrospun biodegradable chitosan based-poly (urethane urea) scaffolds for soft tissue engineering. Mater Sci Eng C 103: 109819
22. Petrova VA, Golovkin AS, Mishanin AI, Romanov DP, Chernyakov DD, Poshina D et al (2020) Cytocompatibility of bilayer scaffolds electrospun from chitosan/alginate-chitin nanowhiskers. Biomedicines 8:305
23. Zarei M, Samimi A, Khorram M, Abdi MM, Golestaneh SI (2021) Fabrication and characterization of conductive polypyrrole/chitosan/collagen electrospun nanofiber scaffold for tissue engineering application. Int J Biol Macromol 168:175–186
24. Fereshteh Z (2018) Freeze-drying technologies for 3D scaffold engineering. In: Functional 3D tissue engineering scaffolds. Woodhead Publishing, pp 151–174
25. Merivaara A, Zini J, Koivunotko E, Valkonen S, Korhonen O, Fernandes FM et al (2021) Preservation of biomaterials and cells by freeze-drying: change of paradigm. J Control Release 336:480–498
26. Collins MN, Ren G, Young K, Pina S, Reis RL, Oliveira JM (2021) Scaffold fabrication technologies and structure/function properties in bone tissue engineering. Adv Funct Mater 31: 2010609

27. Cai SJ, Li CW, Weihs D, Wang GJ (2017) Control of cell proliferation by a porous chitosan scaffold with multiple releasing capabilities. Sci Technol Adv Mater 18:987–996
28. Ahmed S, Ali A, Sheikh J (2018) A review on chitosan centred scaffolds and their applications in tissue engineering. Int J Biol Macromol 116:849–862
29. Zhao P, Gu H, Mi H, Rao C, Fu J, Turng LS (2018) Fabrication of scaffolds in tissue engineering: a review. Front Mech Eng 13:107–119
30. Yadav LR, Chandran SV, Lavanya K, Selvamurugan N (2021) Chitosan-based 3D-printed scaffolds for bone tissue engineering. Int J Biol Macromol 183:1925–1938
31. Soundarya SP, Menon AH, Chandran SV, Selvamurugan N (2018) Bone tissue engineering: scaffold preparation using chitosan and other biomaterials with different design and fabrication techniques. Int J Biol Macromol 119:1228–1239
32. Yousefiasl S, Sharifi E, Salahinejad E, Makvandi P, Irani S (2023) Bioactive 3D-printed chitosan-based scaffolds for personalized craniofacial bone tissue engineering. Eng Regen 4: 1–11
33. Zhou L, Ramezani H, Sun M, Xie M, Nie J, Lv S et al (2020) 3D printing of high-strength chitosan hydrogel scaffolds without any organic solvents. Biomater Sci 8:5020–5028
34. Zulkifli MZA, Nordin D, Shaari N, Kamarudin SK (2023) Overview of electrospinning for tissue engineering applications. Polymers 15:2418
35. Maji K, Pramanik K (2022) Electrospun scaffold for bone regeneration. Int J Polym Mater 71: 842–857
36. Padaki M, Subramanya TM, Prasad D, Deon S, Jadhav AH (2021) Electrospun nanofibers: role of nanofibers in water remediation and effect of experimental variables on their nano topography and application processes. Environ Sci Water Res Technol 7:2166–2205
37. Mudenda S, Chabalenge B, Daka V, Mfune RL, Salachi KI, Mohamed S et al (2023) Global strategies to combat antimicrobial resistance: a one health perspective. Pharmacol Pharm 14: 271–328
38. Ke CL, Deng FS, Chuang CY, Lin CH (2021) Antimicrobial actions and applications of chitosan. Polymers 13:904
39. Confederat LG, Tuchilus CG, Dragan M, Sha'at M, Dragostin OM (2021) Preparation and antimicrobial activity of chitosan and its derivatives: a concise review. Molecules 26:3694
40. Matica MA, Aachmann FL, Tøndervik A, Sletta H, Ostafe V (2019) Chitosan as a wound dressing starting material: antimicrobial properties and mode of action. Int J Mol Sci 20:5889
41. Li J, Zhuang S (2020) Antibacterial activity of chitosan and its derivatives and their interaction mechanism with bacteria: current state and perspectives. Eur Polym J 138:109984
42. Ning LG, Kang ET, Wang YB, Hu XF, Xu LQ (2018) Recent developments in controlled release of antibiotics. Curr Pharm Des 24:911–925
43. Goller S, Turner NJ (2020) The antimicrobial effectiveness and cytotoxicity of the antibiotic-loaded chitosan: ECM scaffolds. Appl Sci 10:3446
44. Alizadeh S, Farshi P, Farahmandian N, Ahovan ZA, Hashemi A, Majidi M, Gholipourmalekabadi M (2023) Synergetic dual antibiotics-loaded chitosan/poly (vinyl alcohol) nanofibers with sustained antibacterial delivery for treatment of XDR bacteria-infected wounds. Int J Biol Macromol 229:22–34
45. Tang S, Zheng J (2018) Antibacterial activity of silver nanoparticles: structural effects. Adv Healthc Mater 7:1701503
46. Tabassum Z, Mohan A, Girdhar M (2024) Insight into recent trends in ZnO nanoparticle reinforced chitosan-based composite films for sustainable packaging: a critical review on its current status, challenges and perspective. Mater Today Proc. https://doi.org/10.1016/j.matpr.2024.05.069
47. Alavi M, Nokhodchi A (2020) An overview on antimicrobial and wound healing properties of ZnO nanobiofilms, hydrogels, and bionanocomposites based on cellulose, chitosan, and alginate polymers. Carbohydr Polym 227:115349

48. Zhang J, Ye X, Li W, Lin Z, Wang W, Chen L et al (2023) Copper-containing chitosan-based hydrogels enabled 3D-printed scaffolds to accelerate bone repair and eliminate MRSA-related infection. Int J Biol Macromol 240:124463
49. Karakeçili A, Topuz B, Ersoy FS, Sahin T, Gunyakti A, Demirtas TT (2022) UiO-66 metal-organic framework as a double actor in chitosan scaffolds: antibiotic carrier and osteogenesis promoter. Biomater Adv 136:212757
50. Ansari-Asl Z, Shahvali Z, Sacourbaravi R, Hoveizi E, Darabpour E (2022) Cu(II) metal-organic framework@polydimethylsiloxane nanocomposite sponges coated by chitosan for antibacterial and tissue engineering applications. Micropor Mesopor Mater 336:111866
51. Zhao C, Liu W, Zhu M, Wu C, Zhu Y (2022) Bioceramic-based scaffolds with antibacterial function for bone tissue engineering: a review. Bioact Mater 18:383–398
52. Tao F, Ma S, Tao H, Jin L, Luo Y, Zheng J et al (2021) Chitosan-based drug delivery systems: from synthesis strategy to osteomyelitis treatment–a review. Carbohydr. Polym. 251:117063
53. El-Sayed SA, Mabrouk M, Khallaf ME, Abd El-Hady BM, El-Meliegy E, Shehata MR (2020) Antibacterial, drug delivery, and osteoinduction abilities of bioglass/chitosan scaffolds for dental applications. J Drug Deliv Sci Technol 57:101757
54. Ghasemi S, Ghomi H (2021) Investigation of applying chitosan coating on antibacterial and biocompatibility properties of bredigite/titanium dioxide composite scaffolds. J. Biomater Appl 36:406–418
55. Rahman MM, Shahruzzaman M, Islam MS, Khan MN, Haque P (2019) Preparation and properties of biodegradable polymer/nano-hydroxyapatite bioceramic scaffold for spongy bone regeneration. J Polym Eng 39:134–142
56. Pańtak P, Czechowska JP, Belcarz A, Zima A (2024) The influence of titanium and cooper on physiochemical and antibacterial properties of bioceramic-based composites for orthopaedic applications. Ceram Int 51:1214–1226
57. Yang Y, Yang S, Wang Y, Yu Z, Ao H, Zhang H et al (2016) Anti-infective efficacy, cytocompatibility and biocompatibility of a 3D-printed osteoconductive composite scaffold functionalized with quaternized chitosan. Acta Biomater 46:112–128
58. Goh YF, Akram M, Alshemary A, Hussain R (2016) Antibacterial polylactic acid/chitosan nanofibers decorated with bioactive glass. Appl Surf Sci 387:1–7
59. Chen S, Galuskova D, Kankova H, Zheng K, Michalek M, Liverani L et al (2020) Electrospun PCL fiber mats incorporating multi-targeted B and Co co-doped bioactive glass nanoparticles for angiogenesis. Materials 13:4010
60. Zhang M, Fan Z, Zhang J, Yang Y, Huang C, Zhang W et al (2023) Multifunctional chitosan/alginate hydrogel incorporated with bioactive glass nanocomposites enabling photothermal and nitric oxide release activities for bacteria-infected wound healing. Int J Biol Macromol 232:123445
61. Tamilmani H, Srinivas BK, Suresh N, Saranya K (2024) Advancing periodontal care: development of a novel collagenchitosan-bioglass scaffold as a substitute for autologous soft tissue grafts. Cureus 16:e68644
62. Manjubaashini N, Bargavi P, Thomas NG, Krishnan N, Balakumar S (2024) Chitosan bioactive glass scaffolds for in vivo subcutaneous implantation, toxicity assessment, and diabetic wound healing upon animal model. Int J Biol Macromol 256:128291
63. Kassem A, Ayoub GM, Malaeb L (2019) Antibacterial activity of chitosan nano-composites and carbon nanotubes: a review. Sci Total Environ 668:566–576
64. Choudhary P, Ramalingam B, Das SK (2020) Fabrication of chitosan-reinforced multifunctional graphene nanocomposite as antibacterial scaffolds for hemorrhage control and wound-healing application. ACS Biomater Sci Eng 6:5911–5929
65. Dubey P, Gopinath P (2016) PEGylated graphene oxide-based nanocomposite-grafted chitosan/polyvinyl alcohol nanofiber as an advanced antibacterial wound dressing. RSC Adv 6:69103–69116

66. Shrestha S, Shrestha BK, Ko SW, Kandel R, Park CH, Kim CS (2021) Engineered cellular microenvironments from functionalized multiwalled carbon nanotubes integrating Zein/Chitosan @Polyurethane for bone cell regeneration. Carbohydr Polym 251:117035
67. Meng D, Garba B, Ren Y, Yao M, Xia X, Li M et al (2020) Antifungal activity of chitosan against Aspergillus ochraceus and its possible mechanisms of action. Int J Biol Macromol 158:1063–1070
68. Bazban-Shotorbani S, Hasani-Sadrabadi MM, Karkhaneh A, Serpooshan V, Jacob KI, Moshaverinia A et al (2017) Revisiting structure-property relationship of pH-responsive polymers for drug delivery applications. J Control Release 253:46–63
69. Ciftci F, Ozarslan AC (2024) Fabrication of polycaprolactone-chitosan/curcumin polymer composite fibers and evaluation of their in vitro release kinetic behavior and antibacterial-antifungal activity. J Sol-Gel Sci Technol 109:192–203
70. Thattaruparambil Raveendran N, Mohandas A, Ramachandran Menon R, Somasekharan Menon A, Biswas R, Jayakumar R (2018) Ciprofloxacin-and fluconazole-containing fibrin-nanoparticle-incorporated chitosan bandages for the treatment of polymicrobial wound infections. ACS Biomater Sci Eng 2:243–254
71. Maita KC, Avila FR, Torres-Guzman RA, Garcia JP, Eldaly AS, Palmieri L et al (2022) Local anti-inflammatory effect and immunomodulatory activity of chitosan-based dressing in skin wound healing: a systematic review. J Clin Transl Res 8:488
72. Placha D, Jampilek J (2021) Chronic inflammatory diseases, anti-inflammatory agents and their delivery nanosystems. Pharmaceutics 13:64
73. Wang Z, Qi F, Luo H, Xu G, Wang D (2022) Inflammatory microenvironment of skin wounds. Front Immunol 13:789274
74. Lebaudy E, Fournel S, Lavalle P, Vrana NE, Gribova V (2021) Recent advances in antiinflammatory material design. Adv Healthc 10:2001373
75. Lazar KM, Shetty S, Chilkoti A, Collier JH (2023) Immune-active polymeric materials for the treatment of inflammatory diseases. Curr Opin Colloid Interface Sci 67:101726
76. Raziyeva K, Kim Y, Zharkinbekov Z, Kassymbek K, Jimi S, Saparov A (2024) Immunology of acute and chronic wound healing. Biomolecules 11:700
77. Dou Y, Li C, Li L, Guo J, Zhang J (2020) Bioresponsive drug delivery systems for the treatment of inflammatory diseases. J Control Release 327:641–666
78. Ahmed S, Ikram S (2016) Chitosan based scaffolds and their applications in wound healing. Achiev Life Sci 10:27–37
79. Lee S, Byun S, Lee C, Park SH, Rudra D, Iwakura Y et al (2022) Resolving the mutually exclusive immune responses of chitosan with nanomechanics and immunological assays. Adv Healthc Mater 13:2102667
80. Li H, Wang X, Shen Y, Tang H, Tang X, Zhang Y (2022) [Chondrogenic and ameliorated inflammatory effects of chitosan-based biomimetic scaffold loaded with icariin]. Sheng Wu Gong Cheng Xue Bao (Chin J Biotechnol) 38:2308–2321
81. Soriente A, Fasolino I, Gomez-Sánchez A, Prokhorov E, Buonocore GG, Luna-Barcenas G et al (2022) Chitosan/hydroxyapatite nanocomposite scaffolds to modulate osteogenic and inflammatory response. J Biomed Mater Res A 110:266–272
82. Sumayya AS, Muraleedhara Kurup G (2021) In vitro anti-inflammatory potential of marine macromolecules cross-linked bio-composite scaffold on LPS stimulated RAW 264.7 macrophage cells for cartilage tissue engineering applications. J Biomater Sci Polym Ed 32:1040–1056
83. Lin HY, Yeh CT (2012) Genipin-crosslinked chitosan scaffolds and its efficacy in releasing anti-inflammatory medicine. Biomed Mater Eng 22:321–332
84. Atmaca H, Oguz F, Ilhan S (2024) Chitosan in cancer therapy: a dual role as a therapeutic agent and drug delivery system. Z Naturforsch C J Biosci 79:95–105
85. Shakil MS, Mahmud KM, Sayem M, Niloy MS, Halder SK, Hossen MS et al (2021) Using chitosan or chitosan derivatives in cancer therapy. Polysaccharides 2:795–816

86. Saberian E, Jenča A, Petrášová A, Zare-Zardini H, Ebrahimifar M (2024) Application of scaffold-based drug delivery in oral cancer treatment: a novel approach. Pharmaceutics 16:802
87. Sachdeva B, Sachdeva P, Negi A, Ghosh S, Han S, Dewanjee S et al (2023) Chitosan nanoparticles-based cancer drug delivery: application and challenges. Mar Drugs 21:211
88. Alsaikhan F, Farhood B (2024) Recent advances on chitosan/hyaluronic acid-based stimuli-responsive hydrogels and composites for cancer treatment: a comprehensive review. Int J Biol Macromol 280:135893
89. Ali M, Mir S, Abid OUR, Ajlouni AW, Ghafoor Alvi S, Bibi S (2023) Applications of chitosan based bionanocomposites in drug-delivery and anticancer treatment—a review. Eur Polym J 201:112576
90. Narmani A, Jafari SM (2021) Chitosan-based nanodelivery systems for cancer therapy: recent advances. Carbohydr Polym 272:118464
91. Florczyk SJ, Kievit FM, Wang K, Erickson AE, Ellenbogen RG, Zhang M (2016) 3D porous chitosan–alginate scaffolds promote proliferation and enrichment of cancer stem-like cells. J Mater Chem B 4:6326–6334
92. Maleki Dana P, Hallajzadeh J, Asemi Z, Mansournia MA, Yousefi B (2021) Chitosan applications in studying and managing osteosarcoma. Int J Biol Macromol 169:321–329
93. Dornjak L, Kovačić M, Ostojić K, Angaits A, Szpunar J, Urlić I et al (2025) Chitosan-boric acid scaffolds for doxorubicin delivery in the osteosarcoma treatment. Polymers (Basel) 14:4753
94. Ul-Islam M, Subhan F, Islam SU, Khan S, Shah N, Manan S et al (2019) Development of three-dimensional bacterial cellulose/chitosan scaffolds: analysis of cell-scaffold interaction for potential application in the diagnosis of ovarian cancer. Int J Biol Macromol 137:1050–1059
95. Bambole V, Yakhmi JV (2016) Tissue engineering: use of electrospinning technique for recreating physiological functions. In: Nanobiomaterials in soft tissue engineering. William Andrew, pp 387–455
96. Bishop CJ, Kim J, Green JJ (2014) Biomolecule delivery to engineer the cellular microenvironment for regenerative medicine. Ann Biomed Eng 42:1557–1572
97. Wang Z, Wang Z, Lu WW, Zhen W, Yang D, Peng S (2017) Novel biomaterial strategies for controlled growth factor delivery for biomedical applications. NPG Asia Mater 9:e435
98. Caballero Aguilar LM, Silva SM, Moulton SE (2019) Growth factor delivery: defining the next generation platforms for tissue engineering. J Control Release 306:40–58
99. Qu M, Jiang X, Zhou X, Wang C, Wu Q, Ren L et al (2020) Stimuli-responsive delivery of growth factors for tissue engineering. Adv Healthc Mater 9:1901714
100. Shan BH, Wu FG (2024) Hydrogel-based growth factor delivery platforms: strategies and recent advances. Adv Mater 36:2210707
101. Gohil SV, Padmanabhan A, Deschamps J, Nair LS (2017) Chitosan-based scaffolds for growth factor delivery. In: Chitosan based biomaterials, vol 2. Woodhead Publishing, pp 175–207
102. Liu Z, Wang K, Peng X, Zhang L (2022) Chitosan-based drug delivery systems: current strategic design and potential application in human hard tissue repair. Eur Polym J 5:110979
103. Palomino-Durand C, Lopez M, Marchandise P, Martel B, Blanchemain N, Chai F (2020) Chitosan/polycyclodextrin (CHT/PCD)-based sponges delivering VEGF to enhance angiogenesis for bone regeneration. Pharmaceutics 12:784
104. Mohandas A, Anisha BS, Chennazhi KP, Jayakumar R (2025) Chitosan-hyaluronic acid/VEGF loaded fibrin nanoparticles composite sponges for enhancing angiogenesis in wounds. Colloids Surf B Biointerfaces 127:105–113
105. Brun P, Zamuner A, Cassari L, D'Auria G, Falcigno L, Franchi S et al (2021) Chitosan covalently functionalized with peptides mapped on vitronectin and bmp-2 for bone tissue engineering. Nanomaterials 11:2784
106. He JM, Zhu PF, Li LH, Wang Z, Li XL, Wang S et al (2021) Silk fibroin/chitosan/TGF-β1-loaded microsphere scaffolds for cartilage reparation. Biomed Mater Eng 32:347–358

107. Liu F, Hao F, Hao P, Zhao W, Gao Y, Duan H et al (2021) bFGF-chitosan scaffolds effectively repair 20 mm sciatic nerve defects in adult rats. Biomed Mater 16:025011
108. Yang S, Lan L, Gong M, Yang K, Li X (2022) An asymmetric wettable PCL/chitosan composite scaffold loaded with IGF-2 for wound dressing. J Biomater Appl 37:577–587
109. Jain A, Jain A, Gulbake A, Shilpi S, Hurkat P, Jain SK (2013) Peptide and protein delivery using new drug delivery systems. Crit Rev Ther Drug Carrier Syst 30:293–329
110. Chen G, Kang W, Li W, Chen S, Gao Y (2022) Oral delivery of protein and peptide drugs: from non-specific formulation approaches to intestinal cell targeting strategies. Theranostics 12:1419
111. Wang L, Wang N, Zhang W, Cheng X, Yan Z, Shao G et al (2022) Therapeutic peptides: current applications and future directions. Signal Transduct Target Ther 7:1–27
112. Khosravimelal S, Chizari M, Farhadihosseinabadi B, Moosazadeh Moghaddam M, Gholipourmalekabadi M (2021) Fabrication and characterization of an antibacterial chitosan/silk fibroin electrospun nanofiber loaded with a cationic peptide for wound-dressing application. J Mater Sci Mater Med 32:1–11
113. Rao F, Wang Y, Zhang D, Lu C, Cao Z, Sui J et al (2020) Aligned chitosan nanofiber hydrogel grafted with peptides mimicking bioactive brain-derived neurotrophic factor and vascular endothelial growth factor repair long-distance sciatic nerve defects in rats. Theranostics 10:1590
114. Can AS, Erdal MS, Güngör S, Özsoy Y (2013) Optimization and characterization of chitosan films for transdermal delivery of ondansetron. Molecules 18:5455–5471
115. Racine L, Costa G, Bayma-Pecit E, Texier I, Auzély-Velty R (2017) Design of interpenetrating chitosan and poly (ethylene glycol) sponges for potential drug delivery applications. Carbohydr Polym 170:166–175
116. Biondi M, Ungaro F, Quaglia F, Netti PA (2008) Controlled drug delivery in tissue engineering. Adv Drug Deliv Rev 60:229–242
117. Vunain E, Mishra AK, Mamba BB (2017) Fundamentals of chitosan for biomedical applications. In: Jennings JA, Bumgardner JD (eds) Chitosan based biomaterials, vol 1, pp 3–30
118. Singh A, Peppas NA (2014) Hydrogels and scaffolds for immunomodulation. Adv Mater 26:6530–6541
119. Chen R, Ma H, Zhang L, Bryers JD (2018) Precision-porous templated scaffolds of varying pore size drive dendritic cell activation. Biotechnol Bioeng 115:1086–1095
120. Fan M, Ma Y, Tan H, Jia Y, Zou S, Guo S et al (2017) Covalent and injectable chitosan-chondroitin sulfate hydrogels embedded with chitosan microspheres for drug delivery and tissue engineering. Mater Sci Eng C 71:67–74
121. Yan J, Chen R, Zhang H, Bryers JD (2019) Injectable biodegradable chitosan-alginate 3D porous gel scaffold for mRNA vaccine delivery. Macromol Biosci 19:1800242
122. Shrestha H, Bala R, Arora S (2014) Lipid-based drug delivery systems. J Pharm 2014:801820
123. Zhao Z, Yao W, Wang N, Liu C, Zhou H, Chen H et al (2019) Synthesis and evaluation of mono- and multi-hydroxyl low toxicity pH-sensitive cationic lipids for drug delivery. Eur J Pharm Sci 133:69–78
124. Saengkrit N, Saesoo S, Srinuanchai W, Phunpee S, Ruktanonchai UR (2014) Influence of curcumin-loaded cationic liposome on anticancer activity for cervical cancer therapy. Colloids Surf B Biointerfaces 114:349–356
125. Gao H, Hui KM (2001) Synthesis of a novel series of cationic lipids that can act as efficient gene delivery vehicles through systematic heterocyclic substitution of cholesterol derivatives. Gene Ther 8:855–863
126. Radwan Almofti M, Harashima H, Shinohara Y, Almofti A, Baba Y, Kiwada H (2003) Cationic liposome-mediated gene delivery: biophysical study and mechanism of internalization. Arch Biochem Biophys 410:246–253
127. Li S, Tseng WC, Beer Stolz D, Wu SP, Watkins SC, Huang L (1999) Dynamic changes in the characteristics of cationic lipidic vectors after exposure to mouse serum: implications for intravenous lipofection. Gene Ther 6:585–594

128. Eliyahu H, Joseph A, Schillemans JP, Azzam T, Domb AJ, Barenholz Y (2007) Characterization and in vivo performance of dextran–spermine polyplexes and DOTAP/cholesterol lipoplexes administered locally and systemically. Biomaterials 28:2339–2349
129. Obata Y, Tajima S, Takeoka S (2010) Evaluation of pH-responsive liposomes containing amino acid-based zwitterionic lipids for improving intracellular drug delivery in vitro and in vivo. J Control Release 142:267–276
130. Zhao W, Zhuang S, Qi X-R (2011) Comparative study of the in vitro and in vivo characteristics of cationic and neutral liposomes. Int J Nanomedicine 6:3087
131. Aoki A, Akaboshi H, Ogura T, Aikawa T, Kondo T, Tobori N et al (2015) Preparation of pH-sensitive anionic liposomes designed for drug delivery system (DDS) application. J Oleo Sci 64:233–242
132. Chen J, Cheng D, Li J, Wang Y, Guo JX, Chen ZP et al (2013) Influence of lipid composition on the phase transition temperature of liposomes composed of both DPPC and HSPC. Drug Dev Ind Pharm 39:197–204
133. Mitragotri S, Burke PA, Langer R (2014) Overcoming the challenges in administering biopharmaceuticals: formulation and delivery strategies. Nat Rev Drug Discov 13:655–672
134. Değim Z, Çelebi N, Alemdaroğlu C, Deveci M, Öztürk S, Özoğul C (2011) Evaluation of chitosan gel containing liposome-loaded epidermal growth factor on burn wound healing. Int Wound J 8:343–354
135. Filipczak N, Yalamarty SSK, Li X, Khan MM, Parveen F, Torchilin V (2021) Lipid-based drug delivery systems in regenerative medicine. Materials 14:5371
136. Wang SS, Yang MC, Chung TW (2008) Liposomes/chitosan scaffold/human fibrin gel composite systems for delivering hydrophilic drugs—release behaviors of tirofiban in vitro. Drug Deliv 15:149–157

Adv Polym Sci (2025) 295: 213–234
https://doi.org/10.1007/12_2024_185

Published online: 10 December 2024

Applications of Chitosan Hydrogel in Drug Delivery

Joel Maria Joju, Sona Sunil, Megha Visakhan, and R. Jayakumar

Contents

1 Introduction 214
2 Drug Delivery Applications of Chitosan Hydrogel 216
2.1 Antibacterial 216
2.2 Antifungal 219
2.3 Anti-inflammatory 221
2.4 Anticancer 223
2.5 Gene Delivery 226
2.6 Peptide and Protein Delivery 228
3 Conclusion 229
References 230

Abstract Chitosan-based hydrogels are water-absorbing three-dimensional networks that are formed through different cross-linking mechanisms. The much-increased interest in using these hydrogels as delivering vehicles arises from their diverse physicochemical properties. The presence of hydrophilic groups like amine and hydroxyl groups contributes to its multitude of properties. The ability of these hydrogels to encapsulate, retain, and release bioactive molecules is made use in designing different drug-delivering systems across various biomedical fields. Recently, the focus has been shifted to fabricating "smart systems" that can respond to external stimuli, including light, temperature, pH, and magnetic field. Diverse molecules like genes, proteins, peptides, antibacterial, anticancer, antifungal, and anti-inflammatory drugs can be delivered using these hydrogel systems. This chapter reviews the recent progress in the fabrication and use of chitosan-based hydrogels in pharmaceutical delivery.

J. M. Joju, S. Sunil, M. Visakhan, and R. Jayakumar (✉)
Polymeric Biomaterials Lab, School of Nanosciences and Molecular Medicine, Amrita Vishwa Vidyapeetham, Kochi, India
e-mail: rjayakumar@aims.amrita.edu

Keywords Antibacterial · Anticancer · Antifungal · Anti-inflammatory · Chitosan · Gene delivery · Hydrogel · Peptide-protein delivery

1 Introduction

Hydrogels are cross-linked matrices that possess superabsorbent characteristics [1]. These extracellular matrix (ECM) mimicking structures can swell by imbibing large quantities of water without getting disintegrated [1, 2]. Hydrogels as drug-delivering vehicles are attracting much interest these days. Among them, the natural polymer chitosan (CS)-based hydrogels are investigated extensively due to their ease of fabrication, availability of raw materials, low cost, low toxicity, biocompatibility, biodegradability, and ability to be applied across various fields. CS is a linear random copolymer of D-glucosamine and N-acetyl-D-glucosamine units [3]. This polycationic polymer is sourced from chitin, which is originally extracted from crustaceans like crabs and shrimps [4]. The innate antibacterial activity of CS stems from the positive charge it possesses [5]. In addition, CS has a multitude of features, including antifungal, anticancer, anti-inflammatory, antioxidant, hemostatic, gelling, and mucoadhesive properties. Furthermore, the presence of functionalities like hydroxyl groups, amine groups, acetamide groups, and glycosidic linkages opens up the possibility of functional group modification, thereby yielding different CS derivatives [6].

The two key factors that influence the effectiveness of CS-based delivering vehicles include deacetylation degree and molecular weight. The solubility and hydrophobicity of the matrix are determined by these factors, which in turn can influence encapsulation efficiency [7]. Another factor that can contribute to the drug-delivering ability of CS-based hydrogel is its ability to open up epithelial tight junctions, arising from its inherent mucoadhesiveness [8]. CS hydrogels with different shapes and geometries can be prepared by making use of different cross-linking mechanisms [9, 10] (Fig. 1). Tunable swelling behavior can be achieved for CS hydrogels prepared via physical cross-linking. In this technique, the hydrogel can be prepared by simply mixing the required components in appropriate conditions. These gels are suitable for applications involving immediate delivery and are safe to use clinically. In order to overcome the demerit of inconsistent in vivo performance due to the lack of control over the pore size, degradation, and chemical functionalization in a physically cross-linked CS hydrogel, the use of small molecules and other polymers came into existence to synthesize chemically cross-linked CS hydrogels [9]. Other types of cross-linking mechanisms, such as self-gelation, double cross-linking, enzymatic cross-linking, and photo cross-linking are also being used in developing advanced hydrogels with versatile features. Thus, the cross-linking mechanism will greatly influence the properties of the prepared CS hydrogel.

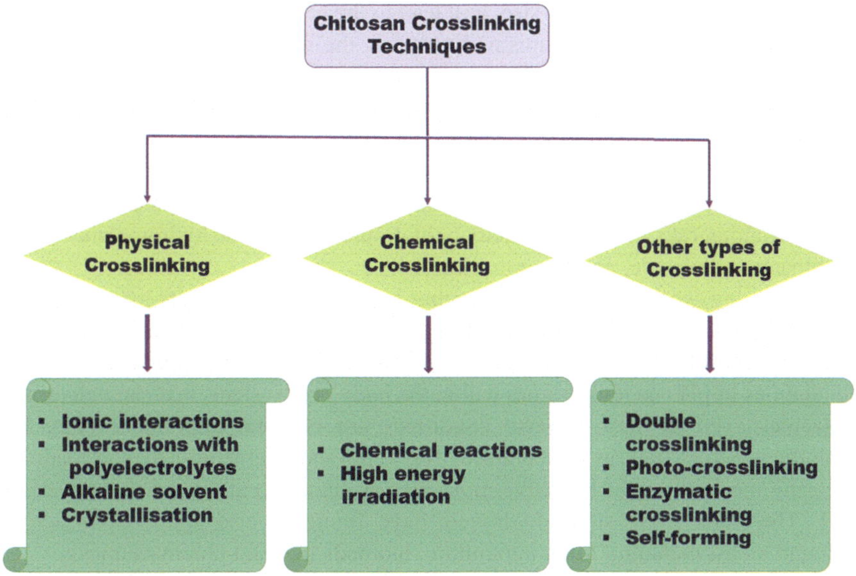

Fig. 1 Different types of cross-linking techniques

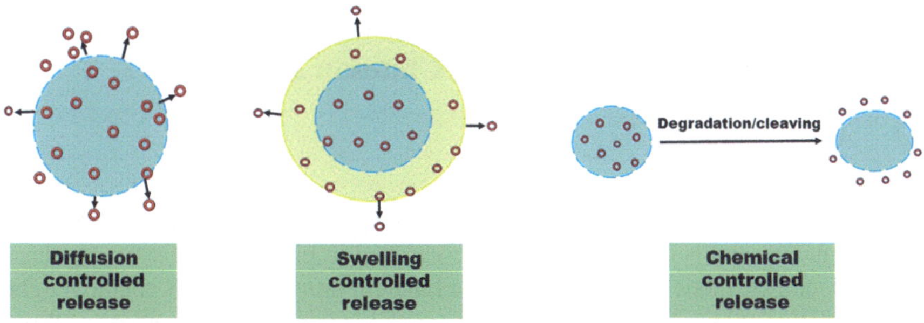

Fig. 2 Drug release mechanisms

The physicochemical properties of the prepared hydrogel will greatly influence its ability to be used as a drug-delivering matrix. In fact, the hydrogel preparation and drug loading mechanism should be carefully chosen based on the properties as well as the method of action of the drug. Permeation, entrapment, and covalent bonding are the three major drug-loading techniques, which possess their own unique advantages and disadvantages [9].

The drug-release-triggering mechanisms are the next factor that should be studied to understand the efficiency of the hydrogel matrix. The three most common release mechanisms are i) diffusion, ii) swelling, and iii) chemical control (Fig. 2) [10]. In diffusion control, the small-size drug will disperse from the hydrogel matrix through the pores which are bigger in size. On the other hand, the swelling of the hydrogel, which results in increased pore size, will lead to the release of the encapsulated drug in the swelling control mechanism. Lastly, in the chemical control mechanism, physical breakup via degradation or enzymatic or hydrolytic cleavage takes place, and this can cause the release of the entrapped drug. Release through diffusion is the most common mechanism, while release via chemical control can be employed if there is a significant interaction between the hydrogel framework and the drug molecule. Generally, the factors that determine bioactive release from the gel matrix are the swelling rate, the hydrophilicity or hydrophobicity of the hydrogel, the size of the bioactive molecule, and the pore size [10].

Present-day works focus on developing highly advanced smart hydrogels like stereocomplexed hydrogels and in situ hydrogels, which can respond to multiple external stimuli [11]. Conditions like inflammation, lesions, and injuries can lead to fluctuations in pH and temperature within the body, and systems that can deliver the molecules even in these adverse conditions are of great importance [12]. By exploiting factors such as temperature, pH, light, and magnetic field, the loading-releasing mechanisms of bioactive molecules are tuned to attain the desired effect [13]. Therefore, these smart hydrogels have found diverse applications across different areas, including food, agriculture, biomedicine, and pharmaceuticals.

This review chapter briefly summarizes the utilization of CS hydrogel in the pharmaceutical delivery of various bioactive molecules across different applications.

2 Drug Delivery Applications of Chitosan Hydrogel

2.1 Antibacterial

CS is a widely used cationic biopolymer that has gained extensive recognition for its versatile nature and is being thoroughly researched in biomedical, tissue engineering, food, and cosmetic industries. The vast inherent properties of CS enable it to be a candidate for exhaustive research. In this era, the increasing development of drug-resistant bacteria poses a threat, hence the current trend shifts to using efficient and effective bioagents that do not induce the conditions that promote drug resistance. The broad-spectrum antimicrobial action of CS makes it an excellent choice for antibacterial applications [14].

There is a rising focus on developing hydrogels based on CS to attain controlled and sustained delivery, especially in the case of antibacterial drug delivery. The advantage of using hydrogels is that they are delivered to the body by a minimally invasive process such as through a syringe, ruling out any need for surgery [15]. Another added advantage is that human cells can use up CS degradation products to produce ECM. The characteristic antibacterial efficacy of CS is solely due to its polycationic nature. The positive charge of CS will facilitate ionic interactions with the negative bacterial cell membrane, initiating membrane disruption, the alteration of membrane permeability, and the discharge of cytoplasmic components. This process is accompanied by the binding of CS to the bacterial

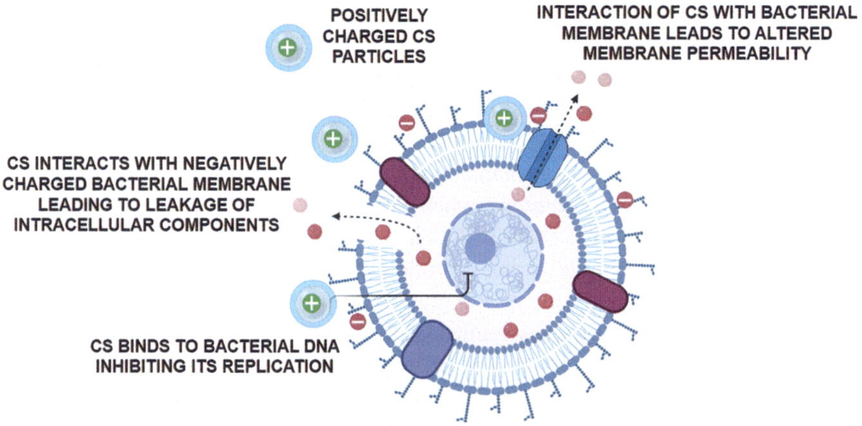

Fig. 3 Diagrammatic representation of the antibacterial mechanism of chitosan

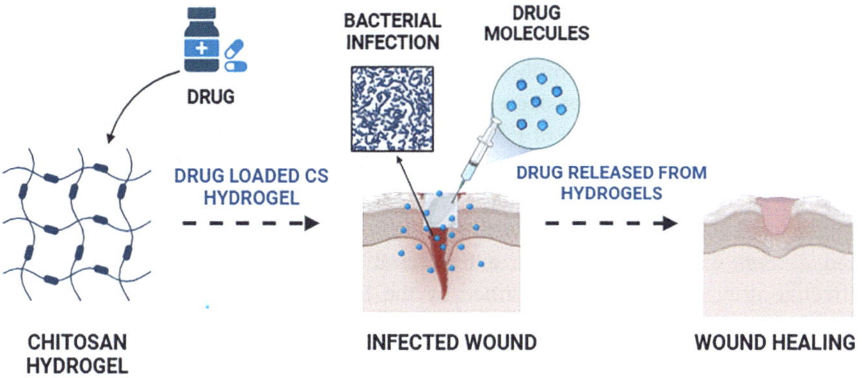

Fig. 4 Schematic representation of antibacterial-drug-loaded chitosan hydrogel in wound healing

DNA, preventing its replication and ultimately leading to cell death (Fig. 3). Therefore, as the deacetylation degree of CS increases, the amine group content in turn increases, which enhances the strength of positive charges, leading to stronger antibacterial action. Another mechanism proposed to indicate the antibacterial activity of CS suggests that its chelating capacity enables it to capture trace metal ions involved in the production of toxins. The intrinsic antibacterial activity of CS is also influenced by factors such as molecular weight, pH, ionic strength, chelating capacity, positive charge density, and the type of microorganism [16]. Even though CS possesses innate antibacterial activity, its broad-spectrum antibiotic action is only capable of controlling the infection and is not useful in eliminating the infection. To solve this obstacle, CS hydrogels have been incorporated with drugs that enhance the antibacterial activity of the hydrogel (Fig. 4). Recent developments include the preparation of composite hydrogels that combine CS with other polymers, which results in improved physical and mechanical properties.

A study showed the drug delivery efficiency and antibacterial activity of a hydrogel system based on *N*-trimethyl chitosan (TMC) and sodium carboxymethyl xanthan gum (CMXG) that encapsulated ciprofloxacin as the model drug. The drug release study indicated that when the drug loading efficiency of the hydrogel was higher, it exhibited faster and higher drug release. Furthermore, the hydrogel showed significant activity against *Escherichia coli*, *Klebsiella pneumoniae*, *Staphylococcus aureus*, and *Streptococcus mutans* when compared to the reference antibiotics (gentamicin and ampicillin). These results indicated that the TMC/CMXG hydrogel system was successful in delivering ciprofloxacin, which can be confirmed by its activity against both gram-positive and gram-negative bacterial strains [17].

Another hydrogel system was synthesized by electron beam irradiation that can release doxycycline (Doxy), where one hydrogel (CA-O-CMCS) was made from *O*-carboxymethyl chitosan (*O*-CMCS) and caffeic acid (CA) and another composite hydrogel which contained polyacrylamide (PAAm) (CA-*O*-CMCS/PAAm) in addition to the components of the former hydrogel. The Doxy release is caused by the diffusion of water molecules into the matrix, causing the hydrogel to swell and release the drug. The minor rate difference of drug release from both hydrogels is due to the fact that CA-*O*-CMCS/PAAm hydrogel has greater chemical potential for drug release and solvent adsorption, which creates a longer diffusion path for drug release because of its higher degree of swelling when compared to CA-*O*-CMCS hydrogels. The incorporation of PAAm resulted in higher drug release as PAAm is highly hydrophilic and enables more drug loading in its structure, resulting in higher swelling and subsequent drug release. By performing the disk diffusion test, the antibacterial action of Doxy-unloaded and Doxy-loaded hydrogels against *S. aureus* and *E. coli* was compared. The antibacterial efficacy of the hydrogels and an effective drug release were confirmed by the formation of the zone of inhibition by Doxy-loaded CA-*O*-CMCS and CA-*O*-CMCS/PAAm hydrogels against *S. aureus* and *E. coli*, where the latter gel had a higher zone of inhibition due to higher drug release when compared to Doxy-loaded CA-*O*-CMCS [18]. Antony et al. proposed a research study that deals with the preparation of an antiseptic CS hydrogel system that incorporates poly(hexamethylene) biguanide (PHMB) for treating infected wounds. The drug release study of the CS-PHMB hydrogel checked for 7 days indicated an initial burst release, followed by a controlled release of PHMB till day 7. The weak interaction between CS and PHMB is the reason for the initial burst release, and afterward, the release became steady, resulting from the weakening of the van der Waals' as well as hydrogen bonding. The antibacterial action of the CS-PHMB system was confirmed by the zone of inhibition the hydrogel showed against *K. pneumoniae*, *E. coli*, *Pseudomonas aeruginos*a, methicillin-resistant *S. aureus* (MRSA), *S. aureus*, and clinical strains of MRSA, *E. coli*, and *K. pneumoniae*. The hydrogel also showed a significant inhibition of biofilm formation for both *E. coli* and *S. aureus* [19]. Another research was conducted to prepare a promising material that enables controlled and sustained antibiotic delivery by cross-linking CS and dialdehyde xanthan gum interpenetrated with hydroxypropyl methylcellulose (CXH). The drug-delivering efficiency of this hydrogel system was studied using the following antibiotics: ampicillin trihydrate (AT), minocycline hydrochloride (MH), and rifampicin (RF). The drug release results of CXH hydrogel showed a burst release of 18–36% of the drug in under 6 h, followed by a gradual release of ≥50% of the drug after 48 h. The antibacterial activity of the hydrogels loaded with the three antibiotics, CXH-AT, CXH-MH, and CXH-RF, was tested against the bacteria *E. coli*, *S. aureus*, and *K. pneumonia*. A higher zone of inhibition was observed for the hydrogels loaded with the drugs (CXH-AT, CXH-MH, and CXH-RF) compared to the control (CXH). All three drug-loaded gels showed the highest activity against *S. aureus*. The control also showed low inhibition, which may be attributed to the antibacterial property of CS [20]. Recent developments aim at inhibiting the drug resistance resulting from the rapid release of the drug from the hydrogels. A group of researchers reported the fabrication of nanocomposite hydrogels using quaternized chitosan (QCS) and clindamycin-loaded hyperbranched nanoparticles via dual cross-linking. The results indicated a sustained and stable release of clindamycin in acidic pH. The controlled drug release allows the prevention of drug resistance, which can occur due to the initial burst release. The in vitro antibacterial studies showed excellent antibacterial activity against *E. coli*, *S. aureus*, and MRSA [21]. The use of smart biomaterials is gaining wide interest in biomedical applications. These types of biomaterials are sensitive to either chemical stimuli, physical stimuli, or biological stimuli, and based on these stimuli, the material interacts and functions with the biological surroundings. The stimuli-responsive hydrogels are well known for their application in controlled and sustained drug delivery [22]. Khan et al. fabricated a smart and pH-responsive hydrogel system that comprises arabinoxylan (ARX), CS, and reduced graphene sheets (rGO), which were cross-linked using tetraethyl orthosilicate (TEOS) to verify its application in skin wound healing. The composite hydrogels were synthesized by varying weight percentage of rGO. Silver sulphadiazine, a well-known antibiotic used for wound-healing applications, was chosen as the model drug to investigate the drug delivery efficiency of the prepared hydrogel. The drug release results showed a controlled release of the drug with 93.1% release (maximum) occurring at a physiological pH of 7.4 and minimum release at an alkaline pH of 8.4. The antibacterial efficacy of the hydrogel studied against *P. aeruginosa*, *S. aureus*, *E. faecalis*, and *E. coli* showed that all the composite hydrogels were effective, while ATC-3 (0.03% rGO) was found to be the best, forming the highest zone of inhibition against *P. aeruginosa*, *E. faecalis*, and *E. coli*. Hence, the studies conducted indicated that this composite hydrogel was a smart hydrogel responsive to pH, having controlled and efficient drug delivery and excellent antibacterial activity to be used for skin wound healing [23]. The above research studies are indications of the utilization of CS-based hydrogels as an antibacterial drug-delivering vehicle, which in turn can lead to the effective healing of wounds.

2.2 Antifungal

Fungal infections are a major health issue globally because of their tendency to recur, their widespread presence, and their resistance to multiple drugs, all of which can significantly impact the quality of life for patients. Traditional treatments often face limitations in their effectiveness due to poor water solubility and low bioavailability. While topical creams and oral medications are commonly used, their efficacy is uncertain and is often associated with numerous side effects [24]. Hydrogel-based delivery systems represent a promising strategy for the effective and controlled release of antifungal drugs. These systems not only enhance drug solubility but also improve drug permeability across the skin barrier [24]. CS can be utilized to develop hydrogels, wherein the CS matrix protects the drug from degradation and enables slow and sustained release [25]. The mucoadhesive property of CS allows it to adhere to mucosal surfaces, such as vaginal and oral mucosal membranes, and thereby increase the drug retention time [26]. Moreover, CS hydrogels can be formulated to encapsulate antifungal drugs like caspofungin, fluconazole, itraconazole, and difluorinated curcumin. The release of these drugs can be tuned by optimizing conditions such as pH, temperature, and enzymatic activity. This approach holds great potential for optimizing the therapeutic efficacy of antifungal treatments [24].

A thermosensitive hydrogel system was developed for treating the fungal infection onychomycosis. The system consists of CS, carboxymethylcellulose (CMC), scleroglucan, and montmorillonite (MMT) clay nanoparticles, with ketoconazole as the antifungal drug. Upon scanning electron microscopy (SEM) characterization, the gel showed a porous and interconnected network structure. Rheology studies demonstrated the gel's potential to transform into a high-viscosity gel from a low-viscosity liquid at a physiological temperature. Furthermore, swelling studies revealed higher swelling in acidic pH (similar to the nail plate) than at neutral pH, depicting the pH-sensitive behavior of the hydrogel. Additionally, the composite hydrogel demonstrated potential antifungal activity against *T. rubrum* and *T. mentagrophytes*. The synergistic effect of CS and MMT and the controlled release of ketoconazole from the hydrogel further enhanced the antifungal activity [27]. Another hydrogel-based drug delivery system was developed by using CS with three halogenated monoaldehydes: 4-fluoro, 4-chloro, and 4-bromo benzaldehyde. This system was then mixed with fluconazole, an antifungal drug. The antifungal activity of this drug delivery system was tested against four strains of *Candida* spp., and it was found that it was most effective against *C. parapsilosis*, showing a high inhibition zone diameter of up to 64 mm. When the untreated *Candida* cells were compared with the treated ones, it was observed that the hydrogel covered and surrounded the yeast cells, preventing them from forming biofilms, while untreated cells formed biofilms. Therefore, this CS-based hydrogel system has the potential to be used for the controlled and sustained release of antifungal drugs [28].

Another study highlighted the potential of caspofungin-loaded CS hydrogel derivatives grafted with [2-(methacryloyloxy)ethyl]dimethyl-(3-sulfopropyl)ammonium hydroxide (SBMA), in varying proportions, which showed very high swelling ability. Caspofungin was loaded at various concentrations, and initially, the hydrogels exhibited a burst release, followed by a sustained release. Additionally, the hydrogels displayed antibacterial activity and are suitable for topical application

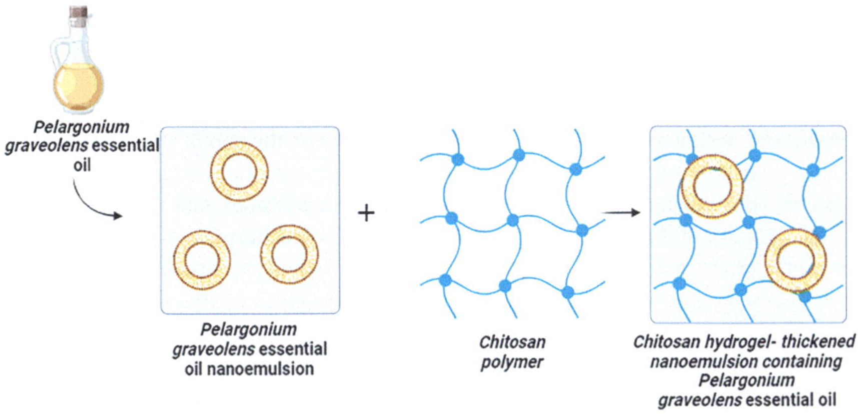

Fig. 5 Diagrammatic representation of *Pelargonium graveolens*-incorporated nanoemulsion developed from chitosan hydrogel for treating vaginal candidiasis

[25]. For treating vaginal candidiasis, *Pelargonium graveolens* essential oil containing nanoemulsion from CS hydrogel was developed (Fig. 5). In vitro antifungal studies tested against five different species of *Candida* spp. suggested that the system exhibited significant antifungal activity, with MIC (minimum inhibitory concentration) values 64 times lower than with the system containing essential oil alone. Stability studies indicated that the system remained stable for over 3 months. Therefore, this system containing CS hydrogel not only increases the stability and mucoadhesive properties of the nanoemulsion but also enhances its antifungal efficiency [29]. CS is a polymer with inherent antibacterial activity and limited antifungal activity. A group of researchers formulated a novel hydrogel based on CS (CS-3) and its nanoparticles (CS-3NP), prepared by cross-linking CS with heterocyclic compounds. The activity of the derivatives was investigated against multiple fungal strains, including *C. albicans*, *A. flavus*, and *A. niger*. The studies revealed that, compared to normal CS gel, CS-3 exhibited a significant zone of inhibition for *A. niger* and *Candida*. Also, in comparison to CS-NP hydrogels, CS-3NPs exhibited antifungal activity. The findings suggest that the modified form of CS has great potential in clinical applications, particularly in treating fungal infections [30].

2.3 *Anti-inflammatory*

Inflammation is a natural reaction of the body toward harmful or foreign particles that disrupt the natural homeostasis of the body. It is a very essential step as it acts as a cleaning process that enables the killing and removal of all the harmful microbes that may have entered through an infected wound. Even though its significance is not much to debate, inflammation can have adverse effects in cases of infected wounds, such as diabetic foot ulcers, leading to delayed healing. Hence, the demand for effective treatment for inflammation is rising worldwide with the increase in reports of patients suffering from chronic wounds. The current treatment methods involved fail to reduce inflammation efficiently due to the irregular delivery of drugs or the subsequent toxicity of the drug. The nonspecificity of the drugs in differentiating normal and inflamed tissue also plays a role in the downfall of current treatment methods. Researchers focus on developing efficient methods that enable the controlled delivery of the drug at the precise location. For this purpose, scientists all over the world are focusing their interest on developing biopolymer-based drug delivery systems. CS is one of the emerging polymers chosen for this application (Fig. 6) [31].

Sami et al. conducted a study that reported the use of CS-guar gum hydrogels as a drug delivery vehicle for the efficient delivery of the common analgesic drug paracetamol. This developed system aims to control the side effects of taking paracetamol orally, such as toxicity in the liver and complications in the gastric tract. The drug release of the transdermal patch was checked in the avian skin model by two methods: direct method and drug release through the chicken skin model. The release profile of paracetamol followed a diffusion mechanism that depended on the porosity of the hydrogel system. The results showed a linear trend, suggesting the sustained and controlled delivery of paracetamol. This drug displays an anti-inflammatory effect by inhibiting the cyclooxygenase (COX) enzymes. It was found that protein denaturation occurs as a result of inflammation. Hence, to check the biological activity of the drug released, inhibition of the denaturation of albumin protein was observed. The results indicated that the drug released showed full antiprotease activity, thereby exhibiting an anti-inflammatory response [32]. Another group of researchers developed a hydrogel system using CS and poly (vinyl pyrrolidone) as the core polymers that were cross-linked using epichlorohydrin for studying diclofenac sodium release. The mechanism of drug diffusion depended greatly on the swelling properties of the gel. It was observed that more than 90% of

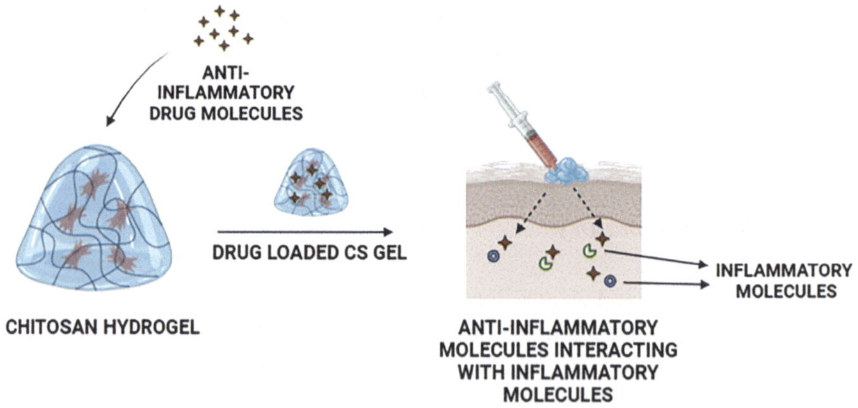

Fig. 6 Schematic representation of chitosan hydrogel-based anti-inflammatory drug delivery

the diclofenac sodium release occurred within 130 min in phosphate-buffered saline (PBS) (pH 7.4). They also observed that the cross-linking density of the hydrogel remained a significant factor affecting the release of drug molecules. As the cross-linking density of the gel increased, slower was the drug release [33]. Ailincai et al. showcased the use of diclofenac-sodium-salt-loaded citryl-imine PEG-ylated CS hydrogels as a fitting choice for controlled and sustained drug delivery. The release profile showed that the higher the swelling of the gel, the more the drug was released. The hydrogels with higher PEG content were the more hydrophilic ones, causing a faster drug release. As expected, the hydrogels with lower PEG content exhibited slower drug release. After around 7 days, the highly hydrophilic hydrogels released all of the drugs, whereas the less hydrophilic drug released only 91%. Hence, the initial burst release observed from the hydrogel makes it suitable for acute pain treatment, whereas its sustained and controlled release observed at the later stage makes it suitable for treating chronic conditions as well. This controlled release can be proved because of the strong interactions occurring between the drug and the imino-PEG-ylated CS hydrogels [34].

It is well known that inflammation is a critical step involved in the healing of a wound. However, when the inflammation stage is prolonged, it leads to delayed wound healing, which is not desirable. Hence the anti-inflammatory signals have to jump at the right time so as to take the wound to the next level of healing, which is proliferation. Olmo et al. proposed CS-genipin hydrogels that exhibited synergistic anti-inflammatory effects and investigated the release profile of acetylsalicylic acid (ASA), cefuroxime (CFX), tetracycline (TCN), and amoxicillin (AMX) from the system. It was found that the drug release differed between different types of hydrogels loaded with different drugs. This can be attributed to the differences in the interactions between different drugs and the hydrogels and other physiochemical properties of the drug and the hydrogels. ASA, an inflammatory drug, had the highest drug loading efficiency owing to its negative charge and low molecular weight. Controlled release was achieved via the interaction between anionic ASA and cationic CS-genipin hydrogels. The anti-inflammatory effect of ASA-loaded hydrogel was observed by checking whether pro-inflammatory cytokines were released by THP-1 cells. The results of the drug release of ASA indicated a reduction in the levels of pro-inflammatory cytokines such as TNF-α, IL-8, and IL-6, thereby confirming the anti-inflammatory response of the ASA-loaded hydrogels [35].

2.4 *Anticancer*

Cancer, which is driven by a combination of various genetic factors and environmental influences, remains a significant health crisis worldwide [36, 37]. Despite the availability of various antineoplastic therapies, such as chemotherapy, immunotherapy, and radiation therapy, these treatments often result in severe side effects due to their nonselective action and radiotoxicity [38, 39]. The requirement of high doses of these drugs due to their poor absorption can be toxic to normal cells, thus limiting its

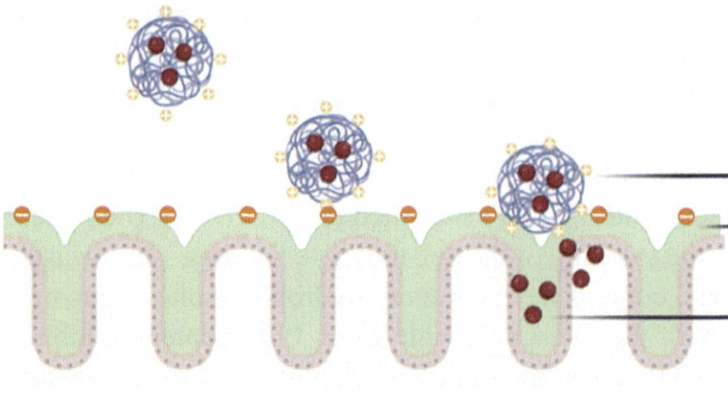

Fig. 7 Release of drug from chitosan hydrogel

potential. Many efforts have been put forward to develop more effective anticancer treatments that address the limitations of conventional methods. Among these innovations, hydrogels as drug-delivering systems are particularly promising due to their potential to incorporate and sustainably deliver chemotherapeutic agents selectively to the tumor site by establishing a diffusion barrier while sparing normal tissues [38]. In situ gelation and high loading capacity make hydrogels more advantageous than other types of drug delivery vehicles, facilitating the prolonged release of the drug with minimum toxicity. CS hydrogel has gained significant attention in this field due to its appealing properties, such as enhanced biodegradability, nontoxicity, and mucoadhesive properties [39, 40]. The mucoadhesive property can be exploited to specifically target the drug to the cancer site, where the production of mucin is ten times more than in normal cells, thereby increasing treatment efficiency by enhancing the contact time (Fig. 7) [38].

CS hydrogel beads were successfully prepared using gamma irradiation and then were copolymerized with L-glutamic acid. These beads showed great potential for delivering the anticancer drug doxorubicin (DOX). The pH-sensitive hydrogel beads release DOX primarily at a pH of 5.8, which is close to cancerous tissues' pH. A cell viability study conducted on the HEK-292 cell line showed about 95% cell viability and good biocompatibility. The anticancer efficiency of the DOX-loaded CS-g-GA hydrogel was tested using a trypan blue assay on the breast cancer cell line (MCF-7 cells), and it demonstrated a cancer cell toxicity of about 78.18%, indicating its potential for localized cancer therapy [38]. A study was conducted that encapsulated DOX inside thermoplastic polyurethane-polylactic-co-glycolic acid nanoparticles (TPU-PLGA-DOX NPs) and incorporated it into CS hydrogel (CTP-DOX NPs) for the sustained release of DOX. This hybrid system of hydrogel and nanoparticles is responsive to pH, which enables the sustained release of DOX and aids in its prolonged retention. The system exploits the mucoadhesive property of CS for treating cervical cancer by binding to the vaginal mucosal membrane to extend its residence time in the tissue. The encapsulation efficiency was calculated to be 71.5% for TPU-PLGA-DOX NPs and 65.8% for CTP-DOX NP hydrogel. Drug release studies on physiological and endolysosomal pH demonstrated the sustained release of the drug from the nanoparticles. Both systems exhibited pH-dependent release and rapid degradation in an acidic pH. CTP-DOX NP hydrogel exhibited more sustained release of the drug than the other one, which might be due to the synergistic effect of nanoparticles and hydrogel. Therefore, this hybrid system holds promise as a drug delivery system for DOX, reducing its cardiotoxicity and enhancing its stability [41].

For the sustained release of disulfiram (DSF), a thermosensitive and injectable CS-based hydrogel was developed via physical cross-linking, which quickly forms a gel at body temperature. Swelling studies showed that the hydrogel exhibited maximum swelling at 37 °C at a pH of 1.2, indicating that the effective delivery of DSF is achieved under physiological conditions. Rheological studies demonstrated the viscoelastic behavior of the injectable gel. Human hepatocellular carcinoma (HCC) cell lines (SMMC-7721 cells) were used for conducting cell cytotoxicity tests for 24 and 48 h. The results showed that toxicity was more dose-dependent compared to the free DSF solution. Confocal laser scanning microscopy (CLSM) and flow cytometer results were consistent and showed higher cellular uptake by CS/DSF hydrogel compared to the free drug [36]. The anticancer properties of curcumin and chrysin were investigated by incorporating them into an alginate-CS hydrogel, which was tested on lung cancer and breast cancer cell lines. In vivo cytotoxicity studies on the hydrogel exhibited dose-dependent and time-dependent antigrowth effects and halted the cell cycle in both A549 and T47D cell lines. Therefore, this alginate-CS hydrogel has the potential to serve as an effective drug delivery system [39]. In another work, methacrylate glycol CS (MGC) hydrogel was conjugated with 6-monodeoxy-6-monoamino-β-cyclodextrin·hydrochloride (6-NH2-β-CD·HCl), polyethylene glycol (PEG), and folic acid (FA) for the site-specific delivery of paclitaxel (PTX) (MGC/CDPF-ic-PTX). Anticancer studies done in xenograft models showed that MGC/CDPF-ic-PTX demonstrated better anticancer effects against breast cancer compared to the reference models MGC/PTX and MGC/CD-ic-PTX. Release profile studies indicated that the MGC-containing sample showed a burst release of PTX on the first day, followed by a sustained release for 7 days. It was also found that the swelling ratios of the samples gradually increased for 3 days before stabilizing. The results suggest that MGC/CDPF-ic-PTX has the potential for clinical applications [42].

Specialized CS hydrogel, modified with covalently bonded organophosphorus derivatives, was found suitable for the oral delivery of camptothecin. This happens because of the interaction between the positive CS and negative mucosal surfaces. The resulting hydrogel demonstrates adhesive properties, allowing for the extended retention of the drug at the administration site and consequently enhancing its absorption [43]. A novel hydrogel system was developed for the localized and pH-regulated delivery of cisplatin (CDDP) and DOX for lung cancer treatment. The hydrogel, developed with CS, nitrosalicylaldehyde, and hyaluronic acid, showed the enhanced inhibition of lung cancer cells when delivering both drugs together as opposed to individually. The study also demonstrated that the hydrogel swelled and released drugs more efficiently at an acidic pH (5.5), similar to the tumor microenvironment. Animal studies indicated that the hydrogel effectively inhibited tumor growth compared to free drugs, suggesting its potential to improve drug delivery and reduce side effects. This system shows promise for lung cancer therapy [44].

2.5 Gene Delivery

The necessity of gene therapy is rising worldwide due to an increase in inherited genetic conditions, viral infections, cancer treatment, etc. Gene therapy is a method that delivers genes into the target cell, which then leads to the expression of a transgene. The direct delivery of genes into the cell faces multiple challenges, such as nucleases present in the plasma, the presence of an anionic plasma membrane that inhibits the entry of negatively charged genes, and the nonspecificity of the genes to the target cell. Hence, efficient delivery vehicles are required for the protected transfection of genes into the target cell. Generally, viral vectors, such as adenovirus, are being used in gene therapeutic treatment because of their high transfection efficiency and high levels of gene expression [45]. The usage of these types of vectors has lost interest in gene delivery applications due to possible toxicity, undesirable immune responses, inflammation, and low loading capacity. Thus, the utilization of nonviral vectors for gene therapy has gained the utmost interest in the research field. However, amid the numerous advantages of nonviral vectors, such as low immune reaction, unrestricted gene material size, and large-scale production possibility, their gene transfection efficacy is lower than that of viral vectors. Cationic lipids and polymers have been well investigated as alternatives for viral vectors. CS, a cationic polymer, is extensively studied for targeted gene delivery due to its biocompatibility, low toxicity, ability to form complexes with negatively charged genes, etc. [46]. The applications of CS hydrogels in targeted gene delivery have been widely researched worldwide (Fig. 8).

A group of researchers reported the use of alginate-CS-based hydrogel embedded with polydeoxyribonucleotide-loaded calcium carbonate nanoparticles (PCNPs) for wound healing. The prepared system acts as an efficient delivery system as it was effective in releasing PCNPs, which were taken up by skin fibroblast cells. PCNPs

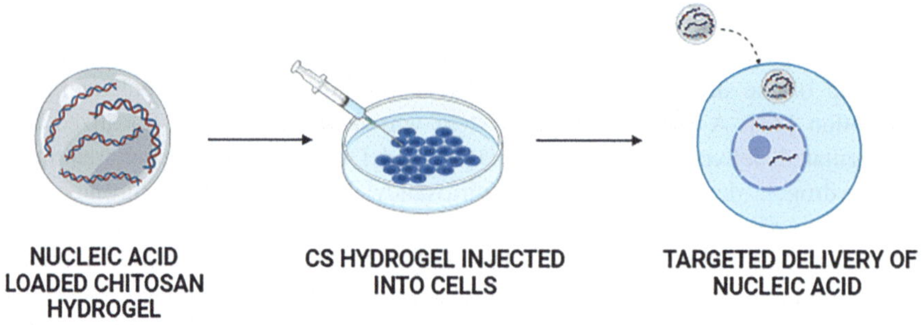

Fig. 8 Schematic representation of chitosan-hydrogel-based gene delivery

also serve as a reservoir of calcium ions, which help in cross-linking and in the gel formation process of alginate and also play a role in wound healing. The sustained release of PCNPs from the hydrogel system enabled the hydrogel to be molded before gelation, making it perfectly fit in the wound site [47]. Steinle et al. developed an injectable CS-alginate hybrid hydrogel system that can deliver synthetic messenger ribonucleic acid (mRNA) to the target site, which allows the site-specific and controlled expression of proteins in cells. This research study focused on encapsulating Cy3-labeled synthetic secretable humanized Gaussia luciferase (hGLuc) mRNA in alginate, CS, and CS-alginate hybrid hydrogels. The release kinetics of Cy3-labeled hGLuc mRNA was analyzed by measuring the fluorescence intensity emitted by it. The results showed a pattern of sustained release of hGLuc mRNA for 21 days for all hydrogel systems. The fastest release was observed for control alginate hydrogels, while the slowest release of Cy3-labeled hGLuc mRNA was observed for CS hydrogels. The most promising result was observed for alginate-CS hydrogel, exhibiting a faster release rate than CS and having a slower release rate than alginate. The interaction between anionic mRNA and cationic CS could be the reason for the slower release of mRNA from the CS-incorporated hydrogel systems. Hence, this study indicates that the release rate of mRNA can be controlled by using differently charged biomaterials [48].

The impact of smart biomaterials for various applications is expanding worldwide. Recent studies have been focusing on developing stimuli-sensitive smart CS hydrogels for localized gene delivery. A study reported the fabrication of an injectable thermoresponsive CS hydrogel system that loads both irinotecan (CPT-11)-loaded graphene oxide and SLP2 shRNA, targeted at the epidermal growth factor receptor (EGFR), which was found successful in local gene delivery in glioblastoma cells. The injectable hydrogel was synthesized by coentrapping SLP2 shRNA, and CPT-11 to cetuximab (CET)-conjugated GO (GO-CET/CPT11) in a thermosensitive CS-g-poly (N-isopropyl acrylamide) (CPN) polymer solution. The system works by getting endocytosed by U87 human glioblastoma cells that overexpress (EGFR) for pH-responsive drug release for increased antitumor efficiency. The CPT-11 and shRNA exhibited controlled and sustained release rates from the CPN hydrogel. Thus, the hydrogel system turned out to be a successful smart delivery system for gene delivery [49]. In another study, a smart thermoresponsive and degradable hydrogel gene delivery system was prepared via free radical polymerization of highly deacetylated CS and *N*-isopropyl acrylamide for delivering nucleic acids. A cationic amphipathic peptide, RALA, and pEGFP-N1 were loaded as the gene vector and nucleic acid, respectively, in the prepared hydrogel system. The synthesized RALA/pEGFP-N1 nanoparticles were directly loaded in the thermoresponsive hydrogel matrix. The release kinetics of RALA/pEGFP-N1 nanoparticles from the hydrogel system were observed and were found to show a rapid release for 8 h. It was then accompanied by a controlled release for up to 3 weeks. The functional activity of the delivered nucleic acid was confirmed by the successful transfection of the NCTC-929 fibroblast cell line [50].

2.6 Peptide and Protein Delivery

Proteins and peptides are abundant in our body and are the important constituents of many molecules, including hormones, antibodies, and enzymes. They have the potential to treat various diseases, such as diabetes and cancer, and are also used in therapies like gene therapy, wound dressings, and regenerative medicine. However, their clinical application is limited due to their inherent instability, poor bioavailability, and limited cell permeability [51]. Advancements in biotechnology have enabled the development of delivery vehicles that allow the controlled and sustained release of peptide molecules. One such system is hydrogels, which offer a promising solution. Hydrogels protect the peptide molecules from degradation and help retain them at the target site for a longer period. CS being a cationic polymer can be easily converted into a hydrogel due to its hydrophilic nature. This makes CS hydrogel a promising platform for delivering peptides, thanks to its biocompatibility, biodegradability, and controlled release capabilities [52].

The release of bovine serum albumin from an enzyme-sensitive CS-glycol hydrogel was studied using 3-(4-hydroxyphenyl)propionic acid (HPP) as the cross-linking agent. They also studied the impact of acetylation on the degradation of the hydrogel to improve its biodegradability. In vitro experiments, for testing the hydrogel's degradation in the presence and absence of lysozyme enzymes (known to degrade CS), showed that hydrogels with higher acetylation levels degraded faster. Furthermore, the in vivo study conducted on mice revealed that different acetylation levels led to varied protein release from the hydrogel when placed subcutaneously. This ability to control the degradation rate and protein release by adjusting the degree of acetylation could be exploited in developing hydrogels as a platform for delivering proteins, peptides, and other molecules [53]. In another study, microparticles from carboxymethyl β-cyclodextrin were synthesized for delivering insulin orally by grafting with carboxymethyl cellulose (CMCS). This system aims to protect insulin from decomposition in an acidic stomach environment and to provide controlled release. The hydrogel exhibits pH-sensitive swelling behavior, which is essential for maintaining the optimal therapeutic level of insulin. Release studies conducted in the simulated stomach and intestinal conditions exhibited that insulin was successfully preserved in matrix microparticles in the stomach environment and gradually released in the intestinal conditions. Intestinal permeability studies using a calcium carbonate ($CaCO_3$) monolayer further confirmed the hydrogel microparticle's enhanced insulin permeability across the intestinal barriers. Furthermore, studies conducted in a diabetic mice model exhibited a decrease in blood glucose levels when compared to the control groups. This type of oral delivery of insulin shows great potential as a treatment option for diabetic patients [54]. A self-healing and injectable hydrogel made of CS and konjac glucomannan, which is responsive to pH, was developed to control the release profile of protein. The release rate of the enclosed protein was assessed under different pH conditions. The hydrogel demonstrated pH-responsive behavior indicated by the higher amount of release at pH 7.2

than at pH 1.2. Furthermore, self-healing analysis showed the hydrogel's ability to restore its structure quickly after being subjected to mechanical damage [55].

Egle et al. fabricated a fucoidan/CS hydrogel for the sustained delivery of platelet-containing fibrin molecules to promote wound healing and tissue engineering. They conducted a protein release study using an enzyme-linked immunosorbent assay (ELISA) for analyzing growth factors, including VEGF, EGF, IL-8, PDGF-BB, and TGF-β1, over 7 days, which showed a gradual and sustained release. Encapsulation within the hydrogel significantly enhanced the release of platelet-rich fibrin (PRF) compared to pure PRF. This controlled release can be utilized for tissue engineering applications [52]. A CS hydrogel containing nanoparticles embedded with MK2 inhibitor YARA peptide (YARAAARQARAKALNRQGLVAA) was developed for treating atopic dermatitis. The focus was on improving localized drug delivery and enhancing anti-inflammatory effects. The inhibitor peptide loaded into the nanoparticles achieved a loading efficiency of greater than 50%. In vitro skin penetration studies comparing hydrogel-encapsulated YARA with nonencapsulated YARA, conducted using a porcine skin model, suggested that the hydrogel delivered two to four times more peptide than the nonencapsulated form. Additionally, in vitro release studies indicated that the YARA-loaded hydrogel provided a sustained release of the peptide over 120 h, demonstrating its effectiveness in treating atopic dermatitis [56].

3 Conclusion

This chapter briefly reviews the various applications in which CS hydrogel has been used as a delivering matrix. CS, due to its various characteristics, such as low toxic effects, great biocompatibility, biodegradability, mucoadhesiveness, inherent antibacterial activity, and anti-inflammatory effects, has been chosen as an ideal candidate for this purpose. The unique water absorption property of hydrogel contributes to designing it as a pharmaceutical carrier. Various constituents (such as antibacterial, antifungal, anticancer, and anti-inflammatory agents; proteins; peptides; and genes) have been incorporated within the CS hydrogel matrix to study their effective release into the site of need and their overall effectiveness in treating the disease. There is a long history of using CS hydrogel as a pharmaceutical delivery vehicle, and this demonstrates their safety aspect as a carrier matrix. In short, CS hydrogel is a smart biomaterial that can be used in delivering various pharmaceutical molecules to the targeted site.

Acknowledgments One of the authors, Joel Maria Joju, is grateful to Amrita Vishwa Vidyapeetham for providing her research fellowship. The authors are also thankful to Amrita Vishwa Vidyapeetham for providing seed grant fund.

References

1. Vashist A, Vashist A, Gupta YK, Ahmad S (2014) Recent advances in hydrogel based drug delivery systems for the human body. J Mater Chem B 2:147–166
2. Ou Y, Tian M (2021) Advances in multifunctional chitosan-based self-healing hydrogels for biomedical applications. J Mater Chem B 9:7955–7971
3. Ali A, Ahmed S (2018) A review on chitosan and its nanocomposites in drug delivery. Int J Biol Macromol 109:273–286
4. Al-Rooqi MM, Hassan MM, Moussa Z, Obaid RJ, Suman NH, Wagner MH, Natto SSA, Ahmed SA (2022) Advancement of chitin and chitosan as promising biomaterials. J Saudi Chem Soc 26:101561
5. Amirthalingam S, Rajendran AK, Mani P, Rangasamy J (2021) Perspectives and challenges of using chitosan in various biological applications. In: Rangasamy J, Prabaharan M (eds) Chitosan for biomaterials III: structure-property relationships. Springer, Cham, pp 1–22
6. Khanna N, Chatterji T, Singh S, Khanna N (2023) A medico-biological approach to chitosan bionanocomposites. In: Swain SK, Biswal A (eds) Chitosan nanocomposites: bionanomechanical applications. Springer, Singapore, pp 55–80
7. Ahsan SM, Thomas M, Reddy KK, Sooraparaju SG, Asthana A, Bhatnagar I (2018) Chitosan as biomaterial in drug delivery and tissue engineering. Int J Biol Macromol 110:97–109
8. Singh MK, Prajapati SK, Mahor A, Rajput N, Singh R (2011) Chitosan: a novel exepient in pharmaceutical formulation. Int J Pharm Sci Res 2:2266–2277
9. Bhattarai N, Gunn J, Zhang M (2010) Chitosan-based hydrogels for controlled, localized drug delivery. Adv Drug Deliv Rev 62:83–99
10. Almajidi YQ, Gupta J, Sheri FS, Zabibah RS, Faisal A, Ruzibayev A, Adil M, Saadh MJ, Jawad MJ, Alsaikhan F, Narmani A, Farhood B (2023) Advances in chitosan-based hydrogels for pharmaceutical and biomedical applications: a comprehensive review. Int J Biol Macromol 253:127278
11. Buwalda SJ, Boere KWM, Dijkstra PJ, Feijen J, Vermonden T, Hennink WE (2014) Hydrogels in a historical perspective: from simple networks to smart materials. J Control Release 190:254–273
12. Tian B, Liu J (2023) Smart stimuli-responsive chitosan hydrogel for drug delivery: a review. Int J Biol Macromol 235:123902
13. Eivazzadeh-Keihan R, Noruzi EB, Mehrban SF, Aliabadi HAM, Karimi M, Mohammadi A, Maleki A, Mahdavi M, Larijani B, Shalan AE (2022) The latest advances in biomedical applications of chitosan hydrogel as a powerful natural structure with eye-catching biological properties. J Mater Sci 57:3855–3891
14. Peers S, Montembault A, Ladavière C (2020) Chitosan hydrogels for sustained drug delivery. J Control Release 326:150–163
15. Domalik-Pyzik P, Chłopek J, Pielichowska K (2019) Chitosan-based hydrogels: preparation, properties, and applications. In: Ibrahim MD, Mondal H (eds) Cellulose-based superabsorbent hydrogels. Springer Nature, Cham, pp 1665–1693
16. Yilmaz Atay H (2019) Antibacterial activity of chitosan-based systems. In: Jana S, Jana S (eds) Functional chitosan: drug delivery and biomedical applications. Springer, Singapore, pp 457–489
17. Hanna DH, Saad GR (2019) Encapsulation of ciprofloxacin within modified xanthan gum-chitosan based hydrogel for drug delivery. Bioorg Chem 84:115–124
18. Hafezi Moghaddam R, Dadfarnia S, Shabani AMH, Amraei R, Hafezi Moghaddam Z (2020) Doxycycline drug delivery using hydrogels of O-carboxymethyl chitosan conjugated with caffeic acid and its composite with polyacrylamide synthesized by electron beam irradiation. Int J Biol Macromol 154:962–973
19. Antony IR, Pradeep A, Pillai AV, Menon RR, Kumar VA, Jayakumar R (2023) Antiseptic chitosan-poly(hexamethylene) biguanide hydrogel for the treatment of infectious wounds. J Funct Biomater 14:528

20. Ngwabebhoh FA, Zandraa O, Patwa R, Saha N, Capáková Z, Saha P (2021) Self-crosslinked chitosan/dialdehyde xanthan gum blended hypromellose hydrogel for the controlled delivery of ampicillin, minocycline and rifampicin. Int J Biol Macromol 167:1468–1478
21. Wei S, Liu X, Zhou J, Zhang J, Dong A, Huang P, Wang W, Deng L (2020) Dual-crosslinked nanocomposite hydrogels based on quaternized chitosan and clindamycin-loaded hyperbranched nanoparticles for potential antibacterial applications. Int J Biol Macromol 155: 153–162
22. Bordbar-Khiabani A, Gasik M (2022) Smart hydrogels for advanced drug delivery systems. Int J Mol Sci 23:3665
23. Khan MUA, Haider S, Raza MA, Shah SA, Razak SIA, Kadir MRA, Subhan F, Haider A (2021) Smart and pH-sensitive rGO/Arabinoxylan/chitosan composite for wound dressing: in-vitro drug delivery, antibacterial activity, and biological activities. Int J Biol Macromol 192:820–831
24. Rouf M, Ahmad Z, Mahmood D, Qavi Y, Shchinar S, Latif R (2024) Hydrogel-based delivery systems for topical antifungal therapy: a review. Cellul Chem Technol 58:425–436
25. Lazaridou M, Nanaki S, Zamboulis A, Papoulia C, Chrissafis K, Klonos PA, Kyritsis A, Vergkizi-Nikolakaki S, Kostoglou M, Bikiaris DN (2021) Super absorbent chitosan-based hydrogel sponges as carriers for caspofungin antifungal drug. Int J Pharm 606:120925
26. Araujo VHS, de Souza MPC, Carvalho GC, Duarte JL, Chorilli M (2021) Chitosan-based systems aimed at local application for vaginal infections. Carbohydr Polym 261:117919
27. Kancı Bozoğlan B, Duman O, Tunç S (2021) Smart antifungal thermosensitive chitosan/carboxymethylcellulose/scleroglucan/montmorillonite nanocomposite hydrogels for onychomycosis treatment. Colloids Surf Physicochem Eng Asp 610:125600
28. Ailincai D, Morariu S, Rosca I, Sandu AI, Marin L (2023) Drug delivery based on a supramolecular chemistry approach by using chitosan hydrogels. Int J Biol Macromol 248:125800
29. dos Santos MK, Kreutz T, Danielli LJ, De Marchi JGB, Pippi B, Koester LS, Fuentefria AM, Limberger RP (2020) A chitosan hydrogel-thickened nanoemulsion containing Pelargonium graveolens essential oil for treatment of vaginal candidiasis. J Drug Deliv Sci Technol 56: 101527
30. Ahmed MELS, Mohamed HM, Mohamed MI, Kandile NG (2020) Sustainable antimicrobial modified chitosan and its nanoparticles hydrogels: synthesis and characterization. Int J Biol Macromol 162:1388–1397
31. Nguyen HTT, Do NHN, Lac HD, Nguyen PLN, Le PK (2023) Synthesis, properties, and applications of chitosan hydrogels as anti-inflammatory drug delivery system. J Porous Mater 30:655–670
32. Sami AJ, Khalid M, Jamil T, Aftab S, Mangat SA, Shakoori AR, Iqbal S (2018) Formulation of novel chitosan guargum based hydrogels for sustained drug release of paracetamol. Int J Biol Macromol 108:324–332
33. Gull N, Khan SM, Butt OM, Islam A, Shah A, Jabeen S, Khan SU, Khan A, Khan RU, Butt MTZ (2020) Inflammation targeted chitosan-based hydrogel for controlled release of diclofenac sodium. Int J Biol Macromol 162:175–187
34. Ailincai D, Mititelu-Tartau L, Marin L (2020) Citryl-imine-PEG-ylated chitosan hydrogels – promising materials for drug delivery applications. Int J Biol Macromol 162:1323–1337
35. Andrade del Olmo J, Pérez-Álvarez L, Sáez-Martínez V, Benito-Cid S, Ruiz-Rubio L, Pérez-González R, Vilas-Vilela JL, Alonso JM (2022) Wound healing and antibacterial chitosan-genipin hydrogels with controlled drug delivery for synergistic anti-inflammatory activity. Int J Biol Macromol 203:679–694
36. Ahsan A, Farooq MA, Parveen A (2020) Thermosensitive chitosan-based injectable hydrogel as an efficient anticancer drug carrier. ACS Omega 5:20450–20460
37. Hou G, Sun T, Qian J, Zhang Y, Guo M, Xu W, Wang J, Suo A (2021) Hydroxyethyl chitosan hydrogels for enhancing breast cancer cell tumorigenesis. Int J Biol Macromol 184:768–775

38. Nisar S, Pandit AH, Nadeem M, Pandit AH, Rizvi MMA, Rattan S (2021) γ-Radiation induced L-glutamic acid grafted highly porous, pH-responsive chitosan hydrogel beads: a smart and biocompatible vehicle for controlled anti-cancer drug delivery. Int J Biol Macromol 182:37–50
39. Abbasalizadeh F, Alizadeh E, Bagher Fazljou SM, Torbati M, Akbarzadeh A (2022) Anticancer effect of alginate-chitosan hydrogel loaded with curcumin and Chrysin on lung and breast cancer cell lines. Curr Drug Deliv 19:600–613
40. Uddin MS, Ju J (2020) Effect of crosslinking agents on drug distribution in chitosan hydrogel for targeted drug delivery to treat cancer. J Polym Res 27:81
41. Manhas P, Sharma R, Wangoo N, Saini R, Saima AJN, Sharma RK (2023) Dox-loaded TPU-PLGA nanoparticles embedded chitosan hydrogel formulation as an effective anti-cancer therapy. J Mol Liq 383:122170
42. Hyun H, Park MH, Jo G, Lee BY, Choi JW, Chun HJ, Kim HS, Yang DH (2021) Injectable glycol chitosan hydrogel containing folic acid-functionalized Cyclodextrin-paclitaxel complex for breast cancer therapy. Nanomaterials 11:317
43. Martínez-Martínez M, Rodríguez-Berna G, Bermejo M, Gonzalez-Alvarez I, Gonzalez-Alvarez M, Merino V (2019) Covalently crosslinked organophosphorous derivatives-chitosan hydrogel as a drug delivery system for oral administration of camptothecin. Eur J Pharm Biopharm 136:174–183
44. Anirudhan TS, Mohan M, Rajeev MR (2022) Modified chitosan-hyaluronic acid based hydrogel for the pH-responsive co-delivery of cisplatin and doxorubicin. Int J Biol Macromol 201: 378–388
45. Chuan D, Jin T, Fan R, Zhou L, Guo G (2019) Chitosan for gene delivery: methods for improvement and applications. Adv Colloid Interf Sci 268:25–38
46. Cao Y, Tan YF, Wong YS, Liew MWJ, Venkatraman S (2019) Recent advances in chitosan-based carriers for gene delivery. Mar Drugs 17:381
47. Jing X, Sun Y, Liu Y, Ma X, Hu H (2021) Alginate/chitosan-based hydrogel loaded with gene vectors to deliver polydeoxyribonucleotide for effective wound healing. Biomater Sci 9:5533–5541
48. Steinle H, Ionescu TM, Schenk S, Golombek S, Kunnakattu SJ, Tutku Özbek M, Schlensak C, Wendel HP, Avci-Adali M (2018) Incorporation of synthetic mRNA in injectable chitosan-alginate hybrid hydrogels for local and sustained expression of exogenous proteins in cells. Int J Mol Sci 19:1313
49. Lu YJ, Lan YH, Chuang CC, Lu WT, Chan LY, Hsu PW, Chen JP (2020) Injectable Thermo-sensitive chitosan hydrogel containing CPT-11-loaded EGFR-targeted graphene oxide and SLP2 shRNA for localized drug/gene delivery in glioblastoma therapy. Int J Mol Sci 21:7111
50. Ziminska M, Wilson JJ, McErlean E, Dunne N, McCarthy HO (2020) Synthesis and evaluation of a thermoresponsive degradable chitosan-grafted PNIPAAm hydrogel as a “smart” gene delivery system. Materials 13:2530
51. Doostmohammadi M, Ameri A, Mohammadinejad R, Dehghannoudeh N, Banat IM, Ohadi M, Dehghannoudeh G (2019) Hydrogels for peptide hormones delivery: therapeutic and tissue engineering applications. Drug Des Devel Ther 13:3405–3418
52. Egle K, Dohle E, Hoffmann V, Salma I, Al-Maawi S, Ghanaati S, Dubnika A (2024) Fucoidan/chitosan hydrogels as carrier for sustained delivery of platelet-rich fibrin containing bioactive molecules. Int J Biol Macromol 262:129651
53. Gohil SV, Padmanabhan A, Kan HM, Khanal M, Nair LS (2021) Degradation-dependent protein release from enzyme sensitive injectable glycol chitosan hydrogel. Tissue Eng Part A 27:867–880
54. Yang Y, Liu Y, Chen S, Cheong KL, Teng B (2020) Carboxymethyl β-cyclodextrin grafted carboxymethyl chitosan hydrogel-based microparticles for oral insulin delivery. Carbohydr Polym 246:116617

55. Lin CX, Yang K, Li PC, Gao LT, Aziz Y, Li JH, Miyatake H, Ito Y, Chen YM (2024) Self-healing and injectable chitosan/konjac glucomannan hydrogel with pH response for controlled protein release. Colloids Surf B Biointerfaces 242:114089
56. Dartora VFC, Passos JS, Osorio B, Hung RC, Nguyen M, Wang A, Panitch A (2023) Chitosan hydrogels with MK2 inhibitor peptide-loaded nanoparticles to treat atopic dermatitis. J Control Release 362:591–605

Adv Polym Sci (2025) 295: 235–260
https://doi.org/10.1007/12_2024_186

Published online: 4 January 2025

Chitosan and Its Derivatives in the Delivery of Peptides and Proteins

Pravas Ranjan Sahoo, Ipsita Kar, Ankita Priyadarshini, Aravindh Sundar, Revanasiddappa Bistuvalli Chandrashekarappa, and Harishkumar Madhyastha

Contents

1 Introduction 236
  1.1 Chitosan Protein Derivatives in Drug Delivery 238
2 Drug Delivery System with Natural Polymer and Chitosan Conjugates 239
  2.1 Chitosan in Drug Delivery Modes 240
3 Chitosan Derivatives in Cancer Biology 242
  3.1 Cancer Drug Delivery Mode 242
  3.2 Tumor Targeting System 243
  3.3 Immune Regulation Modes 243
4 Chitosan Derivatives and Their Breakthrough Applications in Cancer Biology 243
  4.1 Chitosan Derivatives and Their Functionalized Derivatives for Cancer Treatment . 244
5 Chitosan Complexes in Orthopedic Disorders 245

P. R. Sahoo (✉) and A. Priyadarshini
Department of Veterinary Biochemistry, College of Veterinary Science and Animal Husbandry, Odisha University of Agriculture and Technology, Bhubaneshwar, Odisha, India
e-mail: prsahoo@ouat.ac.in

I. Kar
Department of Veterinary Microbiology, ICAR-Indian Veterinary research institute, Bareilly, Uttar Pradesh, India

A. Sundar
Department Veterinary Physiology, College of Veterinary science and Animal Husbandry, Odisha University of Agriculture and Technology, Bhubaneshwar, Odisha, India

R. B. Chandrashekarappa
Department of Pharmaceutical Chemistry, NGSM Institute of Pharmaceutical Sciences, Nitte University, Paneer, Deralakatte Campus, Mangaluru, India

H. Madhyastha (✉)
Department of Cardiovascular Physiology, Faculty of Medicine, University of Miyazaki, Miyazaki, Japan
e-mail: hkumar@med.miyazaki-u.ac.jp

5.1 Treatment of Osteoarthritis and Restoration of Cartilage 246
5.2 Drug Delivery for Orthopedic Disorders and Other Pathologies 246
6 Chitosan and Its Other Pathological Healing Properties 247
6.1 Effects on Wound Healing and Hemostasis 247
6.2 Chitosan in Infectious Diseases 248
6.3 Applications of Chitosan in Tumor Management 248
6.4 Chitosan Protein Conjugates in Neurological Disorders 249
6.5 Mechanisms of Chitosan-Mediated Neuroprotection 251
7 Chitosan-Mediated Delivery of Neurotrophic Factors 252
7.1 Surface Modification Perspectives and Challenges in Chitosan-Based Therapies 252
8 Summary 254
References 254

Abstract Natural polymeric polysaccharides are being used to deliver therapeutic agents like peptides, proteins, deoxyribonucleic acid (DNA), vaccines, and sensitive drugs. Natural polymers like chitosan are being conjugated with different growth factors, proteins, and peptides and are being delivered in the form of microparticles, nanoparticles, nanomeshes, nanogels, and nano sprays to the site of damaged tissues for regeneration purposes. Several strategies are being focused on to refine chitosan properties for the effective delivery of active biomolecules. In this chapter, we summarize the challenges, pitfalls, and prudence of chitosan-mediated drug delivery in the scenario of regenerative disease-reversing strategies.

Keywords Chitosan · Drug delivery system · Protein and peptides

1 Introduction

The transition of the millennium, seen with the inception of microtechnology and a shift to nanotechnology, which opened the knowledge gate to a whole new world, has witnessed many technological innovations in biomedical sciences. The close and cohesive cooperativity of engineers, clinicians, and basic medical scientists is key to innovation sustainability in medicine. Recent applications of nanotechnology-derived biocomponents, like enzymes, peptides, proteins, etc., are seen in the areas of pharmaceutical formulation, agrochemical synthesis, nutraceutical preparation, industrial bioproduct manufacturing, and environmental remediation. Because of the natural polymer's great mechanical strength, biocompatible qualities, and readily modifiable surface chemistry, biopolymeric scientists are particularly interested in this affordable and widely available particle. [1]. Currently, biopolymer materials from renewable sources like plants, animals, fungi, and those of bacterial origins are gaining attention from the biomedical field. Polysaccharides like cellulose, chitosans (CSs), pectins of plant origin, collagen and α-keratins derived from animal sources, phospholipids, exopolysaccharides, polyesters like polyhydroxyalkanoates (PHAs) purified from bacterial sources are being investigated extensively in bioengineering-aided biomedical sciences [2]. Nanomaterials with specific physical and chemical properties, which are easily taken up by cells and penetrate deep into tissues and circulate in the bloodstream can be toxic to the human body [3]. In addition, metallic nanoparticles can also harm the soil, water, and environment, with negative impacts on organisms living in the environment. Therefore, it is a much-needed task to identify suitable and compatible nanoparticles with less toxic effects on human beings and with an effective delivery system. Hence, the concept of biodegradable nanomaterials with surface-coated biopolymeric nanoparticles with higher biocompatibility and biosuitability is gaining importance nowadays.

Among all natural biopolymers of animal origin, Chitosan's suitability for use as a conjugate agent or carrier is now the important subject of area for research. N-acetyl-D-glucosamine with β-1,4-glycosidic bonds as a linker molecule, chitin, a linear macromolecular bio polysaccharide, found abundantly in crustacean shell. The dissolving coefficient and biodegradability of chitin is extremely poor because of large number of acetyl amino, hydroxyl and carbonyl group with extremely complexes inter and intra molecular hydrogen bonds within. However, the presence of a large number of free amino groups increased the positive charges of chitin, and thus, they dissolve only in dilute acid solutions. However, presence of amide group and hydroxyl group at C_6 and C_3 position attracts the phenomenon of chemical modification strategies in chitosan. Chemical reactions like acylation by acetic acid, myristoyl chloride, octanoyl chloride, and linoleic acid with active reaction sites like -OH, -NH_3, and -NH_3/-OH help in drug carrier properties to enhance antimicrobial properties. Next, the carboxylation of chitin by monochloroacetic acid helps in water solubility [4]. Alkylation reaction by dodecanol, docdecyl aldehyde, and alkyl halide improves the hemostatic behavior of chitin. Quaternization and sulfonation reaction by iodomethane and sulfonation conjugation by SO3-pyridine trimethyleamine-sulfur trioxide help in the drug delivery system of chitin [5]. The phosphoric-acid-aided phosphorylation reaction of chitin derivatives helps in the proliferation of MC3T3-E1 cells. Thioglycolic acid bound to the ammonia (NH_3) group of chitins by thiolation reaction helps in tissue engineering technology [6]. In addition, chitosan, after chemical modifications, dwells in different material derivatives, like sponges, nanoparticles, hydrogels, films, scaffolds, beads, and nanofibers, with various biomedical applications [7, 8]. With these excellent physiochemical properties of chitosan derivatives, they have been widely applied in medicine, food, chemical industries, agricultural products, the environment, textiles, and other applications.

Structurally liable and sensitive compounds like therapeutic proteins, peptides, and antigens are administered by traditional routes like the parenteral administration system, such as the intravenous injection system. However, injections have psychological disadvantages for many patients due to pain and high cost. The current research trend is focusing on the mucophilic properties of chitosan, with advantages in noninvasive therapy via administration through buccal, vaginal, oral, and anal routes [9]. In this comprehensive review, we discuss the power, pitfalls, and prudence of chitosan derivatives in proteins, peptides, enzymes, and antigen complexes in various pathophysiological and regenerative diseases.

1.1 Chitosan Protein Derivatives in Drug Delivery

Chitosan has emerged as a pivotal component of the adjuvant drug delivery system, especially for the sustained and controlled release of the candidate drug. It is a natural cationic polymer derived from chitin through the deacetylation process and is composed of β (1–4)-linked 2-acetamido-2-deoxy-D-glucopyranose and 2-amino-2-deoxy-Dglycopyranose units. These biopolymers enhance the drug's bioavailability and stability, making them crucial for optimizing the efficacy and safety of the drug. Chitosan derivates are regarded as an ideal drug delivery material due to their peculiar characteristics, including nontoxicity; biodegradability; availability; resistance to acid, anticoagulants, and ulcers; and the ability to float and expand in an acidic environment [10]. The polymer has good film-forming properties with proper compatibility and permeability. The unique polycationic property of chitosan enables electrostatic interaction with microcapsule surface components (sodium alginate), increasing their stability and optimizing both the loading and release rates of drugs. Chitosan can be modified physically through some techniques, viz., blending, graft copolymerization, and curing [11]. However, chemical modification occurs by altering functional groups using several methods, like radiation, photochemical, plasma-induced, and enzymatic grafting methods. Chitosan derivatives such as N-trimethylenechloride chitosan, chitosan esters, and conjugates are widely utilized in pharmaceutical industries as drug delivery agents. N-trimethylene chloride, a quaternary chitosan derivative, exhibits higher water solubility, intestinal permeability, and reduced transepithelial electrical resistance, significantly facilitating the drug's absorption capacity. Chitosan esters, like glutamate and succinate, enhance sustained release in alkaline environments but reduce their solubility in acidic conditions. Additionally, chitosan conjugates, those linked to methyl pyrrolidinone or 4-thiobutylamidine, have been utilized to deliver active ingredients, like peptides, to improve both absorption and mucosal adhesion. Chitosan-based drug delivery has been formulated for oral, mucosal, ophthalmic, nasal, and topical routes. The release of drugs from a chitosan-based delivery system generally depends on the morphology and physiobiological properties of the drugs. It involves different mechanisms viz. expansion of the polymer, diffusion of adsorbed drug or drug diffusion through the polymeric net-mesh, polymer erosion or degradation and with properties of both erosion and degradation mechanism. The swelling of a polymer depends on its solubility in water. The release of a drug from the surface is done by the rapid dissolution of the adsorbed drug, leading to a burst effect. A linear relationship between the amount of drugs released and time kinetics was demonstrated by Mitra and Dey [12], highlighting the role of the diffusion mechanism in drug release. Chitosan compounds have been used in various dosages, such as chitosan-based mucoadhesive metronidazole tablets for vagina chitosan capsules.

2 Drug Delivery System with Natural Polymer and Chitosan Conjugates

Many regenerative diseases, like cancer, cardiovascular disorders, neuro disorders, and arthritis, originate in the body due to the generation and accumulation of free radicals, which induce oxidative stress, protein misfolding, lipid oxidation, and damage to nucleic acid [5]. However, natural polyphenols, proteins, and peptides have tremendous and promising therapeutic potentials, but their short- and fast-degrading phenomenon limits their use. Chitosan, along with other naturally occurring adjuvants, has synergic effects for effective use in regenerative disorders. Some of the natural polymers (Table 1) have been conjugated, which enhances the solubility and H-donor phenomenon of chitosan.

Table 1 Protein-/peptide-based natural polymers used in regenerative medicine

Natural polymer	Delivery system	Action/result	Reference
Cecropins A and B	Liposomes	Bladder cancer	[13]
Sericin	Chitosan nanoparticles	Antioxidant and anti-inflammatory	[14]
Solenospin	Solid lipid nanoparticles	Antiangiogenic	[15]
Apitoxin	PLGA nanoparticles	Arthritis and dermatitis	[16]
Melittin	Polycaprolactone microparticles	Mitochondria hemostasias in glioblastoma	[17]
Elastin	Solid lipid nanoparticles	Wound healing	[18]
Resilin	Liposomes	Adjuvant for cancer drugs	[19]
Nisin	Liposome	Antimicrobial	[20]
Albumin	Chitosans	Anti-allergic	[21]
Collagen	Scaffold engineering	Tissue repair	[22]
Fucoidan	Microvesicles	Thrombolytic disease	[23]
Hyaluronic acid	Self-polymerizing conjugate agents	Antitumor, wound healing	[24]
Zein	Thymol-incorporated nanocomposites	Antipathogens	[25]
Gliadin	Cross-linker molecules	Antimicrobial	[26]
Vicilin	pH coacervation agents	Kidney protectants	[27]
Platelet-rich fibrin	Chondrocyte-stabilizing agents	OA pathology	[28]
Mucin	Liposomal formulation	Lung ARDS	[29]

2.1 *Chitosan in Drug Delivery Modes*

2.1.1 Colon-Targeted Drug Delivery

Chitosan is a potential natural polymer for colon-targeted drug delivery due to its biodegradable properties [30]. The strategies for formulating colon-targeted drugs involve time-dependent release [31], the pH-responsive polymeric coating phenomenon [32], the prodrug symbiosis approach, microbiota-targeted drug delivery, the conjugation of drugs with polymers, and bioadhesive pharma molecules [33]. The multiarticulate dosage form, like nanoparticles, is efficient for achieving the sustained release of drugs at the target site. Moreover, chitosan is capable of opening tight junctions reversibly between the epithelial cells, promoting paracellular transport [34]. In the case of inflammatory bowel disease, chitosan nanoparticles enhance the retention of the drug by interacting with the anionic sialic acid group of the mucous membrane. Additionally, chitosan oligosaccharide also possesses anti-inflammatory properties through the inhibition of macrophage response and maintains epithelial barrier integrity by assembling tight junctions. A chitosan-based microsphere drug developed for ulcerative colitis was directly compressed into matrices with 5-aminosalicylic acid (5-ASA), and subsequently, the matrices were introduced into tablets or capsules. Some other colon-targeted drugs, like tylophorine malate, mesalazine, ketoprofen, and ascorbic acid, are also encapsulated through chitosan microspheres. CS-based particles, obtained by encapsulating proteins into the granules, were developed through ionic cross-linking to maintain protein vitality during storage and oral administration. Therefore, chitosan-modified drugs could be a great option for treating intestinal diseases.

2.1.2 Mucosal Drug Delivery

Mucosal drug delivery is an advantageous technique for noninvasive drug administration, which is attributed to its easy accessibility with promising action, the abolition of the hepatic first-pass effect in tissues, and better bioavailability. It involves different body surfaces, such as buccal, nasal, vaginal, ocular, rectal, and uterine, for the prolonged release of drugs. Chitosan polymers have excellent mucoadhesive properties, which are beneficial for the delivery of drugs through mucosal surfaces. Buccal drug administration can provide both systemic and local effects via enhanced permeability through a thin epithelial layer, utilized for treating oral mucositis, candidiasis, gingivitis, and periodontitis. Trimethylated and thiolated chitosan derivatives are found to be highly effective in promoting the absorption of hydrophilic macromolecules and peptides. Mucoadhesive tablets of chitosan, in combination with different polymers, such as Carbopol 940 [35] or gelatin [2], were directly compressed for the buccal delivery of carvedilol or propranolol hydrochloride, respectively. Furthermore, thiolated-triethyl-chitosan-based nanoparticles, embedded in mucoadhesive CS films, are utilized for enhanced buccal delivery. Ovalbumin-containing CS microparticles showed an increased uptake of microparticles by the Peyer patches of the gut-associated lymphoid tissue (GALT).

2.1.3 Nasal Drug Delivery

The nasal route offers a large mucosa for drug absorption, prevents first-pass metabolism, and is suitable for molecules that are unstable in the gastrointestinal (GI) tract [36]. Microparticulates and nanoparticle-based chitosan derivatives, functionalized with mannose, have been developed for the innovative delivery of various vaccines [37]. CS nanoparticles have been utilized for the carriage of drugs in central nervous system (CNS) diseases [38]. In addition, carboxymethyl-CS nanoparticles were developed for the intranasal delivery of carbamazepine for epilepsy management [39]. Similarly, galantamine-loaded thiolated-CS nanoparticles were developed for the treatment of Alzheimer's disease [40]. Chitosan-based insitu forming hydrogels was developed for the intranasal delivery of rasagiline mesylate drugs to cure Parkinson's disease [41].

2.1.4 Ocular Drug Delivery

Chitosan, an efficient carrier for ophthalmological drug delivery, enhances retention and bioavailability. Additionally, good ocular tolerance, bioadhesion, and high permeability make it an ideal ocular drug delivery vehicle. Similarly, CS-sodium tripolyphosphate nanoparticles were designed for the topical ocular delivery of acetazolamide for treating glaucoma and were reported to significantly have higher hypotensive activity in vivo ($p < 0.05$) as compared to a simple drug solution [42]. Shinde et al. [43] prepared mucoadhesive trimethyl-CS nanoparticles loaded with flurbiprofen against conjunctivitis, which showed an increment in both drug solubility and in situ residence time. Chitosan-coated liposomes transport triamcinolone acetonide for topical administration in posterior eye diseases [44]. Hence, these studies suggested that chitosan derivatives could be useful for delivering ocular drugs to achieve their therapeutic efficacy.

2.1.5 Topical Drug Delivery

This delivery is an elite pathway for enhancing the therapeutic effects and reducing the side effects of administered drugs. Chitosan can promote the dermal absorption of small polar molecules, peptides, and proteins. Silver nanocrystalline chitosan dressing (SNC) was developed against wounds [45]. Chitosan-associated lidocaine anesthesia improved efficiency and absorbability compared to normal lidocaine anesthesia. However, a study reported the increased chemical stability and permeation of the antiviral acyclovir, used in skin drug delivery systems, when it was incorporated with chitosan-tripolyphosphate nanoparticles [46]. These studies suggested that chitosan is a potential natural polymer, widely utilized in pharmaceutical industries. It has been applied in drug dosage formulation for the sustained release of drugs at their specific target site, thus improving efficacy in various diseases.

3 Chitosan Derivatives in Cancer Biology

Cancer is one of the principal causes of death worldwide, irrespective of advancements in restorative approaches. Certain challenges, like ineffective drug delivery and drug antagonism, constitute formidable struggles to combat. However, chitosan and its derivatives have emerged as a favorable tool in cancer biology. Derived from the deacetylation of chitin, found in the exoskeletons of insects and crustaceans, the biocompatible polymer chitosan has nontoxic properties and biodegradability, making it a rightful prospect in many scientific applications. These alterations elevate its solubility in the synthesis of nanoparticles, which can target tumor cells more efficiently. The derivatives have enriched bioavailability, stability, and target specificity, widening their usage in tumor-targeted treatments [47]. Originating from natural sources, its derivatives are considered safe. According to toxicological studies, chitosan nanoparticles have minimal adverse effects. There is an easy conjugation with different bioactive compounds, such as anticancer agents, due to the cationic properties of chitosan. The expression of proapoptotic and antiapoptotic proteins can be regulated by chitosan, leading to cell cycle detention following apoptosis. Apoptotic pathways in tumor cells are triggered by reactive oxygen species (ROS), which are produced by chitosan nanoparticles. Such increased ROS can hinder the performance of mitochondria, thus releasing cytochrome c and promoting caspase cascade activation. This causes programmed cell death [48].

3.1 Cancer Drug Delivery Mode

The derivatives of chitosan are considered useful anticancer drug transporters. Chitosan mucoadhesive properties facilitate absorption and drug release at the targeted site. According to research, chitosan nanoparticles significantly enhance the bioavailability of medications like paclitaxel and doxorubicin and facilitate their therapeutic effects in numerous cancer prototypes [49]. Folate receptor α (FRα) has been the limelight and hallmark target for cancer treatment for many decades, including for ovarian, lung, and breast cancers [14]. Functionalized folic acid conjugated with poly(styrene-*alt*-maleic anhydride (SMA) with biological linker 2,4-diaminobutyric acid (DABA) as a copolymer has been studied extensively to reduce tumor spheroid volume in breast and prostate cancers. Here, the investigator [50] found that SMA-conjugated DABA increases cell surface FRα and later disturbs the folic-acid-dependent signaling pathway by reducing the Notch signaling network in prostate cancer.

3.2 Tumor Targeting System

In cancer treatments, it is quite a challenge to ensure that therapeutic agents significantly target the tumor cells instead of damaging the healthy tissues. Modifications can be employed in chitosan compounds so that tumor targeting would be more accurate. This could be performed by the conjugation of specific ligands, which bind selectively to receptors overexpressed in most cancer cells. The cepropin family and folate, a well-understood concentrated ligand, and its conjugated chitosan nanoparticles have demonstrated a high uptake of folate receptors, which overexpress the majority of cancer cells. This has led to a significant reduction in the size of bladder tumors in vivo. [13]. This selective targeting not only complements therapeutic efficacy but also reduces the side effects related to conventional chemotherapy.

3.3 Immune Regulation Modes

Chitosan derivatives regulate immune reactions to tumors. The immunoactivity of macrophages and dendritic cells can be stimulated by chitosan, enhancing their ability to recognize and eliminate cancer cells. By improving immune reaction, these derivatives could potentially improve the efficacy of immunotherapies and result in improved outcomes of cancer therapy. N-acetyl chitosan, a chitosan derivative, is used as an adjuvant in cancer vaccines by stimulating the immune system, increasing antigen presentation, promoting the maturation of dendritic cells, and activating T cells. Chitosan derivatives also play a significant role in immune checkpoint blockade, when incorporated into anti-PD-1 or anti-CTLA-4 antibodies, and help in barring inhibitory signals, preventing immune cells from attacking tumors.

4 Chitosan Derivatives and Their Breakthrough Applications in Cancer Biology

Chitosan compounds have a promising potential in the treatment of cancer, including as anticancer drugs. Chitosan derivatives have been proven in studies to cause apoptosis while also inhibiting tumor growth in colorectal and prostate malignancies. Due to such properties, chitosan is considered a dual-function agent by delivering other therapeutic agents and exerting its effects directly on cancer cells. Chitosan derivatives show synergistic effects in the combination. According to studies, a widespread increase in the efficacy of radiotherapy is concurred as they sensitize cancer cells [51]. Due to this, there is a decline in the dosage of radiation, resulting in the minimization of toxic effects and the maximization of therapeutic effects. Further, antitumor immunity is enhanced by incorporating chitosan-based drug delivery with immunotherapeutic agents. Chitosan derivatives act as carriers for photothermal agents, so they are used in photothermal treatments. Chitosan nanoparticles convert light energy into heat, leading to the destruction of cancer cells [52]. Such a process is usually less invasive and provides high advantages. Many studies have been executed to analyze chitosan nanoparticles to attain controlled release and targeted delivery, enriching therapeutic results. In a study by Bajpai and others [25], it was found that there is increased cytotoxicity against breast tumor cells due to chitosan-coated nanoparticles with paclitaxel. A multifaceted strategy with the encapsulation of small interfering ribonucleic acid (siRNA) in chitosan nanoparticles is being used against cancer, which results in the restoration of tumor suppressor genes [39]. Chitosan-based prototypes deliver the genes to various cancer cell types with minimal cytotoxicity due to the proficiency of chitosan to constitute stable complexes with RNA or deoxyribonucleic acid (DNA) and characterize them as convenient carriers for various therapeutic genes.

4.1 Chitosan Derivatives and Their Functionalized Derivatives for Cancer Treatment

Chitosan, in its native form, is more active, but its biological activity and solubility can be enhanced by modification. So to improve its efficiency in cancer biology, some chemical modifications, like carboxylation, alkylation, and acylation, are made in its structure. This improves their therapeutic efficiency, solubility, and targeting abilities.

4.1.1 Carboxymethyl Chitosan (CMCS)

Water-soluble derivatives, having enhanced mucoadhesive properties, are most suitable for oral delivery in cancer treatment. The enriched biodegradability and bioavailability of such compounds lead to the development of nanoparticles with encapsulating drugs. Another modification of chitosan, namely, N-acetyl chitosan with enhanced solubility, enhances the bioavailability of insoluble drugs and acts as a vaccine adjuvant in cancer therapy.

4.1.2 Chitosan Oligosaccharides (COSs)

With low molecular weight and anti-inflammatory and antioxidant properties, chitosan oligosaccharides help in the enhancement of cellular uptake, as well as in the effective delivery of drugs to tumor cells [47]. They cause apoptosis by regulating different signaling pathways, including the NF-κB and MAPK pathways. N-trimethyl modification with higher solubility of chitosan is enhanced on a broader range of pH, making it highly effective for oral drug delivery. It alleviates the administration of different chemotherapeutic reagents and siRNA in the treatment of cancer.

5 Chitosan Complexes in Orthopedic Disorders

Chitosan is a biofriendly, decomposable, and harmless biopolymer derived from converting chitin found in crustaceans' exoskeletons through deacetylation. The solubility, nucleic acid binding ability, biological properties, and biomolecular interactions of chitosan compounds are greatly influenced by molecular weight (MW) and the degree of deacetylation (DD). In the scope of tissue engineering, chitosan can create intricate architectures with various biologically active constituents, rendering it a multiple functional material. Chitosan is a type of polysaccharide that contains acid; in acidic conditions, it exhibits a positive charge, which causes chitosan to attach to negatively charged substances such as drugs and various growth factors for bone reconstruction and healing. There is significant progress in the increased utilization of chitosan-based complexes for treating and handling orthopedic conditions like osteoporosis, osteoarthritis, bone fractures, and bone tumors. Usually, conventional methods like synthetic polymers and metal implants exhibit many constraints, including mechanical mismatch and bioincompatibility, so such chitosan complexes deliver a unique direction in this area. Chitosan develops porous scaffolds, which support the regeneration of tissue. This behavior of chitosan is important in orthopedic disorders and their treatments. It helps in cell growth, adhesion, and differentiation as it structurally mimics the extracellular matrix (ECM). These properties make chitosan an excellent candidate for repairing and regenerating damaged bone and cartilage in orthopedic applications. Biocompatibility delivers minimal immunogenicity and is considered a suitable compound for scaffold construction [53]. So chitosan helps in the regeneration of damaged bones, cartilages, and related tissues, which are a significant focus area in orthopedic tissue engineering. It acts as a potential carrier for bioactive molecules, including growth factors or drugs (e.g., bisphosphonates for osteoporosis). Such compounds, when delivered straight to the site of bone injury, help in the stimulation of bone growth and improve healing rates.

With its low immunogenicity and nontoxicity, chitosan favors bone and cartilage repair. It also naturally degrades into nontoxic by-products, hence avoiding surgical removal after implantation. The infection risk in postoperative orthopedic procedures is lessened due to its antibacterial activity. In the regeneration of bones, scaffolding is important because it mimics the natural bone ECM and advocates bone growth, i.e., osteo-conduction and the stimulation of progenitor cells to become bone-forming cells, i.e., osteo-induction. Hydroxyapatite (HA), calcium phosphate, or bioactive glass is often combined with chitosan scaffolds, which form three-dimensional (3D) structures, supporting the differentiation and proliferation of cells, hence augmenting osteo-conduction. It also assists in the regeneration of large bone defects by facilitating bone cell adhesion and the deposition of matrices [54]. Cell infiltrations, integration with enfold tissues, and the transport of nutrients and waste materials usually rely on the porosity of scaffolds [5]. For efficient load-bearing procedures, chitosan scaffolds can be simulated with regulated pore sizes, which influence biological functionality and mechanical strength.

5.1 Treatment of Osteoarthritis and Restoration of Cartilage

Chitosan hydrogels play a major role in osteoarthritis (OA) in cartilage tissue engineering. A supple and hydrated matrix is provided, along with the growth of chondrocytes. It usually shows biological features that assist in the regeneration of tissues [55]. Chitosan scaffolds are incorporated with certain growth factors, like bone morphogenetic proteins (BMPs) or mesenchymal stem cells (MSCs), which are beneficial for cartilage repair. A transforming growth factor-beta (TGF-β) is used in injectable chitosan hydrogels for the effective repair of cartilage tissue during osteoarthritis. Chitosan scaffolds deliver a conducive environment for cartilage regeneration due to mechanical properties [56]. Chitosan compounds not only have mechanical and biological properties but also possess anti-inflammatory properties, which are quite helpful in lessening local inflammation in injured joints and in fostering improved repair strategies [57]. Chitosan coatings in metal implants enhance biocompatibility, which reduces the chances of postsurgery infections. Metallic materials like nails, plates, or screws, utilized for fixing bone fractures, lead to infection and inadequate osteointegration. However, chitosan coatings on them have shown enriched osteointegration. Other benefits of chitosan include reduced infection at the implantation site due to antibacterial features and localized treatment directly at the injury site as chitosan complexes encapsulate antibiotics or other therapeutic agents.

5.2 Drug Delivery for Orthopedic Disorders and Other Pathologies

Chitosan is converted to gels, nanoparticles, or beads, which renders it a suitable platform for the controlled release of drugs (i.e., delivery of anti-inflammatory agents for osteoarthritis or antibiotics across bone infections). For instance, chitosan nanoparticles containing dexamethasone, which is an anti-inflammatory medication, demonstrated decreased inflammation in osteoarthritis cases. In the same way, chitosan delivery systems containing bisphosphonates, which are medications for treating osteoporosis, have led to stronger bones and lower chances of fractures. Chitosan-hydroxyapatite composites have been used to achieve optimal bone regeneration by mixing chitosan with hydroxyapatite or similar calcium-based materials to improve the mechanical strength and osteo-conductivity of the scaffold. These materials imitate the mineralized structure of natural bone and help osteoblasts grow and attach [58]. Nanofiber chitosan scaffolds are created through electrospinning to imitate the ECM found in living organisms. These nanofibers also aid in providing a larger surface area to assist in tissue regeneration, such as repairing bones or cartilage and promoting cell attachment.

6 Chitosan and Its Other Pathological Healing Properties

Chitin, a naturally occurring polysaccharide, is prevalent in crab shells, fungal cuticles, and crustacean exoskeletons. Its derivative, chitosan, comprises randomly distributed β (1 → 4)-linked D-glucosamine and N-acetyl-d-glucosamine units. Extraction involves multiple steps: pretreatment, deproteination, demineralization, decoloration, and purification [59]. Chitosan's unique properties are influenced by acetylation levels as well as pH, which helps in its tremendous use in biological and technical applications. As the sole natural polycation, chitosan exhibits antimicrobial, antioxidant, anti-inflammatory, anticancerous, hemostatic, and wound-healing potentials [60]. The quest for innovative biomaterials drives research, fueled by the search for effective treatments and sustainable solutions. Chitosan's low toxicity, biodegradability, and biocompatibility make it an attractive biopolymer for medical applications. This chapter explores chitosan's role in addressing underlying pathological conditions, leveraging its potential in drug delivery and nanoparticle technology.

6.1 Effects on Wound Healing and Hemostasis

The numerous and varied health benefits of chitosan are far-reaching, attending to a wide range of body requirements. Its hemostatic properties are attributed to its diverse molecular weights and degrees of deacetylation, enabling tailored applications [61]. This process occurs through the ionic interaction between chitosan's cationic glucosamine units and the anionic surface of red blood cells (RBCs), leading to enhanced blood clotting [13]. Chitosan's unique properties include forming a stable, protein-based barrier through interpolymer complexation. This process leverages covalent and noncovalent interactions between chitosans and plasma proteins, generating a physical obstacle that hastens hemostasis. Importantly, this mechanism occurs through traditional coagulation cascades, providing an advantageous solution to coagulopathy patients [62]. Researchers explored innovative methods to minimize surgical-site infections associated with hernia repair mesh. Utilizing a rat model, they simulated mesh infection through simultaneous and delayed *Staphylococcus aureus* inoculation and monitored the outcomes over 8 days. The results demonstrated that meshes treated with triclosan-loaded chitosan gel exhibited significant protective effects against graft infection, offering the potential for improved therapeutic outcomes. Recent advancements help in the development of chemically modified chitosan derivatives, enhancing their biological efficacy. Novel polymers, such as ammonium and carboxymethyl chitosan, have reaped significant attention. Effective wound care necessitates infection prevention, moisture retention, wound protection, and rapid healing with minimal scarring.

6.2 Chitosan in Infectious Diseases

Chitosan causes membrane permeability and permits chitosan nanoparticles (CNPs) to pass through the bacterial cell wall. Following their penetration, CNPs mix with DNA to impair DNA replication, which ceases the bacteria's survivability [63]. Additionally, it targets the electron transport chain in the microbe, resulting in effective therapeutic applications [64]. Chitosan-based nanoparticles demonstrated clear antibacterial activity against many *Staphylococcus aureus* and *Escherichia coli*. According to studies, *Streptococcus mutans* biofilm development is inhibited in vitro by chitosan nanoparticles, particularly those made with low molecular chitosan [65]. Drug resistance is well established in certain bacterial diseases, like *Helicobacter pyroli*, in which CNPs can target this. Since the development of mucosal immunity is the first line of defense, the induction of immunity is investigated in BALB/c (Bagg albino) mice following nasal intervention using three *Brucella abortus* recombinant proteins: malate dehydrogenase (rMdh), outer membrane proteins (rOmp) 10 and 19, and mucoadhesive CNPs [66]. One of the main health problems that affect women is thought to be vulvovaginal candidiasis. An antibiotic called clotrimazole is frequently used to treat the illness. The antibiotic was incorporated into O-palmitoyl chitosan nanoparticles for topical administration against the disease to improve treatment efficacy [67].

Viral-mediated transmission is also hindered by the application of chitosan. For instance, PEGylated chitosan (LPC) nanoparticles loaded with lamivudine have demonstrated a greater degree of potency in preventing human immunodeficiency virus (HIV) replication by restricting the generation of HIV virions [68]. Likewise, human hepatoma cell lines validated that curcumin-encapsulated CNPs had an antiviral impact against the hepatitis C virus [69].

6.3 Applications of Chitosan in Tumor Management

Cancer is one of the foremost causes of death around the globe due to the abnormal increase of cells, which later spreads to other parts of the human body [70]. Cancer patients rely on chemotherapeutic drugs to reduce the malignancy and to increase their odds of survival. However, patients often experience profound drug-induced side effects due to higher dose requirements, poor bioavailability, and resistance [71]. Therefore, to counteract these rising concerns, biopolymers are being employed as anticancer agents or drug delivery vehicles. Among various other biopolymers, chitosans stand apart, owing to their enhanced efficacy, potency, and biocompatibility, compared to other derivatives [72]. Chitosans have excellent anticancer compounds for chemotherapeutic treatment and are often used as a delivery vehicle for nucleic acids [73]. Their derivatives, including chitosan-thymine conjugate (CTC), carboxymethyl chitosan (CMC), sulfated chitosan (SCS), sulfated benzaldehyde chitosan (SBCS), and polypyrene-chitosan (PPC), have intriguing anticancer activity against various cancer cells. Furthermore, in combination with nanocomposite or other chemical agents, more modifications can be done in the anticancerous activities of chitosan derivatives. Chemically modified chitosan specifically traverses through tumor cells and carry out antiproliferative activities through immunoenhancing, antiangiogenic, apoptotic, and enzymatic regulation [74]. Additionally, they are utilized in various formulations (nanoformulation, hydrogel emulsion, and surfactants), thereby improving the biocompatibility and efficacy of anticancer agents and leading to a reduction in therapeutic doses [75]. Besides, the role of chitosan and their derivates revealed prospective applications in immunotherapy, photodynamic therapy, gene therapy, and chemotherapeutic drug delivery via ligand-mediated chitosans. Therefore, chitosan-based biomaterials have a promising future in the formulation of therapeutic cancer vaccines as they potentiate humoral and cellular immunity [47]. The most practical use of nanomedicine that is targeted to tumors is designing the adoptive method to prevent degradation by proteolytic enzymes. Tumor mass is densely packed with extracellular matrix, making it difficult for tumor medicine and delivery devices to penetrate. Collagenase nano capsules degrade this matrix, allowing nanocarrier penetration [76]. However, the use of enzymes as delivery agents is debated. In order to overcome this constraint, acid degradable material, like ethyleneglycol dimethacrylate (EGDMA) as a pH-sensitive cross-linker, has been developed, which has a high degree of tumor zone penetration and, ultimately, cancer nanodrug release (Fig. 1).

6.4 Chitosan Protein Conjugates in Neurological Disorders

Neurodegenerative disorders (NDDs) have become an appalling context around the globe, attributable to trivial drug delivery systems. It paves the foundation for numerous motor and cognitive dysfunctions, which include Alzheimer's and Parkinson's diseases. The blood-brain barrier (BBB) and the lack of lymphatic vessels in tissues cause immune injury to the CNS [77]. However, in regard to growing lesions, the CNS could not elicit an adaptive immune response. Furthermore, additional pathways cause massive calcium influx, and high glutamate

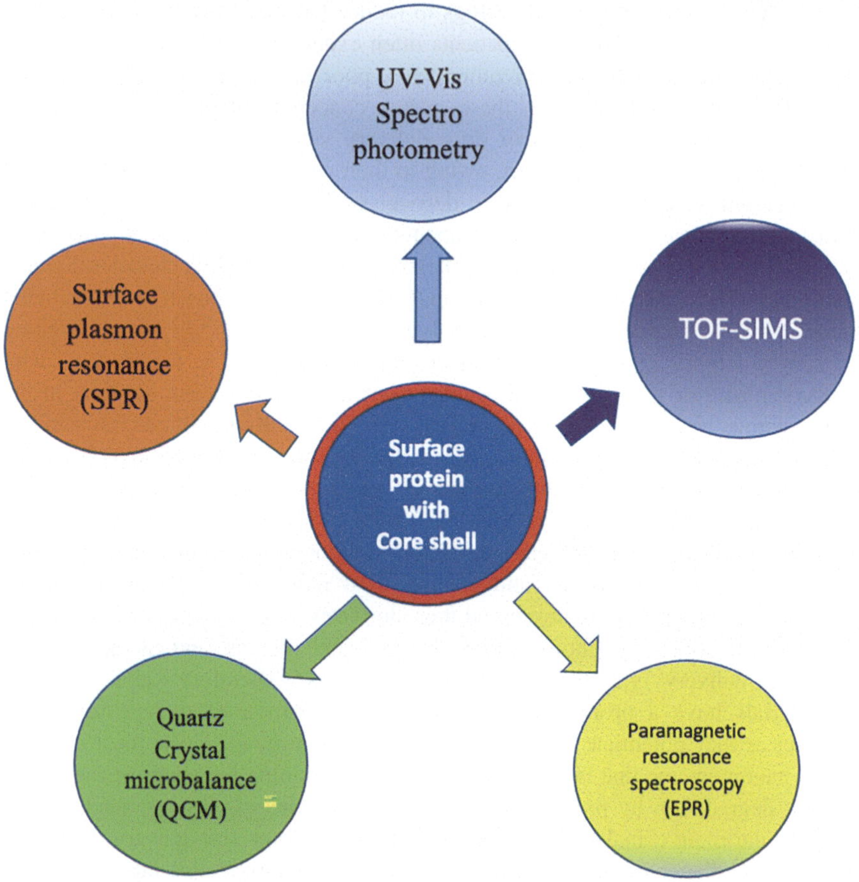

Fig. 1 Protein and peptide delivery toward tumor mass

concentrations are claimed to be the primary factors for neuronal deterioration [78]. Extrapolating several such degenerative mechanisms and pathways bids hope for targeting NDDs with novel therapeutics. Therefore, the drug material possesses limited access to the CNS, hampering modulatory functions, which cues the difficulty of medication for NDDs.

Recently, chitosan, a modified β (1 → 4)-linked carbohydrate, turned out to be a breakthrough in drug delivery as it is safe, nontoxic, and proficient in treating neurological illness [59]. Chitin, a polysaccharide, is technically present in crab shells, fungal cuticles, and outer skeletons of crustaceans [60]. It is an alluring idea in the frame of the pharmaceutical industry due to its plastic and distinguishing biological characteristics.

6.5 Mechanisms of Chitosan-Mediated Neuroprotection

6.5.1 Antiapoptosis

It is an essential feature in all living neuronal and nonneuronal cells. However, certain diseases in the CNS hasten apoptosis during the course of the disease and instigate further impairment. Therapeutic approaches targeting apoptotic substances could mitigate pathological outcomes by intervening in the key pathways. Chitosan has provable effects as an apoptotic inhibitor as it reduces the release of lactate dehydrogenase (LDH) and also lessens the disruption of MMP (mitochondrial membrane potential), which is induced by the neurotoxicant dibutyltin [48]. It also tolerates the serum starvation of astrocytes and shelters cells against the damage induced by glutamate [79]. Most neurodegenerative diseases frequently have oxidative stress as a consequence, which is predominantly posed by free radicals. Further, chitosan is able to hinder oxidative-stress-induced amyloid β production in NT2 neuron cells through antioxidant mechanisms [80].

6.5.2 Antineuroinflammatory

Some significant studies showed a pledging link between inflammatory responses to chronic neurological disease. Alzheimer's disease shows a persistent inflammatory response that is mediated through interleukin-1beta (IL-1beta) and beta-amyloid (Abeta). Pretreatment with water-soluble carbohydrate (WSC) in human astrocytoma cells markedly reduced the release and production of pro-inflammatory cytokines, such as TNF-alpha and IL-6, as well as the expression of inducible nitric oxide synthase (iNOS) [81].

6.5.3 Antiexcitotoxic

Among the main endogenous excitatory neurotransmitters, glutamate mediates the central nervous system's physiology. However, increased glutamate levels are poised to cause neuronal death by either provoking oxidative stress or causing a massive extrusion of calcium ions into the cells. Chito oligosaccharide (COS), a byproduct of chitosan, shields hippocampus neuronal cells from glutamate-induced death. Calcium concentration is downregulated by COS, which also causes the activation of caspase-3 mediated by glutamate [82].

6.5.4 Diminished Impact on Amyloid Beta Consolidation

It is thought that the principal constituent in Alzheimer's patients is amyloid peptide (Aβ), which causes neurodegeneration. Mutations in amyloid precursor protein

(APP) ignite the accumulation of APP oligomers in the brain. For example, the Swedish mutation such as (K.M670/671 N > L) and A673 > V, occured in peptide are leading cause of the disease. According to a study, it was observed that treating cells with conjugate oxyscavenger (COS) nanoparticles prevents them from causing Aβ deposition and producing reactive oxygen species (ROS) [83]. A sensitive nanoparticle-aided opit-biobarcode sensor was developed to estimate amyloid-derived diffusible ligand levels in the central nervous system [84].

7 Chitosan-Mediated Delivery of Neurotrophic Factors

The hunt for novel therapeutic strategies to combat CNS diseases has been fueled by the rising prevalence of neurodegenerative diseases and the dearth of effective treatments. The grueling concern for developing pharmaceutical drugs is attributed to the BBB, which permits only certain endogenous hormones and nutrients. Consequently, a futuristic strategy for enabling the delivery of therapeutic compounds to the brain is the use of carriers/vehicles and through nontraditional administration routes.

Poly(lactic-co-glycolic acid) PLGA and poly(lactic acid) PLA are two nanocarrier materials whose surfaces can be effectively capped with modified chitosan to enhance BBB penetration [48]. The development of protein-polysaccharide drug delivery vehicles has been reallocated to chitosan-collagen nanocomposites due to their exceptional scaffolds [85].

N-trimethyl chitosan (TMC), a derivative, has been targeted in brains affected with Alzheimer's due to interactions between positively charged TMC and anionic sialic acid residues of glycoprotein found on the BBB [86]. Tween 80-coated chitosan, PEGylated chitosan, and zwitterionic chitosan derivatives are further forms of chitosan derivatives [87]. DNA forms complexes with chitosan, which are used in gene therapy [24]. They extend the shelf life of the biological protein and protect the DNA from nuclease.

7.1 Surface Modification Perspectives and Challenges in Chitosan-Based Therapies

Establishing a long adequate shelf life for chitosan formulations causes more advancements in practical utility. Stability can be enhanced by modifying the chitosan's structure to ensure adequate purity and deacetylation levels, blending chitosan with other polymers, and controlling environmental factors, especially temperature. It is interesting to guise into the depolymerization of chitosan, which has the potential to generate free radicals, favoring oxidative damage [59]. As a whole, the surface modification of proteins with polymers like chitosan is evident

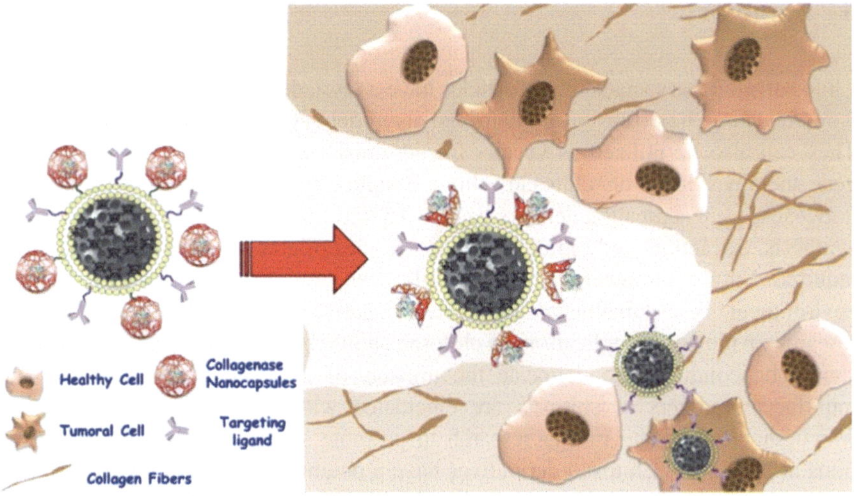

Fig. 2 Additional ultra-high technologies to identify the adsorbed protein on the surface of a core shell nanoparticle (adopted from DOI: https://doi.org/10.3390/molecules23051008). This article is an open-access article distributed under the terms and conditions of the Creative Commons Attribution (CC BY) license

and can be identified by various ultra-high techniques. Surface-adsorbed protein on nanoparticle core shell can be analyzed by a variety of methods to examine the amount of protein adsorbed, adsorption kinetics, and structural modification. A thematic explanation is depicted in Fig. 2.

Common methods like ultraviolet-visible (UV-Vis) spectrophotometry have less accuracy [88, 89]. Surface-adsorbed protein can also be measured with extra accuracy by a variety of high-end instruments like time-of-flight secondary ion mass spectrometry (ToF-SIMS), electron paramagnetic resonance (EPR) [90], quartz crystal microbalance (QCM), and surface plasmon resonance (SPR) for the real-time determination of adsorbed protein, which enable gaining a comprehensive understanding of surface protein interactions [91]. Chitosan-based biomedical application raises the boundaries in tissue engineering, scaffolding, and regenerative medicine. Besides, the credible aspect of the biopolymer extends to elevating the bars in novel drug delivery for treating NDDs. Recently, commendable studies have been crafted on the therapeutic management of Alzheimer's disease and brain toxicity by niruriflavone-loaded and piracetam-loaded magnetic chitosan nanoparticles [92]. Furthermore, microspheres loaded with brain-derived neurotropic factors for improving axonal development are elucidated [93, 94]. Chitosan nanoparticles improve oral and nasal medication delivery because they facilitate better mucoadhesion by relaxing tight connections. Further studies are recommended to evaluate the purity level and moisture-holding ability and alter biochemical characteristics. To achieve translation into the clinical level, future studies utilizing chitosan that target the central nervous system should combine previously proposed methodologies with new multidisciplinary approaches.

8 Summary

Chitosan and its associated nanoparticles have a myriad of biological uses, including wound healing, hemostasis, antimicrobial agent delivery, and antitumorous activity. They can also be utilized as vectors for administering DNAs and for undergoing gene therapy. However, chitosan's limited solubility in water is one of its main drawbacks as a medication delivery carrier. Therefore, comprehensive studies on the synthesis, extraction, and biological characteristics of chitosan ought to be urgently required. Natural biomaterials have recently been promoted more because they are environmentally sustainable and have fewer negative impacts on the body. In-depth research on the novel applications of chitosan should be undertaken, and animal tests should be conducted to delineate the impacts of the material on humans. The multifaceted targets of chitosan are elucidated within the biomedical industry, which should instigate further research to promote several advantageous applications in the future. Chitosan derivatives have a potential impact on cancer biology; however, certain challenges still prevail. There are problems with inefficient large-scale production, consistency, and quality and a lack of research on drug release mechanisms, drug target strategies, and regulatory issues seen in this area. In the future, chitosan derivatives incorporated with highly advanced cancer therapies like personalized medicines and immunotherapy carry an incredible potential to provide more efficient treatments in cancer biology. Further research is required to ensure the efficacy, safety, and reproducibility of chitosan-based treatments. 3D-printing techniques in chitosan-based scaffolds and their fabrication still need more optimization. For long-term performance in tissue repair, there should be proper control of chitosan degradation. The versatility of chitosan compounds has been used in various preclinical studies. However, translation applications require further research to combat the limitations.

References

1. Singh AK, Malviya R, Prajapati B, Singh S, Yadav D, Kumar A (2023) Nanotechnology-aided advancement in combating the cancer metastasis. Pharmaceuticals 16:899–903
2. Abruzzo A, Cerchiara T, Bigucci F, Gallucci MC, Luppi B (2015) Mucoadhesive buccal tablets based on chitosan/gelatin microparticles for delivery of propranolol hydrochloride. J Pharm Sci 104:4365–4372
3. Mathur P, Sathishkumar K, Chaturvedi M, Das P, Sudarshan KL, Santhappan S, Nallasamy V, John A, Narasimhan S, Roselind FS (2020) ICMR-NCDIR-NCRP Investigator Group. Cancer statistics, Report from National Cancer Registry Programe, India. JCO Glob Oncol 6:1063–1075
4. Schatz C, Bionaz A, Lucas JM, Pichot C, Viton C, Domard A, Delair T (2005) Formation of polyelectrolyte complex particles from self-complexation of N-sulfated chitosan. Biomacromolecules 6(3):1642–1647
5. Prabaharan M (2008) Chitosan derivatives as promising materials for controlled drug delivery. J Biomater Appl 23(1):5–36

6. Bratskaya S, Privar Y, Nesterov D, Modin E, Kodess M, Slobodyuk A, Marinin D, Pestov A (2019) Chitosan gels and cryogels cross-linked with diglyceryl ethers of ethylene glycol and polyethylene glycol in acidic media. Biomacromolecules 20(4):1635–1643
7. EzEldeen M, Loos J, Mousavi Nejad Z, Cristaldi M, Murgia D, Braem A, Jacobs R (2021) 3D-printing-assisted fabrication of chitosan scaffolds from different sources and cross-linkers for dental tissue engineering. Eur Cell Mater 5(41):485–501
8. Kazemi-Aghdam F, Jahed V, Dehghan-Niri M, Ganji F, Vasheghani-Farahani E (2021) Injectable chitosan hydrogel embedding modified halloysite nanotubes for bone tissue engineering. Carbohydr Polym 269:118–131
9. Takka S, Gürel A (2010) Evaluation of chitosan/alginate beads using experimental design: formulation and in vitro characterization. AAPS PharmSciTech 11(1):460–478
10. Sonia TA, Sharma CP (2011) Chitosan and its derivatives for drug delivery perspective. Adv Polym Sci 243:23–53
11. Shukla SK, Mishra AK, Arotiba OA, Mamba BB (2013) Chitosan-based nanomaterials: a state-of-the-art review. Int J Biol Macromol 59:46–58
12. Mitra A, Dey B (2011) Chitosan microspheres in novel drug deliver system. Indian J Pharm Sci 73(4):355–366
13. Suttmann H, Retz M, Paulsen F, Harder J, Zwergel U, Kamradt J (2008) Antimicrobial peptides of the Cecropin-family show potent antitumor activity against bladder cancer cells. BMC Urol 8:59–76
14. Mumtaz S, Ali S, Pervaiz A, Qureshi MZ, Kanwal K, Saleem T (2023) Apoptotic and antiproliferative effects of silk protein sericin conjugated-$AgNO_3$ nanoparticles in human breast cancer cells. Saudi J Biol Sci 30:103551
15. Kang B, Park H, Kim B (2019) Anticancer activity and underlying mechanism of phytochemicals against multiple myeloma. Int J Mol Sci 20:23–32
16. Ullah A, Aldakheel FM, Anjum SI, Raza G, Khan SA, Tlak Gajger I (2023) Pharmacological properties and therapeutic potential of honey bee venom. Saudi Pharm J 31:96–109
17. Ertilav K, Nazıroğlu M (2023) Honey bee venom melittin increases the oxidant activity of cisplatin and kills human glioblastoma cells by stimulating the TRPM2 channel. Toxicon 222: 345–359
18. Daamen WF, Veerkamp JH, van Hest JCM, van Kuppevelt TH (2007) Elastin as biomaterial for tissue engineering. Biomaterials 28:4378–4398
19. Mark B, Eldijk V, Christopher LM, Kristi LK, Jan CM (2012) Elastomeric polypeptides. Top Curr Chem 310:71–116
20. Romero GB, Arntjen A, Keck CM, Müller RH (2016) Amorphous cyclosporin a nanoparticles for enhanced dermal bioavailability. Int J Pharm 498:217–224
21. Sleep D (2015) Albumin and its application in drug delivery. Expert Opin Drug Deliv 12(5): 793–812
22. Báez J, Olsen D, Polarek JW (2005) Recombinant microbial systems for the production of human collagen and gelatin. Appl Microbiol Biotechnol 69:245–252
23. Vazquez-Prada KX, Moonshi SS, Wu Y, Akther F, Tse BRC, Sokolowaski KA, Peter K, Wang X, Xu G, Ta HT (2023) A spiky silver-iron oxide nanoparticle for highly efficient targeted photothermal therapy and multimodal imaging of thrombosis. Small 11:567–582
24. Sarvanakumar G, Deepagan VG, Jayakumar R, Hyung PJ (2014) Hyaluronic acid-based conjugates for tumor targeted drug delivery and imaging. J Biomed Nanotech 10(1):17–31
25. Bajpai SK (2015) Chitosan-coated nanoparticles for effective delivery of paclitaxel in breast cancer therapy. J Nanobiotechnol 13(1):48–59
26. Fajardo P, Balaguer MP, Gomez-Estaca J, Gavara R, Hernandez-Munoz P (2014) Chemically modified gliadins as sustained release systems for lysozyme. Food Hydrocoll 41:53–59
27. Ezpeleta I, Rache J, Gueguen J, Orecchioni A (1997) Properties of glutaraldehyde cross-linked vicilin nano-and microparticles. J Microencapsul 14(5):557–565

28. Mort JS, Beaudry F, Theroux K, Emmott A, Richard H, Fisher WD, Lee R, Poole AR, Laverty S (2016) Early cathepsin K degradation of type II collagen *in vitro* and *in vivo* in articular cartilage. Osteoarthritis Cartilage 46:461–469
29. Valente SA, Silva LM, Guido RL, Sarmento B, Coimbra MA, Passos CP (2021) Polysaccharide-based formulations as potential carriers for pulmonary delivery – a review of their properties and fates. Carbohydr Polym 277:418–434
30. Chourasia MK, Jain SK (2003) Pharmaceutical approaches to colon targeted drug delivery systems. J Pharm Pharm Sci 6(1):33–66
31. Kotla NG, Rana S, Sivaraman G, Sunnapu O, Vemula PK, Pandit A, Rochev Y (2019) Bioresponsive drug delivery systems in intestinal inflammation: state-of-the-art and future perspectives. Adv Drug Deliv Rev 146:248–266
32. Sinha VR, Kumria R (2003) Microbially triggered drug delivery to the colon. Eur J Pharm Sci 18(1):3–18
33. Vass P, Démuth B, Hirsch E, Nagy B, Andersen SK, Vigh T, Marosi G (2019) Drying technology strategies for colon-targeted oral delivery of biopharmaceuticals. J Control Release 296:162–178
34. Du H, Liu M, Yang X, Zhai G (2015) The design of pH-sensitive chitosan-based formulations for gastrointestinal delivery. Drug Discov Today 20(8):1004–1011
35. Fathima N, Das P, Kuchana V (2015) Formulation and evaluation of mucoadhesive buccal tablets of carvedilol using natural binders. Int J Res Pharm Chem 5:699–707
36. Touitou E, Illum L (2013) Nasal Drug Delivery. Drug Deliv Transl Res 3:1–13
37. Cho Y, Zhang S, Yang Y, Chen J (2020) Enhancing the efficacy of radiotherapy using chitosan nanoparticles. Int J Nanomedicine 15:7873–7885
38. Erdő F, Bors LA, Farkas D, Bajza Á, Gizurarson S (2018) Evaluation of intranasal delivery route of drug administration for brain targeting. Brain Res Bull 143:155–170
39. Liu Y (2017) Chitosan-based gene delivery systems for cancer therapy. Biotechnol Adv 35(5): 771–782
40. Sunena S, Singh SK, Mishra DN (2018) Nose to brain delivery of Galantamine loaded nanoparticles: *in-vivo* Pharmacodynamic and biochemical study in mice. Curr Drug Deliv 16: 51–58
41. Ravi PR, Aditya N, Patil S, Cherian L (2015) Nasal *in-situ* gels for delivery of rasagiline mesylate: improvement in bioavailability and brain localization. Drug Deliv 22:903–910
42. Manchanda S, Sahoo PK (2017) Topical delivery of acetazolamide by encapsulating in mucoadhesive nanoparticles. Asian J Pharm Sci 2:550–557
43. Shinde UA, Joshi PN, Jain DD, Singh K (2019) Preparation and evaluation of N-trimethyl chitosan nanoparticles of flurbiprofen for ocular delivery. Curr Eye Res 44:575–582
44. Li J, Cheng T, Tian Q, Cheng Y, Zhao L, Zhang X, Qu Y (2019) A more efficient ocular delivery system of triamcinolone acetonide as eye drop to the posterior segment of the eye. Drug Deliv 26:188–198
45. Lu SY, Gao WJ, Gu HY (2008) Construction, application and biosafety of silver nanocrystalline chitosan wound dressing. Burns 34:623–631
46. Hasanovic A, Reznicek G, Zehl M, Valenta C (2009) Chitosan-TPP nanoparticles as possible skin drug delivery system for acyclovir with enhanced stability. J Pharm Pharmacol 61:1609–1616
47. Shakil MS, Mahmud KM, Sayem M, Niloy MS, Halder SK, Hossen MS, Uddin MF, Hasan MA (2021) Using chitosan or chitosan derivatives in cancer therapy. Polysaccharides 2:795–816
48. Huang G, Liu Y, Chen L (2017) Chitosan and its derivatives as vehicles for drug delivery. Drug Deliv 24(2):108–113
49. Prajapati VD (2017) Nanoparticles of chitosan for drug delivery. Drug Deliv Syst 16(2):61–76
50. DeCarlo A, Jugroot CM, Szewczuk MR (2021) Folic acid functionalized nanomedicine: Folic acid conjugated copolymer and folate receptor interactions disrupt receptor functionality resulting in dual therapeutic anti-cancer potential in breast and prostrate cancer. Bioconjug Chem 32(3):512–522

51. Cao S, Xu G, Li Q, Zhang S, Yang Y, Chen J (2022) Double crosslinking chitosan sponge with antibacterial and hemostatic properties for accelerating wound repair. Compos Part B 234:109–127
52. Zhang H, Liu Z, Liu C, Zhou X (2017) Chitosan nanoparticles for cancer therapy: a review. Carbohydr Polym 173:11–19
53. Jayakumar R, Prabaharan M, Kumar PTS, Nair SV, Tamura H (2010) Chitosan scaffolds in tissue engineering: review. J Biomed Mater Res B Appl Biomater 93:507–521
54. Khor E, Lim LY (2003) Implantable applications of chitin and chitosan. Biomaterials 24:2339–2349
55. Freier T, Koh HS, Kazazian K, Shoichet MS (2005) Chitosan-based hydrogel scaffolds for cartilage tissue engineering. Biomaterials 26:3215–3225
56. Kim IY, Seo SJ, Moon HS, Yoo MK, Park IY, Kim BC, Cho CS (2008) Chitosan and its derivatives for tissue engineering applications. Biotechnol Adv 26:1–21
57. Dash M, Chiellini F, Ottenbrite RM, Chiellini E (2011) Chitosan: a versatile biomaterial for tissue engineering applications. J Biomater Sci Polym Ed 22:51–62
58. Bhumiratana S, Bernhard JC, Alfi DM, Yeager K, Eton RE, Bova J, Shah F, Gimble JM, Lopez MJ, Eisig SB, Vunjak-Novakovic G (2014) Tissue engineering: a new approach to bone and cartilage repair. Nat Mater 13:676–679
59. Jiménez-Gómez CP, Cecilia A (2020) Chitosan: a natural biopolymer with a wide and varied range of applications. Molecules 23:17–28
60. Iyer M, Elangovan A, Sennimalai R, Babu HW, Thiruvenkataswamy S, Krishnan J, Yadav MK, Gopalakrishnan AV, Narayanasamy A, Vellingiri B (2024) Chitosan–An alternative drug delivery approach for neurodegenerative diseases. Carbohydr Polym Technol Appl 67:2341–2356
61. Lim C, Hwang DS, Lee DW (2021) Intermolecular interactions of chitosan: degree of acetylation and molecular weight. Carbohydr Polym 259:117–7122
62. Lee VK, Lee T, Ghosh A, Saha T, Bais MV, Bharani KK, Chag M, Parikh K, Bhatt P, Namgung B, Venkataramanan G (2024) An architecturally rational hemostat for rapid stopping of massive bleeding on anticoagulation therapy. Proc Natl Acad Sci 121(5):316170121
63. Murugesan C, Chun KKD (2020) Antibacterial activity of chitosan nanoparticles: a review. Processes 8:11–73
64. Raghuwanshi D, Mishra V, Das D, Kaur K (2012) Dendritic cell targeted chitosan nanoparticles for nasal DNA immunization against SARS CoV nucleocapsid protein. Mol Pharmacol 9:946–956
65. de Paz LEC, Resin A, Howard KA, Sutherland DS, Wejse PL (2011) Antimicrobial effect of chitosan nanoparticles on streptococcus mutans biofilms. Appl Environ Microbiol 77(11): 3892–3895
66. Shim S, Soh SH, Im YB, Park HE, Cho CS, Kim S, Yoo HS (2020) Elicitation of Th1/Th2 related responses in mice by chitosan nanoparticles loaded with *Brucella abortus* malate dehydrogenase, outer membrane proteins 10 and 19. Int J Med Microbiol 310:151–162
67. Facchinatto WM, Galante J, Mesquita L, Silva DS, Martins Dos Santos D, Moraes TB, Campana-Filho SP, Colnago LA, Sarmento B, das Neves J. (2021) Clotrimazole-loaded N-(2- hydroxy)-propyl-3-trimethylammonium, O-palmitoyl chitosan nanoparticles for topical treatment of vulvovaginal candidiasis. Acta Biomater 4(42):34–41
68. Aghasadeghi MR, Heidari H, Sadat SM, Irani S, Amini S, Siadat SD, Fazlhashemy ME, Zabihollahi R, Atyabi SM, Momen SB, Khosraavy MS, Davari M, Mohammadi T, Doroudian M, Ardestani MS (2013) Lamivudine-PEGylated chitosan: a novel effective nanosized antiretroviral agent. Curr HIV Res 11:309–320
69. Loutfy SA, Elberry MH, Farroh KY, Mohamed HT, Mohamed AA, Mohamed EB, Faraag AHI, Mousa SA (2020) Antiviral activity of chitosan nanoparticles encapsulating curcumin against hepatitis C virus genotype 4a in human hepatoma cell lines. Int J Nanomedicine 15:2699–2715
70. Ward RA, Fawell S, Floch N, Flemington V, McKerrecher D, Smith PD (2021) Challenges and opportunities in cancer drug resistance. Chem Rev 121:3297–3351

71. Senapati S, Mahanta AK, Kumar S, Maiti P (2018) Controlled drug delivery vehicles for cancer treatment and their performance. Signal Transduct Target Ther 3:7–21
72. Hasnain MS, Ahmed SA, Alkahtani S, Milivojevic M, Kandar CC, Dhara AK, Nayak AK (2020) Biopolymers for drug delivery. In: Hasnain MS, Nayak AK (eds) Advanced biopolymeric systems for drug delivery. Springer International Publishing, Cham, pp 1–29
73. Park JH, Saravanakumar G, Kim K, Kwon IC (2010) Biocompatible nano-structured chitosan-based carriers for gene delivery. Carbohydr Polym 79:403–411
74. Franconetti A, López Ó, Fernandez-Bolanos JG (2020) Carbohydrates: potential sweet tools against cancer. Curr Med Chem 27:1206–1242
75. Adhikari HS, Yadav PN (2018) Anticancer activity of chitosan, chitosan derivatives, and their mechanism of action. Int J Biomater 30:295–315
76. Villeges MR, Baeza A, Regi MV (2018) Nanotechnological strategies for protein delivery. Molecule 23(5):1008–1021
77. Engelhardt B, Carare RO, Bechmann I, Flügel A, Laman JD, Weller RO (2016) Vascular, glial, and lymphatic immune gateways of the central nervous system. Acta Neuropathol 132:17–338
78. Hao C, Wang W, Wang S, Zhang L, Guo Y (2017) An overview of the protective effects of chitosan and acetylated chitosan oligosaccharides against neuronal disorders. Mar Drugs 15(4): 348–367
79. Wang X, Miao J, Yan C, Ge R, Liang T, Liu E (2016) Chitosan attenuates dibutyltin-induced apoptosis in PC12 cells through inhibition of the mitochondria dependent pathway. Carbohydr Polym 151:996–1005
80. Koo HN, Jeong HJ, Hong SH, Choi JH, An NH, Kim HM (2002) High molecular weight water-soluble chitosan protects against apoptosis induced by serum starvation in human astrocytes. J Nutr Biochem 13(4):245–249
81. Khodagholi F, Eftekharzadeh B, Maghsoudi N, Rezaei PF (2010) Chitosan prevents oxidative stress-induced amyloid beta formation and cytotoxicity in NT2 neurons: involvement of transcription factors Nrf2 and NF-kappaB. Mol Cell Biochem 337:39–51
82. Kim MS, Sung MJ, Seo SB, Yoo SJ, Lim WK, Kim HM (2002) Water-soluble chitosan inhibits the production of pro-inflammatory cytokine in human astrocytoma cells activated by amyloid β peptide and interleukin-1β. Neurosci Lett 321(1):105–109
83. Zhou S, Yang Y, Gu X, Ding F (2008) Chito oligosaccharides protect cultured hippocampal neurons against glutamate-induced neurotoxicity. Neurosci Lett 444(3):270–274
84. Dai X, Chang P, Zhu Q, Liu W, Sun Y, Zhu S (2013) Chitosan oligosaccharides protect rat primary hippocampal neurons from oligomeric β-amyloid 1-42-induced neurotoxicity. Neurosci Lett 32(554):64–69
85. Georganopoulou DG, Chang L, Nam J-M, Thaxton CS, Mufson EJ, Klein WL, Mirkin CA (2005) Proc Natl Acad Sci U S A 102(7):2273–2276
86. Rezaii M, Oryan S, Averi A (2019) Curcumin nanoparticles incorporated collagen-chitosan scaffold promotes cutaneous wound healing through regulation of TGF-β1/Smad7 gene expression. Mater Sci Eng C 98:347–357
87. Sarvaiya J, Agrawal YK (2015) Chitosan as a suitable nanocarrier material for anti-Alzheimer drug delivery. Int J Biol Macromol 72:454–465
88. Duceppe N, Tabrizian M (2010) Advances in using chitosan-based nanoparticles for *in vitro* and *in vivo* drug and gene delivery. Expert Opin Drug Deliv 7(10):1191–1207
89. Vulpoi A, Gruian C, Vanea E, Baia L, Steinhoff HJ, Göller G, Simon V (2012) Bioactivity and protein attachment onto bioactive glasses containing silver nanoparticles. J Biomed Mater Res A 100:1179–1186
90. Wagener V, Boccaccini AR, Virtanen S (2017) Protein adsorption and CaPO4 formation on chitosan bioactive glass composite coatings. Appl Surf Sci 416:454–460
91. Gruian CM, Rickert C, Nicklisch SC, Vanea E, Steinhoff HJ, Simon S (2017) Conformational changes and competitive adsorption between serum albumin and hemoglobin on bio-ceramic substrates. Chem Phys Chem 18:634–664

92. Höök F, Vörös J, Rodahl M, Kurrat R, Böni P, Ramsden JJ, Textor M, Spencer ND, Tengvall P, Gold J, Kasemo BA (2002) Comparative study of protein adsorption on titanium oxide surfaces using in situ ellipsometry, optical waveguide light-mode spectroscopy, and quartz crystal microbalance/dissipation. Colloids Surf B: Biointerfaces 24:155–170
93. Abomosallam M, Hendam BM, Abdallah AA, Refaat R, Elshatory A, Gad El Hak HN (2023) Neuroprotective effect of piracetam-loaded magnetic chitosan nanoparticles against thiacloprid-induced neurotoxicity in albino rats. Inflammopharmacology 34:11–23
94. Aranaz I, Acosta N, Revuelta J, Bastida A, Gomez-Casado V, Civera C (2023) Fast and sustained axonal growth by BDNF released from chitosan microspheres. Mar Drugs 21(2): 91–104

Adv Polym Sci (2025) 295: 261–284
https://doi.org/10.1007/12_2024_177

Published online: 14 November 2024

Chitosan Nanoparticles: Drug Delivery Carriers in Cancer Therapy

Kiran Joy, D. Sathya Seeli, and M. Prabaharan

Contents

1 Introduction 262
2 Chitosan NPs in Tumor-Specific Targeted Drug Delivery 263
2.1 Chitosan NPs in Passive Drug Delivery Systems 263
2.2 Chitosan NPs in Active Drug Delivery Systems 265
3 Applications of Chitosan NPs in Different Types of Cancer 271
3.1 In Breast Cancer 272
3.2 In Prostate Cancer 274
3.3 In Lung Cancer 275
3.4 In Colon Cancer 277
3.5 In Other Types of Cancer 278
4 Summary 279
References 280

Abstract Chitosan nanoparticles (NPs) have garnered significant interest in cancer therapy due to their unique properties, including biodegradability, biocompatibility, and mucoadhesive characteristics. This chapter provides a comprehensive overview of the applications of chitosan NPs in tumor-specific targeted drug delivery, emphasizing both passive and active delivery systems. Passive drug delivery systems utilize the enhanced permeability and retention (EPR) effect to accumulate chitosan NPs in tumor tissues, while active drug delivery systems utilize ligand-receptor interactions to enhance targeting specificity. The chapter delves into various strategies for active drug delivery using chitosan NPs, including folic acid-based, nucleic acid-based, peptide-based, and antibody-based targeting. Additionally, the chapter discusses the applications of chitosan NPs in different cancer categories, such as breast cancer, prostate cancer, lung cancer, colon cancer, and other malignancies like

K. Joy, D. Sathya Seeli, and M. Prabaharan (✉)
Department of Chemistry, Hindustan Institute of Technology and Science, Chennai, India

bladder cancer and gliomas. These discussions underscore the versatility of chitosan NPs in encapsulating a wide range of therapeutic agents and delivering them with increased precision to tumor sites, resulting in improved treatment outcomes and reduced systemic toxicity.

Keywords Cancer therapy · Chitosan · Cytotoxicity · Drug delivery · Nanoparticles

1 Introduction

Nanoparticles (NPs) are fabricated from various materials and play a pivotal role in biomedical applications. Their creation involves top-down techniques such as sonication, high-pressure treatments, and homogenization, as well as bottom-up approaches like solvent displacement and reactive precipitation [1]. At the atomic scale, NPs exhibit distinctive properties that differ from their bulk counterparts. In biomedicine, NPs are crucial in diagnostics, imaging, therapy, and targeted drug delivery. By serving as contrast agents in magnetic resonance imaging (MRI), computed tomography (CT) scans, and fluorescence imaging methods, NPs enhance tissue visualization and facilitate early disease detection. Moreover, they function as carriers for therapeutic agents to improve treatment efficacy while minimizing adverse effects [2]. By engineering NPs specifically for targeted drug delivery, they can selectively attach to cancer cells. This selective binding allows them to optimize drug concentration at the intended site, minimizing unintended impacts elsewhere in the body. The integration of both diagnostic and therapeutic functions within theranostic NPs represents a significant advancement in personalized medicine. This integration underscores the substantial potential of NPs to revolutionize biomedical research and healthcare practices [3].

Chitosan, composed of *N*-acetyl-*D*-glucosamine and *D*-glucosamine units, stands out for its remarkable properties due to the presence of amino and hydroxyl groups within each glycosidic unit [4]. Chitosan is derived from chitin, a polysaccharide abundantly found in crustacean shells, through a deacetylation process [5]. As a biomaterial, chitosan exhibits antimicrobial properties, minimal immunogenic response, and outstanding biodegradability and biocompatibility. The primary amino groups within chitosan are accountable for various functions such as muco-adhesion, controlled drug release, and targeted delivery to specific areas of the human body [6]. Chitosan NPs hold immense potential in biomedical applications due to their unique properties. These NPs possess a high surface area-to-volume ratio and can be easily functionalized, allowing precise control and adaptation to meet specific biomedical requirements [7]. One key advantage of chitosan NPs is their ability to encapsulate a variety of therapeutics, including proteins, drugs, and nucleic acids. Moreover, their biocompatibility and biodegradability make them suitable for cancer interventions, such as targeted drug delivery to tumor sites and controlled release of anticancer agents. Additionally, the muco-adhesive properties of chitosan

NPs enhance drug absorption across mucosal surfaces, offering potential applications in oral cancer drug delivery and mucosal vaccination strategies [8]. Furthermore, chitosan NPs can be engineered to target specific tissues to prevent non-specific interactions, improve local drug concentration, and reduce the toxicity and undesired effects of systemic administration. Targeted drug delivery can be accomplished by changing chitosan NPs with antibodies, peptides, aptamers, or small molecules. These targeting approaches not only reduce the required drug dosage but also enhance receptor binding efficiency [9]. This review aims to provide insight into various chitosan NPs designed for tumor-targeted drug delivery, with a special focus on different targeting strategies.

2 Chitosan NPs in Tumor-Specific Targeted Drug Delivery

Cancer is a foremost reason of mortality worldwide. The drawbacks of current cancer treatments include painful side effects, limited effectiveness, and considerable expenses. In response to these challenges, nanotechnology has emerged as a promising pathway for cancer therapy. In recent years, chitosan NPs have received significant attention as potential drug carriers for cancer therapy due to their biodegradability, biocompatibility, improved drug loading, and targetability. The versatile utility of chitosan NPs as both drug and imaging agent carriers enhances their potential for applications in imaging, diagnosis, and treatment techniques [10]. Although a limited number of chitosan-based nanoformulations have transitioned to large-scale applications, further research and combination with other compounds hold promise for expanding their utilization in cancer therapy [11]. Chitosan NPs can deliver therapeutic agents to tumor sites through passive and active-targeting mechanisms. In passive targeting, chitosan-based nanocarriers capitalize on the enhanced permeability and retention (EPR) effect, allowing them to accumulate preferentially in cancerous tissues. This approach reduces systemic toxicity and minimizes adverse effects, making it a critical aspect of modern cancer therapy. The active targeted drug delivery approach combines ligand-receptor binding with the unique physicochemical traits of NPs. This approach offers precision in delivering drugs to cancer cells. By using specific targeting strategies, it is possible to direct treatments with greater accuracy. This reduces the risk of damaging healthy tissue and increases the effectiveness of cancer therapies. As a result, targeted drug delivery can enhance drug efficacy and lower the potential for side effects.

2.1 *Chitosan NPs in Passive Drug Delivery Systems*

Passive drug delivery systems play a crucial role in cancer treatment by capitalizing on the unique microenvironment of tumors to selectively deliver therapeutic agents to malignant cells. In the context of cancer therapy, passive targeting exploits the

EPR effect, where the leaky vasculature and poor lymphatic drainage in tumors permit NPs to gather preferentially within the tumor microenvironment [12]. This phenomenon enables the passive delivery of chemotherapeutic drugs, NPs, or other therapeutic agents directly to tumor tissues while minimizing exposure to healthy tissues. Through passive targeting mechanisms, the internalization efficiency of drug-loaded NPs into cancer cells can increase by up to 50 times compared to normal cells [13].

Recently, chitosan-based NPs, with dimensions ranging from 10 to 200 nm, have been designed to efficiently pass through the fenestrated blood vessels commonly found in and around cancerous tissues. The porous nature of these blood vessels, along with inadequate lymphatic drainage and compromised venous structure, allows these nanocarriers to easily infiltrate tumor sites and be taken up by cancer cells. Furthermore, chitosan-based NPs can linger in tumors for extended periods due to the weak venous network and insufficient lymphatic clearance, resulting in sustained accumulation within the tumor microenvironment [14]. Tumor environments typically have leaky blood vessels and weak lymphatic drainage, which allows chitosan NPs to concentrate in tumor tissues. This characteristic is useful for the targeted delivery of chemotherapeutic agents and reducing the impact of the drugs on normal tissues. Since the passive targeting approach leads to fewer side effects for the patient, it opens the possibility of administering higher doses of chemotherapy to improve therapeutic effects [15]. The schematic representation of chitosan NPs in passive drug delivery systems is exhibited in Fig. 1.

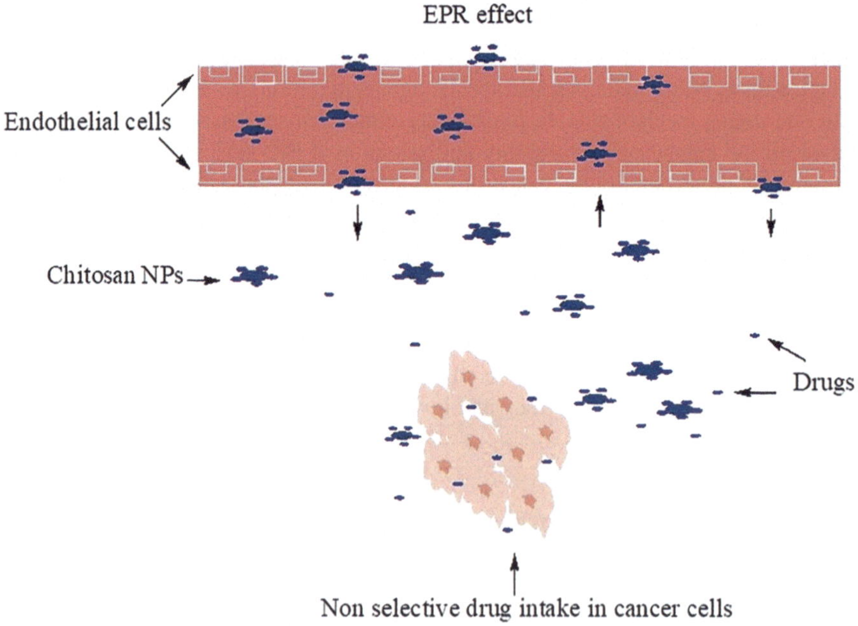

Fig. 1 Chitosan NPs in passive drug delivery systems

2.2 Chitosan NPs in Active Drug Delivery Systems

Passive drug delivery systems often lack targeting precision as they depend on the body's natural circulatory system, which can be inaccurate. This inaccuracy can result in drug distribution to non-target tissues, causing side effects and reducing therapeutic effectiveness. Tumor heterogeneity is another complicating factor common in many cancers, as tumor vascularization can vary widely, affecting drug distribution. Similarly, differences in interstitial pressure within a tumor can lead to inconsistent accumulation of chitosan NPs. These variations can have a direct impact on the effectiveness of the treatment, making it less reliable. Additionally, passive delivery systems generally have limited control over drug release, which can result in under-dosing or overdosing. Under-dosing may render the treatment ineffective, while overdosing can lead to severe side effects. Because of these risks, passive systems might not be suitable for all cancer types or therapeutic needs, highlighting the importance of exploring other approaches such as active drug delivery systems [16]. Active drug delivery systems offer precision in delivering drugs to cancer cells from chitosan-based NPs through receptor-mediated endocytosis mechanisms. The schematic representation of chitosan NPs in active drug delivery systems is shown in Fig. 2. So far, a variety of chitosan-based NPs conjugated with different types of targeting ligands, such as folic acid (FA), nucleic acids, peptides, and proteins, have been developed for active tumor-targeted drug delivery.

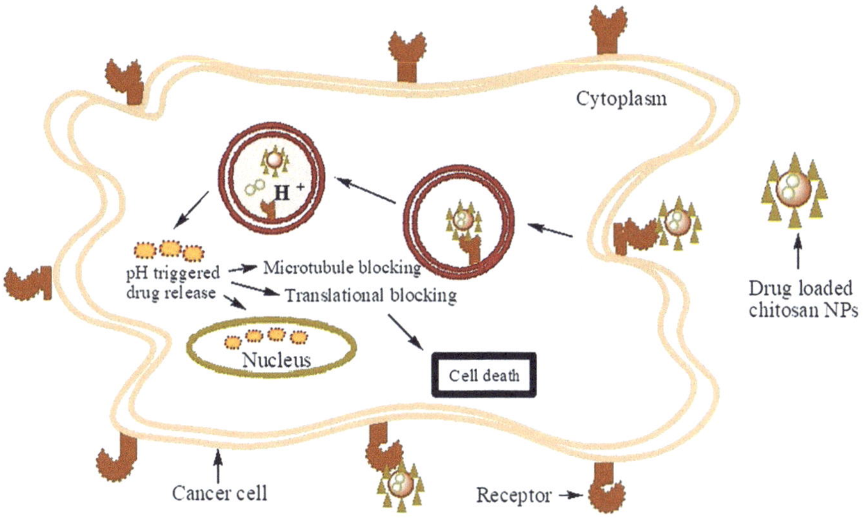

Fig. 2 Chitosan NPs in active drug delivery systems

2.2.1 FA-Conjugated Chitosan NPs

FA is a water-soluble vitamin essential for various biological functions, including DNA synthesis, repair, and methylation. FA is important for cell division and growth and is recommended for pregnant women to support fetal development. In targeted drug delivery, FA is valuable due to its affinity for folate receptors, which are often overexpressed on certain cancer cells. This feature makes FA useful for developing targeted cancer therapies, as it can deliver therapeutic agents directly to cancer cells, potentially reducing side effects and improving treatment effectiveness.

A study by Ullah et al. focused on formulating FA-conjugated chitosan NPs loaded with 5-fluorouracil for colon cancer treatment [17]. The developed FA-conjugated chitosan NPs had an average size of 235 nm, while the FA-free chitosan NPs were about 208 nm. Both types of NPs had a zeta potential of around +20 mV. The drug content yield and entrapment efficiency of these NPs were within acceptable ranges. In vitro drug release studies showed that the FA-conjugated chitosan NPs presented controlled release of the encapsulated drug up to 96%, suggesting these NPs are suitable for sustained drug delivery. Cytotoxicity studies indicated that the FA-conjugated chitosan NPs were more effective against cancer cells, suggesting that FA decoration on chitosan NPs could offer a potentially more effective and safer approach for colon cancer treatment. Recently, Kumbhar et al. synthesized chitosan NPs tagged with FA and loaded with dasatinib using tripolyphosphate (TPP) as an ionic coupling agent [18]. The reaction scheme for the preparation of FA-tagged chitosan NPs and schematic representation of FA-tagged chitosan NPs loaded with dasatinib are shown in Fig. 3. The NPs produced in this study exhibited smaller diameters, greater drug encapsulation, and loading efficiency. They also had a consistent spherical shape and released their contents in vitro under controlled conditions, leading to a noticeable decrease in cytotoxicity. The chitosan NPs developed in this study could be useful in targeting folate receptor-positive cancer cells and consequently could be a more promising option for treating multidrug-resistant cancers.

San et al. developed turmeric oil-loaded FA-conjugated chitosan-alginate nanocapsules for targeted breast cancer therapy [19]. In this study, chitosan was grafted with FA through an amide bond, resulting in a 12.86% degree of substitution. These NPs had a consistent size of 189 nm, an encapsulation efficiency of 35.9%, and a loading capacity of 1.82%. An in vitro release study indicated that the developed drug-loaded nanocapsules provided a sustained release of tamoxifen at pH levels of 5.5 and 7.4. Furthermore, these nanocapsules demonstrated a lower IC50 compared to both the FA-free nanocapsules and tamoxifen-free nanocapsules when tested using MDA-MB-231 and MCF-7 cancer cell lines. Ince et al. developed radioiodinated FA-chitosan conjugated thymoquinone NPs aimed at improving targeted treatment for ovarian cancer cells [20]. The loading efficiency of these NPs was measured at $20.0 \pm 1.0\%$, and FA bonding on the NP surface was found to be $46.0 \pm 0.5\%$. The developed NPs showed improved uptake by the SKOV3 (human ovarian adenocarcinoma) cell line compared to free thymoquinone.

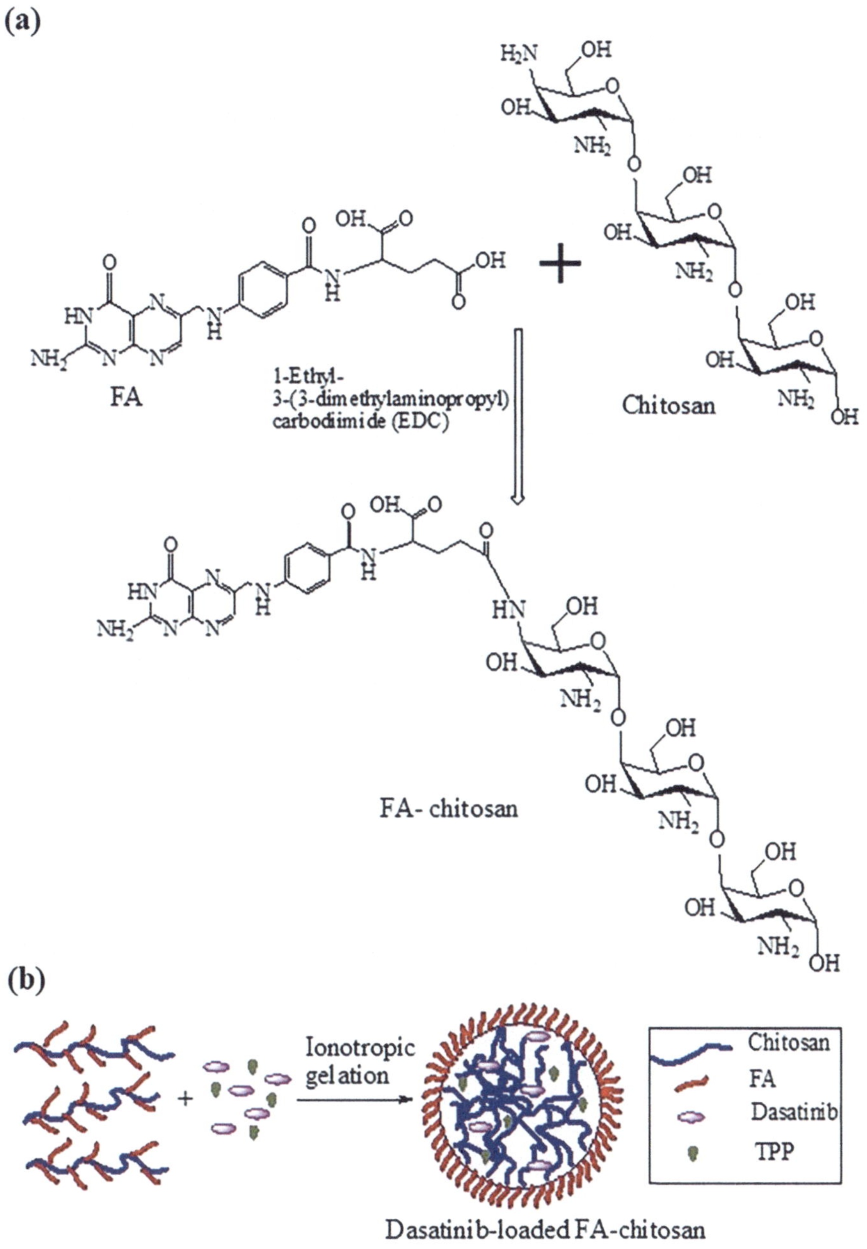

Fig. 3 (**a**) Synthesis of FA-conjugated chitosan and (**b**) schematic representation of FA-conjugated chitosan loaded with dasatinib

In another study, Mazzotta et al. developed redox-responsive FA-conjugated chitosan NPs for intracellular methotrexate delivery using an ionotropic gelation technique [21]. Due to the presence of L-cysteine and FA, the developed NPs exhibited redox responsiveness and tumor-targeted drug delivery via a folate-receptor-mediated endocytosis mechanism. The unique design of these NPs allowed for tumor specificity and controlled drug release. In in vitro studies, the drug release from the redox-responsive NPs was influenced by glutathione. In a normal environment, the NPs remained stable. However, in a reducing environment, such as that found in tumor cells, the encapsulated methotrexate was quickly released. The redox-responsive FA-conjugated chitosan NPs exhibited a greater inhibitory impact on HeLa cell growth when compared to FA-free chitosan NPs.

2.2.2 Nucleic Acid-Conjugated Chitosan NPs

Tumor-targeted drug delivery using nucleic acid-based nanocarriers for cancer treatment is an innovative approach aimed at enhancing the precision of cancer therapies. This method employs nucleic acids like DNA or RNA to deliver therapeutic agents directly to cancer cells, showing promise for developing more effective and less invasive cancer treatments. Ultimately, it could improve patient outcomes and quality of life [22]. A major drawback of nucleic acid-based drug carriers is the instability and degradation of mRNA. To address this issue, Maiyo and Singh developed chitosan-coated selenium NPs with a FA-targeting group to deliver Fluc mRNA to cancer cells. The prepared NPs were steady, uniformly dispersed, and in the size range of 59 to 102 nm. These NPs effectively shielded mRNA from ribonuclease deprivation and showed low toxicity against nasopharyngeal (KB), human embryonic kidney (HEK293), and breast adenocarcinoma (MCF-7) cancer cell lines. In colorectal carcinoma (Caco-2) and colon carcinoma (HT-29) cell lines, the chitosan-coated selenium NPs showed moderate cytotoxicity due to selenium-induced apoptosis. Selenium uptake studies supported the transfection results, demonstrating significant expression of the transgene in KB cells that overexpress FA receptors, compared to other type of cells without FA receptor expression [23].

Liang et al. developed hyaluronic acid (HA) dialdehyde-conjugated chitosan NPs loaded with siRNA for targeting the Bcl2 oncogene [24]. The developed NPs ranged in size from 100 to 120 nm, were stable, and had a high siRNA encapsulation capability. These NPs exhibited low toxicity and were blood-compatible. The siRNA present in the NPs could enter T24 cells through CD44 receptor-ligand-mediated endocytosis, facilitating targeted gene silencing by delivering cy3-siRNA. In vivo results showed that the siRNA-loaded NPs accumulated at tumor sites, effectively reducing the activity of the targeted oncogene and inhibiting tumor growth.

Recently, Yan et al. developed pH-responsive self-assembled carboxymethyl chitosan (CMC) and fluorescein isothiocyanate (FITC)-labeled chitosan hydrochloride NPs loaded with siRNA as an epithelial surfaces-targeted or mucosa-targeted delivery system (Fig. 4) [25]. At an optimal ratio of CMC and FITC-labeled chitosan

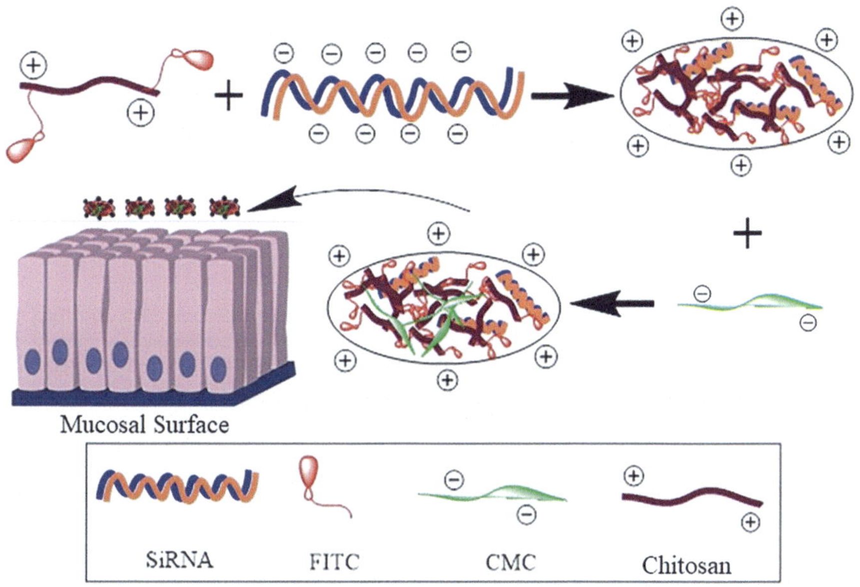

Fig. 4 Schematic representation of the formation and epithelial-targeting of siRNA-loaded CMC and FITC-labeled chitosan hydrochloride NPs

hydrochloride, the encapsulation efficiency of siRNA in the developed NPs was 76.7 ± 3.7%. These NPs protected siRNA from disintegration in acidic environments and allowed for sustained delivery of siRNA when exposed to ultrasound in the suitable pH conditions due to its pH-responsive behavior. The results of this study demonstrated that the siRNA-loaded CMC and FITC-labeled chitosan hydrochloride nanoplatform could effectively deliver siRNA into colon cancer cells and target β-catenin protein expression in HT-29 cells for gene silencing.

2.2.3 Peptide-Conjugated Chitosan NPs

Chitosan NPs, with their unique properties and peptide-based targeting, offer a promising approach to cancer treatment. Yadav et al. prepared raloxifene-encapsulated chitosan and raloxifene-encapsulated arginylglycylaspartic acid (RGD)-conjugated chitosan NPs using an ionic gelation method for breast cancer treatment [26]. These NPs exhibited greater firmness, increased zeta potential, along with improved cellular uptake at the acidic pH levels typical of tumor microenvironment compared to physiological pH. Moreover, the coupling of the RGD peptide enhanced the in vitro cellular uptake of chitosan NPs in breast cancer cells expressing αvβ3 integrin, leading to higher cellular apoptosis.

In another study, Bicak et al. designed tyrosyllysylthreonine peptide-loaded chitosan NPs using a modified ionic gelation method to enhance the effectiveness of a tripeptide while minimizing its adverse effects [27]. In this study, tripeptide and peptide-loaded chitosan NPs were tested on a prostate cancer cell line (MAT-Lylu) in vitro. The results showed that tripeptide and tripeptide-loaded chitosan NPs exhibited higher toxicity against the cancer cell line compared to control chitosan NPs. A controlled in vitro release study conducted at pH 7.4 found that the peptide-loaded chitosan NPs released 34% of the tripeptide by the end of the first day and 97% by the end of the eleventh day.

2.2.4 Antibody-Conjugated Chitosan NPs

In recent years, chitosan NPs have emerged as a platform for antibody-based targeted drug delivery in cancer treatment due to their biocompatibility and versatility. By functionalizing chitosan NPs with specific antibodies, a wide variety of tumor-targeted drug delivery carriers have been developed. These carriers can precisely deliver anticancer drugs to tumor sites, thereby reducing side effects and enhancing therapeutic efficacy [28]. Recently, Helmi et al. synthesized chitosan NPs to selectively deliver the anti-cancer drug doxorubicin to the breast tumor microenvironment [29]. To increase their circulation time in the bloodstream, the surface of the chitosan NPs was modified with polyethylene glycol (PEG). To enhance drug selectivity, chitosan NPs were conjugated with anti-human mammaglobin (Anti-hMAM) and anti-human epidermal growth factor receptor 2 (Anti-HER2) (Fig. 5). These chitosan-based NPs showed greater cytotoxicity against MCF-7 cancer cells compared to L-929 normal cells, in contrast to free doxorubicin. Moreover, these NPs exhibited pH-responsive doxorubicin release. In the acidic breast tumor environment, the rate and amount of doxorubicin release increased because of the protonation of amino groups of chitosan and doxorubicin. In normal physiological conditions, the release of the drug from the NPs was controlled. Due to the presence of monoclonal antibodies, the developed chitosan NPs demonstrated improved cellular uptake and higher rates of intracellular accumulation, allowing for targeted drug release and reduced tumor cell viability.

In another study, Sabra et al. developed mucoadhesive cetuximab-conjugated citrus pectin-chitosan NPs for the targeted delivery of curcumin in the treatment of colorectal cancer (Fig. 6) [30]. In vitro drug release studies revealed that this formulation is suitable for colon delivery. Additionally, these NPs demonstrated improved cellular uptake and anticancer activity against the Caco2 cancer cell line. This increased uptake can be attributed to a combination of factors, including the optimal size, positive charge, active targeting, and mucoadhesive properties of the NPs.

Recently, Manivasagam et al. prepared a multifunctional NPs made up of thiol chitosan-layered gold nanoshells conjugated with anti-epidermal growth factor receptor (EGFR) and loaded with paclitaxel for fluorescence/photoacoustic dual-imaging-guided chemophotothermal therapy [31]. This system demonstrated near-infrared (NIR) laser-triggered drug release combined with a chemophotothermal effect guided by photoacoustic imaging. The nanocarrier exhibited outstanding

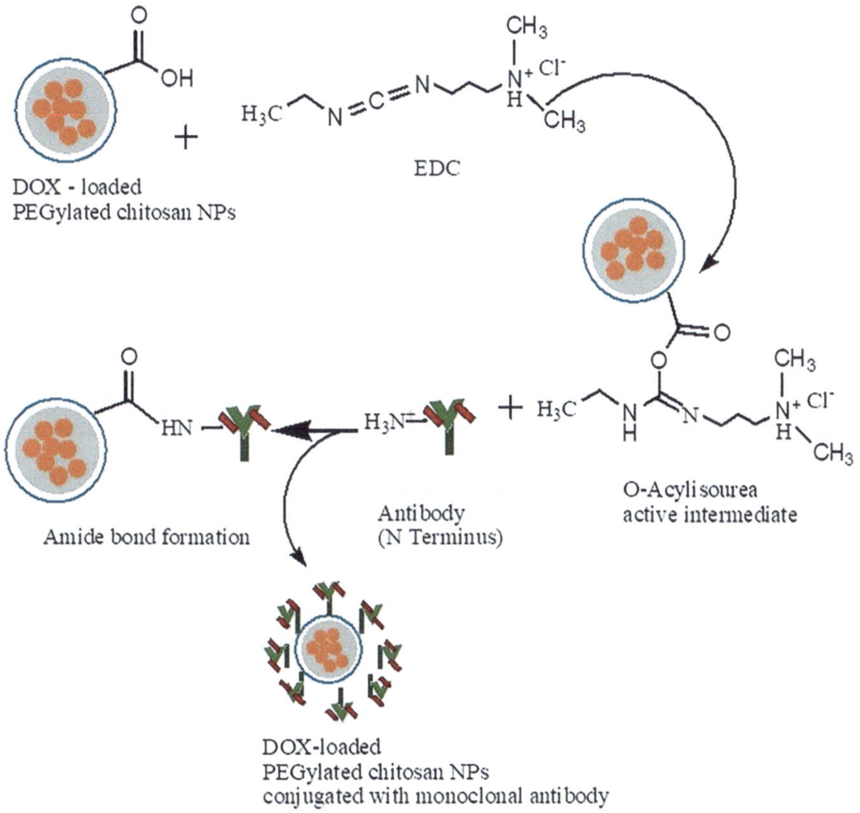

Fig. 5 The schematic representation for the preparation of doxorubicin-loaded chitosan NPs surface modified with PEG and antibodies

photostability, biosafety, stability, biocompatibility, NIR absorbance, and targeting ability. Moreover, it showed significant drug loading capacity and responsiveness to both pH and NIR laser. In vivo chemophotothermal treatment with the multifunctional nanocarrier and laser irradiation resulted in substantial tumor ablation with no signs of recurrence. Additionally, the developed nanocarrier demonstrated excellent biosafety and biocompatibility.

3 Applications of Chitosan NPs in Different Types of Cancer

Chitosan NPs have been employed in the dealing of breast, prostate, lung, and colon cancers. In breast cancer, chitosan NPs have been utilized to deliver chemotherapeutics and target specific markers like HER2. For prostate cancer, these NPs can

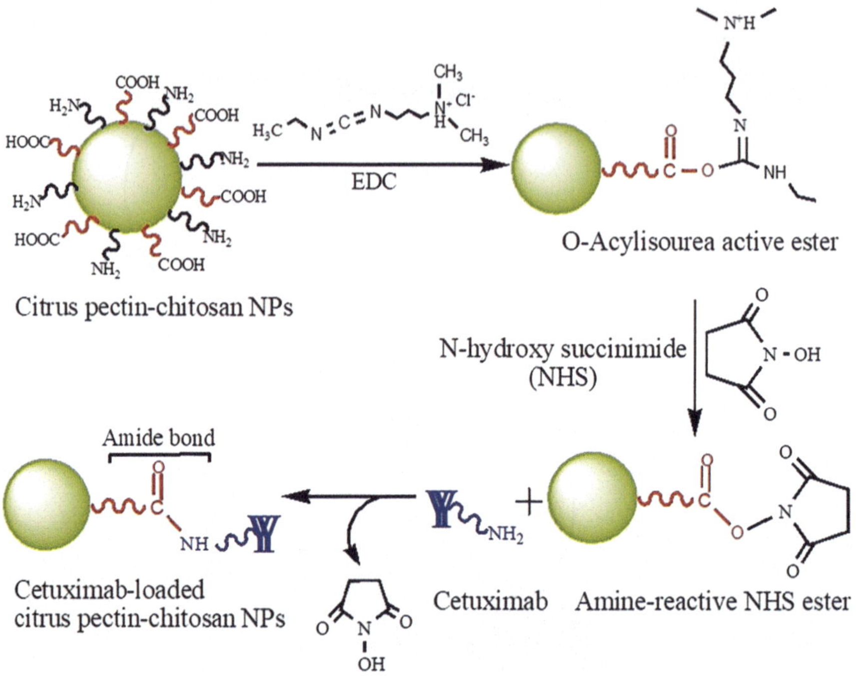

Fig. 6 Schematic representation for the preparation of cetuximab-conjugated citrus pectin-chitosan NPs

target androgen receptors to achieve targeted drug delivery. Beyond drug delivery, chitosan NPs are being explored for diagnostic purposes and combination therapies, such as photothermal therapy and immunotherapy. As research progresses, the applications of chitosan NPs in cancer treatment continue to expand, offering the potential for more effective and targeted therapies with fewer side effects.

3.1 In Breast Cancer

Breast cancer is one of the most common cancers globally, and traditional treatments often bring significant side effects. In this context, chitosan NPs have shown great potential as targeted drug delivery carriers for breast cancer therapy [32]. Chitosan NPs with various functionalities have been used to deliver chemotherapeutics, gene therapy, and immunotherapy with fewer side effects [33]. Recently, Afrifa et al. developed chitosan modified with PEG NPs for triple-negative breast cancer treatment [34]. In this study, various chitosan-PEG copolymers were prepared through formaldehyde-mediated grafting of PEG onto chitosan. Different formaldehyde ratios were used to vary the physicochemical properties of the copolymer. In vivo studies indicated that the auranofin loaded in the chitosan-PEG NPs had a stronger necrotic effect on triple-negative breast cancer cells than free auranofin, highlighting the improved potency of the drug-encapsulated chitosan-PEG NPs. Olmos et al. explored the sensitizing effect of chitosan-based NPs on reversine-treated breast cancer cells [35]. They measured cell metabolic activity to evaluate cell viability, used the alkaline comet assay to assess DNA damage, and employed immunofluorescence with anti-P-H3 as a marker for mitosis. The findings indicated that chitosan NPs trigger cellular responses that disturb mitosis and cell viability and can sensitize breast cancer cells to X-ray radiation. Furthermore, it was found that the sensitization effect was not due to direct DNA damage caused by the NPs.

In another study, black pomegranate peel extract-loaded core-shell chitosan magnetic NPs were synthesized by Taherian et al. to serve as drug carriers for cancer therapy [36]. The synthesized NPs demonstrated good drug-loading efficiency and an impressive release rate in a tumor-like environment. Cytotoxicity tests indicated that the NPs were biocompatible, showing no significant toxicity to both normal and cancerous cells. Additionally, these NPs exhibited toxicity toward cancerous cells without affecting normal cells. Gounden et al. developed cisplatin-loaded silver (Ag) NPs functionalized with chitosan for breast cancer therapy [37]. The developed chitosan-Ag NPs encapsulated over 80% of cisplatin. Additionally, they demonstrated targetability toward breast cancer cells, causing over 80% cell demise at doses below 10 μg, while showing nominal cytotoxicity in normal cells. Hassanpour et al. developed salicylic acid-loaded chitosan NPs using TPP as a cross-linker [38]. These NPs showed an encapsulation efficiency of 84%. The release pattern of salicylic acid from chitosan NPs was examined using the Korsmeyer-Peppas and Higuchi models, indicating that the drug delivery was primarily ruled by the Korsmeyer-Peppas model. Additionally, the toxicity of free salicylic acid and salicylic acid-loaded chitosan NPs was assessed against the MDA-MB-231 breast cancer cell line using the MTT method. The results disclosed that salicylic acid-loaded chitosan NPs were more cytotoxic than free salicylic acid.

Mehata et al. prepared docetaxel-loaded D-α-tocopherol PEG succinate-conjugated chitosan NPs with or without trastuzumab [39]. The average size and entrapment efficacy of these NPs ranged between 126–186 nm and 74–78%, respectively. The results of in vitro studies conducted using SK-BR-3 cells revealed that the developed NPs improved cytotoxicity and cellular uptake. The IC50 values of non-targeted and targeted NPs were found to be 43 and 223 times lower than those of the control conventional formulation (Docel™). In vivo study demonstrated a 2.33-fold increase in the docetaxel relative bioavailability for non-targeted NPs and a 2.82-fold increase for HER-2 receptor-targeted NPs. After intravenous administration, the non-targeted and targeted NPs achieved half-lives that were 3.48 and 5.94 times higher, respectively, than the control. The relative bioavailability, area under the curve, and mean residence time were also increased for non-targeted and targeted NPs when related to control. Furthermore, histopathology studies showed that non-targeted and targeted NPs caused fewer toxicity to liver, lungs, and kidneys.

Recently, Gogoi et al. developed phosphorylated chitosan NPs loaded with curcumin and (±) α-lipoic acid using the ionic gelation method (Fig. 7) [40]. The developed NPs demonstrated improved drug loading efficacy with favorable

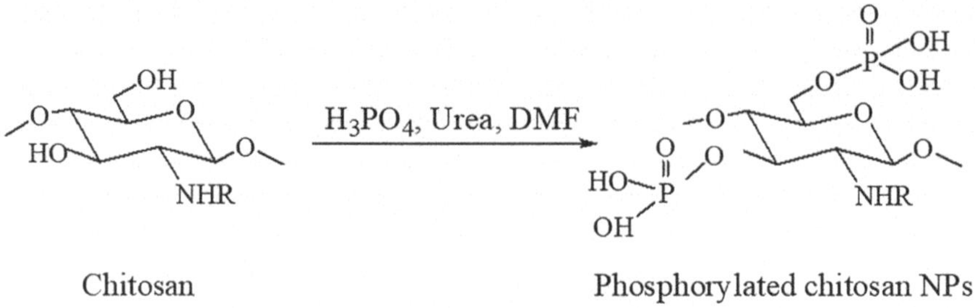

Fig. 7 Synthesis of phosphorylated chitosan

swelling and drug delivery character at pH 5.0. Morphological characterization revealed that the prepared NPs had a rough and dense surface with particle sizes ranging from 82.66 to 142.64 nm. The prepared NPs were not found to be cytotoxic to normal human lymphocytes; rather, they promoted cell proliferation. However, when exposed to MDA-MB-231 cells, the NPs exhibited anti-breast cancer activity. The successful preparation of drug-loaded phosphorylated chitosan NPs demonstrated the ability to obstruct growth and increase the amount of dead MDA-MB-231 breast cancer cells when related to drug-free NPs.

3.2 *In Prostate Cancer*

Prostate cancer is a significant health challenge among men worldwide, necessitating innovative therapeutic approaches. Traditional treatments, including surgery, radiation therapy, and chemotherapy, are often effective but come with considerable side effects and risks [41, 42]. In recent years, chitosan NPs have emerged as a promising avenue in prostate cancer treatment, offering a more targeted and less invasive approach. Chitosan NPs can serve as vehicles for drug delivery, enabling targeted therapies that could revolutionize prostate cancer treatment by reducing side effects and enhancing therapeutic efficacy [43].

Saravanakumar et al. developed TM2-loaded chitosan NPs (TM2 is a metabolite from T. atroviride with anti-prostate cancer properties) for prostate cancer therapy [44]. Anticancer assays demonstrated that TM2-loaded chitosan NPs induced a higher rate of prostate cancer cell death compared to drug-free NPs by promoting reactive oxygen species (ROS) production, mitochondrial membrane loss, nuclear damage, and apoptosis-related protein expression. This study showed that chitosan NPs can boost the therapeutic efficacy of TM2 by facilitating targeted delivery, leading to more effective prostate cancer treatments. Recently, Zamanvaziri et al. developed chitosan NPs conjugated with PEG and folic acid (FA) as a promising drug delivery system for sodium butyrate to prostate cancer cells [45]. These NPs showed an adequate particle size, higher drug entrapment efficiency, and controlled drug release at physiological pH. MTT assay, gene expression analysis, and apoptosis-related cell cycle arrest results showed the potential of these NPs to suppress prostate cancer cells. The results suggest that chitosan NPs conjugated with PEG and FA hold promise for effective prostate cancer treatment, as their components are biodegradable, biocompatible, non-toxic, and effective in combating cancer. Due to the presence of FA, the developed NPs can be suitable candidates for targeted drug delivery in cancers with high FA receptor expression.

3.3 In Lung Cancer

Lung cancer continues to be a significant health challenge, accounting for a large proportion of cancer-related deaths globally [46, 47]. In this context, chitosan NPs offer a promising new approach to lung cancer treatment. They can encapsulate therapeutic agents and deliver them directly to lung cancer cells, reducing collateral damage to healthy tissues and potentially improving treatment efficacy. This unique capability has sparked interest in using chitosan NPs for targeted drug delivery, gene therapy, and other innovative approaches to lung cancer treatment [48]. Almutairi et al. designed a drug delivery system for lung and liver cancer treatments based on a HA and chitosan complex to improve the half-life and activity of raloxifene, as shown in Fig. 8 [49]. This study investigated the pro-apoptotic and cytotoxic effects of drug-loaded HA and chitosan complex on non-small cell lung cancer A549 cells and liver cancer cells, including HepG2 and Huh-7. The developed nanocomplex showed 92% entrapment efficiency and demonstrated improved cytotoxicity against A549 cells compared to the control formulations. The suppression of A549 cell viability was linked to a reduction in glucose uptake, leading to decreased cellular bioenergetics and the activation of apoptosis through elevated nitric oxide levels.

Viswanadh et al. prepared docetaxel-encapsulated chitosan-FA and chitosan-FA-EGFR NPs [50]. In vitro analysis of docetaxel-loaded chitosan-FA-EGFR NPs showed significantly greater cytotoxicity against A-549 cells, with an IC50 nearly 34 times lower than that of the control docetaxel. An in vivo pharmacokinetic study with Wistar rats demonstrated that all NP formulations had improved relative bioavailability compared to the drug control. These results suggest that chitosan NPs targeting both EGFR and FA receptors enhanced docetaxel's cytotoxicity against A-549 lung cancer cells and considerably improved its pharmacokinetic profile in rats.

Photodynamic therapy (PDT) is a potential treatment method for non-small cell lung cancer, but the limited tumor selectivity of current photosensitizers has restricted its broader clinical use. To address this, Zhang et al. synthesized erlotinib-modified chitosan NPs containing indocyanine green for combined targeted chemotherapy and PDT [51]. These NPs exhibited an oval shape with a diameter of 267.5 nm, a zeta potential of -15.2 mV, and notable photostability. The in vitro drug delivery studies indicated that erlotinib and indocyanine green were gradually released from chitosan NPs under acidic conditions with lysozyme. Under NIR laser irradiation, these NPs could generate ROS for PDT. The combination of

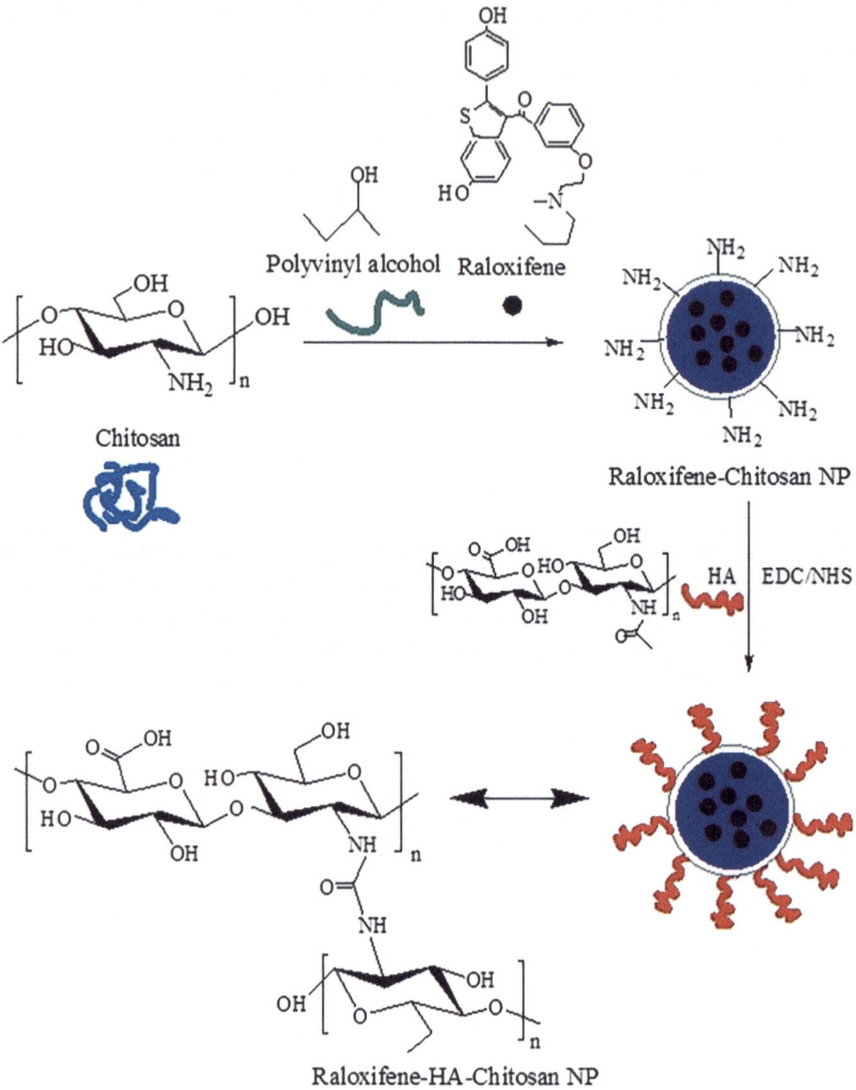

Fig. 8 Schematic diagram of the preparation of raloxifene-loaded HA and chitosan complex

targeted molecular therapy and PDT provided by chitosan NPs demonstrated synergistic effects in inhibiting cell development and encouraging apoptosis in lung cancer cells.

Recently, Kuen et al. developed protocatechuic acid-encapsulated chitosan-palmitoyl NPs for lung cancer therapy [52]. The inclusion of the hydrophobic palmitoyl group in chitosan NPs increased protocatechuic acid encapsulation by 54.5% compared to unmodified chitosan NPs. The cytotoxic efficacy of chitosan-palmitoyl NPs was assessed using the MTT test on lung cancer A549 cells. The developed NPs demonstrated controlled release and enhanced therapeutic efficacy, evidenced by lower IC50 values over 72 h of treatment. Using the hydrophobic chitosan-palmitoyl NPs, protocatechuic acid's efficacy was considerably increased at 24, 48, and 72-h intervals compared to a single administration of the compound or unmodified chitosan NPs.

3.4 In Colon Cancer

Colon cancer ranks among the most prevalent cancers worldwide, leading to significant morbidity and mortality [53]. In recent years, chitosan NPs have emerged as a promising tool in the fight against colon cancer, offering the potential to enhance drug delivery, improve targeted therapy, and reduce side effects [14, 54]. Chitosan NPs can encapsulate chemotherapeutic agents, enabling controlled and targeted drug release directly to colon cancer cells. This precision in drug delivery could lead to more effective treatments with fewer adverse effects, representing a significant advancement in colon cancer therapy.

Recently, Abuelmakarem et al. studied the luminescent properties of chitosan NPs and investigated light propagation in colon cancer and normal cells at wavelengths of 650 nm and 808 nm [55]. The goal of this work was to evaluate the ability of chitosan NPs to differentiate cancerous cells from normal cells using a Monte Carlo simulation model. The optical fluence rate spreading on the surface of the tissue was utilized to recreate images through the diffuse equation using the finite element method. The results showed that chitosan NPs exhibit a luminescent effect, suggesting their potential application in imaging and differentiating cancerous from normal cells. In another study, Ciro et al. developed methotrexate-loaded phytic acid-chitosan NPs for colon cancer therapy (Fig. 9) [56]. The anticancer activity of the NPs was assessed using the MTT test on colon adenocarcinoma cells (HT-29). The results of this study indicated that the methotrexate-loaded phytic acid-chitosan NPs had positive zeta potential values, sizes less than 300 nm, and low polydispersity. These NPs demonstrated good physical stability and the ability to modulate the release of methotrexate via a non-Fickian mechanism.

In another study, Chen et al. developed chitosan NPs encapsulated with a photothermal agent (IR780) and a photodynamic agent (5-aminolevulinic acid) for photothermally enhanced PDT [57]. These NPs were more firm in acidic pHs, such as the stomach, which substantially enhanced drug incorporation and local accumulation in intravenous colon cancer cells in mice. The results of this study showed that the developed NPs could enhance PDT through photothermal effects by increasing oxidative stress, leading to elevated levels of ROS, superoxide, and singlet oxygen creation. Moreover, these NPs demonstrated considerable therapeutic effect for cancer therapy in vivo after oral drug management without causing any noticeable adverse effects. Similarly, Rashedi et al. prepared 5-fluorouracil-loaded chitosan NPs crosslinked with quercetin using the ionotropic gelation method for colon cancer therapy in Wistar rats. The in vitro drug release results indicated that the

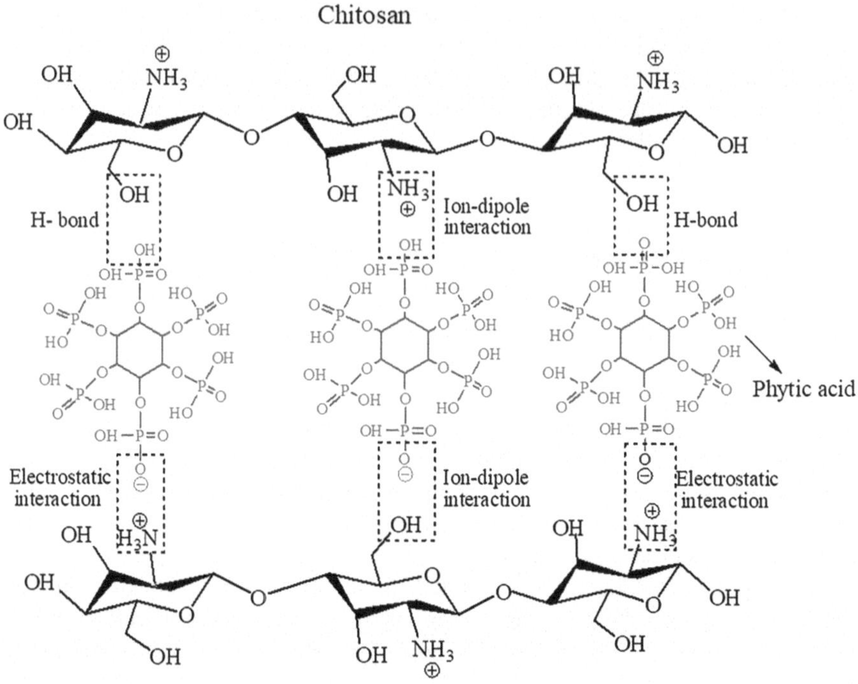

Fig. 9 The chemical structure of phytic acid-chitosan NPs

developed NPs released a lower amount of drug in simulated colonic fluid and a higher amount of drug in simulated tumor microenvironment fluid due to their greater swelling capacity compared to the simulated colonic fluid.

3.5 *In Other Types of Cancer*

Chitosan NPs have also garnered significant attention for treating bladder cancer, oral cancer, and brain cancer. Bladder cancer, the most prevalent urinary system cancer [58, 59], despite having a high survival rate, poses a considerable risk of recurrence and disease progression. Standard therapies involve the intravesical administration of anticancer drugs following the transurethral resection of bladder tumors. However, chemotherapeutic drugs often exhibit insufficient efficiency due to the development of chemoresistance and cytotoxicity. Additionally, the frequent emptying of the bladder and the low permeability of the bladder urothelium further reduce drug retention, weakening the antitumor response. To address these challenges, chitosan NPs have recently been explored as carriers for delivering anticancer drugs to bladder cancer sites. They can enhance drug penetration into deeper cell layers of the bladder by transiently disrupting the urothelium barrier, offering diverse biomedical applications in urinary bladder treatment [60].

Oral cancer is a significant cause of mortality, particularly in developing countries [61, 62]. Recently, Mariadoss et al. developed a promising anticancer formulation using phloretin-loaded chitosan NPs for oral cancer treatment [63]. These NPs demonstrated enhanced cellular uptake, sustained drug release, and improved bioavailability of phloretin. Moreover, the cationic nature of phloretin-loaded chitosan NPs contributed to higher anticancer activity through electrostatic interactions with negatively charged groups present in tumor cells.

Gliomas represent a prevalent type of brain tumor, constituting about 80% of all malignant primary brain tumors [64, 65]. Glioblastoma multiforme, an aggressive astrocytic tumor, accounts for over 60% of adult gliomas, primarily affecting individuals aged 55 to 60 years. Despite being classified as rare, glioblastoma multiforme is a highly fatal disease with a dismal prognosis. In an effort to combat this, Ahmad et al. developed poly-lactide-co-glycolic acid NPs coated with carmustine-loaded chitosan using a solvent vaporization method [66]. The results of drug release and permeation experiments demonstrated controlled drug release and increased carmustine absorbency. The area under the curve for Albino Wistar rat brain tissue obtained from carmustine-loaded NPs was 2.8 times higher than that from intranasal carmustine suspension and 14.7 times higher than from intravenous carmustine. Additionally, the MTT test on the U87 MG cells revealed a substantial reduction in the IC50 value of chitosan-coated NPs (71.23 μg/ml) when related to the drug suspension (90.02 μg/ml). These results underscore the potential of chitosan-coated NPs as a promising delivery system for carmustine via intranasal administration to treat glioblastoma multiforme.

4 Summary

Chitosan NPs have emerged as a versatile and effective platform for targeted drug delivery in cancer therapy, offering a unique combination of biodegradability, biocompatibility, and mucoadhesive properties. Their ability to encapsulate a wide range of therapeutic agents, from chemotherapeutics to genetic materials, allows for precision in targeting tumor cells while minimizing adverse effects on healthy tissues. The adaptability of chitosan NPs for both passive and active drug delivery systems, as well as their capability to be functionalized with specific ligands or antibodies, enhances their potential in cancer treatment.

The use of chitosan NPs in different types of cancer has demonstrated significant improvements in drug delivery and therapeutic outcomes. Studies have shown that these NPs can increase drug stability, solubility, and bioavailability, resulting in sustained drug release and improved treatment efficacy. The ability to target specific receptors on cancer cells enables chitosan NPs to achieve more localized and effective drug delivery, reducing the risk of systemic toxicity and enhancing patient outcomes. Moreover, the development of chitosan NPs has extended beyond traditional drug cancers, showing promise in treating other types of cancer such as gliomas, bladder cancer, and oral cancer. The application of chitosan NPs in these areas has illustrated their potential to revolutionize targeted cancer therapy, with significant reductions in tumor size and improved survival rates in experimental models. Despite these promising results, challenges remain in translating chitosan NPs from laboratory research to clinical applications. Issues such as scale-up production, long-term stability, and potential immunogenic responses must be addressed. Additionally, further studies are needed to explore the long-term safety and efficacy of chitosan NPs in human subjects.

In summary, chitosan NPs represent a transformative approach to cancer treatment, with the potential to improve targeted drug delivery, reduce side effects, and enhance patient outcomes. As research in this field continues to progress, chitosan NPs could become a cornerstone in the development of more effective and personalized cancer therapies, offering new hope to patients and healthcare providers alike.

References

1. Nienhaus K, Wang H, Nienhaus GU (2020) Nanoparticles for biomedical applications: exploring and exploiting molecular interactions at the nano-bio interface. Mater Today Adv 5:100036
2. Han X, Xu K, Taratula O, Farsad K (2019) Applications of nanoparticles in biomedical imaging. Nanoscale 11(3):799–819
3. Dang Y, Guan J (2020) Nanoparticle-based drug delivery systems for cancer therapy. Smart Mater Med 1:10–19
4. Abourehab MA, Pramanik S, Abdelgawad MA, Abualsoud BM, Kadi A, Ansari MJ, Deepak A (2022) Recent advances of chitosan formulations in biomedical applications. Int J Mol Sci 23(18):10975
5. Aranaz I, Alcántara AR, Civera MC, Arias C, Elorza B, Heras Caballero A, Acosta N (2021) Chitosan: an overview of its properties and applications. Polymers 13(19):3256
6. Harugade A, Sherje AP, Pethe A (2023) Chitosan: a review on properties, biological activities and recent progress in biomedical applications. React Funct Polym:105634
7. Gao Y, Wu Y (2022) Recent advances of chitosan-based nanoparticles for biomedical and biotechnological applications. Int J Biol Macromol 203:379–388
8. Rostami E (2020) Progresses in targeted drug delivery systems using chitosan nanoparticles in cancer therapy: a mini-review. J Drug Del Sci Technol 58:101813
9. Bashir SM, Ahmed Rather G, Patrício A, Haq Z, Sheikh AA, Shah MZUH, Fonte P (2022) Chitosan nanoparticles: a versatile platform for biomedical applications. Materials 15(19):6521
10. Shanmuganathan R, Edison TNJI, Lewis Oscar F, Kumar P, Shanmugam S, Pugazhendhi A (2019) Chitosan nanopolymers: an overview of drug delivery against cancer. Int J Biol Macromol 130:727–736
11. Garg U, Chauhan S, Nagaich U, Jain N (2019) Current advances in chitosan nanoparticles based drug delivery and targeting. Adv Pharm Bulletin 9(2):195
12. Tewabe A, Abate A, Tamrie M, Seyfu A, Abdela Siraj E (2021) Targeted drug delivery – from magic bullet to nanomedicine: principles, challenges, and future perspectives. J Mult Healthcare:1711–1724
13. Batlle R, Andrés E, Gonzalez L, Llonch E, Igea A, Gutierrez-Prat N, Berenguer-Llergo A, Nebreda AR (2019) Regulation of tumor angiogenesis and mesenchymal–endothelial transition by p38α through TGF-β and JNK signaling. Nat Commun 10(1):3071

14. Ding J, Guo Y (2022) Recent advances in chitosan and its derivatives in cancer treatment. Front Pharmacol 13:888740
15. Bashir SM, Ahmed Rather G, Patrício A, Haq Z, Sheikh AA, Shah MZ, Singh H, Khan AA, Imtiyaz S, Ahmad SB, Nabi S (2022) Chitosan nanoparticles: a versatile platform for biomedical applications. Materials 15(19):6521
16. Garg U, Chauhan S, Nagaich U, Jain N (2019) Current advances in chitosan nanoparticles based drug delivery and targeting. Adv Pharm Bull 9(2):195
17. Ullah S, Azad AK, Nawaz A, Shah KU, Iqbal M, Albadrani GM, Al-Joufi FA, Sayed AA, Abdel-Daim MM (2022) 5-fluorouracil-loaded folic-acid-fabricated chitosan nanoparticles for site-targeted drug delivery cargo. Polymers 14(10):2010
18. Kumbhar ST, Patil RY, Bhatia MS, Choudhari PB, Gaikwad VL (2022) Synthesis and characterization of chitosan nanoparticles decorated with folate and loaded with dasatinib for targeting folate receptors in cancer cells. OpenNano 7:100043
19. San HH, Alcantara KP, Bulatao BP, Sorasitthiyanukarn FN, Nalinratana N, Suksamrarn A, Vajragupta O, Rojsitthisak P, Rojsitthisak P (2022) Folic acid-grafted chitosan-alginate nanocapsules as effective targeted nanocarriers for delivery of turmeric oil for breast cancer therapy. Pharmaceutics 15(1):110
20. İnce İ, Yıldırım Y, Güler G, Medine Eİ, Ballıca G, Kuşdemir BC, Göker E (2020) Synthesis and characterization of folic acid-chitosan nanoparticles loaded with thymoquinone to target ovarian cancer cells. J Radioanal Nucl Chem 324:71–85
21. Mazzotta E, De Benedittis S, Qualtieri A, Muzzalupo R (2019) Actively targeted and redox responsive delivery of anticancer drug by chitosan nanoparticles. Pharmaceutics 12(1):26
22. Karayianni M, Sentoukas T, Skandalis A, Pippa N, Pispas S (2023) Chitosan-based nanoparticles for nucleic acid delivery: technological aspects, applications, and future perspectives. Pharmaceutics 15(7):1849
23. Maiyo F, Singh M (2019) Folate-targeted mRNA delivery using chitosan-functionalized selenium nanoparticles: potential in cancer immunotherapy. Pharmaceuticals 12(4):164
24. Liang Y, Wang Y, Wang L, Liang Z, Li D, Xu X, Chen Y, Yang X, Zhang H, Niu H (2021) Self-crosslinkable chitosan-hyaluronic acid dialdehyde nanoparticles for CD44-targeted siRNA delivery to treat bladder cancer. Bioact Mater 6(2):433–446
25. Yan L, Gao S, Shui S, Liu S, Qu H, Liu C, Zheng L (2020) Small interfering RNA-loaded chitosan hydrochloride/carboxymethyl chitosan nanoparticles for ultrasound-triggered release to hamper colorectal cancer growth *in vitro*. Int J Biol Macromol 162:1303–1310
26. Yadav AS, Radharani NN, Gorain M, Bulbule A, Shetti D, Roy G, Baby T, Kundu GC (2020) RGD functionalized chitosan nanoparticle mediated targeted delivery of raloxifene selectively suppresses angiogenesis and tumor growth in breast cancer. Nanoscale 12(19):10664–10684
27. Bicak B, Budama-Kilinc Y, Kecel-Gunduz S, Zorlud T, Akman G (2021) Peptide based nano-drug candidate for cancer treatment: preparation, characterization, *in vitro* and in silico evaluation. J Mol Struct 1240:130573
28. Petrilli R, Pinheiro DP, de Cássia Evangelista de Oliveira F, Galvão GF, Marques LG, Lopez RF, Pessoa C, Eloy JO (2021) Immunoconjugates for cancer targeting: a review of antibody-drug conjugates and antibody-functionalized nanoparticles. Curr Med Chem 28(13):2485–2520
29. Helmi O, Elshishiny F, Mamdouh W (2021) Targeted doxorubicin delivery and release within breast cancer environment using PEGylated chitosan nanoparticles labeled with monoclonal antibodies. Int J Biol Macromol 184:325–338
30. Sabra R, Billa N, Roberts CJ (2019) Cetuximab-conjugated chitosan-pectinate (modified) composite nanoparticles for targeting colon cancer. Int J Pharm 572:118775
31. Manivasagan P, Jun SW, Hoang G, Mondal S, Kim H, Doan VH, Kim J, Kim CS, Oh J (2019) Anti-EGFR antibody conjugated thiol chitosan-layered gold nanoshells for dual-modal imaging-guided cancer combination therapy. J Control Release 311:26–42
32. Leon-Ferre RA, Goetz MP (2023) Advances in systemic therapies for triple negative breast cancer. BMJ:381

33. Debela DT, Muzazu SG, Heraro KD, Ndalama MT, Mesele BW, Haile DC, Kitui SK, Manyazewal T (2021) New approaches and procedures for cancer treatment: current perspectives. SAGE Open Med 9:20503121211034366
34. Afrifa MA, Kim JH, Pitton KA, Olelewe C, Arojojoye AS, Strachan DR, Suckow MA, Awuah SG (2024) Auranofin-loaded chitosan nanoparticles demonstrate potency against triple-negative breast cancer. ACS Appl Bio Mater 7(3):2012–2022
35. Pina Olmos S, Diaz Torres R, Elbakrawy E, Hughes L, Mckenna J, Hill MA, Kadhim M, Ramirez Noguera P, Bolanos-Garcia VM (2019) Combinatorial use of chitosan nanoparticles, reversine, and ionising radiation on breast cancer cells associated with mitosis deregulation. Biomol Ther 9(5):186
36. Taherian A, Esfandiari N, Rouhani S (2021) Breast cancer drug delivery by novel drug-loaded chitosan-coated magnetic nanoparticles. Cancer Nanotechnol 12:1–20
37. Gounden S, Daniels A, Singh M (2021) Chitosan-modified silver nanoparticles enhance cisplatin activity in breast cancer cells. Biointerface Res Appl Chem 11:10572–10584
38. Hassanpour M, Jafari H, Sharifi S, Rezaie J, Lighvan ZM, Mahdavinia GR, Gohari G, Akbari A (2021) Salicylic acid-loaded chitosan nanoparticles (SA/CTS NPs) for breast cancer targeting: synthesis, characterization and controlled release kinetics. J Mol Struct 1245:131040
39. Mehata AK, Bharti S, Singh P, Viswanadh MK, Kumari L, Agrawal P, Singh S, Koch B, Muthu MS (2019) Trastuzumab decorated TPGS-g-chitosan nanoparticles for targeted breast cancer therapy. Colloids Surf B Biointerfaces 173:366–377
40. Gogoi P, Dutta A, Ramteke A, Maji TK (2020) Preparation, characterization and cytotoxic applications of curcumin-(±) α-lipoic acid coloaded phosphorylated chitosan nanoparticles in MDA MB 231 breast cancer cell line. Polym Adv Technol 31(11):2827–2841
41. Zhou Z, Yue F, Jin L, Zhai T, Zhang J, Gu W, Liu S, Luo M, Peng B, Yao X, Ye L (2021) Characteristics and risk differences of different tumor size on localized prostate cancer: a retrospective cohort study in the SEER database. Cancer Med 10(8):2763–2773
42. Wang J, Wang L, Ye W, Xing N, Liu Y, Guan Y, Gao X (2022) National guidelines for diagnosis and treatment of prostate cancer 2022 in China (English version) National Health Commission of the People's Republic of China. Chin J Cancer Res 34:270–288
43. Amiri H, Javid H, Einafshar E, Ghavidel F, Rajabian A, Hashemy SI, Hosseini H (2024) Development and evaluation of PLGA nanoparticles surfaced modified with chitosan-folic acid for improved delivery of resveratrol to prostate cancer cells. BioNanoScience 14(2):1–11
44. Amiri H, Javid H, Einafshar E, Ghavidel F, Rajabian A, Hashemy SI, Hosseini H (2021) pH-sensitive release of fungal metabolites from chitosan nanoparticles for effective cytotoxicity in prostate cancer (PC3) cells. BioNanoScience 14(2):1–11
45. Zamanvaziri A, Meshkat M, Alazmani S, Khaleghi S, Hashemi M (2023) Targeted PEGylated chitosan nano-complex for delivery of sodium butyrate to prostate cancer: an *in vitro* study. Technol Cancer Res Treat 22:15330338231159223
46. Liu X, Shi D, Guo L, Zhou X, Shang M, Sun X, Li J (2021) Echogenic, ultrasound-sensitive chitosan nanodroplets for spatiotemporally controlled DKK-2 gene delivery to prostate cancer cells. Int J Nanomedicine 16:421–432
47. Faria N, Lacerda C, Lopes J, Viana C, Sucena M (2023) PET-CT SUVmax and endobronchial ultrasound features for prediction of malignancy: a prospective study. Clin Lung Cancer 24(8): 753–760
48. Kuen YC, Masarudin MJ (2022) Chitosan nanoparticle-based system: a new insight into the promising controlled release system for lung cancer treatment. Molecules 27(2):473
49. Almutairi FM, Abd-Rabou AA, Mohamed MS (2019) Raloxifene-encapsulated hyaluronic acid-decorated chitosan nanoparticles selectively induce apoptosis in lung cancer cells. Bioorg Med Chem 27(8):1629–1638
50. Vikas VMK, Mehata AK, Sharma V, Priya V, Varshney N, Mahto SK, Muthu MS (2021) Bioadhesive chitosan nanoparticles: dual targeting and pharmacokinetic aspects for advanced lung cancer treatment. Carbohydr Polym 274:118617

51. Zhang L, Wang J, Zhang Y, Ke L, Lin X, Li Z, Chen H, Gao Y (2019) Indocyanine green-encapsulated erlotinib modified chitosan nanoparticles for targeted chemo-photodynamic therapy of lung cancer cells. Dyes Pigments 170:107588
52. Kuen YC, Galen T, Fakurazi S, Othman SS, Masarudin MJ (2020) Increased cytotoxic efficacy of protocatechuic acid in A549 human lung cancer delivered via hydrophobically modified-chitosan nanoparticles as an anticancer modality. Polymers 12(9):1951
53. Choukaife H, Seyam S, Alallam B, Doolaanea AA, Alfatama M (2022) Current advances in chitosan nanoparticles based oral drug delivery for colorectal cancer treatment. Int J Nanomedicine 17:3933–3966
54. Shakil S, Mahmud KM, Sayem M, Niloy MS, Halder SK, Hossen S, Uddin F, Hasan A (2021) Using chitosan or chitosan derivatives in cancer therapy. Polysaccharides 2:795–816
55. Abuelmakarem HS, Hamdy O, Sliem MA, El-Azab J, Om-Hashem MA, Ahmed WA (2020) Colonic carcinoma diagnosis using chitosan nanoparticles based on the optical properties. J Phys Conf Ser 1472:012001
56. Ciro Y, Rojas J, Di Virgilio AL, Alhajj MJ, Carabali GA, Salamanca CH (2020) Production, physicochemical characterization, and anticancer activity of methotrexate-loaded phytic acid-chitosan nanoparticles on HT-29 human colon adenocarcinoma cells. Carbohydr Polym 243: 116436
57. Chen G, Zhao Y, Xu Y, Zhu C, Liu T, Wang K (2020) Chitosan nanoparticles for oral photothermally enhanced photodynamic therapy of colon cancer. Int J Pharm 589:119763
58. Rashedi J, Ghorbani Haghjo A, Mesgari Abbasi M, Dastranj Tabrizi A, Yaqoubi S, Sanajou D, Ashrafi Jighe Z, Namvaran A, Mohammadi A, Mohammadi Khoshraj J, Baradaran B (2019) Anti-tumor effect of quercetin loaded chitosan nanoparticles on induced colon cancer in wistar rats. Adv Pharm Bull 9(3):409–415
59. Yu C, Wang S, Lai WF, Zhang D (2023) The progress of chitosan-based nanoparticles for intravesical bladder cancer treatment. Pharmaceutics 15(1):211
60. Harsanyi S, Novakova ZV, Bevizova K, Danisovic L, Ziaran S (2022) Biomarkers of bladder cancer: cell-free DNA, epigenetic modifications and non-coding RNAs. Int J Mol Sci 23(21): 13206
61. Irani S (2020) New insights into oral cancer – risk factors and prevention: a review of literature. Int J Prev Med 11:202
62. Ali K (2022) Oral cancer-the fight must go on against all odds. J Evid Based Dent 23:4–5
63. Das JM, Hoang S, Mesfin FB (2023) Intramedullary spinal cord tumors. In: StatPearls [Internet]. StatPearls Publishing
64. Bray F, Laversanne M, Sung H, Ferlay J, Siegel RL, Soerjomataram I, Jemal A (2024) Global cancer statistics 2022: GLOBOCAN estimates of incidence and mortality worldwide for 36 cancers in 185 countries. CA. CA Cancer J Clin 74(3):229–263
65. Ilhan B, Lin K, Guneri P, Wilder-Smith P (2020) Improving oral cancer outcomes with imaging and artificial intelligence. J Dent Res 99(3):241–248
66. Ahmad S, Khan I, Pandit J, Emad NA, Bano S, Dar KI, Rizvi MMA, Ansari MD, Aqil M, Sultana Y (2022) Brain targeted delivery of carmustine using chitosan coated nanoparticles via nasal route for glioblastoma treatment. Int J Biol Macromol 221:435–445

Adv Polym Sci (2025) 295: 285–318
https://doi.org/10.1007/12_2025_190

Published online: 19 February 2025

Chitosan-Based Formulations for Enhanced Topical, Transdermal, and Ocular Drug Delivery

Victor Hmingthansanga, Lalit Kumar, Neha, Rajkumar Malayandi, and Subramanian Natesan

Contents

1 Introduction 286
2 Chitosan 288
3 Dermal and Transdermal Drug Delivery 289
3.1 Barriers 290
3.2 Epidermis 290
3.3 Dermis 292
3.4 Hypodermis 292
4 Role of Chitosan in Topical and Transdermal Drug Delivery 292
4.1 Nanoparticles 292
4.2 Nanocapsules 295
4.3 Microneedles 296
4.4 Transdermal Patches 297
4.5 Hydrogels 298
5 Ocular Drug Delivery System 299
5.1 Barriers to the Ocular Drug Delivery System 301
6 Application of Chitosan in Ocular Drug Delivery 304
6.1 Chitosan Nanoparticles 304
6.2 Chitosan Solutions 305
6.3 In Situ Gelling System 306
6.4 Chitosan-Coated Nanosystems 307
6.5 Contact Lenses 308

V. Hmingthansanga, L. Kumar, Neha, and S. Natesan (✉)
Advanced Formulation Laboratory, Department of Pharmaceutics, National Institute of Pharmaceutical Education and Research Kolkata, Kolkata, West Bengal, India
e-mail: snatesan@niperkolkata.ac.in

R. Malayandi
Department of Pharmaceutics, National Institute of Pharmaceutical Education and Research Hajipur, Export Promotion Industrial Park (EPIP), Hajipur, Bihar, India

7 Regulatory Status 309
8 Conclusion 310
References 310

Abstract Chitosan, a naturally occurring cationic polymer, has various pharmaceutical utilizations, including drug delivery. The primary amino groups present in the chitosan trigger many vital functions, such as mucoadhesion and mucopenetration, controlled drug release, in situ gelation, transfection ability, and inhibition of efflux pump. Transdermal drug delivery systems have emerged as promising approaches for delivering drugs with poor oral bioavailability and hydrophobic nature than traditional oral delivery systems. It gained popularity due to its noninvasiveness, simple access and removal, sustained drug delivery for prolonged periods, bypassing hepatic first-pass metabolism, higher bioavailability, etc. Furthermore, chitosan has been incorporated as an adjuvant to enhance skin penetration. Moreover, chitosan-based drug delivery systems are demonstrated as a potential candidate for transcending the barriers of the eye, improving bioavailability by its mucoadhesive property, enhancing corneal permeability, and prolonging residence time at the target site. This chapter has emphasized topics such as the chitosan and its physicochemical properties; barriers related to topical, transdermal, and ocular delivery; applications of various chitosan-based formulations, which include nanoparticles, nanocapsules, gels, microneedles, patches, and contact lenses; and lastly its regulatory status.

Keywords Chitosan · Drug delivery · Hydrogel · Nanoparticle · Ocular · Topical · Transdermal

1 Introduction

Chitin is the most plentiful nature-derived polymer after cellulose. It is a white, solid, elastic, and nitrogenous polysaccharide derived from plants and animals, especially in coastal regions [1]. The majority of polysaccharides found in nature are acidic. However, chitin's unique characteristic is its strong basicity, which makes it superior to other naturally occurring polysaccharides [2]. Chitosan, a chitin derivative, is composed of repeating linear β-(1-4)-2-amino-2-deoxy-d-glucopyranose units. N-Acetylglucosamine residues in the chitin macromolecular chain are completely deacetylated to form N-glucosamine residues. N-Glucosamine residues in the chitosan are obtained by deacetylation of N-acetylglucosamine residues in the macromolecular chitin. Typically, N-glucosamine and N-acetylglucosamine units are copolymerized to form chitosan dispersed randomly throughout the polymer chain [3]. The solubility of chitosan in diluted acid solutions provided a way to process it into various forms in milder processing conditions than its parent polymer,

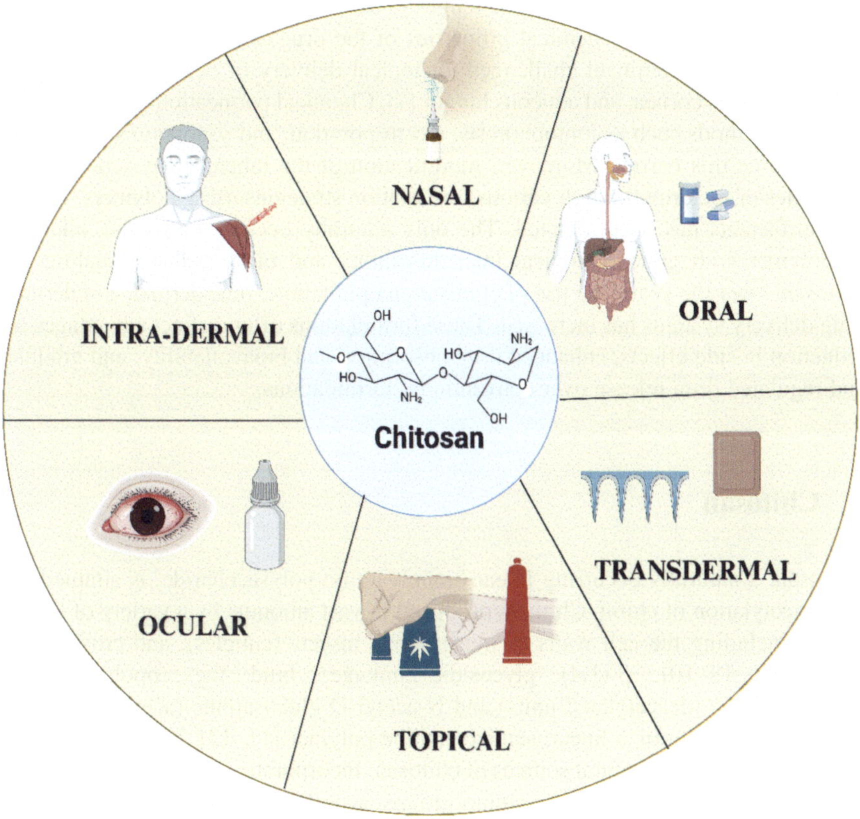

Fig. 1 Application of chitosan in various forms of drug delivery system

making it a more appealing biopolymer for various applications. The interesting chemical and biological characteristics of chitosan compelled its use in several pharmaceutical areas, particularly in drug delivery, as shown in Fig. 1.

Because oral drug delivery is convenient, safe, and economical, it is the preferred method of drug administration [4]. However, most of the drugs do not fulfill the requirements because of low solubility and degradation within the gastrointestinal system. Topical or transdermal drug delivery has provided a promising alternative, specifically for poorly water-soluble and oral bioavailable drugs and for treating local ailments. The noninvasiveness, ease of administration and removal, regulated drug release, avoidance of hepatic first-pass metabolism, and enhanced bioavailability of these two methods make them acceptable [5].

The two sections of the eye, anterior and posterior, work together as a sensory organ to provide vision. Ocular drug delivery often uses topical, systemic, intraocular, and periocular drug delivery techniques because of their many benefits, which include less systemic toxicity and improved ocular pharmacokinetics. Compared to other delivery systems, topical delivery of ocular drugs is less intrusive and dangerous [6]. The predominant challenges in dermal or transdermal drug delivery are the inherent physicochemical properties of the drug and the stratum corneum (SC) [5]. The predominant challenges for topical delivery of ocular drugs include precorneal loss, cornea, and aqueous humor [7]. Chemical permeation enhancers and physical methods such as iontophoresis, electroporation, and sonophoresis are used to overcome this barrier. Moreover, modification of the inherent physicochemical properties of the drug through various formulation strategies using polymers, lipids, and surfactants has been adopted. The only naturally occurring cationic, alkaline biopolymer with good biological inter-miscibility and biodegradable qualities is chitosan. Over the years, the use of chitosan in ophthalmic, transdermal, and dermal drug delivery systems has increased. These formulations offer distinct advantages of reduction in side effects, enhanced drug absorption and bioavailability, and gradual and regulated drug release over conventional formulations.

2 Chitosan

Chitosan, a naturally occurring linear biopolyamino polysaccharide, is attained by the deacetylation of chitin. Chitin is produced in vast amounts by a variety of living things, including the cell walls of mushrooms, insects (cuticles), and crustaceans (skeletons) [8–10]. β-(1,4) glycosidic linkages bind the copolymers of D-glucosamine (deacetylated units) and N-acetyl-D-glucosamine (acetylated units) that make up chitosan, a linear, semicrystalline polymer [11, 12]. In addition, fungi and yeast are also the natural sources of chitosan. Incorporating additional functional groups in chitosan, such as free amino groups, acetyl amine groups, and hydroxyl groups, imparts distinctive properties from cellulose. The physicochemical composition can affect its functional characteristics, including reactivity to chemicals, solubility, and biological activity. Molecular weight and the deacetylation degree (DD) are the primary factors significantly impacting this biopolymer's numerous physical, chemical, and biological characteristics. Because of its substantial biological activity and uses in the food, pharmaceutical, and agricultural sectors, chitosan has drawn increased interest. Till now, it has been assessed that chitosan derivatives showed applications in more than 200 diverse fields. It is commercially viable to produce chitosan from the shells of crustaceans. The carotenoid astaxanthin, found in crustacean shells, is added to fish feed in aquaculture. Chitosan can also be chemically extracted straight from fungus. To produce chitin and chitosan more profitably, several scientists have developed techniques to deproteinize crustacean chitin using proteolytic enzyme bacteria. The standard process involves deacetylating chitin in an alkaline solution to produce chitosan. It is also crucial to isolate chitin from various sources by crushing the raw material and washing it with water and detergent. Although there is a lot of interest in using chitosan as a pharmaceutical excipient, no approved medications are currently being used except topical. Nevertheless, it is being investigated for usage in various pharmaceutical formulations, such as injectables and ophthalmic, nasal, dermal, transdermal, and oral drug delivery applications. Moreover, chitosan has been shown as a potential hemostatic, antimicrobial, anti-ulcerogenic, anti-tumor, immune-altering, osteogenic, and renal protective agent and used in tissue engineering and is backed by several in vivo and in vitro investigations looking into potential efficacy.

3 Dermal and Transdermal Drug Delivery

Topical application of formulations or pharmaceuticals to the skin for localized effects is known as dermal drug delivery. The viable layer of the skin is the primary target of dermal medication administration. Medications are typically administered directly to the afflicted area using a variety of formulations, including creams, ointments, gels, and lotions. These formulations may contain active pharmaceutical ingredients (APIs) and excipients that enhance drug penetration, stability, and skin compatibility [13]. The dermal drug delivery mechanism involves diffusion across the stratum corneum (SC) and subsequent distribution within the layers of skin. Factors influencing drug penetration include the physicochemical properties of the drug (e.g., molecular weight, lipophilicity), formulation characteristics, skin condition, and, most importantly, application site [14, 15]. Dermal drug delivery finds widespread applications in dermatology and cosmetology for treating various skin conditions and enhancing skin appearance [16]. Standard dermatological formulations include corticosteroids for eczema and psoriasis, antibiotics for acne, antifungals for fungal infections, and emollients for dry skin. Cosmetic formulations such as antiaging creams, sunscreens, and moisturizers also utilize dermal drug delivery technologies to deliver active ingredients to the skin. These formulations help improve skin hydration, texture, and overall appearance. The absorption-enhancing effects of chitosan have been intensively investigated in terms of its dermal application. Chitosan can facilitate the skin's absorption of peptides, proteins, and tiny polar compounds. Chitosan improves drug transport by promoting bioadhesion and temporarily opening the tight junction of the cell membrane to allow polar medications to pass through. Chitosan's action is probably caused by its positive charge, which interacts with the dermal cells' tight connection. However, due to its cationic polyelectrolyte nature, it has a significant electrostatic interaction with negatively charged dermal cell surfaces and other macromolecules like DNA. Furthermore, dermal applications are also essential for genetic immunization since the skin is a substantial immunological site home to antigen-presenting cells like Langerhans cells. Various techniques, including gene guns, needle-free devices, and particulate systems, have been used to deliver pDNA. Chitosan particles have been used successfully to transport DNA to Langerhans cells.

Administration of medications through the skin for systemic drug effects is known as transdermal drug delivery. Unlike dermal delivery, which primarily targets the local application site, transdermal delivery aims to deliver drugs into the bloodstream, bypassing the gastrointestinal tract and liver metabolism [17–21]. Transdermal drug delivery offers several advantages for giving drugs with systemic effects, including improved patient compliance, reduced gastrointestinal side effects, and steady blood medication levels. Pain management, hormone replacement therapy, cardiovascular diseases, quitting smoking, and disorders of the central nervous system are among the most popular therapeutic areas for transdermal medication administration. Many transdermal preparations, including hydrogels, transdermal patches and membranes, emulsions, transdermal microneedles, nanocapsules, and nanoparticles, contain chitosan and its derivatives and are commonly used to transfer active compounds to the systemic circulation. Mechanistically, the drug diffuses into the skin from the system, where circulation absorbs and circulates it. Molecular weight, polymer concentration, particle size and shape, drug concentration, and skin permeability are some variables that affect the rate of drug delivery.

The mechanism of transdermal drug delivery involves several processes, including drug release from the system, diffusion through the stratum corneum, absorption into the viable epidermis and dermis, and subsequent transport into the systemic circulation [14, 15]. The formulation design is crucial in ensuring optimal drug delivery kinetics, stability, and patient compliance. Chitosan and its derivatives' mode of action in transdermal drug delivery systems have been thoroughly investigated, and new applications are constantly being created [22]. A summary of the most recent developments in chitosan-related transdermal preparations will be provided in the next part, covering the mechanism of chitosan in sustained release and transdermal delivery.

The dermal and transdermal drug delivery systems represent versatile strategies for administering medications through the skin, offering advantages such as targeted delivery, improved patient compliance, and enhanced therapeutic outcomes. While both approaches have their unique characteristics, ongoing research and technological innovations continue to drive advancements in drug delivery science, paving the way for developing more effective and patient-friendly formulations in the future.

3.1 Barriers

Skin barriers mainly hinder the transdermal delivery of various drugs. Skin covers around 2 m^2 in an average adult, and considering the disadvantages of transdermal drug delivery, it also acts as a protective barrier against undesired external influences with self-healing capabilities [23]. The anatomical skin structure is shown in Fig. 2.

3.2 Epidermis

Under the epidermis there are several layers such as stratum corneum, stratum lucidum, stratum germinativum, stratum granulosum, and stratum spinosum. In addition, there are some Langerhans cells, dendritic cells, keratinocytes, and melanocytes. The overall thickness of SC is around 15 μm with a layer of corneocytes

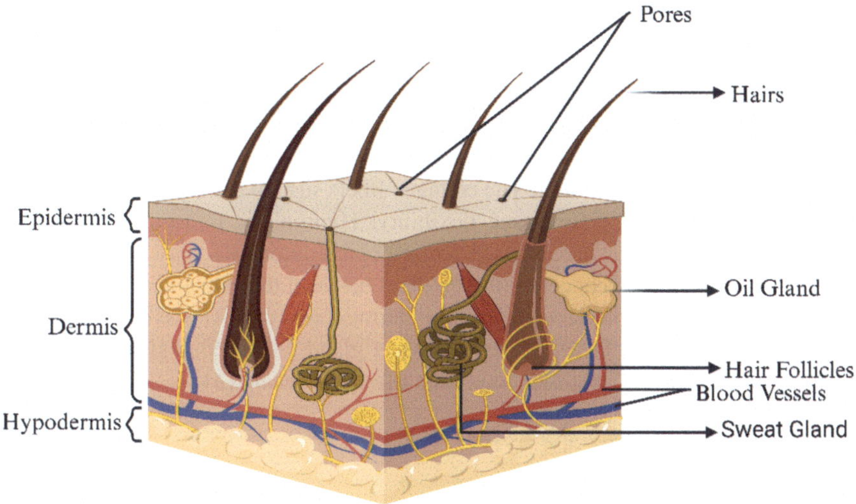

Fig. 2 Anatomical structure of the skin

Fig. 3 Structure of the epidermis layer

separated by an intercellular lipid domain. Figure 3 represents the structure of the epidermis. The various lipids that comprise the SC layer are free fatty acids, triglycerides, ceramides, neutral lipids, phospholipids, glycosphingolipids, and cholesterol. The corneocytes and lipidic domain in the SC are organized in a brick wall formation in which the corneocytes are called the brick and the lipids are called mortar. Moreover, the corneocytes are pulled together with the help of corneodesmosomes. The thick overlapping corneocytes are held together by corneodesmosomes, which promote the rigidity and impermeability features of the skin and form a major barrier for hydrophilic drugs having a molecular mass higher than 200–350 Da [24].

3.3 Dermis

The thickness of the dermis layer is around 0.1–0.5 cm, consisting of hydrophilic deposits such as collagen fibers and elastin, which are deep-rooted in the mucopolysaccharide matrix that consists of sweat glands, pilosebaceous units, blood vessels, and lymphatic and nerve ending. The collagen fibers sustain the layers of skin, while the elastin is responsible for tissue flexibility. Hydrophilic molecules can easily reach the dermis layer, whereas very hydrophobic compounds may find it to be a significant obstacle. The concentration gradient between the skin and dermis layer is preserved because the blood vessels in the dermis layer remove the chemicals that penetrate the epidermis layer. Moreover, sweat glands, sebaceous glands, and hair follicles create a "shunt" pathway favoring a few permeants [24].

3.4 Hypodermis

The hypodermis is the skin's innermost layer, and it helps absorb any shocks that are brought to the body. Connective tissue, adipose tissue (fat cells), fibroblasts, blood vessels, and macrophages are among the several cells that make up the hypodermis [24].

4 Role of Chitosan in Topical and Transdermal Drug Delivery

Chitosan and its derivatives are used for topical and transdermal delivery of drugs because of their unique properties, comprehensively discussed in the preceding section. Figure 4 depicts some of the applications for chitosan and its derivatives, which are detailed below.

4.1 Nanoparticles

Nanoparticles are solid colloidal particles having a size in the nanometer range (0.1–100 nm) in 3D space, also called nanopowders or nano-dust. Because of their minuscule size, nanoparticles possess exceptional permeability and a large specific surface area, making them highly valuable for transdermal drug administration.

A scaffold with curcumin-loaded chitosan nanoparticles (NCs/Cur) was prepared using polycaprolactone (PCL) and gelatin (GE). In a full-thickness excisional animal model, human endometrial stem cells incorporated with a hybrid scaffold demonstrated favorable biocompatibility and wound-healing capabilities. The hybrid fiber

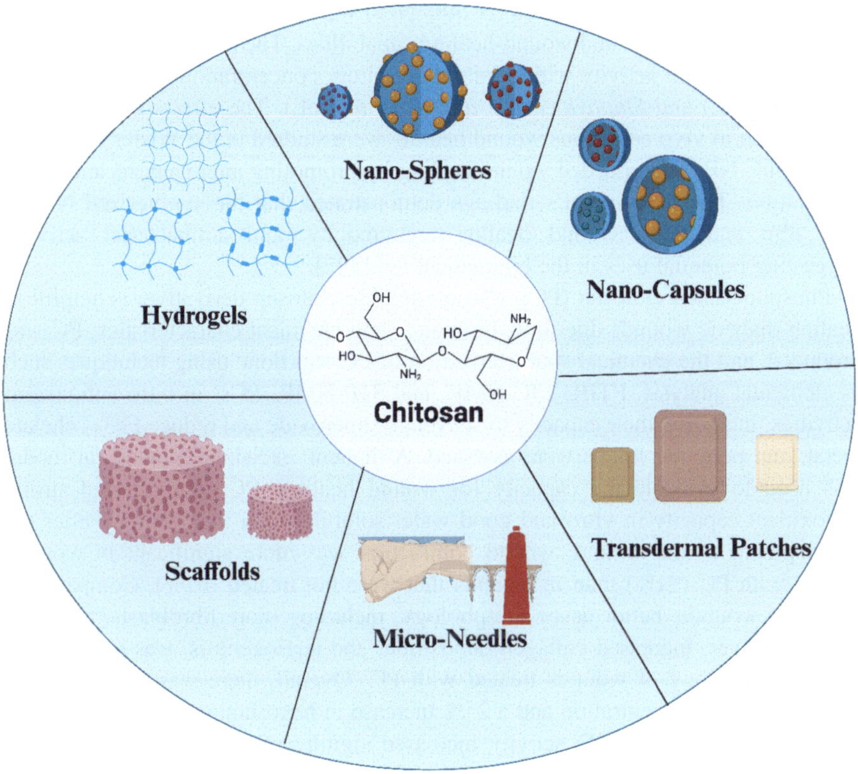

Fig. 4 Application of chitosan in topical and transdermal drug delivery

scaffolds had excellent mechanical and structural qualities, such as homogeneous shape, uniform pore size, porosity, perfect contact angle, wettability, and degradability. In addition, hybrid fiber scaffolds (PCL/Gela/NCs/Cur) showed a greater rate of cell attachment and proliferation compared to PCL and PCL/Gela scaffolds. The innovative tissue-engineered construct may act as a skin substitute for healing damaged skin and for use in regenerative medicine [25]. Similarly, chitosan nanoparticles loaded with epigallocatechin 3-gallate (NCs-EGCG) were electrospun into collagen and polyvinyl alcohol-conjugated tyramine fibrous mat. The composite was then used to assess the created wound patch's in vivo wound-healing capabilities. In an animal model, the prepared fibers loaded with stem cells demonstrated suitable biocompatibility and the potential to heal wounds. The ideal hydrophilic/hydrophobic ratio, as well as strong mechanical and structural qualities, including porosity and shape homogeneity, was demonstrated by the results. The prepared fibers containing NCs-EGCG exhibited a noteworthy increase in cellular adhesion and proliferation compared to the control samples. The histological examination of the hybrid fibrous patch may indicate that this structure can be used as a good skin substitute to heal wounds on the skin [26]. In a different research direction, films of

N-succinyl chitosan nanoparticles (N-SuC NPs) were made and evaluated for their cytotoxic, antibacterial, and wound-healing capabilities. The produced films showed potent antibacterial activity with minimal inhibitory concentrations for *Escherichia coli* (6 mg/mL) and *Staphylococcus aureus* (8 mg/mL). The effects of the N-SuC NPs film on in vivo cutaneous wound healing were studied in the Wister rat model. The N-SuC NPs film speeded wound healing by promoting angiogenesis and tissue granulation. The experiment's findings demonstrated that the synthesized N-SuC NPs film had good wound healing, cytotoxicity, and antibacterial activity, suggesting potential uses in the biomedical field [27].

Phosphorylated chitosan (PC), a water-soluble chitosan derivative, is helpful in healing diabetic wounds due to its biological and chemical characteristics. PC was produced, and the chemical characterization of PC was done using techniques such as elemental analysis, FTIR, 13C-NMR, and 31P-NMR. PCs' in vitro antioxidant activities, including their capacity to scavenge superoxide and reduce Fe3+, chelate metal, and peroxide lipids, were assessed. A diabetic excisional wound rat model was used to evaluate the capacity for wound healing. PC demonstrated strong antioxidant capacity in vitro and good water solubility. On the 14th day after the development of the wound, wound contraction was more significant in wounds treated with PC (91%) than in wounds that were not treated (67%). Compared to untreated wounds, better tissue morphology, including more fibroblasts, a thicker epithelial layer, increased collagen deposition, and angiogenesis, was observed in the histopathology of wounds treated with PC. Overall, there was a 57% rise in hydroxylamine concentration and a 25% increase in hexosamine content compared to untreated wounds. SOD activity increased significantly ($P \leq 0.05$), and lipid peroxides decreased significantly ($P \leq 0.05$) in PC-treated wounds compared to untreated wounds. These findings showed that PC can be a valuable treatment for diabetic wound healing [28]. Tacrolimus is a naturally occurring macrolide that inhibits T lymphocytes to have an antiproliferative effect. Consequently, it was investigated as a possible topical therapy to alleviate and manage psoriatic plaques. This study used a modified ionic gelation process to achieve the appropriate reaction and skin retention of tacrolimus, a lipophilic medication, for the first time in chitosan nanoparticles. Tacrolimus, a hydrophobic drug, was successfully encapsulated into positively charged particles measuring 140.8 nm ± 50.0 with an EE of 65.5% ± 1.3. With 82.0% ± 0.6 of the medication remaining in the skin, local skin deposition of the drug was dramatically improved compared to 34.0% ± 0.9 from tacrolimus ointment. After 3 days of therapy, the animals showed an amazing response to the produced formula: increased hair growth rates. This indicates that the skin is recovering from the induced psoriatic plaques [29]. Various other studies have been conducted using chitosan-based nanoparticles for the dermal or transdermal delivery of drugs such as amphotericin B [30], tacrolimus and nicotinamide [31], methotrexate [32], etoricoxib [33], methotrexate [34], conotoxin [35], and ginkgo biloba extract [36].

4.2 Nanocapsules

Drugs can be contained within the capsule wall of polymeric materials to create a core-shell structure by using microencapsulation technology. This keeps the medication from degrading and inactive while releasing it at the desired location. Nanocapsules are generally defined as submicron-size capsules with a particle size of less than 1 μm. Nanocapsules can enhance stability, solubility, and dispersibility in addition to their embedding, targeting, and sustained release capabilities. Imiquimod-loaded chitosan nanocapsules having a particle size of 200 nm and a PDI of less than 0.3 are prepared for transdermal drug delivery. Under physiological circumstances, the nanocapsules exhibited good stability for at least 48 h. Pig skin was used to test the nanocapsules' permeability. In addition, it was evident that chitosan was crucial in promoting permeability and isolation. According to Raman spectrum analysis, the imiquimod band was detected at a depth of more than 20 μm. This indicates that the nanocapsules could enter the stratum corneum and provide the perfect platform for imiquimod's transdermal distribution [37].

Chitosan-coated nanostructured lipid carriers (NLCs) loaded with tetrahydrocurcumin were prepared for breast cancer therapy which had drug loading of $0.28 \pm 0.01\%$, zeta potential of 17.5 ± 0.5 mV, particle size of 244 ± 18 nm, and an entrapment efficiency of $76.6 \pm 0.2\%$. A sustained release at varying pH is demonstrated by the in vitro release research. Additionally, when compared to unencapsulated THC, Ch-NLCs showed significant cytotoxicity toward breast cancer cells, enhanced in vitro skin permeability, and cellular absorption, indicating that they would be effective transdermal nanocarriers of THC for the treatment of breast cancer [38]. In a similar work, they loaded 5-fluorouracil (5-FU) in an emulsion and coated with chitosan. The release rates of the emulsions containing Span 20, sodium lauryl sulfate, Tween 80 (T80), and a combination of Span 20 and polyethylene glycol (PEG) were 88%, 86%, 90%, and 92%, respectively. In contrast to the 5-FU solution, chitosan-coated emulsion shows controlled release. The drug permeation study across rat skin suggests that the combination of T80 and PEG has a good penetration profile with a possible mechanism of fluidizing lipids and proteins of the SC derived through ATR-FTIR spectrograms. The study demonstrated that 5-FU emulsions coated with chitosan offer a great chance to administer a model medication via transdermal distribution. Mechanical approaches with nanoparticles have gained much interest in transdermal drug delivery [39]. Hen egg-white lysozyme (HEL) was successfully delivered to hair follicles with the help of PLGA nanoparticles coated with chitosan hydroxypropyltrimonium chloride using iontophoresis (IP). The developed formulation has a size of 98.4 ± 36.8 nm with a surface charge number density of 1.28×10^{-2} M. Results of the cellular uptake study in dendritic cells suggest that the nanoparticles were internalized successfully. The ex vivo skin permeation study with IP also revealed that the nanoparticles were firmly accumulated within the skin membranes, increasing the viable concentration of HEL. In vivo, percutaneous immunoreactivity analysis was used to determine the HEL-specific Ig G1 and IgG2a titers, which were 7.4×10^4 and 6.6×10^2, respectively. Combining IP and nanoparticle carriers could effectively target hair follicles [40].

4.3 Microneedles

Transdermal microneedles use micro- or nanotechnology to open the skin pipeline. As long as the microneedles are adhered to the skin, the microneedles can penetrate the cuticle layer of the epidermis, which serves as a barrier for medications. Drugs can be applied quantitatively and continuously by enabling them to slowly penetrate the epidermis through a patch applied to the skin. Microneedles inserted into the skin might be a local drug reservoir to ensure continuous medication administration. On the other hand, transdermal microneedles improve patient tolerance by avoiding the pain associated with skin punctures, which is a drawback of standard injection. Additionally, the danger of infection from injection is much decreased because there is no major skin wound.

High quantities of cell-free DNA primarily cause psoriasis, and because of its limited penetration, immunological suppression, and skin irritation, it has proven challenging to cure. Biguanide chitosan microneedles (BGC-MNs) were created to circumvent the skin barrier and extract cfDNA from the dermis to address these issues. The formulation containing 20% bisguanidine (BGC2) was effective with high DNA-binding capacity, biocompatibility, and inflammation inhibition. The formulation also showed good therapeutic activity in an in vivo study, markedly clearing the cfDNA and lowering the dermal concentration of inflammatory cytokines. Their study results indicate that BGC-MNs could treat psoriasis by scavenging cfDNA and exerting anti-inflammatory effects [41]. In another study, dydrogesterone (DYD)-loaded microneedle patch (MNP) was fabricated using polyvinyl acetate and thiolated chitosan. Compared to pure drugs, the MNP showed sustained release in an in vitro dissolution study, and the skin penetration study showed better penetration. Moreover, a histological study of the mice's skin reveals that the MNP has good penetration into the skin. The fabricated MNP holds great potential for an effective transdermal delivery system for DYD [42]. Similarly, tacrolimus, an immunosuppressant agent, was loaded in a microneedle patch (MNP) as an alternative to the oral delivery system. Thiolated chitosan was utilized in this study to create MNP, and its mechanical strength, skin penetration, drug release, and skin irritation were assessed both in vitro and in vivo. The developed MNP with 225 needles and 575 μm demonstrated good mechanical qualities with 84% drug skin penetration. The MNP showed in vitro drug release of 82.5% of the medication [43]. Recently, 3D printing technology has been extensively employed in the area of drug delivery due to its ease, accuracy, and less time-consuming process.

Dual temperature-sensitive and photosensitive microneedles were developed using digital light processing (DLP) 3D printing and smart chitosan biomaterial hydroxybutyl methacrylated chitosan (HBCMA). Microneedles exhibited an extension of 4D properties when exposed to temperature with good resolution, sharpened needles, and mechanical strength. In addition, the 3D-printed microneedles were used to deliver a small molecular drug that could penetrate soft tissue [44]. To effectively transfer antigens into the skin's viable epidermis and dermis region – explicitly targeting the antigen-presenting cells – an embeddable chitosan microneedle was created on a poly(L-lactide-co-D,L-lactide) (PLA) supporting array. The developed microneedle was evaluated in vivo for its performance. It came up with an exciting outcome, such as prolonged antigen release for 14 days after a gradual degradation process. The antigen-antibody response lasted for 6 weeks in comparison to intramuscular injection. According to these study findings, embeddable chitosan microneedles hold promise as a depot for prolonged release of encapsulated antigens to enhance immunogenicity [45]. Chitosan can be used to facilitate the healing of skin wounds because of its inherent antibacterial and anti-inflammatory qualities. A microneedle patch made of chitosan was created to aid in the healing of skin wounds. Planar chitosan membranes, microneedle-free patches, chitosan microneedle patches, and PBS were applied to the wound models. The findings demonstrated that the chitosan microneedle patch group had the lowest expression levels of two proinflammatory factors (IL-6 and TNFα), suggesting that the chitosan microneedle patch had the most significant impact on skin wound healing when compared to other groups [46].

4.4 Transdermal Patches

Transdermal patches are thin-film preparations applied topically to the skin, allowing medications to permeate the epidermis and reach the bloodstream. Transdermal patches comprise four layers: adhesive, backing, drug reservoir, and protective layer. The backing layer, removed before usage, is situated on the outermost edge of the delivery system from the external environment. The middle is where the drug storage is located and equipped with membranes to ensure a continuous and regulated release of pharmaceuticals. The adhesive layer, which secures the skin by making intimate contact with it, is the deepest. The benefit of the patch is that the medication can remain firmly applied to the skin for an extended period without losing its effectiveness, guaranteeing the stability and sustainability of drug administration.

Wound healing is an intricate body process that requires our immune system to repair the damaged tissue. Transdermal patches fabricated with biocompatible material with an intrinsic immune activation capacity have been used for wound healing. A transdermal patch was prepared using chitosan and polyvinyl alcohol loaded with a GO/CuO nanocomposite (NC). The GO/CuO NC primarily aided in forming and stabilizing extracellular matrix skin proteins and the angiogenesis process. Therefore, by promoting cell proliferation, antibacterial properties, and the quick onset of inflammation, GO/CuO NC enhances the wound healing mechanism [47]. Clobetasol propionate (CP)-loaded chitosan (CS) patches were created for the topical management of long-term oral conditions. The optimized structures demonstrated good mechanical qualities and a water uptake of 200–360%, enabling the creation of flexible patches in wet conditions. The in vitro release was 80% in the first 2 h, and the CP loading was adjusted by varying the CP concentration in the deposition bath between 0.002 and 0.12 mg/cm^2. The characteristics of the developed patch made it interesting for the local oral release of CP. Burn and other skin injuries are now being treated with biocompatible polymers. It has been demonstrated that polymers like chitosan and polyvinyl alcohol are biocompatible and least harmful when used to heal wounds with little to no adverse effects [48]. The main ingredient in turmeric, curcumin, has very limited bioavailability but strong anti-inflammatory and antibacterial effects. Curcumin's bioavailability improved dramatically when it was converted to a nano-form, which made it possible for it to be significant for the healing of wounds. As demonstrated by the outcomes of cell line tests and the MTT experiment, this PVA/Chi/Cur patch boosted cell proliferation. Its water retention capabilities and antibacterial potential against four bacterial strains frequently found in wound sites suggest it is an ideal material for wound therapy. The patch outperformed commercial ointment in healing a surgically inflicted wound, according to the findings of in vivo trials on Wistar rats. This method of treating epidermal wounds speeds up the healing process and decreases the need to replace the patch as often [49].

4.5 Hydrogels

Skin injuries often accompany microbial contamination, leading to antimicrobial resistance, a growing concern today. To fix this problem, various novel treatment approaches are commonly used to eradicate biofilm formation and enhance the intricate process of wound curing. Chlorhexidine-loaded chitosomes were prepared and incorporated into a matrix of chitosan hydrogel. The developed nanoparticles have a size of ($\approx$250 nm) and good biological and antibacterial activity when incorporated in chitosan hydrogel. Moreover, both systems showed a better antimicrobial effect than formulations devoid of chitosan. This formulation could be a basis for preventing infection and bacterial eradication in acute wounds [50]. A chitosan composite hydrogel, in combination with functionalized graphene oxide (CGO), was prepared to promote wound healing. Adding CGO increases the mechanical properties due to the appropriate interaction between the two components. The compressive stress and tensile strength were also increased in CS-CGO composite hydrogel compared to CS-GO composite hydrogel, and structural analysis revealed a porous structure with a smaller pore size as the concentration of the CGO increased. The prepared chitosan composite hydrogel was biocompatible and could promote adhesion and proliferation of fibroblasts. Lastly, in an in vivo study, the CS-CGO composite hydrogel significantly accelerates wound healing, and the wound closure rate reaches up to 92.2% after 21 days [51]. The antibacterial properties of CS-GT (chitosan-gentamicin)-based scald dressings were investigated to create a novel system for wound healing. In vitro cell culture study demonstrated that cell viability exhibits a declining inclination as CS-GT concentration increases and the culture period is extended. Moreover, CS-GT exhibits strong hemocompatibility and cytocompatibility. In an in vivo wound-healing study, CS-GT hydrogel showed the best rate of skin healing (99.61%) followed by matrix (77.39%), GT hydrogel (87.50%), and the positive control (wet burn ointment) (at 94.98%), while the blank control group had the last healing rate. A subsequent study revealed that CS-GT hydrogel could stimulate skin granulation tissue's total protein (TP) synthesis, increasing the hydroxyproline (HYP) content. This, in turn, promoted healing of wounds. In conclusion, the hydrogel CS-GT shows promise as a scald dressing [52].

5 Ocular Drug Delivery System

The eye is the primary site of vision due to its unique anatomy and physiology (Fig. 5). It is a highly complex, isolated, and specialized organ [53, 54]. The eye is divided into two segments: anterior and posterior. The anterior segment is the frontal area of the eye that includes the pupil, cornea, iris, ciliary body, aqueous fluid, and lens. In contrast, the posterior segment covers two-thirds of the eye. It contains the anterior hyaloid membrane and all the optical structures behind it, such as the sclera, retina, choroid, vitreous humor, and optic nerve [53]. The most common ocular diseases affecting the anterior segment include conjunctivitis, trauma, dry eye syndrome, cataract, and glaucoma, and the posterior segment includes age-related macular degeneration (AMD), diabetic retinopathy (DR), proliferative vitreoretinopathy, cytomegalovirus (CMV), Stargardt disease, and retinoblastoma [55, 56]. Numerous medications have been licensed for treating various disorders. Still, due to the particular structure and physiology of the eye, it poses a challenge to get these drugs to the target tissue of the eye. Therefore, different modes of

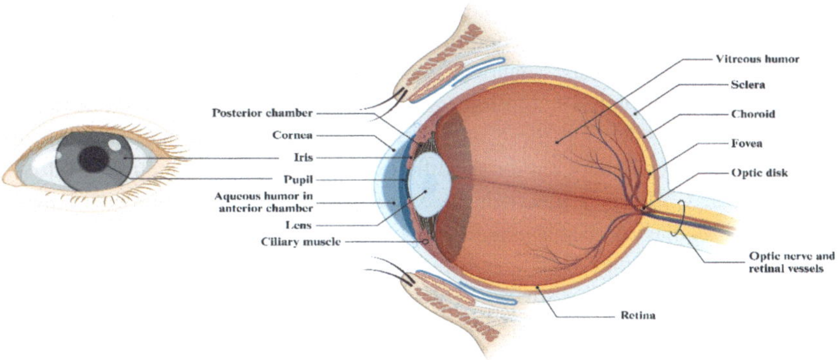

Fig. 5 Anatomy of the eye

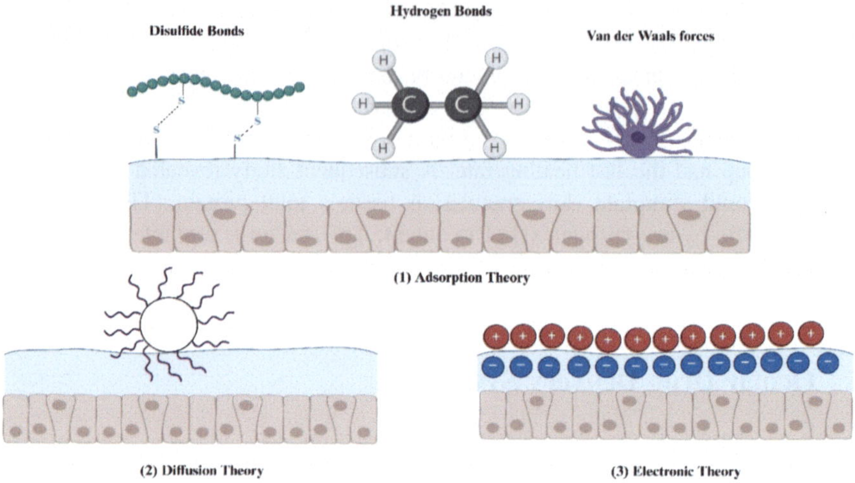

Fig. 6 Mechanisms of mucoadhesion for ocular drug delivery vehicles

administration have been adopted in preclinical and clinical settings, including topical, systemic, intraocular, and periocular administration [53, 57]. Drug delivery to the eye is challenging due to the anatomical location of tissues and accompanying barriers.

Topical instillation of eye drops is the most popular type of ocular treatment due to its intrinsic benefits, such as ease of administration and noninvasiveness, which result in high patient compliance and quick effect. It solves systemic delivery concerns such as first-pass metabolism, blood-ocular barriers, low eye vascularization, adverse effects, etc. However, frequent installation of concentrated solutions is required to achieve the therapeutic outcome. To increase the ocular bioavailability of topical formulation, there has been a quest to identify a more efficient drug delivery system than the existing ones. The most common strategy is increasing the ocular surface's residence time and reducing drug drainage. To boost ocular bioavailability, many delivery systems have been utilized, such as drug-impregnated contact lenses, patches, colloidal carriers (nanoparticles, different kinds of liposomes, etc.), microneedles, films, gels, or solutions of mucoadhesive polymers. Various materials have been used to develop these intelligent delivery systems: natural or synthetic polymeric matrices, which can be modified for a particular application. Chitosan is one of them. Chitosan's mucoadhesive and mucopenetration properties make it suitable for ocular medication delivery systems. The mechanistic properties are depicted in Figs. 6 and 7. The polycationic groups of chitosan interact with the mucin on the eye's anterior surface. The electrostatic forces will retain the chitosan-based drug delivery system at the application site, increasing drug penetration to the posterior region of the eye.

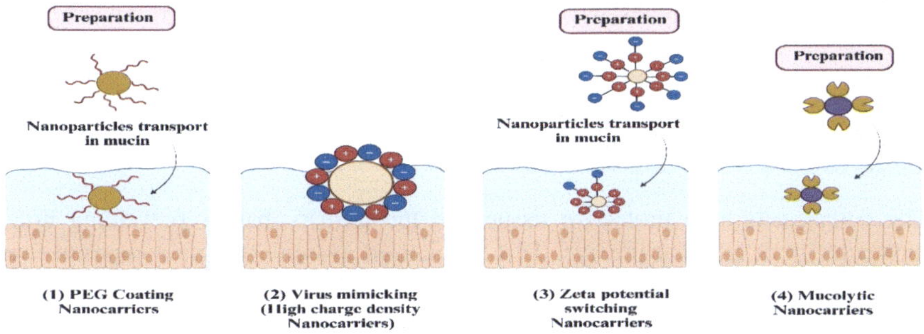

Fig. 7 Mechanisms of mucopenetration for ocular drug delivery vehicles

5.1 Barriers to the Ocular Drug Delivery System

5.1.1 Precorneal Barriers

Precorneal barriers refer to the barriers encountered by topically given medicines before they contact the cornea. These include tear film barriers, limited lacrimal volume and leakage, tear production, reflex blinking, and nasolacrimal drainage [58].

5.1.2 Capacity of Cul-de-sac

A cul-de-sac is described as a small-depth portion in the lower eyelid that connects the two parts of the conjunctiva, and its capacity offers various ocular barriers. In humans, a cul-de-sac can hold up to 30 μL. Returning the lower eyelid to its former position will limit the capacity by 70–80%. Inflammation and allergies can reduce cul-de-sac capacity. The duration and concentration of a drug's residency determine its activity. Cul-de-sac's poor capacity reduces drug concentration in the eye, limiting its therapeutic effectiveness [59].

5.1.3 Tear Film Barrier

The tear film is the first barrier to drug transport in the eye, consisting of three layers: an exterior thin lipid layer, a middle aqueous layer, and an innermost mucous layer [60]. The lipophilic layer keeps water from evaporating and lowers drug absorption in the cornea and sclera. In the middle aqueous layer, endogenous proteins such as globulin, albumin, and lactoferrin might bind and metabolize the administered drug, reducing its bioavailability [61]. Essential components such as water, enzymes, salts, lipids, and mucins are present in the inner lining of the mucous layer. The structural lining of the mucous layer is thick at the epithelial apex, and as it moves toward the tear fluid, it becomes thinner. It acts as a barrier to drug delivery by trapping and adhering to foreign particles, which are then washed away by mucus turnover before reaching the cornea [62, 63].

5.1.4 Cornea

The cornea acts as a protective tissue and the primary barrier preventing drug molecules from entering the eye. This specialized avascular tissue comprises five layers: epithelium, Bowman's layer, stroma, Descemet's membrane, and endothelium [64, 65]. The corneal epidermis comprises basal and squamous epithelial cells connected by zonula occludens. The corneal epithelium is highly lipophilic and acts as a rate-limiting barrier for most hydrophilic drugs and few lipophilic drugs. Lipophilic drugs cross the corneal epithelium through the intracellular route. Meanwhile, hydrophilic drugs or low molecular weight ions follow the normal paracellular route. The permeability of the cornea is temporally altered by a class of vehicles called penetration enhancers. Collagen fibers in the stroma make it hydrophilic, preventing the diffusion of lipophilic substances. The corneal endothelium comprises a single layer of cells equipped with fluid pumps, which are adhered to the membranes to transport fluid to the anterior portion of the eye. This monolayer creates a strong barrier between the stroma and aqueous humor [66].

Various factors influence the effectiveness of topically administered ophthalmic drugs in penetrating the cornea. The corneal composition offers an equivalent to a fat-water-fat structure. Therefore, the oil-water partition coefficients of nonpolar substances are essential for their penetration. Drug binding to ocular tissues can inhibit transcorneal penetration. Topically applied drugs are retained in the corneal tissue, making up a drug reservoir that slowly releases the drug to other parts of the eye and is primarily removed by aqueous humor turnover and venous blood flow. Scleral-conjunctival ocular penetration may be faster for hydrophilic drugs than transcorneal ocular penetration [66].

5.1.5 Conjunctiva

The conjunctiva is a thin, transparent membrane covering the sclera and extending from the eyelid to the limbus. The conjunctiva has three sections: bulbar conjunctiva, conjunctival fornix, and palpebral conjunctiva. It has two layers: epithelium and stroma. Conjunctival cells are connected by tight junctions of occludin, claudins, tricellulin, and JAM-A, attached to the cytoskeleton by ZO-1, 2, 3. The conjunctival epithelium has a lower resistance (about 1,500 W/cm^2) and a 15-fold larger surface area than the cornea. The conjunctival epithelium is more significant than the corneal epithelium; its bulbar and palpebral conjunctiva diameters are 3.2 and 4.8 nm, respectively, with wider intercellular spacing and higher drug permeability. Hence, larger molecules (up to 40 KDa molecular weight) can permeate through this tissue [67].

5.1.6 Blood-Aqueous Barriers (BAB)

The BAB consists of altered vascular endothelium that connects neighboring epithelial cells of the nonpigmented ciliary body epithelium. The aqueous humor is supplied to the posterior part of the eye by the ciliary body. The ciliary body extends posteriorly from the iris root to the retina, creating a ring around the globe [68]. It stops drugs from entering the aqueous humor from the plasma. However, the BAB is not a total barrier because of fenestrated capillaries in the ciliary body stroma. Fenestrated capillaries are a secondary source of plasma protein leakage to the iris, allowing tiny molecules to enter the circulation. The BRB limits drug diffusion into the retina after systemic circulation [69]. If medications are administered systemically, it is probable that the paracellular transport of those substances through the blood-ocular barrier probably has to be further constrained. BAB has a lower barrier function than BRB and BBB, and it is more permeable to hydrophilic molecules. Even though BAB has reduced barrier function, the absorption of high-lipophilic substances may appear to be the same as that of BBB and BRB due to the transcellular transport of these substances [70]. The BAB enables the quick clearance of small lipophilic medicines from the anterior chamber by allowing them to enter the uveal blood circulation. Conversely, more significant hydrophilic medicines are removed through the BAB's aqueous humor [71].

5.1.7 Melanin Binding

The delivery of drugs in the eyes may be impacted by melanin. The interaction with melanin may impact the number of free drugs at the intended site and reduce the pharmacological action. Melanin is commonly found in the uvea and RPE regions of the eye. It binds to medicines and free radicals through simple charge transfer or electrostatic and van der Waals forces. Most basic and lipophilic medicines are bound to melanin. Whereas a medicine linked to melanin is typically unavailable for receptor binding, higher doses must be given. The amount of drug absorbed into the posterior part of the eye after a transscleral or systemic administration is also influenced by melanin in the choroid and RPE. The penetration of hydrophobic drug in bovine choroid and RPE was greater than that of hydrophilic drug due to melanin binding. Similar evidence has been found that melanin increases the binding of lipophilic substances to the choroid-Bruch's membrane in bovine. As a result, the choroid-Bruch's membrane has a higher resistance to solute absorption than the sclera, which lacks melanin [72].

5.1.8 Nasolacrimal Drainage

Approximately 95% of the given amount of drug is removed systemically through the nasolacrimal duct and/or conjunctiva. The movement of tears from the ocular region to nasal cavity is facilitated by the lacrimal drainage. It comprises of lacrimal sac, canaliculi, puncta, and nasolacrimal duct. Post topical instillation of the eye drop, it comes in contact with the lacrimal fluid. The drug has a contact time of around 1–2 min with ocular tissues. Roughly half of the drug enters to both canaliculus of the lacrimal sac. After that, the flow widens and enters the nasolacrimal duct, which empties into the nose. The concentration of a topically injected medication depends on a number of factors, including the patient's age, reflex blinking, and the amount of drug solution given. Smaller amounts are quickly expelled from the lacrimal sac, but larger injected volumes from the nasolacrimal sac easily enter the nose. Because it exposes the medications to the systemic circulation, drug loss from the nasolacrimal duct via transconjunctival or transnasal absorption is undesirable [73, 74].

5.1.9 Reflex Blinking

Eyelid closure during reflex blinking protects against external issues. Tears cover the eye's surface. Lysozyme and immunoglobulin in the eye have an anti-infectious effect on the eye. Additionally, the medication is rendered inactive by interacting with the proteins in tears and conjunctival mucin [75]. A standard eyedropper measures 25–56 μL of the topical formulation, averaging 39 μL. However, an eye can hold up to 30 μL, and the remaining is also eliminated by reflex blinking (5–7 blinks/min), reducing the total concentration of the drug intended for therapeutic action [72, 76].

6 Application of Chitosan in Ocular Drug Delivery

Chitosan has been extensively exploited as a base material for the preparation of various formulations for the treatment of various ocular diseases, and it is also a fundamental material in sustained drug delivery. This section will specifically elaborate on research on chitosan-based formulation and its application in ocular drug delivery. The chemical structure of chitosan gives enormous opportunity to enhance its properties through the conjugation of other polymers or drugs. With its intrinsic properties such as adhesiveness, penetration, and antimicrobial, chitosan prevails to be a preferable candidate for preparing formulations for ocular drug delivery.

6.1 Chitosan Nanoparticles

Solid particles composed of polymeric materials that range in size from 10 to 1,000 nm are known as polymeric nanoparticles. Using polymeric nanoparticles for drug delivery aims to maximize therapeutic efficacy while reducing side effects. Because of its small size, surface charge, sustained release, and ability to cross cell and tissue barriers, it has received a lot of interest in ocular medication delivery. Furthermore, the polymers used to make the nanoparticles are biocompatible, biodegradable, nontoxic, and immunosuppressive, making them better suited for ocular applications. Despite these advantages, there are drawbacks, such as particle-particle agglomeration, low drug loading, and burst release. Research in this area of periocular drug delivery has gained enormous attention. For instance, miltefosine-loaded polymeric nanoparticles (MF-CS-NPs) were prepared using chitosan to treat *Acanthamoeba* keratitis. MF-CS-NPs had a high EE of 86.3% when synthesized by the ionic gelation process. The MF-CS-NPs are 46.61 ± 18.16 nm in size and exhibit a 210 ± 3.2 mV zeta potential. The present study found it less cytotoxic to Vero cells than free MF, with no hemolytic activity in vitro or ocular discomfort in rabbits. The MF-CS-NPs significantly reduced trophozoite viability relative to free miltefosine and were somewhat effective against cysts [77]. In another study, sulfobutylether-β-cyclodextrin/chitosan NPs (Nag-CD/CS-NPs) were developed and loaded with naringenin and their efficacy was tested in topical ophthalmic administration. The particle size was 446.4 ± 112.8 nm with a zeta potential of +22.5 ± 4.91 mV and spherical shape. An in vivo study in rabbit eyes showed that the NPs were nonirritating and capable of prolonging the residence time of naringenin, hence increasing its bioavailability in the aqueous humor. The chitosan NPs were found to be a promising alternative for the ocular delivery of poorly soluble drugs [78]. Timolol maleate-loaded galactosylated chitosan (GC) nanoparticles were formulated for the treatment of glaucoma. The particle size, EE, and DL were 213.3 ± 6.83 nm, 38.58 ± 1.31%, and 17.72 ± 0.28%, respectively. TM-GC-NPs have longer retention than timolol eye drops, and the in vivo study proposed that TM-GC-NPs effectively reduced intraocular pressure (IOP) and had a longer working period than commercial TM eye drops [79]. Many other studies have been conducted by using polymeric nanoparticles for subconjunctival delivery of various drugs such as tedizolid phosphate [80], meloxicam [81], triamcinolone acetonide [82], dexamethasone [83], resveratrol [84], bevacizumab [85], gentamycin [86], chloramphenicol [87], resveratrol and quercetin [88], and resveratrol [89].

6.2 Chitosan Solutions

The most popular way to administer drug for anterior segment illness is through eye drops. More than 90% of marketed ophthalmic formulations are eye drop solutions due to their convenience, patient compliance, ease of production, and dose uniformity. To improve ocular drug delivery, many strategies have focused on increasing contact time between the formulations and the corneal/conjunctival epithelium. Additionally, the presence of mucoadhesive characteristics in chitosan polymer enhances the effect. Studies have found that a combination of medium molecular weight chitosan and hydroxypropyl guar gum in ophthalmic agents resulted in optimal physicochemical, rheological, and mucoadhesive properties for olopatadine (0.1% w/v) eye drop formulations. In vitro drug permeability was evaluated in HCE-T cell-based model and PAMPA model with chitosan acting as a permeation enhancer. The MTT cytotoxicity test showed that the studied formulations were safe and well tolerated [90]. Fluconazole was administered using chitosan solution and a combination poloxamer and chitosan in situ gel. The study found a poloxamer/chitosan in situ gel reduced the release of fluconazole relative to a chitosan solution. Ex vivo experiments on porcine cornea showed that chitosan solution delivered more fluconazole than in situ gel-forming. The in vivo investigation found that both formulations had similar sustained release characteristics. Chitosan solution with an in situ gelling formulation can treat fungal keratitis. These solutions are cost-effective and easy to prepare with commercially available polymers [91]. In another study, a chitosan film was prepared using a solution casting method in which an antiglaucoma drug, brimonidine tartrate (BT), was loaded. The prepared chitosan film was transparent, stable, and mucoadhesive in nature. The drug was dispersed homogeneously in the chitosan film and exhibited high permeability in the cornea [92].

6.3 In Situ Gelling System

A temperature-sensitive hydrogel loaded with moxifloxacin hydrochloride (MOX) was prepared for enhanced ocular administration. The temperature-sensitive hydrogel demonstrated optimum gelation temperature at 35°C with a gelation time within 2 min. Structural analysis of the optimized formulation showed a porous structure, and the rheological analysis showed good elasticity. In addition, biological assessment showed a better antibacterial activity compared to plain drug and marketed product. Moreover, histological evaluation of ocular tissues validated the antibacterial activity after topical application. The developed CS/β-GP thermosensitive hydrogel shows potential for improving ocular administration of MOX for treating bacterial infections [93]. Another study developed the hydrogel using dialdehyde starch (DAS) and chitosan (CS) to deliver betamethasone. Their findings showed that an increase in the concentration of DAS might induce quick gelation, good mechanical strength, and larger pore size with rapid release of the drug. Furthermore, in an in vitro cell model, the composite hydrogel was highly biocompatible [94]. A nanoparticle-embedded in situ gelling system was developed using carboxymethyl chitosan (CMCh) and poloxamer as gelling materials. The nanoparticle-loaded gel showed increased antifungal efficacy against *Candida albicans*. When compared to VCZ nanosuspension, the optimized batch containing CMCh demonstrated 2.86-fold improvement in mucoadhesion. An ex vivo investigation found that the drug release was sustained for up to 8 h and the study on the hen's egg chorioallantoic membrane showed no sign of toxicity [95].

A temperature-responsive gel comprising poloxamer and chitosan was developed for topical administration of tacrolimus (TCS). In vitro, testing for the optimized gel showed prolonged TCS release and enhanced corneal penetration compared to the TCS solution. The TCS release follows the Higuchi model, including Fickian diffusion transport. Furthermore, histopathology and HET-CAM investigations demonstrated that the developed gel was nonirritating and suitable for ocular delivery [96].

A conjugated form of chitosan and maleic acid was used to study the mucoadhesiveness and antibacterial activity for treating eye infections. The degree of conjugation was quantified using 2,4,6-trinitrobenzene sulfonic acid (TNBS). The conjugate showed no toxicity or irritation and a degree of modification of 50.5%. The mucoadhesive properties were increased by 2.68- and 1.81-fold for both elastic and viscous modulus, respectively. In addition, the developed formulation showed better antibacterial activity. In vivo experiments demonstrated that hydrogels were well tolerated, indicating that the proposed conjugation had potential properties. The compound demonstrated potential antibacterial activity, long-lasting mucoadhesive properties, and enhanced pharmacokinetics, which revolutionized the treatment of eye bacterial infections [97].

6.4 Chitosan-Coated Nanosystems

The ocular delivery system utilizes various delivery platforms comprising different polymers and lipids. However, these nanosystems exhibit a short interaction time with the ocular membrane, leading to quick washout and low ocular bioavailability. Therefore, it is commonly coated with chitosan or a chitosan derivative to enhance the overall ocular bioavailability. Chitosan-coated liposomes loaded with timolol maleate successfully decreased intraocular pressure compared to normal eye drops. Moreover, increased particle size when mixed with mucin indicates the ability to bind with mucin and increase the residence time. Additionally, the apparent diffusion coefficient increased three times compared to eye drops [98]. Flurbiprofen-loaded liposomes coated with chitosan increased the transcorneal penetration in an in vitro condition. The apparent permeability coefficient of chitosan-coated liposomes was 1.29 times greater than that of the uncoated liposomes and 4.59 times higher than that of the 0.03% flurbiprofen solution. A significant improvement was also shown by examining the residence time in vivo, which was more than twice as high as the outcomes of liposomes without chitosan [99]. Recently, triamcinolone acetonide has been encapsulated in chitosan-coated liposomes. According to a study, coated liposomes showed longer-lasting drug release than solution or uncoated liposomes. Moreover, the results of a histological study after the formulation administration showed no sign of toxicity. Additionally, the authors used coumarin-6 (C6) as a fluorescent probe to observe the penetration of chitosan-coated liposomes into the anterior and posterior segment ocular tissues following topical treatment. C6 was carried in solution and in liposomes with and without chitosan. It was found that

HCEC and ARPE-19 cells absorbed chitosan-coated liposomes more readily, which raised their fluorescence levels. Additionally, C57BL/6 mice received topical administration of 5 μL of these formulations. The findings demonstrated that C6 encapsulated in chitosan-coated liposomes had increased fluorescence intensity in both the anterior and posterior segments [100]. Liposome-trimethyl chitosan (TMC) nanoparticles that co-deliver insulin and vascular endothelial growth factor small interfering RNA (siVEGF) were prepared to treat corneal alkali burns by inhibiting ferroptosis. TIL had a higher positive charge (+15.52 ± 4.78 mV) and a greater diameter (183.23 ± 12.61 nm) than insulin-liposomes (124.22 ± 6.04 nm), indicating that TMC was successfully coated on insulin-liposomes. TIL, as positive NPs, can adsorb negative siRNA. Therefore, siVEGF-TIL had a lesser zeta potential (12.40 ± 0.40) and greater diameter (216.63 ± 4.51) than TIL, indicating siVEGF loading in TIL. Moreover, si-VEGF-TIL treatment had the optimal qualities of NPs, such as good biosafety, low toxicity, ease of preparation, adhesion, and prolonged release, making it a promising delivery platform for the cornea [101]. Many other drugs have been loaded in lipid nanoparticles for ocular delivery examples including diclofenac [102], cyclosporine A [103], triamcinolone acetonide [82], atorvastatin calcium [104], sirolimus [105], baicalin [106], sirolimus [107], sodium butyrate [108], clotrimazole [109], ciprofloxacin [110], brimonidine [111], ibuprofen [112], ganciclovir [113], resveratrol [114], and 5-fluorouracil [115].

6.5 Contact Lenses

In recent years, therapeutic contact lenses have received a lot of attention. In this research work, chitosan-conjugated poly(2-hydroxyethyl methacrylate) (PHEMA) were employed for contact lens application. It aimed to improve the affinity of ascorbic acid, a small molecule anionic medication, via chitosan-conjugated PHEMA. Chitosan immobilization increases drug loading efficiency and promotes prolonged ascorbic acid release. The chitosan hydrogel also inhibits the biofouling of tear fluid components. The findings demonstrated that chitosan surface modification prevents protein and bacterial accumulation on contact lenses. According to the results, chitosan surface modification prevents proteins and bacterial accumulation on contact lenses. The results of protein absorption studies showed that tear proteins were absorbed at 28.4 ± 4.4 μg/cm^2 in plain PHEMA and 18.5 ± 1.8 μg/cm^2 in chitosan-conjugated hydrogel. The PHEMA bacterial adhesion showed a mean CFU value of 273 ± 27. However, the chitosan group had a much lower number of bacterial colonies (9 ± 6) [116]. Bacterial keratitis (BK) and other inflammatory diseases cause irreversible corneal tissue damage. This study constructed a new antibacterial and antioxidant contact lens using a novel polyacrylamide semi-interpenetrating network hydrogel containing quaternized chitosan and tannic acid (PAM-QCS-TA). The developed hydrogels exhibited a high-water content (>85%), swelling resistance, light transmittance (>90%), and variable mechanical properties. Antibacterial quantitative and qualitative tests exhibited a high degree of sterilization, particularly against almost 100% of *Escherichia coli* and *Staphylococcus aureus*. Because of the presence of tannic acid, it showed strong antioxidant activities, reducing oxidative stress and protecting cells from reactive oxygen species-induced cytotoxicity. Animal studies also showed that BK had a shorter treatment period (just 3 days) and protected the ocular tissue structure. As a result, medication-free antibacterial and antioxidant contact lenses that prevent drug resistance are a promising therapy for treating ocular infections and inflammatory illnesses [117]. In another work, a unique electrofabrication technology (i.e., electrodeposition) for producing hydrogel contact lenses that may concurrently induce covalent cross-linking between chitosan and epichlorohydrin was applied. Without affecting their oxygen permeability, electrochemical cross-linking improves hydrogel transmittance and durability. Furthermore, the electrode template can also be used to modify the shape of the chitosan-based hydrogel contact lenses, and electrical signals and electrochemical cross-linking can modulate their properties. The electrodeposited hydrogel contact lenses show good optical and mechanical properties and are biocompatible, and their anti-*Staphylococcus aureus* properties are comparable to commercial contact lenses. This study explains the process of electrochemical cross-linking between chitosan and epichlorohydrin and proposes an alternate way for contact lens production [118]. Ocular infections, such as keratitis and conjunctivitis, are the leading causes of vision loss. Getting a sufficient concentration of antibiotics to be administered topically to prevent or treat corneal infections is difficult. A drug-eluting polymeric contact lens was prepared using chitosan, glycerol, and polyethylene glycol and loaded with dexamethasone (DM) and moxifloxacin (MF). The physical properties of the polymeric contact lens were comparable to the commercially available contact lens. After 24 h of incubation, the drug-loaded contact lens demonstrated increased corneal drug distribution ($P < 0.05$) compared to drug solutions administered topically. The MF-loaded contact lens has more antibacterial activity in vitro and in vivo than the standard drug solution. In vivo, drug distribution experiments showed that contact lens application resulted in a higher concentration of MF in the cornea, sclera, and aqueous humor than in drug solutions. Overall, the polymeric contact lens effectively delivered MF and DM at therapeutic amounts. Drug-eluting contact lenses might be employed in postoperative settings to prevent ocular infections [119].

7 Regulatory Status

Exposure of chitosan to humans has an established record but not as a pharmaceutical excipient [120, 121]. Chitosan has been used as a dietary supplement for hypercholesterolemia and obesity due to its capability to adsorb body fats. Undeniably, chitosan supplements are sold and branded as fat trappers and magnets [122]. Certain chitosan-containing medical devices have been used to treat bleeding, such as HemCon or QuikClot, which are marketed across Europe and the USA [123]. Chitosan also has been used as a coating material for contact lenses or as a base material for contact itself [120]. Chitosan has been extensively explored as a material for wound healing, and it also has an established product called Tegasorb, which was marketed in the past but seems now to have been withdrawn [120]. In the food industry, chitosan has gained the status of Generally Recognized as Safe (GRAS) in the USA for usage in foods in general [124]. Additionally, chitosan is registered as a food additive in Japan, Italy, and Finland [125, 126]. In the case of the cosmetic industry, chitosan is used as a base material. Chitosan has industrial use as a flocculant and chelating agent in the clarification of beverages and as a fungicide for crop protection [120].

8 Conclusion

The book chapter on chitosan-based formulations for enhanced topical, transdermal, and ocular drug delivery provides an insightful understanding of the unique properties of chitosan and its effectiveness as a drug delivery system. The chapter highlights chitosan's high degree of deacetylation and molecular weight that can enhance epithelial permeability, particularly for polar drugs, by opening tight junctions-associated proteins in the cell membrane and improving transport across epithelial surfaces. Additionally, the polycationic nature of chitosan allows it to bind to negatively charged surfaces such as mucosal membranes, increasing adhesion and extending contact time for drug penetration. Incorporating active agents in chitosan-based formulations can also protect the biologically active agents from degradation. The transportability of chitosan increases with charge, making it a suitable pH-dependent medication carrier. Chitosan coating can prevent the initial burst of ceramide and degrade by weakly acidic skin for controlled ceramide release. It concludes that chitosan-based formulations have emerged as an effective strategy for enhanced topical drug delivery and ocular drug delivery systems.

References

1. Honarkar H, Barikani M (2009) Applications of biopolymers I: chitosan. Monatsh Chem 140: 1403–1420
2. Rinaudo M (2006) Chitin and chitosan: properties and applications. Prog Polym Sci 31(7): 603–632
3. Steinbach M, Karypis G, Kumar V (2000) A comparison of document clustering techniques
4. Alqahtani MS, Kazi M, Alsenaidy MA, Ahmad MZ (2021) Advances in oral drug delivery. Front Pharmacol 12:618411
5. Hmingthansanga V, Singh N, Banerjee S, Manickam S, Velayutham R, Natesan S (2022) Improved topical drug delivery: role of permeation enhancers and advanced approaches. Pharmaceutics 14(12):2818
6. Gaudana R, Ananthula HK, Parenky A, Mitra AK (2010) Ocular drug delivery. AAPS J 12: 348–360

7. Huang D, Chen YS, Rupenthal ID (2018) Overcoming ocular drug delivery barriers through the use of physical forces. Adv Drug Deliv Rev 126:96–112
8. Anraku M, Gebicki JM, Iohara D, Tomida H, Uekama K, Maruyama T (2018) Antioxidant activities of chitosans and its derivatives in in vitro and in vivo studies. Carbohydr Polym 199: 141–149
9. Kumari S, Annamareddy SK, Abanti S, Rath K (2017) Physicochemical properties and characterization of chitosan synthesized from fish scales, crab and shrimp shells. Int J Biol Macromol 104:1697–1705
10. Singh R, Shitiz K, Singh A (2017) Chitin and chitosan: biopolymers for wound management. Int Wound J 14:1276–1289
11. Hamed I, Zogul FO, Regenstein JM (2016) Industrial applications of crustacean by products (chitin, chitosan, and chitooligosaccharides): a review. Trends Food Sci Technol 48:40–50
12. Ways TM, Lau W, Khutoryanskiy K, Ways TMM, Lau WM, Khutoryanskiy VV (2018) Chitosan and its derivatives for application in mucoadhesive drug delivery systems. Polymers 10(30):267
13. Lee AY (2020) Molecular mechanism of epidermal barrier dysfunction as primary abnormalities. Int J Mol Sci 21:1194
14. Donnelly RF, Singh TRR, Morrow DI, Woolfson AD (2012) Microneedle-mediated transdermal and intradermal drug delivery. Wiley, Hoboken
15. Kretsos K, Kasting GB (2007) A geometrical model of dermal capillary clearance. Math Biosci 208:430–453
16. Lauterbach A, Christel C, Goymann M (2015) Applications and limitations of lipid nanoparticles in dermal and transdermal drug delivery via the follicular route. Eur J Pharm Biopharm 97:152–163
17. Chen MC, Mi FL, Liao ZX, Hsiao CW, Sonaje K, Chung MF, Hsu LW, Sung HW (2013) Recent advances in chitosan-based nanoparticles for oral delivery of macromolecules. Adv Drug Deliv Rev 65:865–879
18. Shi Y, Xue J, Xu S, You Y, Yan XQ, Zhao X, Cao J (2018) Polyelectrolyte complex nanoparticles based on chitosan and methoxy poly (ethylene glycol) methacrylate-co-poly (methylacrylic acid) for oral delivery of ibuprofen. Colloids Surf B Biointerfaces 165:235–242
19. Ahad A, Al-Jenoobi FI, Al-Mohizea AM, Akhtar N, Raish M, Aqil M (2015) Systemic delivery of β-blockers via transdermal route for hypertension. Saudi Pharm J 23:587–602
20. Ahad A, Al-Saleh AA, Akhtar N, Al-Mohizea AM, Al-Jenoobi FI (2015) Transdermal delivery of antidiabetic drugs: formulation and delivery strategies. Drug Discov Today 20: 1217–1227
21. Lim JI, Lee WK (2014) Preparation of macroporous chitosan patch using co-solvent as a transdermal drug-delivery system. J Porous Mater 21:783–788
22. Amjadi M, Sheykhansari S, Nelson BJ, Sitti M (2018) Recent advances in wearable transdermal delivery systems. Adv Mater 30(7):1704530
23. Niehues H, Bouwstra JA, El Ghalbzouri A, Brandner JM, Zeeuwen PL, Van Den Bogaard EH (2018) 3D skin models for 3R research: the potential of 3D reconstructed skin models to study skin barrier function. Exp Dermatol 27:501–511
24. Shastri VP, Lee PJ, Ahmad N, Langer R, Mitragotri S (2006) Evaluation of chemical enhancers in the transdermal delivery of lidocaine. Int J Pharm 331:308
25. Zahiri M, Khanmohammadi M, Goodarzi A, Ababzadeh S, Farahani MS, Mohandesnezhad S, Bahrami N, Nabipour I, Ai J (2020) Encapsulation of curcumin loaded chitosan nanoparticle within poly (ε-caprolactone) and gelatin fiber mat for wound healing and layered dermal reconstitution. Int J Biol Macromol 153:1241–1250
26. Badali E, Goodarzi A, Khodayari H, Khodayari S, Habibi A, Hasanzadeh S, Khanmohammadi M (2022) Layered dermal reconstitution through epigallocatechin 3-gallate loaded chitosan nanoparticle within enzymatically crosslinked polyvinyl alcohol/collagen fibrous mat. J Biomater Appl 37(3):502–516

27. Thao NT, Wijerathna HM, Kumar RS, Choi D, Dananjaya SH, Attanayake AP (2021) Preparation and characterization of succinyl chitosan and succinyl chitosan nanoparticle film: in vitro and in vivo evaluation of wound healing activity. Int J Biol Macromol 193:1823–1834
28. Anushree U, Punj P, Vasumathi BS (2023) Phosphorylated chitosan accelerates dermal wound healing in diabetic Wistar rats. Glycoconj J 40(1):19–31
29. Fereig SA, El-Zaafarany GM, Arafa MG, Abdel-Mottaleb MM (2021) Tacrolimus-loaded chitosan nanoparticles for enhanced skin deposition and management of plaque psoriasis. Carbohydr Polym 268:118–238
30. Riezk A, Van Bocxlaer K, Yardley V, Murdan S, Croft SL (2020) Activity of amphotericin B-loaded chitosan nanoparticles against experimental cutaneous leishmaniasis. Molecules 25(17):4002
31. Yu K, Wang Y, Wan T, Zhai Y, Cao S, Ruan W, Wu C, Xu Y (2018) Tacrolimus nanoparticles based on chitosan combined with nicotinamide: enhancing percutaneous delivery and treatment efficacy for atopic dermatitis and reducing dose. Int J Nanomedicine 22:129–142
32. Barbosa AI, Lima SA, Reis S (2019) Development of methotrexate loaded fucoidan/chitosan nanoparticles with anti-inflammatory potential and enhanced skin permeation. Int J Biol Macromol 124:1115–1122
33. Noshi SH, Ibrahim MS, Salama A, Fathy IA, Elsayyad NM (2023) Chondroitin Sulphate-Chitosan polyelectrolyte complexes for etorocoxib transdermal delivery: in silico, in vitro and in vivo studies. Pharm Dev Technol 28(8):785–798
34. Al-Nemrawi N, Wahsheh Y, Alzoubi KH (2024) Transdermal delivery of methotrexate loaded in chitosan nanoparticles to treat rheumatoid arthritis. Curr Drug Deliv 1(3):451–460
35. Li H, Yao J, Guo Y, Huo J, Zhang H, Zhang Z, Zhao J, Zhang C (2023) Preparation of conotoxin-encapsulated chitosan nanoparticles and evaluation of their skin permeability. AAPS PharmSciTech 24(1):53
36. Owczarek M, Herczynska L, Sitarek P, Kowalczyk T, Synowiec E, Sliwinski T, Krucinska (2023) Chitosan nanoparticles-preparation, characterization and their combination with Ginkgo biloba extract in preliminary in vitro studies. Molecules 28(13):4950
37. Alvarez-Figueroa MJ, Narvaez-Araya D, Armijo-Escalona N, Carrasco-Flores EA, Gonzalez-Aramundiz JV (2020) Design of Chitosan nanocapsules with Compritol 888 ATO(R) for Imiquimod transdermal administration. Evaluation of their skin absorption by Raman microscopy. Pharm Res 37:1–10
38. Truong TH, Alcantara KP, Bulatao BP, Sorasitthiyanukarn FN, Muangnoi C, Nalinratana N, Vajragupta O, Rojsitthisak P, Rojsitthisak P (2022) Chitosan-coated nanostructured lipid carriers for transdermal delivery of tetrahydrocurcumin for breast cancer therapy. Carbohydr Polym 288:119–401
39. Khan TA, Azad AK, Fuloria S, Nawaz A, Subramaniyan V, Akhlaq M, Safdar M, Sathasivam KV, Sekar M, Porwal O, Meenakshi DU (2021) Chitosan-coated 5-fluorouracil incorporated emulsions as transdermal drug delivery matrices. Polymers 13(19):3345
40. Takeuchi I, Suzuki T, Makino K (2021) Iontophoretic transdermal delivery using chitosan-coated PLGA nanoparticles for transcutaneous immunization. Colloids Surf A Physicochem Eng Asp 608:125–607
41. Liu Z, Wang Y, Zhang Y, Hu L, Chen B, Li Y, Guo X, Yu B, Xu FJ (2024) Biguanide chitosan microneedles with cell-free DNA scavenging ability for psoriasis therapy. Bioact Mater 33: 497–505
42. Khalid A, Sarwar HS, Sarfraz M, Sohail MF, Jalil A, Jardan YA, Arshad R, Tahir I, Ahmad Z (2023) Formulation and characterization of thiolated chitosan/polyvinyl acetate-based microneedle patch for transdermal delivery of dydrogesterone. Saudi Pharm J 31(5):669–677
43. Ahmad Z, Khan MI, Siddique MI, Sarwar HS, Shahnaz G, Hussain SZ, Bukhari NI, Hussain I, Sohail MF (2020) Fabrication and characterization of thiolated chitosan microneedle patch for transdermal delivery of tacrolimus. AAPS PharmSciTech 21:1–12

44. Che QT, Seo JW, Charoensri K, Nguyen MH, Park HJ, Bae H (2024) 4D-printed microneedles from dual-sensitive chitosan for non-transdermal drug delivery. Int J Biol Macromol 261: 129638
45. Chen MC, Huang SF, Lai KY, Ling MH (2013) Fully embeddable chitosan microneedles as a sustained release depot for intradermal vaccination. Biomaterials 34(12):3077–3086
46. Chi J, Zhang X, Chen C, Shao C, Zhao Y, Wang Y (2020) Antibacterial and angiogenic chitosan microneedle array patch for promoting wound healing. Bioact Mater 5:253–259
47. Venkataprasanna KS, Prakash J, Vignesh S, Bharath G, Venkatesan M, Banat F, Sahabudeen S, Ramachandran S, Venkatasubbu GD (2020) Fabrication of chitosan/PVA/GO/CuO patch for potential wound healing application. Int J Biol Macromol 143:744–762
48. Ghalayani Esfahani A, Altomare L, Varoni EM, Bertoldi S, Fare S, De Nardo L (2019) Electrophoretic bottom-up design of chitosan patches for topical drug delivery. J Mater Sci Mater Med 30:1–8
49. Niranjan R, Kaushik M, Prakash J, Venkataprasanna KS, Arpana C, Balashanmugam P, Venkatasubbu GD (2019) Enhanced wound healing by PVA/chitosan/curcumin patches: in vitro and in vivo study. Colloids Surf B Biointerfaces 182:110–339
50. Hemmingsen LM, Julin K, Ahsan L, Basnet P, Johannessen M, Skalko-Basnet N (2021) Chitosomes-in-chitosan hydrogel for acute skin injuries: prevention and infection control. Mar Drugs 19(5):269
51. Wang Y, Liu S, Yu W (2021) Functionalized graphene oxide-reinforced chitosan hydrogel as biomimetic dressing for wound healing. Macromol Biosci 21(4):2000432
52. Yan T, Kong S, Ouyang Q, Li C, Hou T, Chen Y, Li S (2020) Chitosan-gentamicin conjugate hydrogel promoting skin scald repair. Mar Drugs 18(5):233
53. Natesan S, Krishnaswami V, Thekkila Veedu S, Pathayappurakkal Mohanan D, Ruckmani K, Palanichamy R (2019) Chitosan-based ocular drug delivery systems. In: Functional chitosan, pp 107–134
54. Li S, Chen L, Fu Y (2023) Nanotechnology-based ocular drug delivery systems: recent advances and future prospects. J Nanobiotechnol 21(1):232
55. Zamboulis A, Nanaki S, Michailidou G, Koumentakou I, Lazaridou M, Ainali NM, Xanthopoulou E, Bikiaris DN (2020) Chitosan and its derivatives for ocular delivery formulations: recent advances and developments. Polymers 12(7):1519
56. Cabrera FJ, Wang DC, Reddy K, Acharya G, Shin CS (2019) Challenges and opportunities for drug delivery to the posterior of the eye. Drug Discov Today 24(8):1679–1684
57. Gote V, Sikder S, Sicotte J, Pal D (2019) Ocular drug delivery: present innovations and future challenges. J Pharmacol Exp Ther 370(3):602–624
58. Faria MJ, Gonzalez-Meijome JM, Oliveira ME, Carracedo G, Lucio M (2024) Recent advances and strategies for nanocarrier-mediated topical therapy and theranostic for posterior eye disease. Adv Drug Deliv Rev 26:115321
59. Ahmed S, Amin MM, Sayed S (2023) Ocular drug delivery: a comprehensive review. AAPS PharmSciTech 24(2):66
60. Amrutkar CS, Patil SB (2023) Nanocarriers for ocular drug delivery: recent advances and future opportunities. Indian J Ophthalmol 71(6):2355–2366
61. Liu LC, Chen YH, Lu DW (2023) Overview of recent advances in nano-based ocular drug delivery. Int J Mol Sci 24(20):15352
62. Tsung TH, Chen YH, Lu DW (2023) Updates on biodegradable formulations for ocular drug delivery. Pharmaceutics 15(3):734
63. Dosmar E, Walsh J, Doyel M, Bussett K, Oladipupo A, Amer S, Goebel K (2022) Targeting ocular drug delivery: an examination of local anatomy and current approaches. Bioengineering 19(1):41
64. Han H, Li S, Xu M, Zhong Y, Fan W, Xu J, Zhou T, Ji J, Ye J, Yao K (2023) Polymer-and lipid-based nanocarriers for ocular drug delivery: current status and future perspectives. Adv Drug Deliv Rev 196:114770

65. Onugwu AL, Nwagwu CS, Onugwu OS, Echezona AC, Agbo CP, Ihim SA, Emeh P, Nnamani PO, Attama AA, Khutoryanskiy VV (2023) Nanotechnology based drug delivery systems for the treatment of anterior segment eye diseases. J Control Release 354:465–488
66. Sanchez-Lopez E, Espina M, Doktorovova S, Souto EB, Garcia ML (2017) Lipid nanoparticles (SLN, NLC): overcoming the anatomical and physiological barriers of the eye–part I–barriers and determining factors in ocular delivery. Eur J Pharm Biopharm 110: 70–75
67. Abla KK, Mehanna MM (2023) Lipid-based nanocarriers challenging the ocular biological barriers: current paradigm and future perspectives. J Control Release 362:70–96
68. Lee J, Pelis RM (2016) Drug transport by the blood–aqueous humor barrier of the eye. Drug Metab Dispos 44(10):1675–1681
69. Guidi L, Cascone MG, Rosellini E (2024) Light-responsive polymeric nanoparticles for retinal drug delivery: design cues, challenges and future perspectives. Heliyon
70. Cunha-Vaz JG (1997) The blood-ocular barriers: past, present, and future. Doc Ophthalmol 93:149–157
71. Toda R, Kawazu K, Oyabu M, Miyazaki T, Kiuchi Y (2011) Comparison of drug permeabilities across the blood-retinal barrier, blood-aqueous humor barrier, and blood-brain barrier. J Pharm Sci 100(9):3904–3911
72. Gaudana R, Ananthula HK, Parenky A, Mitra AK (2010) Ocular drug delivery. AAPS J 12: 348–360
73. Bachu RD, Chowdhury P, Al-Saedi ZH, Karla PK, Boddu SH (2018) Ocular drug delivery barriers-role of nanocarriers in the treatment of anterior segment ocular diseases. Pharmaceutics 10(1):28
74. Wu KY, Joly-Chevrier M, Akbar D, Tran SD (2023) Overcoming treatment challenges in posterior segment diseases with biodegradable nano-based drug delivery systems. Pharmaceutics 15(4):1094
75. Bourlais C, Acar L, Zia H, Sado PA, Needham T, Leverge R (1998) Ophthalmic drug delivery systems-recent advances. Prog Retin Eye Res 17(1):33–58
76. Sikandar MK, Sharma PK, Visht S (2011) Ocular drug delivery system: an overview. Int J Pharm Sci Res 2(5):1168
77. Latifi A, Esmaeili F, Mohebali M, Yasami-Khiabani S, Rezaeian M, Soleimani M, Kazemirad E, Amani A (2024) Chitosan nanoparticles improve the effectivity of miltefosine against Acanthamoeba. PLoS Negl Trop Dis 18(3):0011976
78. Zhang P, Liu X, Hu W, Bai Y, Zhang L (2016) Preparation and evaluation of naringenin-loaded sulfobutylether-β-cyclodextrin/chitosan nanoparticles for ocular drug delivery. Carbohydr Polym 149:224–230
79. Zhao R, Li J, Wang J, Yin Z, Zhu Y, Liu W (2017) Development of timolol-loaded galactosylated chitosan nanoparticles and evaluation of their potential for ocular drug delivery. AAPS PharmSciTech 18:997–1008
80. Kalam MA, Iqbal M, Alshememry A, Alkholief M, Alshamsan A (2022) Development and evaluation of chitosan nanoparticles for ocular delivery of tedizolid phosphate. Molecules 27(7):2326
81. Mohamed HB, Attia Shafie MA, Mekkawy AI (2022) Chitosan nanoparticles for meloxicam ocular delivery: development, in vitro characterization, and in vivo evaluation in a rabbit eye model. Pharmaceutics 14(5):893
82. Dandamudi M, McLoughlin P, Behl G, Rani S, Coffey L, Chauhan A, Fitzhenry L (2021) Chitosan-coated PLGA nanoparticles encapsulating triamcinolone acetonide as a potential candidate for sustained ocular drug delivery. Pharmaceutics 13(10):1590
83. Yu A, Shi H, Liu H, Bao Z, Dai M, Lin D, Wang Y (2020) Mucoadhesive dexamethasone-glycol chitosan nanoparticles for ophthalmic drug delivery. Int J Pharm 575:118943
84. Saha M, Saha DR, Ulhosna T, Sharker SM, Shohag MH, Islam MS, Reza HM (2021) QbD based development of resveratrol-loaded mucoadhesive lecithin/chitosan nanoparticles for prolonged ocular drug delivery. J Drug Deliv Sci Technol 63:102480

85. Pandit J, Sultana Y, Aqil M (2021) Chitosan coated nanoparticles for efficient delivery of bevacizumab in the posterior ocular tissues via subconjunctival administration. Carbohydr Polym 267:118217
86. Alruwaili NK, Zafar A, Imam SS, Alharbi KS, Alotaibi NH, Alshehri S, Elmowafy M (2020) Stimulus responsive ocular gentamycin-ferrying chitosan nanoparticles hydrogel: formulation optimization, ocular safety and antibacterial assessment. Int J Nanomed:4717–4737
87. Huang C, Peng L, Xu X, Lu Y, Wang X, Lan Z, Zhou Y (2023) Preparation and characteristics of a thermosensitive in situ gel loaded with chitosan nanoparticles for optimal ocular delivery of chloramphenicol. J Drug Deliv Sci Technol 89:104962
88. Natesan S, Pandian S, Ponnusamy C, Palanichamy R, Muthusamy S, Kandasamy R (2017) Co-encapsulated resveratrol and quercetin in chitosan and peg modified chitosan nanoparticles: for efficient intra ocular pressure reduction. Int J Biol Macromol 104:1837–1845
89. Pandian S, Jeevanesan V, Ponnusamy C, Natesan S (2017) RES-loaded pegylated CS NPs: for efficient ocular delivery. IET Nanobiotechnol 11(1):32–39
90. Racic A, Calija B, Milic J, Dukovski BJ, Lovric J, Dobricic V, Micov A, Vukovic M, Stepanovic-Petrovic R, Krajisnik D (2021) Formulation of olopatadine hydrochloride viscous eye drops–physicochemical, biopharmaceutical and efficacy assessment using in vitro and in vivo approaches. Eur J Pharm Sci 166:105906
91. Gratieri T, Gelfuso GM, de Freitas O, Rocha EM, Lopez RF (2011) Enhancing and sustaining the topical ocular delivery of fluconazole using chitosan solution and poloxamer/chitosan in situ forming gel. Eur J Pharm Biopharm 79(2):320–327
92. Li B, Wang J, Gui Q, Yang H (2020) Drug-loaded chitosan film prepared via facile solution casting and air-drying of plain water-based chitosan solution for ocular drug delivery. Bioactive Materials 5(3):577–583
93. Asfour MH, Abd El-Alim SH, Awad GEA, Kassem AA (2021) Chitosan/β-glycerophosphate in situ forming thermo-sensitive hydrogel for improved ocular delivery of moxifloxacin hydrochloride. Eur J Pharm Sci 167:106041
94. Aslzad S, Savadi P, Abdolahinia ED, Omidi Y, Fathi M, Barar J (2022) Chitosan/dialdehyde starch hybrid in situ forming hydrogel for ocular delivery of betamethasone. Mater Today Commun 33:104873
95. Pardeshi SR, More MP, Patil PB, Mujumdar A, Naik JB (2022) Statistical optimization of voriconazole nanoparticles loaded carboxymethyl chitosan-poloxamer based in situ gel for ocular delivery: in vitro, ex vivo, and toxicity assessment. Drug Deliv Transl Res 12(12):3063–3082
96. Modi D, Mohammad Warsi MH, Garg V, Bhatia M, Kesharwani P, Jain GK (2021) Formulation development, optimization, and in vitro assessment of thermoresponsive ophthalmic pluronic F127-chitosan in situ tacrolimus gel. J Biomater Sci Polym Ed 32(13):1678–1702
97. Armengol ES, Grassiri B, Piras AM, Zambito Y, Fabiano A, Laffleur F (2024) Ocular antibacterial chitosan-maleic acid hydrogels: in vitro and in vivo studies for a promising approach with enhanced mucoadhesion. Int J Biol Macromol 254:127939
98. Tan G, Yu S, Pan H et al (2017) Bioadhesive chitosan-loaded liposomes: a more efficient and higher permeable ocular delivery platform for timolol maleate. Int J Biol Macromol 94:355–363
99. Chen H, Pan H, Li P, Wang H, Wang X, Pan W, Yuan Y (2016) The potential use of novel chitosan-coated deformable liposomes in an ocular drug delivery system. Colloids Surf B Biointerfaces 143:455–462
100. Cheng T, Li J, Cheng Y, Zhang X, Qu Y (2019) Triamcinolone acetonide-chitosan coated liposomes efficiently treated retinal edema as eye drops. Exp Eye Res 188:107805
101. Xiong X, Jiang H, Liao Y, Du Y, Zhang Y, Wang Z, Zheng M, Du Z (2023) Liposome–trimethyl chitosan nanoparticles codeliver insulin and siVEGF to treat corneal alkali burns by inhibiting ferroptosis. Bioeng Transl Med 2:e10499

102. Li N, Zhuang C, Wang M, Sun X, Nie S, Pan W (2009) Liposome coated with low molecular weight chitosan and its potential use in ocular drug delivery. Int J Pharm 379(1):131–138
103. Li N, Zhuang CY, Wang M et al (2012) Low molecular weight chitosan-coated liposomes for ocular drug delivery: in vitro and in vivo studies. Drug Deliv 19(1):28–35
104. Arafa MG, Girgis GN, El-Dahan MS (2020) Chitosan-coated PLGA nanoparticles for enhanced ocular anti-inflammatory efficacy of atorvastatin calcium. Int J Nanomed:1335–1347
105. Suri R, Neupane YR, Mehra N, Nematullah M, Khan F, Alam O, Kohli K (2021) Sirolimus loaded chitosan functionalized poly (lactic-co-glycolic acid) (PLGA) nanoparticles for potential treatment of age-related macular degeneration. Int J Biol Macromol 191:548–559
106. Li N, Zhao Z, Ma H, Liu Y, Nwafor EO, Zhu S, Pan W (2022) Optimization and characterization of low-molecular-weight chitosan-coated baicalin mPEG-PLGA nanoparticles for the treatment of cataract. Mol Pharm 19(11):3831–3845
107. Suri R, Nag TC, Mehra N, Neupane YR, Shafi S, Sharma D, Kohli K (2023) Sirolimus loaded chitosan functionalized PLGA nanoparticles protect against sodium iodate-induced retinal degeneration. J Drug Deliv Sci Technol 82:104369
108. Dos Reis JS, Teixeira ADR, Quaresma ADV, Almeida TC, Arribada RG, Neto JT, Da Silva GR (2022) Sodium butyrate-loaded nanoparticles coated with chitosan for the treatment of neovascularization in age-related macular degeneration: ocular biocompatibility and antiangiogenic activity. Eur J Pharm Biopharm 179:26–36
109. Al-mahallawi AM, Ahmed D, Hassan M, El-Setouhy DA (2021) Enhanced ocular delivery of clotrimazole via loading into mucoadhesive microemulsion system: in vitro characterization and in vivo assessment. J Drug Deliv Sci Technol 64:102561
110. Onugwu AL, Attama AA, Nnamani PO, Onugwu SO, Onuigbo EB, Khutoryanskiy VV (2022) Development and optimization of solid lipid nanoparticles coated with chitosan and poly (2-ethyl-2-oxazoline) for ocular drug delivery of ciprofloxacin. J Drug Deliv Sci Technol 74: 103527
111. Yadav S, Verma S, Tyagi S, Kumar H, Kumar S, Ashique S (2022) Preparation and in-vitro characterization of brimonidine encapsulated polymer coated microemulsion for ocular drug delivery for the management of glaucoma. J Harbin Inst Technol 54(5)
112. Dukovski BJ, Juretic M, Bracko D, Randjelovic D, Savic S, Moral MC, Lovric J (2020) Functional ibuprofen-loaded cationic nanoemulsion: development and optimization for dry eye disease treatment. Int J Pharm 576:118979
113. Choudhari M, Nayak K, Nagai N, Nakazawa Y, Khunt D, Misra M (2023) Role of mucoadhesive agent in ocular delivery of ganciclovir microemulsion: cytotoxicity evaluation in vitro and ex vivo. Int Ophthalmol 43(4):1153–1167
114. El-Haddad ME, Hussien AA, Saeed HM, Farid RM (2021) Down regulation of inflammatory cytokines by the bioactive resveratrol-loaded chitoniosomes in induced ocular inflammation model. J Drug Deliv Sci Technol 66:102787
115. Sharma DS, Wadhwa S, Gulati M, Kumar B, Chitranshi N, Gupta VK, Singh SK (2023) Chitosan modified 5-fluorouracil nanostructured lipid carriers for treatment of diabetic retinopathy in rats: a new dimension to an anticancer drug. Int J Biol Macromol 224:810–830
116. Kazemi Ashtiani M, Zandi SP, Ehsani M, Baharvand H (2020) Chitosan surface modified hydrogel as a therapeutic contact lens. Polym Adv Technol 31(4):741–748
117. Jiao Z, Huo Q, Lin X, Chu X, Deng Z, Guo H, Wang B (2022) Drug-free contact lens based on quaternized chitosan and tannic acid for bacterial keratitis therapy and corneal repair. Carbohydr Polym 286:119314
118. Yang C, Wang M, Wang W, Liu H, Deng H, Du Y, Shi X (2022) Electrodeposition induced covalent cross-linking of chitosan for electrofabrication of hydrogel contact lenses. Carbohydr Polym 292:119678
119. Gade SK, Nirmal J, Garg P, Venuganti VVK (2020) Corneal delivery of moxifloxacin and dexamethasone combination using drug-eluting mucoadhesive contact lens to treat ocular infections. Int J Pharm 591:120023

120. He P, Davis SS, Illum L (1998) In vitro evaluation of the mucoadhesive properties of chitosan microspheres. Int J Pharm 166(1):75–88
121. Gades MD, Stern JS (2003) Chitosan supplementation and fecal fat excretion in men. Obes Res 11(5):683–688
122. Barnard J, Millner R (2009) A review of topical hemostatic agents for use in cardiac surgery. Ann Thorac Surg 88(4):1377–1383
123. Kojima K, Okamoto Y, Kojima K, Miyatake K, Fujise H, Shigemasa Y, Minami S (2004) Effects of chitin and chitosan on collagen synthesis in wound healing. J Vet Med Sci 66(12): 1595–1598
124. Tago K, Naito Y, Nagata T, Morimura T, Furuya M, Seki T, Ohara N (2007) A ninety-day feeding, subchronic toxicity study of oligo-N-acetylglucosamine in Fischer 344 rats. Food Chem Toxicol 45(7):1186–1193
125. Shigemasa Y, Minami S (1996) Applications of chitin and chitosan for biomaterials. Biotechnol Genet Eng Rev 13(1):383–420
126. Dornish M, Kaplan DS, Arepalli SR (2012) Regulatory status of chitosan and derivatives. In: Chitosan-based systems for biopharmaceuticals. Wiley, pp 463–481

Adv Polym Sci (2025) 295: 319–338
https://doi.org/10.1007/12_2024_176

Published online: 21 November 2024

Chitosan in Oral Drug Delivery

Makoto Anraku and Masaki Otagiri

Contents

1 Introduction ... 320
2 Clinical Oral Application ... 320
2.1 Obesity ... 321
2.2 Diabetes Mellitus ... 322
2.3 Renal Failure ... 324
2.4 Intestinal Disease ... 326
2.5 Cancer ... 327
2.6 Respiratory Disease ... 329
2.7 Other Diseases ... 330
3 Conclusion ... 330
References ... 331

Abstract Chitosan, a natural polymer obtained by alkaline deacetylation of chitin, is non-toxic, biocompatible, and biodegradable. Various chitosan molecular weights (MW) have been designed, from high MW to low MW chitosan oligosaccharides (COS). Chitosan has excellent antibacterial, moisture retention, biocompatibility, safety, and chelating properties, and numerous chitosan products are currently being developed to take advantage of these properties. These properties have made chitosan an excellent carrier for drug delivery and a functional food. This chapter presents examples of oral administration for chitosan, COS, and their derivatives, especially in the medical and pharmaceutical fields and as oral drug delivery carriers.

Keywords Clinical applications · Drug delivery carriers · Medical and pharmaceutical materials · Oral administration

M. Anraku (✉) and M. Otagiri
Faculty of Pharmaceutical Sciences, Sojo University, Kumamoto, Japan
e-mail: anraku@ph.sojo-u.acjp; otagirim@ph.sojo-u.ac.jp

1 Introduction

Chitosan is biocompatible, biodegradable, and cationic, making it more advantageous in medical and pharmaceutical applications than other negatively charged or neutral polysaccharides. Indeed, many recent papers have described the functional properties and various applications of chitosan and its derivatives and complexes [1–5]. The structural and functional properties and multiple applications of chitosan oligosaccharides (COS) and their derivatives, which are hydrolyzed chitosan to low MW, have also been described in many review articles [6–10]. Chitosan is defined as having a MW of 10–1,000 kDa and a degree of deacetylation (DD) in the 50–95% range, while COS is described as having a smaller MW. Depending on the DD and MW, chitosan and COS have different viscosities, solubilities, and applications. When considering the oral administration of chitosan or COS, the size of the MW is very important, and its application is expected to vary greatly depending on the MW. For example, COS with a small MW exerts various effects by being absorbed from the intestinal tract and distributed to tissues [11, 12]. COS is also expected to play a role as a dissolution aid by forming complexes with water-insoluble pharmaceuticals and improving the intestinal tract's absorption of peptides by forming complexes with peptides. Even if chitosan is not absorbed through the intestinal tract due to its large MW, it has been applied to various pathological conditions due to its multiple functions, including antioxidant and anti-inflammatory effects limited to the intestinal tract [3, 5]. This chapter presents the applications of oral administration for chitosan and its derivatives with different properties in the medical and pharmaceutical field, as well as examples of recently developed chitosan, such as nanoparticles, nanofibers, and nanogels, as oral drug delivery carriers.

2 Clinical Oral Application

When considering the oral administration of chitosan, the pharmacokinetics of chitosan itself after oral administration is important. From numerous studies, it has been shown that in rats, chitosan and COS with MW of approximately 30 kDa or less undergo intestinal absorption and are mainly distributed in the liver, spleen, and kidney (1–1, 1–2 in Fig. 1) [11, 12]. Furthermore, chitosan and COS are degraded by lysozyme in plasma, liver, kidney, and urine, which have been demonstrated in vivo and in vitro studies (1–3 in Fig. 1). Subsequently, it has been shown that the kidney plays an important role in the excretion of chitosan and COS [13, 14]. In particular, a study in rats showed that more than 80% of chitosan and COS were excreted in urine within 15 days after administration (1–4 in Fig. 1) [15, 16]. On the other hand, chitosan with a MW of 100 kDa or greater is not absorbed from the intestinal tract. It is considered to play a limited role in the intestinal tract. Therefore, when considering the oral administration of chitosan and COS, the size of their MW is very important, and their kinetics and activity are expected to vary greatly depending

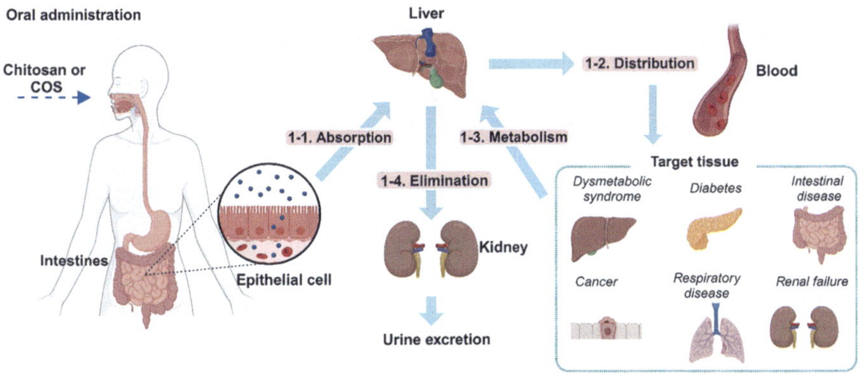

Fig. 1 The pharmacokinetics of chitosan and COS after oral administration

on their MW. Therefore, this section introduces the effects of each MW of chitosan and COS in each disease state and the role of chitosan and COS in pharmaceuticals as carriers.

2.1 *Obesity*

Obesity is usually known to induce dyslipidemia, such as hypercholesterolemia and hypertriglyceridemia, as well as to increase the risk of other diseases, such as cardiovascular disease and diabetes [17]. Chitosan and its derivatives have received significant attention as a drug replacement therapy for obesity and dyslipidemia. As a mechanism by which chitosan effectively reduces body weight and plasma lipids, human clinical trials and animal studies have reported that chitosan effectively reduces body weight and plasma lipids by promoting fecal fat excretion through lipid adsorption in the intestinal tract [18–21]. For example, in a human study, Pokhis et al. reported significant weight loss after 24 weeks of chitosan (MW: 125–145 kDa) intake [18]. In a human study, Trivedi et al. reported significant weight loss after 90 days of chitosan capsule intake [19]. We also reported significant plasma cholesterol and lipid level reductions in a human study of chitosan (Mw: 100 kDa, DD: 90%) as a functional food [22]. The mechanism by which chitosan promotes fat excretion in feces is reported to be the binding of chitosan to lipids and bile acids in the intestinal tract, as well as the inhibition of fatty acid and glycerol reabsorption by chitosan's inhibition of pancreatic lipase activity (2–1 in Fig. 2) [23–25]. On the other hand, COS, once absorbed from the intestinal tract and distributed from the systemic circulatory system to various tissues, has been reported to inhibit adipocyte differentiation by inhibiting PPAR-γ expression and to reduce hepatic lipid metabolic disorders by regulating carnitine palmitoyl transferase one and PPARα in the liver (2–2 in Fig. 2) [20, 26–28]. Few examples of pharmaceutical

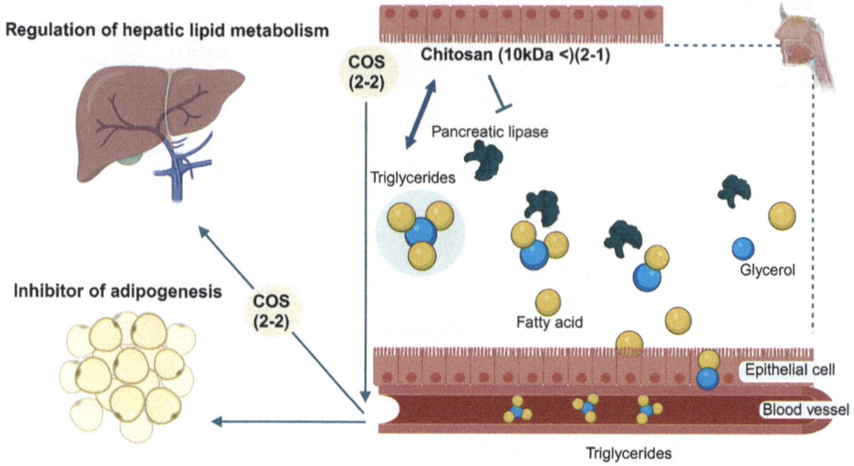

Fig. 2 The possible mechanisms for regulating obesity and dyslipidemia by chitosan and COS

applications of chitosan as a carrier for obesity exist, and most of the reports on the effects of chitosan and COS alone, as described above, are limited.

2.2 *Diabetes Mellitus*

Diabetes is a disease in which insufficient insulin causes an increase in the amount of glucose flowing through the blood [29]. Chitosan and its derivatives have been investigated for several applications in diabetes as an alternative strategy for diabetes treatment [30]. For example, it was proposed that COS effectively reduces postprandial blood glucose levels in diabetes [31]. In addition, a significant reduction in blood glucose levels was observed in a human clinical trial in which pre-diabetic subjects were given COS for 12 weeks [32]. As a mechanism for these effects, it has been suggested that chitosan may reduce hyperglycemia through decreased glycogenesis in the liver, increased glucose uptake and use in skeletal muscle, and suppression of lipid accumulation in adipose tissue (Fig. 3) [33]. Prajapati also proposed that chitosan (Mw: 480 kDa, DD: 75–80%) exerts its anti-diabetic effects through its ability to regulate gut microbiota, reduce insulin resistance and inflammatory cytokine levels, and increase insulin sensitivity [34]. Although many unknowns remain regarding whether chitosan and COS have direct or indirect anti-diabetic effects, further elucidation of these mechanisms will lead to their potential application as therapeutic agents for diabetes.

On the other hand, in addition to using chitosan alone, its use as a carrier in combination with other polysaccharides and the like is also flourishing in treating diabetes mellitus. For example, insulin-encapsulated alginate-chitosan (Mw: 150 kDa, DD: 85%) microspheres and insulin-encapsulated mucin-chitosan (Mw:

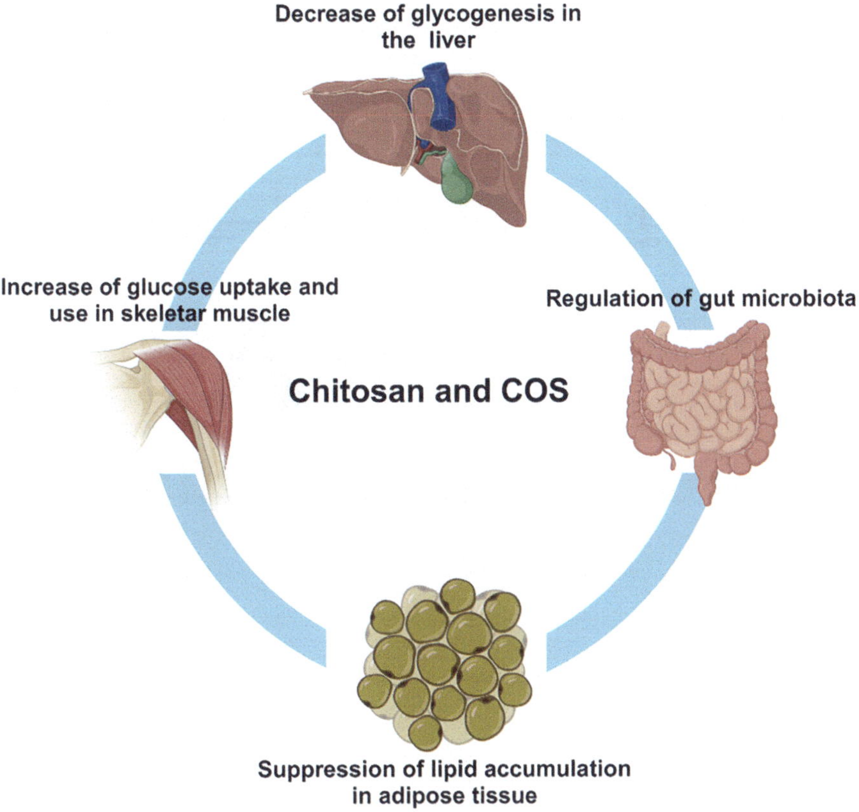

Fig. 3 The possible mechanism of the effect of chitosan and COS on regulating DM

200–300 kDa, DD: 85%) nanoparticles have been prepared and reported to have sustained hypoglycemic effects [35, 36]. Insulin-containing carboxymethyl chitosan (Mw: 530 kDa, DD: 89%)/chitosan (Mw: 400 kDa, DD: 95%) nanoparticles using chitosan derivatives have also been prepared, and similar hypoglycemic effects have been reported [37]. Chitosan-modified nanoparticles coexisting with dipeptidyl peptidase-4 (DPP4), an enzyme inhibitor, and glucagon-like peptide-1 (GLP-1), an anti-diabetic peptide other than insulin, have also been prepared, and reported to exhibit sustained hypoglycemic effects through remarkable uptake of DPP4 and GLP-1 [38]. The mechanism of these effects is that the electrostatic interaction of positively charged chitosan with negatively charged intestinal membrane proteins causes dephosphorylation of claudin-4, which forms a tight junction, and the temporary opening of the tight junction allows peptides, including insulin, to be incorporated. Subsequently, the binding of insulin to the insulin receptor and the promotion of chitosan's gluconeogenesis is thought to be responsible for its hypoglycemic effect (Fig. 4). Insulin-containing cell-permeable peptide/chitosan (Mw: 200 kDa, DD: 75–85%) nanoparticles utilizing cationic peptides that enhance

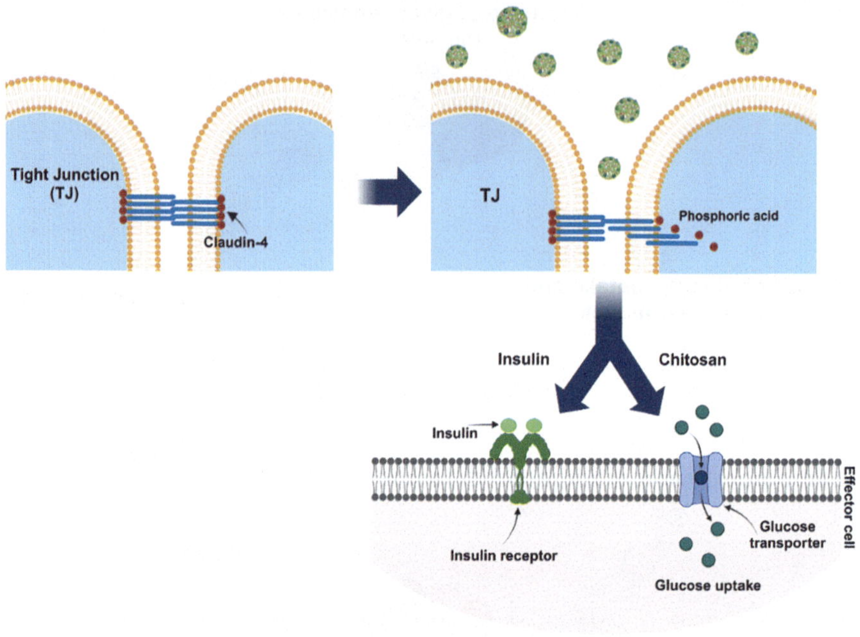

Fig. 4 Mechanism of insulin uptake by insulin-containing chitosan complexes from the intestinal tract

intestinal permeability have also been prepared, and further insulin permeability has been confirmed due to the effect of cell-permeable peptides [39]. We have also prepared guanidylated chitosan (GCS) mimicking membrane-permeable peptides and confirmed the safety of GCS and the high membrane permeability of albumin, a protein similar to insulin [40, 41]. In the future, we plan to develop an oral insulin delivery formulation using this GCS. In summary, COS are expected to have anti-diabetic effects in each tissue of distribution due to their high solubility and absorption. On the other hand, chitosan with a medium MW of 150–500 kDa may be suitable as a substrate for nanoparticle preparation and other carrier applications.

2.3 Renal Failure

Chronic renal failure (CRF) is a gradual decline in kidney function to the point where the filtration capacity is less than 30% of normal, and the body cannot maintain a normal environment. As CRF progresses, renal function cannot be restored. In cases of severe renal impairment, it often progresses to end-stage renal failure (kidney filtration capacity less than 15%), eventually requiring dialysis or a kidney transplant [42, 43]. Currently, more than two million people worldwide are treated with kidney dialysis or kidney transplantation, but how to prevent dialysis, i.e., prevention of

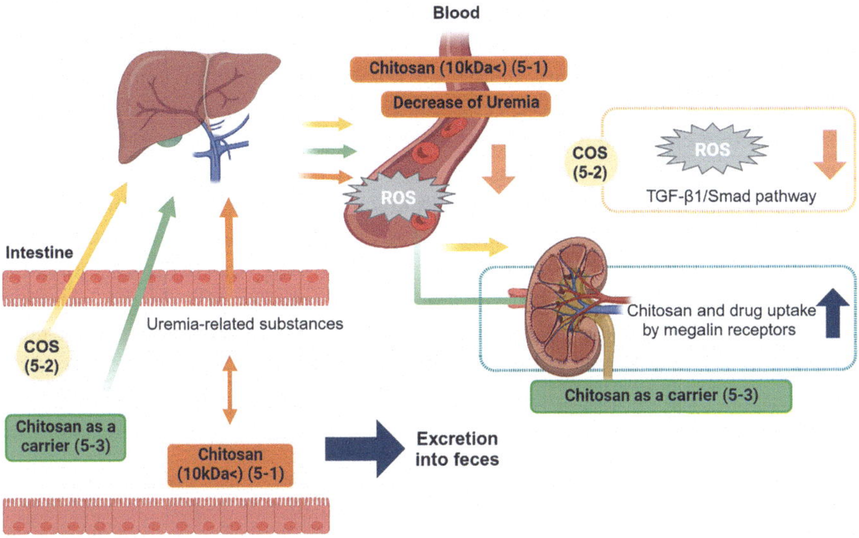

Fig. 5 The possible mechanisms of the inhibitory effects of chitosan and COS on renal failure

CRF in its preliminary stages, is very important [44, 45]. On the other hand, most of the drugs for renal failure are symptomatic treatments, and several applications of chitosan and its derivatives for CRF have been investigated as alternative strategies for the treatment of CRF. Chitosan (Mw: 27 kDa, DD: 89%) ingestion in dialysis patients showed an increase in hemoglobin levels and a significant decrease in cholesterol levels, which inhibited the progression of renal failure [46]. We have also observed that chitosan (Mw:100 kDa, DD:90%) ingestion for 12 weeks in dialysis patients significantly reduced blood uremic substances and serum phosphorus levels [47]. In several animal studies using renal failure models, the protective effect of chitin on renal failure has also been confirmed. For example, oral intake of surface deacetylated chitin nanofibers (DD: 20%) has been suggested to reduce uremic substance concentrations and the associated increase in oxidative stress in 5/6 nephrectomized rats [48]. Chitosan as cat food also improved renal function in cats diagnosed with CRF [49]. The mechanism of these effects is thought to be that chitosan binds to uremic-like toxins and is excreted out of the body, improving renal function through a synergistic effect with the antioxidant effect of excreting uremic substances, which are oxidation-promoting substances (5–1 in Fig. 5) [50].

In a rat model of acute renal failure as well as CRF, COS (Mw: 10 kDa, DD: >90%, Mw: <1.5 kDa) have been shown to inhibit renal failure progression via oxidative stress suppression [51, 52]. As a mechanism for these effects, it has been proposed that COS (Mw: 4. 21 kDa, DD: 92.8%) has a renoprotective effect by reducing oxidative stress damage that occurs during renal failure and by regulating the TGF-β1/Smad pathway to reduce renal fibrosis induced by model mice (5–2 in Fig. 5) [53].

Several chitosans have also been reported as carriers for kidney-targeted drug delivery systems. For example, Zhang et al. prepared chitosan (Mw: 20 kDa, DD: 85%) nanoparticles to efficiently deliver the renal fibrosis drug salvianolic acid B to the kidney. They showed renal-specific migration of these nanoparticles and their renal protective effects in a mouse model of renal fibrosis [54]. Liu et al. prepared telmisartan/canagliflozin-containing chitosan (Mw: <2 kDa) nanoparticles and showed that they are effective carriers with low toxicity and effective renal-specific migration [55]. Thus, the renal-specific transfer of the above drugs was made possible by the increased uptake of chitosan itself and drugs due to the tight junction opening of chitosan (<20 kDa) with a somewhat smaller MW in the intestinal membrane [11] and the increased drug uptake of chitosan by recognition of megalin receptors in the renal tubular epithelial cells after its transfer (5–3 in Fig. 5) [56, 57].

2.4 *Intestinal Disease*

Ulcerative colitis (UC), an inflammatory bowel disease (IBD), is a chronic and highly recurrent intestinal inflammation [58]. Clinical manifestations of UC include varying degrees of systemic symptoms such as persistent or recurrent diarrhea, mucus, pus, bloody stool, and abdominal pain, which are likely to progress to colorectal cancer eventually, and UC occurs more frequently in young people [59, 60]. Analysis of epidemiological data over the past decade shows that the incidence of UC is increasing year by year, and this increase in morbidity has placed a significant burden on the medical and social systems [61–63]. However, because a complete cure for inflammation and prevention of recurrence is difficult, there is a clinical shortage of specific agents. First-line drugs, such as glucocorticoids, 5-aminosalicylic acid, and immunosuppressants, can only reduce symptoms [64, 65]. Functional COS has recently been suggested as a prebiotic that can reduce UC [66, 67]. For example, Azuma et al. orally administered COS (Mw:<1 kDa) to a mouse model of DSS-induced colitis by inhibiting the activation of myeloperoxidase, NF-kB, cyclooxygenase-2, and inducible nitric oxide synthase, found to inhibit inflammation of the colon mucosa [68]. Yusuf et al. also orally administered COS (Mw: 5–10 kDa, DD: >90%) to colitis mouse models and found that COS suppressed colitis progression by decreasing TNF-α and IL-6 levels and MPO activity in the colon tissue [69]. More recently, Mei et al. suggested that oral administration of COS (Mw: 1.5–2.5 kDa, DD: 91%) can significantly enhance short-chain fatty acids and restore intestinal barrier function by increasing the expression of intestinal adhesion-binding proteins [70]. These results strongly suggest that COS may not only play a specific role in preventing colitis by restoring imbalances in the intestinal microbiota, activating mucosal immunity, and suppressing various inflammatory pathways but may also be a therapeutic approach to preventing the development of colitis-associated colorectal cancer (Fig. 6). Regarding this effect on colitis, protection of the intestinal barrier function, anti-

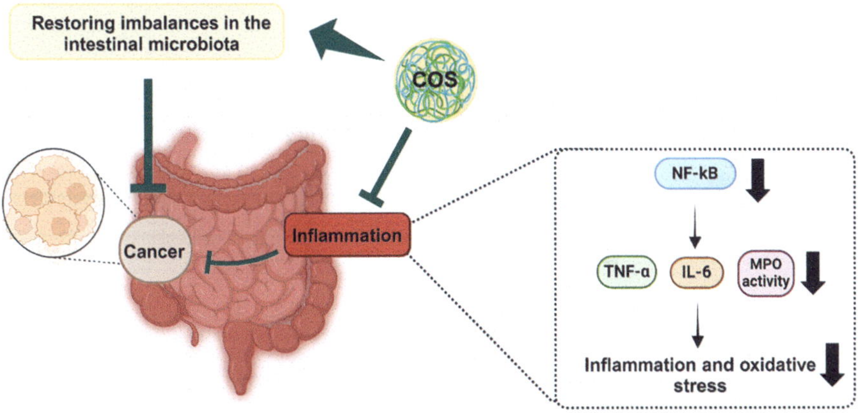

Fig. 6 The possible mechanism of colitis inhibitory effects of chitosan and COS

inflammatory effects, and antifibrotic effects have also been reported in microparticles and nanofibers composed of chitin [71].

Chitosan has also been studied as a drug delivery carrier formulation, such as nanoparticles, microspheres, and beads, for various colon-targeting therapeutic agents in IBD [72]. Omwancha et al. conducted a fascinating study in which they orally administered 5-ASA-containing chitosan beads. These beads were designed to release drugs in the colon, utilizing the colonic microflora's unique degradation process of chitosan [73]. Palma et al. prepared mesalazine-containing chitosan (Mw: 50–190 kDa, DD: 75–85%) microcapsules (average size 1–4 μm). They demonstrated therapeutic efficacy at twofold lower drug concentration than commercial preparations by oral administration in a colitis mouse model [74]. Zhang et al. showed that oral administration of curcumin-embedded chitosan (Mw: 80–120 kDa, DD: 84.1%)/alginate hydrogel achieved better therapeutic results against colitis compared to curcumin alone in a colitis mouse model [75]. Orally administered TNF-α siRNA-containing galactosylated chitosan-modified nanoparticles to a mouse model of colitis not only confirmed the delivery of therapeutic molecules to colitis tissue but also their medical efficacy [76]. Considering the anti-inflammatory and immunomodulatory properties of COS alone against IBD mentioned above, chitosan and COS are expected to have synergistic effects as both a single therapeutic agent and a colon transport carrier.

2.5 *Cancer*

Cancer is the leading cause of the global burden of disease, and cancer-related mortality remains the leading cause of death [77, 78]. Chemotherapy is the most essential treatment for many cancer patients, but it still faces many problems related to numerous side effects that limit its therapeutic efficacy [79]. Several applications

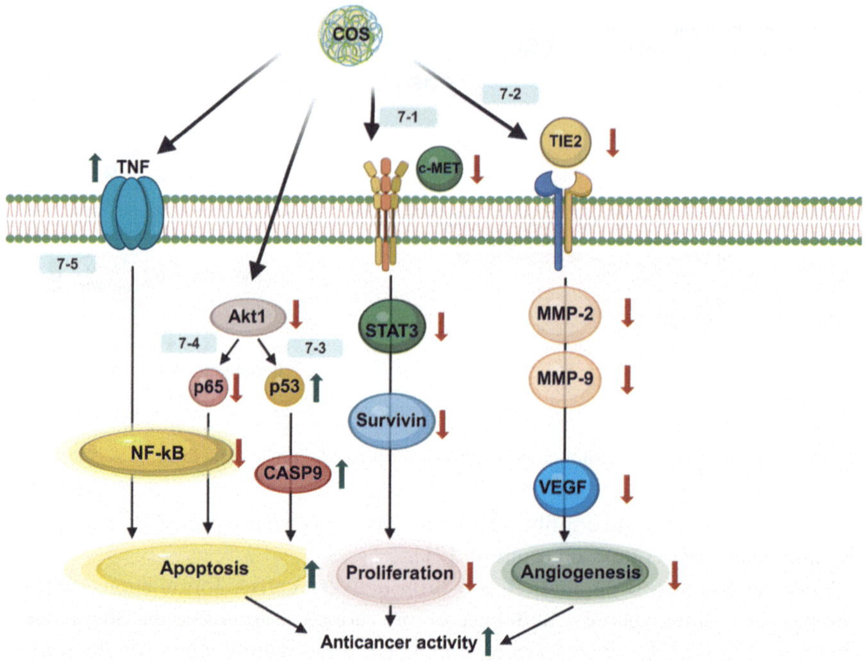

Fig. 7 The signaling pathways of COS with anticancer activity

of chitosan and its derivatives in anticancer therapy have been investigated as alternative strategies for cancer treatment. For example, Lima et al. observed an increase in immune cells in lung cancer patients treated with radiotherapy and oral chitosan (Mw: 4.1 kDa, DD: 93.5%) compared to controls, suggesting that oral chitosan could be used as adjuvant therapy [80]. Studies in mice have also suggested the efficacy of chitosan and COS against cancer. For example, Yeh et al. found that oral administration of low MW chitosan (Mw: 86 kDa) and COS (Mw: 8 kDa) improved survival in a mouse model of lung metastasis [81]. Zou et al. demonstrated that oral administration of COS (Mw:$<$1 kDa) inhibited tumor growth by promoting apoptosis in a mouse model of colon cancer [82]. As a mechanism of these effects, oral administration of low MW chitosan and COS has been reported to inhibit primary augmentation via PI3K/Akt, p38 MAPK, p53, and NF-κB pathways that regulate mitochondrial apoptosis signaling pathways [83–86]. Specifically, the following mechanisms have been proposed. COS downregulates protein expression of proliferation-related molecules such as c-Met, STAT3, and Survivin in HepG2 (7–1 in Fig. 7). COS inhibits the expression of MMP-2, MMP-9, VEGF, and Tie2 in HepG2 (7–2 in Fig. 7). COS induces further downstream CASP9 expression by repressing Akt1, which promotes p53 expression, thereby causing apoptosis in HepG2 (7–3 in Fig. 7). COS inactivates p65 in HepG2, decreasing NF-κB activity and promoting apoptosis in tumor cells (7–4 in Fig. 7). COS enhances anti-tumor activity in LM3 cells by inhibiting NF-κB activity and promoting TNF-α and other inflammatory cytokines (7–5 in Fig. 7). In the oral administration of chitosan and COS for cancer, the key to treatment is their efficient absorption from the intestinal tract and ability to reach cancer cells [86].

Many oral anticancer drugs containing chitosan complexes have been reported using chitosan as a carrier. For example, Almeida et al. suggested that chitosan micelles containing camptothecin may exhibit anticancer effects by improving drug solubility without side effects in colorectal cancer chemotherapy using mouse models of colorectal cancer [87]. Lin et al. also found that ingesting catechin-containing chitosan derivative nanoparticles in a mouse gastric cancer model suppressed gastric inflammatory response [88]. Although the cancers reported above are limited to the intestinal tract, Jayakumar et al. found that doxorubicin-containing chitosan nano gels had a significant cancer-inhibitory effect on several types of liver cancer cells, and Wang et al. also suggested that metformin-containing chitosan nanoparticles are effective against multiple cystic kidneys [89, 90]. As a mechanism for drug-containing chitosan complexes to reach major sites such as the liver and kidney from oral administration, the improvement of intestinal absorption of anticancer drugs using chitosan is a major factor, as the increased bioavailability of chitosan complexes containing doxorubicin by oral administration has been confirmed by Feng et al. [91]. In other words, the adhesive properties and tight junction opening of chitosan in the intestinal membrane are thought to be responsible for increased drug absorption (Fig. 4) [11–15].

2.6 Respiratory Disease

Asthma is a chronic inflammatory disease characterized by various tracheal obstructions. The inflammatory response is initiated by several cells, including helper T-type (Th2) cells [92, 93]. Chung et al. have shown that COS is effective against allergic diseases in a mouse model of asthma [94]. The mechanism suggests that COS suppresses asthma symptoms through its anti-inflammatory effects by downregulating Th2 cytokines and pro-inflammatory cytokines [94]. Airway infection is also one of the various respiratory diseases, especially prominent in children [95]. Singh et al. prepared chitosan nanoparticles containing gingerol, an antimicrobial agent with a short half-life, and administered these nanoparticles orally to rabbits. Due to the chitosan's tight junction opening, the nanoparticles significantly increased gingerol's bioavailability and enhanced its antimicrobial activity [96].

Inhalation is presented in this section as an example of inhalation, with the broad recognition that chitosan administration is currently administered orally. Inhalation is the pulmonary targeted drug delivery system of chitosan for treating respiratory diseases because it is highly effective, non-invasive, avoids first-pass metabolism, reduces the frequency of administration, and has fewer side effects [97–99]. Chitosan's excellent muco-adhesion properties, low viscosity, easy conversion to chemical and biomolecules, and gel-forming properties will allow drugs to remain in the airways longer and enhance therapeutic action. Therefore, various nano pharmaceuticals containing chitosan have been used to treat respiratory diseases with the expectation of a pulmonary pathway that outweighs high therapeutic efficacy and low side effects [100–102]. As a leading example of pulmonary diseases, tuberculosis is the second disease with the highest mortality and morbidity worldwide, after acquired immunodeficiency syndrome [103]. Rawal et al. also prepared nanoparticles containing the antituberculosis drug rifampicin using chitosan of similar MW. They confirmed their lung accumulation and therapeutic efficacy compared to conventional dry powder inhalation formulations in an animal model [104, 105]. Rosière et al. prepared a folate-grafted chitosan derivative containing paclitaxel and demonstrated its effectiveness as an inhaled chemotherapy for treating lung tumors in an animal model [106]. Despite the different pathologies, the MW of these chitosan are all comparable, and some degree of unification of chitosan MW may be necessary to demonstrate their effectiveness as carriers.

2.7 Other Diseases

Chitosan and COS have been extensively studied for their beneficial effects on neurological diseases, especially Alzheimer's. Low MW COS (Mw: <1 kDa, DD: 91.3%) have been reported to be involved in inhibiting amyloid-β aggregation [107]. Chitosan, COS, and nanofibers have also recently been reported to modulate the intestinal microbiota. Still, the detailed mechanism is unknown [108–110]. Chitosan and COS have also been reported to have many applications as antihypertensive, antibacterial, and antioxidant agents. The wide range of chitosan and COS biological activities are expected to be used to treat various diseases [111–113].

3 Conclusion

One of the characteristics of chitosan is that the smaller the MW and the higher the degree of deacetylation, the more water-soluble it is and the wider its application range. After oral administration, chitosan and COS with a MW of 30 kDa or less are absorbed from the intestinal tract. After distribution to the systemic circulatory system, they exert multifaceted effects such as anti-inflammatory and antioxidant actions depending on each target tissue. Even if the MW of chitosan is more than 30 kDa, chitosan and COS are cationic polysaccharides, and their interaction with biological membranes, including the intestinal membrane, promotes the uptake of chitosan and COS themselves and opens tight junctions by adhesion to the membrane; it is also widely used as a transport carrier for drugs and proteins. More recently, the latest technology has enabled the development of chitosan in a nanofibrous form with better membrane adhesion, which is expected to have further applications in the medical field. In the future, chitosan and COS are expected to be used for their effects and in combination with drugs and proteins for various medical and pharmaceutical applications, not only through oral administration but also through various administration sites.

References

1. Pillai CKS, Paul W, Sharma CP (2009) Chitin and chitosan polymers: chemistry, solubility and fiber formation. Adv Polym Sci 34:641–678
2. Rinaudo M (2006) Chitin and chitosan polymers: chitin and chitosan: properties and applications. Prog Polym Sci 31:603–632
3. Jayakumar R, Menon D, Manzoor K, Nair SV, Tamura H (2010) Biomedical applications of chitin and chitosan-based nanomaterials. A short review. Carbohydr Polym 82:227–232
4. Zou P, Yang X, Wang J, Li Y, Yu H, Zhang Y, Liu G (2016) Advances in characterization and biological activities of chitosan and chitosan oligosaccharides. Food Chem 190:1174–1181
5. Anraku M, Gebicki JM, Iohara D, Tomida H, Uekama K, Maruyama T, Hirayama F, Otagiri M (2018) Antioxidant activities of chitosans and its derivatives in in vitro and in vivo studies. Carbohydr Polym 199:141–149
6. Liaqat F, Eltem R (2018) Chitooligosaccharides and their biological activities: a comprehensive review. Carbohydr Polym 184:243–259
7. Naveed M, Phil L, Sohail M, Hasnat M, Baig MMFA, Ihsan AU, Shumzaid M, Kakar MU, Mehmood Khan T, Akabar MD, Hussain MI, Zhou QG (2019) Chitosan oligosaccharide (COS): an overview. Int J Biol Macromol 129:827–843
8. Muanprasat C, Chatsudthipong V (2017) Chitosan oligosaccharide: biological activities and potential therapeutic applications. Pharmacol Ther 170:80–97
9. Lodhi G, Kim YS, Hwang JW, Kim SK, Jeon YJ, Je JY, Ahn CB, Moon SH, Jeon BT, Park PJ (2014) Chitooligosaccharide and its derivatives: preparation and biological applications. Biomed Res Int 2014:654913
10. Guan G, Azad MAK, Lin Y, Kim SW, Tian Y, Liu G, Wang H (2019) Biological effects and applications of chitosan and Chito-oligosaccharides. Front Physiol 7(10):516
11. Chae SY, Jang MK, Nah JW (2005) Influence of molecular weight on oral absorption of water soluble chitosans. J Control Release 102:383–394
12. Zeng L, Qin C, Wang W, Chi W, Li W (2008) Absorption and distribution of chitosan in mice after oral administration. Carbohydr Polym 71:435–440
13. Kean T, Thanou M (2010) Biodegradation, biodistribution and toxicity of chitosan. Adv Drug Deliv Rev 62:3–11
14. Onishi H, Machida Y (1999) Biodegradation and distribution of water-soluble chitosan in mice. Biomaterials 20:175–182
15. Dong W, Han B, Feng Y, Song F, Chang J, Jiang H, Tang Y, Liu W (2010) Pharmacokinetics and biodegradation mechanisms of a versatile carboxymethyl derivative of chitosan in rats: in vivo and in vitro evaluation. Biomacromolecules 11:1527–1533
16. Shao K, Han B, Dong W, Song F, Liu W, Liu W (2015) Pharmacokinetics and biodegradation performance of a hydroxypropyl chitosan derivative. J Ocean Univ China 14:888–896
17. Lavie CJ, Milani RV, Ventura HO (2009) Obesity and cardiovascular disease: risk factor, paradox, and impact of weight loss. J Am Coll Cardiol 53:1925–1932
18. Pokhis K, Bitterlich N, Cornelli U, Cassano G (2015) Efficacy of polyglucosamine for weight loss-confirmed in a randomized double-blind, placebo-controlled clinical investigation. BMC Obes 2:25

19. Trivedi VR, Satia MC, Deschamps A, Maquet V, Shah RB, Zinzuwadia PH (2016) Single-blind, placebo-controlled randomised clinical study of chitosan for body weight reduction. Nutr J 15:3
20. Egan AM, O'Doherty JV, Vigors S, Sweeney T (2016) Prawn shell chitosan exhibits anti-obesogenic potential through alterations to appetite, affecting feeding behaviour and satiety signals in vivo. PLoS One 11:e0149820
21. Huang L, Chen J, Cao P, Pan H, Ding C, Xiao T, Zhang P, Guo J, Su Z (2015) Anti-obese effect of glucosamine and chitosan oligosaccharide in high-fat diet-induced obese rats. Mar Drugs 13:2732–2756
22. Anraku M, Fujii T, Kondo Y, Kojima E, Hata T, Tabuchi N, Tsuchiya D, Goromaru T, Tsutsumi H, Kadowaki D, Maruyama T, Otagiri M, Tomida H (2011) Antioxidant properties of high molecular weight dietary chitosan in vitro and in vivo. Carbohydr Polym 83:501–505
23. Sumiyoshi M, Kimura Y (2006) Low molecular weight chitosan inhibits obesity induced by feeding a high-fat diet long-term in mice. J Pharm Pharmacol 58:201–207
24. Wydro P, Krajewska B, Hac-Wydro K (2007) Chitosan as a lipid binder: a langmuir monolayer study of chitosan-lipid interactions. Biomacromolecules 8:2611–2617
25. Ebihara K, Schneeman BO (1989) Interaction of bile acids, phospholipids, cholesterol and triglyceride with dietary fibers in the small intestine of rats. J Nutr 119:1100–1106
26. Walsh AM, Sweeney T, Bahar B, O'Doherty JV (2013) Multi-functional roles of chitosan as a potential protective agent against obesity. PLoS One 8:e53828
27. Cho EJ, Rahman MA, Kim SW, Baek YM, Hwang HJ, Oh JY, Hwang HS, Lee SH, Yun JW (2008) Chitosan oligosaccharides inhibit adipogenesis in 3T3-L1 adipocytes. J Microbiol Biotechnol 18:80–87
28. Bahar B, O'Doherty JV, O'Doherty AM, Sweeney T (2013) Chito-oligosaccharide inhibits the de-methylation of a 'CpG' Island within the leptin (LEP) promoter during adipogenesis of 3T3-L1 cells. PLoS One 8. Article e60011
29. Karadeniz F, Kim SK (2014) Antidiabetic activities of chitosan and its derivatives: a mini review. In: Kim S (ed) Marine carbohydrates: fundamentals and applications, part B. Elsevier Inc, Oxford, pp 15–31
30. Zhao L, Sun T, Wang L (2017) Chitosan oligosaccharide improves the therapeutic efficacy of sitagliptin for the therapy of Chinese elderly patients with type 2 diabetes mellitus. Ther Clin Risk Manag 13:739–750
31. Kim HJ, Ahn HY, Kwak JH, Shin DY, Kwon YI, Oh CG, Lee JH (2014) The effects of chitosan oligosaccharide (GO2KA1) supplementation on glucose control in subjects with prediabetes. Food Funct 5:2662–2669
32. Liu SH, Chang YH, Chiang MT (2010) Chitosan reduces gluconeogenesis and increases glucose uptake in skeletal muscle in streptozotocin-induced diabetic rats. J Agric Food Chem 58:5795–5800
33. Hsieh YL, Yao HT, Cheng RS, Chiang MT (2012) Chitosan reduces plasma adipocytokines and lipid accumulation in liver and adipose tissues and ameliorates insulin resistance in diabetic rats. J Med Food 15:453–460
34. Prajapati B, Rajput P, Jena PK, Seshadri S (2015) Investigation of chitosan for prevention of diabetic progression through gut microbiota alteration in sugar rich diet induced diabetic rats. Curr Pharm Biotechnol 17:173–184
35. Zhang Y, Wei W, Lv P, Wang L, Ma G (2011) Preparation and evaluation of alginate-chitosan microspheres for oral delivery of insulin. Eur J Pharm Biopharm 77:11–19
36. Mumuni MA, Kenechukwu F, Ofokansi KC, Attama AA, Díaz DD (2020) Insulin-loaded mucoadhesive nanoparticles based on mucin-chitosan complexes for oral delivery and diabetes treatment. Carbohydr Polym 229:115506
37. Wang J, Kong M, Zhou Z, Yan D, Yu X, Cheng X, Feng C, Liu Y, Chen X (2017) Mechanism of surface charge triggered intestinal epithelial tight junction opening upon chitosan nanoparticles for insulin oral delivery. Carbohydr Polym 157:596–602

38. Li L, Yang L, Li M, Zhang L (2017) A cell-penetrating peptide mediated chitosan nanocarriers for improving intestinal insulin delivery. Carbohydr Polym 174:182–189
39. Lee JH, Sahu A, Choi WI, Lee JY, Tae G (2016) ZOT-derived peptide and chitosan functionalized nanocarrier for oral delivery of protein drug. Biomaterials 103:160–169
40. Izawa H, Yagi A, Umemoto R, Ifuku S (2023) Water-soluble guanidinylated chitosan: a candidate material for protein delivery systems. Polym J 55:885–895
41. Khan NF, Nakamura H, Izawa H, Ifuku S, Kadowaki D, Otagiri M, Anraku M (2023) Evaluation of the safety and gastrointestinal migration of guanidinylated chitosan after oral administration to rats. J Funct Biomater 14:340
42. Eddy AA, Neilson EG (2006) Chronic kidney disease progression. J Am Soc Nephrol 17: 2964–2966
43. Jha V, Garcia-Garcia G, Iseki K (2013) Chronic kidney disease: global dimension and perspectives. Lancet 382:208
44. Wada J, Makino H (2013) Inflammation and the pathogenesis of diabetic nephropathy. Clin Sci (Lond) 124:139–152
45. de Boer IH, Khunti K, Sadusky T, Tuttle KR, Neumiller JJ, Rhee CM, Rosas SE, Rossing P, Bakris G (2022) Diabetes management in chronic kidney disease: a consensus report by the American Diabetes Association (ADA) and kidney disease: improving global outcomes (KDIGO). Diabetes Care 45:3075–3090
46. Jing SB, Li L, Ji D, Takiguchi Y, Yamaguchi T (1997) Effect of chitosan on renal function in patients with chronic renal failure. J Pharm Pharmacol 49:721–723
47. Anraku M, Tanaka M, Hiraga A, Nagumo K, Imafuku T, Maezaki Y, Iohara D, Uekama K, Watanabe H, Hirayama F, Maruyama T, Otagiri M (2014) Effects of chitosan on oxidative stress and related factors in hemodialysis patients. Carbohydr Polym 112:152–157
48. Anraku M, Tabuchi R, Ifuku S, Nagae T, Iohara D, Tomida H, Uekama K, Maruyama T, Miyamura S, Hirayama F, Otagiri M (2017) An oral absorbent, surface-deacetylated chitin nano-fiber ameliorates renal injury and oxidative stress in 5/6 nephrectomized rats. Carbohydr Polym 161:21–25
49. Biasibetti E, Martello E, Bigliati M, Biasato I, Cocca T, Bruni N, Capucchio MT (2018) A long term feed supplementation based on phosphate binders in feline chronic kidney disease. Vet Res Commun 42:161–167
50. Yoshimoto H, Jing S, Yamaguchi T (1995) Oral sorbents for acidic substances in uremic toxins. J Chem Soc Jpn 6:486–489
51. Yoon HJ, Moon ME, Park HS, Kim HW, Im SY, Lee JH, Kim YH (2008) Effects of chitosan oligosaccharide (COS) on the glycerol-induced acute renal failure in vitro and in vivo. Food Chem Toxicol 46:710–716
52. Yin S, Liu J, Zhao X, Dong H, Cao Y, Zhang S, Dong X, Zhang G, Jin S, Shi W, Xiang F, Gong J, Wang Z, Han B, Zhang N, Tan X, Wang Z (2023) Chitosan oligosaccharide attenuates acute kidney injury and renal interstitial fibrosis induced by ischemia-reperfusion. Ren Fail 45: 2238831
53. Wu J, Xu Y, Geng Z, Zhou J, Xiong Q, Xu Z, Li H, Han Y (2022) Chitosan oligosaccharide alleviates renal fibrosis through reducing oxidative stress damage and regulating TGF-beta1/ Smads pathway. Sci Rep 12:19160
54. Zhang Q, Li Y, Wang S, Gu D, Zhang C, Xu S, Fang X, Li C, Wu H, Xiong W (2024) Chitosan-based oral nanoparticles as an efficient platform for kidney-targeted drug delivery in the treatment of renal fibrosis. Int J Biol Macromol 256(Pt 1):128315
55. Liu C, Pang M, Wang Q, Yan M, Zhou Y, Yao H, Du B (2024) Intestinal absorption of nanoparticles to reduce oxidative stress and vasoconstriction for treating diabetic nephropathy. ACS Biomater Sci Eng 10:1517–1529
56. Yuan ZX, Zhang ZR, Zhu D, Sun X, Gong T, Liu J, Luan CT (2009) Specific renal uptake of randomly 50% N-acetylated low molecular weight chitosan. Mol Pharm 6:305–314
57. Yuan ZX, Sun X, Gong T, Ding H, Fu Y, Zhang ZR (2007) Randomly 50% N-acetylated low molecular weight chitosan as a novel renal targeting carrier. J Drug Target 15:269–278

58. Glassner KL, Abraham BP, Quigley EMM (2020) The microbiome and inflammatory bowel disease. J Allergy Clin Immunol 145:16–27
59. Ungaro R, Mehandru S, Allen PB, Peyrin-Biroulet L, Colombel JF (2017) Ulcerative colitis. Lancet 389:1756–1770
60. Gibson PR (2017) Use of the low-FODMAP diet in inflammatory bowel disease. J Gastroenterol Hepatol 32:40–42
61. Ng SC, Shi HY, Hamidi N, Underwood FE, Tang W, Benchimol EI, Panaccione R, Ghosh S, Wu JCY, Chan FKL, Sung JJY, Kaplan GG (2017) Worldwide incidence and prevalence of inflammatory bowel disease in the 21st century: a systematic review of population-based studies. Lancet 390:2769–2778
62. GBD 2017 Inflammatory bowel disease collaborators, Alatab S, Sepanlou SG, Ikuta K et al (2020) The global, regional, and national burden of inflammatory bowel disease in 195 countries and territories, 1990–2017: a systematic analysis for the global burden of disease study 2017. Lancet Gastroenterol Hepatol 5:17–30
63. Molodecky NA, Soon IS, Rabi DM, Ghali WA, Ferris M, Chernoff G, Benchimol EI, Panaccione R, Ghosh S, Barkema HW, Kaplan GG (2012) Increasing incidence and prevalence of the inflammatory bowel diseases with time, based on systematic review. Gastroenterology 142:46–54
64. Peyrin-Biroulet L, Sandborn W, Sands BE, Reinisch W, Bemelman W, Bryant RV, D'Haens G, Dotan I, Dubinsky M, Feagan B, Fiorino G, Gearry R, Krishnareddy S, Lakatos PL, Loftus Jr EV, Marteau P, Munkholm P, Murdoch TB, Ordás I, Panaccione R, Riddell RH, Ruel J, Rubin DT, Samaan M, Siegel CA, Silverberg MS, Stoker J, Schreiber S, Travis S, Van Assche G, Danese S, Panes J, Bouguen G, O'Donnell S, Pariente B, Winer S, Hanauer S, Colombel JF (2015) Selecting therapeutic targets in inflammatory bowel disease (STRIDE): determining therapeutic goals for treat-to-target. Am J Gastroenterol 110:1324–1338
65. Chen M, Ding Y, Tong Z (2020) Efficacy and safety of Sophora flavescens (Kushen) based traditional Chinese medicine in the treatment of ulcerative colitis: clinical evidence and potential mechanisms. Front Pharmacol 11:603476
66. Liu Y, Alookaran JJ, Rhoads JM (2018) Probiotics in autoimmune and inflammatory disorders. Nutrients 10:1537
67. Jadhav P, Jiang Y, Jarr K, Layton C, Ashouri JF, Sinha SR (2020) Efficacy of dietary supplements in inflammatory bowel disease and related autoimmune diseases. Nutrients 12: 2156
68. Azuma K, Osaki T, Kurozumi S, Kiyose M, Tsuka T, Murahata Y, Imagawa T, Itoh N, Minami S, Sato K, Okamoto Y (2015) Anti-inflammatory effects of orally administered glucosamine oligomer in an experimental model of inflammatory bowel disease. Carbohydr Polym 115:448–456
69. Yousef M, Pichyangkura R, Soodvilai S, Chatsudthipong V, Muanprasat C (2012) Chitosan oligosaccharide as potential therapy of inflammatory bowel disease: therapeutic efficacy and possible mechanisms of action. Pharmacol Res 66:66–79
70. Mei Z, Huang X, Zhang H, Cheng D, Xu X, Fang M, Hu J, Liu Y, Liang Y, Mei Y (2022) Chitin derivatives ameliorate DSS-induced ulcerative colitis by changing gut microbiota and restoring intestinal barrier function. Int J Biol Macromol 202:375–387
71. Azuma K, Osaki T, Ifuku S, Saimoto H, Tsuka T, Imagawa T, Okamoto Y, Minami S (2012) α-Chitin nanofibrils improve inflammatory and fibrosis responses in inflammatory bowel disease mice model. Carbohydr Polym 90:197–200
72. Hejazi R, Amiji M (2003) Chitosan-based gastrointestinal delivery systems. J Control Release 89:151–165
73. Omwancha WS, Mallipeddi R, Valle BL, Neau SH (2013) Chitosan as a pore former in coated beads for colon specific drug delivery of 5-ASA. Int J Pharm 441:343–351
74. Palma E, Costa N, Molinaro R, Francardi M, Paolino D, Cosco D, Fresta M (2019) Improvement of the therapeutic treatment of inflammatory bowel diseases following rectal

administration of mesalazine-loaded chitosan microparticles vs asamax®. Carbohydr Polym 212:430–438
75. Zhang X, Ma Y, Ma L, Zu M, Song H, Xiao B (2019) Oral administration of chondroitin sulfate-functionalized nanoparticles for colonic macrophage-targeted drug delivery. Carbohydr Polym 223:115126
76. Huang Y, Guo J, Gui S (2018) Orally targeted galactosylated chitosan poly(lactic-co-glycolic acid) nanoparticles loaded with TNF-α siRNA provide a novel strategy for the experimental treatment of ulcerative colitis. Eur J Pharm Sci 125:232–243
77. Cao M, Li H, Sun D, Chen W (2020) Cancer burden of major cancers in China: a need for sustainable actions. Cancer Commun (Lond) 40:205–210
78. Miller KD, Nogueira L, Mariotto AB, Rowland JH, Yabroff KR, Alfano CM et al (2019) Cancer treatment and survivorship statistics. CA Cancer J Clin 69:363–385
79. Harvey AL, Edrada-Ebel RA, Quinn RJ (2015) The re-emergence of natural products for drug discovery in the genomics era. Nat Rev Drug Discov 14:111–129
80. Ma JX, Qian L, Zhou Y (2015) Stimulation effect of chitosan on the immunity of radiotherapy patients suffered from lung cancer. Int J Biol Macromol 72:195–198
81. Yeh MY, Wu MF, Shang HS, Chang JB, Shih YL, Chen YL, Hung HF, Lu HF, Yeh C, Wood WG, Hung FM, Chung JG (2013) Effects of chitosan on xenograft models of melanoma in C57BL/6 mice and hepatoma formation in SCID mice. Anticancer Res 33:4867–4873
82. Zou P, Yuan S, Yang X, Zhai X, Wang J (2018) Chitosan oligosaccharides with degree of polymerization 2-6 induces apoptosis in human colon carcinoma HCT116 cells. Chem Biol Interact 279:129–135
83. Mattaveewong T, Wongkrasant P, Chanchai S, Pichyangkura R, Chatsudthipong V, Muanprasat C (2016) Chitosan oligosaccharide suppresses tumor progression in a mouse model of colitis-associated colorectal cancer through AMPK activation and suppression of NF-κB and mTOR signaling. Carbohydr Polym 145:30–36
84. Masuda S, Azuma K, Kurozumi S, Kiyose M, Osaki T, Tsuka T, Itoh N, Imagawa T, Minami S, Sato K, Okamoto Y (2014) Anti-tumor properties of orally administered glucosamine and N-acetyl-d-glucosamine oligomers in a mouse model. Carbohydr Polym 111:783–787
85. Maeda Y, Kimura Y (2004) Antitumor effects of various low-molecular-weight chitosans are due to increased natural killer activity of intestinal intraepithelial lymphocytes in sarcoma 180-bearing mice. J Nutr 134:945–950
86. Zou P, Yuan S, Yang X, Guo Y, Li L, Xu C, Zhai X, Wang J (2019) Structural characterization and antitumor effects of chitosan oligosaccharides against orthotopic liver tumor via NF-κB signaling pathway. J Funct Foods 57:157–165
87. Almeida A, Castro F, Resende C, Lúcio M, Schwartz Jr S, Sarmento B (2022) Oral delivery of camptothecin-loaded multifunctional chitosan-based micelles is effective in reduce colorectal cancer. J Control Release 349:731–743
88. Lin YH, Chen ZR, Lai CH, Hsieh CH, Feng CL (2015) Active targeted nanoparticles for oral administration of gastric cancer therapy. Biomacromolecules 16:3021–3032
89. Jayakumar R, Nair A, Rejinold SN, Nair MSV (2012) Doxorubicin-loaded pH-responsive chitin nanogels for drug delivery to cancer cells. Carbohydr Polym 87:2352–2356
90. Wang J, Chin D, Poon C, Mancino V, Pham J, Li H, Ho PY, Hallows KR, Chung EJ (2021) Oral delivery of metformin by chitosan nanoparticles for polycystic kidney disease. J Control Release 329:1198–1209
91. Feng C, Wang Z, Jiang C, Kong M, Zhou X, Li Y, Cheng X, Chen X (2013) Chitosan/o-carboxymethyl chitosan nanoparticles for efficient and safe oral anticancer drug delivery: in vitro and in vivo evaluation. Int J Pharm 457:158–167
92. Cohn L, Elias JA, Chupp GL (2004) Asthma: mechanisms of disease persistence and progression. Annu Rev Immunol 22:789–815
93. Medoff BD, Thomas SY, Luster AD (2008) T cell trafficking in allergic asthma: the ins and outs. Annu Rev Immunol 26:205–232

94. Chung MJ, Park JK, Park YI (2012) Anti-inflammatory effects of low-molecular weight chitosan oligosaccharides in IgE-antigen complex-stimulated RBL-2H3 cells and asthma model mice. Int Immunopharmacol 12:453–459
95. Adhikari A, Sen K, Biswas S, Sengupta M (2012) Role of imunomet in upper respiratory tract diorder: a randomized double blind placebo controlled clinical trial. Int J Res Ayurveda Pharm 3:505–507
96. Singh RP, Gangadharappa HV, Mruthunjaya K (2018) Phytosome complexed with chitosan for gingerol delivery in the treatment of respiratory infection: in vitro and in vivo evaluation. Eur J Pharm Sci 122:214–229
97. Pandey A, Raja AN (2020) Recent development in chitosan-based electrochemical sensors and its sensing application. Int J Biol Macromol 164:4231–4244
98. Berezin AS, Skorik YA (2015) Chitosan-isoniazid conjugates: synthesis, evaluation of tuberculostatic activity, biodegradability and toxicity. Carbohydr Polym 127:309–315
99. Del Prado-Audelo ML, Caballero-Florán IH, Sharifi-Rad J, Mendoza-Muñoz N, González-Torres M, Urbán-Morlán Z, Florán B, Cortes H, Leyva-Gómez G (2020) Chitosan-decorated nanoparticles for drug delivery. J Drug Deliv Sci Technol 59:101896
100. Zhao J, Li J, Jiang Z, Tong R, Duan X, Bai L, Shi J (2020) Chitosan, N, N, N-trimethyl chitosan (TMC) and 2-hydroxypropyltrimethyl ammonium chloride chitosan (HTCC): the potential immune adjuvants and nano carriers. Int J Biol Macromol 154:339–348
101. Rasul RM, Tamilarasi Muniandy M, Zakaria Z, Shah K, Chee CF, Dabbagh A, Rahman NA, Wong TW (2020) A review on chitosan and its development as pulmonary particulate anti-infective and anti-cancer drug carriers. Carbohydr Polym 250:116800
102. Liu X, Ma L, Mao Z, Gao C (2011) Chitosan-based biomaterials for tissue repair and regeneration. Adv Polym Sci 244:81–128
103. du Toit LC, Pillay V, Danckwerts MP (2006) Tuberculosis chemotherapy: current drug delivery approaches. Respir Res 7:118
104. Oliveira PM, Matos BN, Pereira PAT, Gratieri T, Faccioli LH, Cunha-Filho MSS, Gelfuso GM (2017) Microparticles prepared with 50-190 kDa chitosan as promising non-toxic carriers for pulmonary delivery of isoniazid. Carbohydr Polym 174:421–431
105. Rawal T, Parmar R, Tyagi RK, Butani S (2017) Rifampicin loaded chitosan nanoparticle dry powder presents an improved therapeutic approach for alveolar tuberculosis. Colloids Surf B Biointerfaces 154:321–330
106. Rosière R, Van Woensel M, Gelbcke M, Mathieu V, Hecq J, Mathivet T, Wauthoz N (2018) New folate-grafted chitosan derivative to improve delivery of paclitaxel-loaded solid lipid nanoparticles for lung tumor therapy by inhalation. Mol Pharm 15:899
107. Dai X, Hou W, Sun Y, Gao Z, Zhu S, Jiang Z (2015) Chitosan oligosaccharides inhibit/disaggregate fibrils and attenuate amyloid beta-mediated neurotoxicity. Int J Mol Sci 16(5):10526–10536
108. Goto M, Iohara D, Michihara A, Ifuku S, Azuma K, Kadowaki D, Maruyama T, Otagiri M, Hirayama F, Anraku M (2020) Effects of surface-deacetylated chitin nanofibers on non-alcoholic steatohepatitis model rats and their gut microbiota. Int J Biol Macromol 164:659–666
109. Chiu CY, Feng SA, Liu SH, Chiang MT (2017) Functional comparison for lipid metabolism and intestinal and fecal microflora enzyme activities between low molecular weight chitosan and chitosan oligosaccharide in high-fat-diet-fed rats. Mar Drugs 15:234
110. Zhang X, Yang H, Zheng J, Jiang N, Sun G, Bao X, Lin A, Liu H (2021) Chitosan oligosaccharides attenuate loperamide-induced constipation through regulation of gut microbiota in mice. Carbohydr Polym 253:117218
111. Park PJ, Je JY, Kim SK (2003) Angiotensin I converting enzyme (ACE) inhibitory activity of hetero-chitooligosaccharides prepared from partially different deacetylated chitosans. J Agric Food Chem 51:4930–4934

112. Pal K, Bharti D, Sarkar P, Anis A, Kim D, Chałas R, Maksymiuk P, Stachurski P, Jarzębski M (2021) Selected applications of chitosan composites. Int J Mol Sci 22(20):10968
113. Tomida H, Fujii T, Furutani N, Michihara A, Yasufuku T, Akasaki K, Maruyama T, Otagiri M, Gebicki JM, Anraku M (2009) Antioxidant properties of some different molecular weight chitosans. Carbohydr Res 344:1690–1696

Adv Polym Sci (2025) 295: 339–366
https://doi.org/10.1007/12_2024_189

Published online: 17 January 2025

Chitosan as Nutraceuticals: Transforming Pharmaceutical Research

Anjali K. I., Aparna Ann Mathew, Swati Gupta, and Sabitha M.

Contents

1 Introduction 340
2 Biological Application of Chitosan 341
2.1 Biomedical Application 341
2.2 Drug Delivery 342
3 Nutraceutical Values of Chitosan 343
3.1 Antimicrobial Properties 343
3.2 Anti-Inflammatory and Immunomodulatory Properties 345
3.3 Antioxidant Effect 347
3.4 Hypolipidemic and Hypocholesterolemic 348
3.5 Antitumor Activity 349
3.6 Antihypertensive Action 356
3.7 Antihypersensitive 358
4 Current Status of Chitosan-Based Nutraceuticals 359
5 Future Prospective 359
6 Summary 360
References 361

Abstract Chitosan, a natural biopolymer derived from chitin, has emerged as a promising nutraceutical with numerous pharmacological applications. Current developments in pharmaceutical research have focused on its ability to improve health and wellness through anti-inflammatory, immunomodulatory, antioxidant, hypolipidemic, hypocholesterolemic, antitumor, and antihypersensitivity properties. These properties make chitosan an ideal constituent for the development of nutraceutical agents with a variety of health benefits. Its biocompatibility,

Anjali K. I., Aparna Ann Mathew, Swati G., and Sabitha M. contributed equally to this work.

A. K. I., A. A. Mathew, S. Gupta, and Sabitha M. (✉)
Amrita School of Pharmacy, Amrita Institute of Medical Sciences and Research Centre, Amrita Vishwa Vidyapeetham, Kochi, India
e-mail: sabitham@aims.amrita.edu

biodegradability, and ability to form multiple complexes enhance its potential as a carrier material in drug delivery and use as an adjuvant in therapies. Furthermore, the increased demand for natural, sustainable, and functional food supplements has increased interest in chitosan-based formulations. This chapter discusses chitosan's numerous health benefits as a nutraceutical, its underlying mechanisms of action, and its transformative impact on pharmaceutical research and product development. The study of chitosan's bioactivities opens up new possibilities for nutraceutical innovation, propelling future research into targeted therapies and functional foods.

Keywords Biodegradable · Biomedical application · Carrier material · Chitosan · COVID-19 · Health benefits · Nutraceutical innovation · Product development

1 Introduction

Nutraceuticals are products derived from various sources, including flora and fauna, inorganic compounds, and microorganisms, that offer additional health benefits beyond the basic nutritional value [1]. These products may include dietary supplements, fortified foods, herbal products, isolated nutrients, diets based on genetically modified foods, processed foods, beverages, marine products, and other substances that contain concentrated forms of biologically active components [2]. One of the major limitations of nutraceuticals is the uncertain bioavailability or the body's ability to access nutrients following digestion, absorption, and transportation [3]. Identifying the genuine source of raw materials, ensuring purity, detecting the presence of other active compounds, assessing quality, addressing the lack of experimental evidence, combating false advertising, preventing heavy metal contamination, understanding interactions between supplements and drugs, and mitigating risks of self-prescription, misuse, or overuse are some of the reported challenges that must be addressed before they are used [3, 4]. Nutraceuticals typically contribute positively to improving the health and well-being of individuals [5].

The impact of nutraceuticals, probiotics, and supplements on reducing the risk of SARS-CoV-2 infection or alleviating COVID-19 symptoms has also been extensively investigated [6]. During the COVID-19 pandemic, the global nutraceutical market experienced significant growth, rising from USD 320 billion in 2020 to USD 352.92 billion by 2021 (Lordan 2021). It is projected to reach USD 658.11 billion by the end of 2028, with a compound annual growth rate exceeding 9%. The Asia-Pacific region, China, North America, South America, and India are anticipated to be the largest markets for nutraceuticals in the coming era [7, 8]. Nutraceuticals with immune-boosting, antiviral, antioxidant, and anti-inflammatory properties offer potential alternatives for both preventive and therapeutic support alongside standard COVID-19 treatments for adults and children [9–11]. Nutraceuticals are a leading trend in the healthcare industry. Various sources of nutraceuticals include products obtained from plants, animals, minerals, and microbes. Recently, marine sources have garnered significant interest from nutraceutical researchers. This is because marine life, which accounts for nearly 80% of the world's biota, is rich in bioactive compounds and secondary metabolites derived from marine invertebrates and many other marine organisms, which can contribute significantly to nutraceuticals in the coming era [12].

Chitosan has attracted considerable interest as a key nutraceutical component due to its biomedical applications and its role as a carrier for various nutraceutical and medicinal compounds. It is created by deacetylating chitin, a biopolymer made up of 2-acetamido-2-deoxy-D-glucose residues linked together by ß-(1-4) bonds. Chitin is sourced from the hard exoskeletons of crustaceans such as crabs, lobsters, and shrimp, which are often considered marine waste.

2 Biological Application of Chitosan

2.1 Biomedical Application

Chitosan possesses multiple biomedical applications like bone regeneration, wound healing, anti-inflammatory, antioxidant antibacterial, and antifungal activity. Chitosan and its derivatives mostly contribute to wound healing during the first three phases such as hemostasis, inflammation, and proliferation [13]. Furthermore, its significant antimicrobial characteristics, capacity to supply moisture and warmth to the affected region, cytocompatibility, immunomodulatory effects, promotion of angiogenesis, and facilitation of the release of various biological mediators such as platelet-derived growth factor (PDGF), transforming growth factor-beta (TGF-β), and vascular endothelial growth factor (VEGF) enhance the cellular proliferation and migration, which augment the advantageous effects of chitosan in the wound-healing process. Numerous chitosan-derived platforms have been developed and are utilized in wound-healing applications, either alone or in combination with various natural or synthetic biomaterials [14]. Furthermore, chitosan stimulates the formation of granulation tissue and the activity of fibroblast proliferation and suppresses fibrosis [15]. The chitosan-based hydrogels possess various properties like swelling, porosity, and mechanical characteristics that make them a great option for use as a scaffold for tissue regeneration [15]. Chitosan is highly suited for tissue engineering due to its low toxicity and nonimmunogenicity. Chitosan has been shown to boost all three crucial phases of tissue regeneration: cell adhesion, proliferation, and differentiation [13]. In case of bone regeneration, chitosan increases the formation of extracellular matrix (ECM) components, including collagen, which facilitates the bone growth and healing. Furthermore, its antibacterial characteristics lessen the risk of infection, and its biodegradable nature makes it a suitable option for long-term use [14].

The antibacterial and antifungal activity of chitosan is mediated by its interaction with the nucleic acid that interrupts the synthesis of vital proteins and enzymes followed by the damage of bacteria and fungi [15]. Antimicrobial studies have revealed that chitosan is effective against a wide range of microorganisms, including bacteria, yeasts, molds, viruses, and pathogenic microbes, including *Salmonella*, *Candida albicans*, *Escherichia coli*, *Staphylococcus aureus*, and *Aspergillus niger* [13]. The antioxidant activity of chitosan may promote the anti-inflammatory activity of chitosan. The antioxidant activity of chitosan results from its ability to scavenge reactive oxygen species (ROS), which have a role in both the onset and continuation of inflammation process [13]. It also exhibits antidiabetic, anticancer, and antilipidemic properties through various mechanisms, as outlined in Table 2. These characteristics enhance its potential for nutraceutical applications.

2.2 Drug Delivery

Drug delivery is a crucial aspect of modern medicine as it determines the efficacy and safety of a therapeutic substance [14]. Because of its biocompatibility and biodegradability, chitosan is certified by the FDA as "generally recognised as safe (GRAS)" and is approved for tissue engineering and drug delivery applications. LD_{50} value of chitosan was reported to be greater than 16 g/kg in mice after oral administration [14]. Additionally, its mucoadhesive nature and mechanical strength facilitate the application of chitosan in both conventional and novel drug delivery systems [16]. Chitosan has been the subject of extensive research as an excipient within the pharmaceutical sector, intended for applications in direct tablet compression, as a tablet disintegrant, facilitating the fabrication of controlled release solid dosage forms, or enhancing the dissolution properties of pharmaceutical compounds [17]. The mucoadhesive nature of chitosan promotes its residence time in the nose with improved bioavailability and facilitates target-specific drug delivery. Alsarra et al. developed mucoadhesive hydrogels using chitosan, polyvinyl pyrrolidone, and carbopol for the intranasal administration of acyclovir. The mucoadhesive evaluation on ovine nasal mucosa indicated that the chitosan with polyvinyl pyrrolidone has superior adhesion to the mucosal tissue. Chitosan-based therapeutic system is used for the delivery of multiple classes of drugs like anti-inflammatory, antibiotics, proteins, peptide-based drugs, and anticancer agents [16]. Chitosan-based nanoformulations were not only used for the invasive application like parenteral route but also for the noninvasive routes including mucosal (oral, buccal, periodontal), nasal, pulmonary, rectal, vaginal, and transdermal. Vesicular and nanoparticulate drug delivery systems are commonly used by parenteral routes either to transport the medication to the target region or to improve the survivability of these systems from phagocytosis. In the case of noninvasive methods, the goal is to improve mucosal adherence or maintain drug release. The cationic nature and limited solubility of chitosan in aqueous media facilitate the applications in developing drug delivery systems in both invasive parenteral and noninvasive mucosal routes, such as oral, buccal, periodontal, nasal, pulmonary, rectal, vaginal, and transdermal [16]. Drug-loaded chitosan can release pharmaceuticals via long-term sustained release, and thereby achieve the continuous drug release required to maintain the therapeutic window [18]. Chitosan has excellent qualities for application in oral delivery systems. Its cationic nature facilitates the interaction with negatively charged cell membranes, hence increasing cellular absorption. Unlike many water-insoluble polymers, chitosan may partially dissolve in water and form a hydrogel in watery settings. Its particular ability to mucoadhesion makes it a good candidate for oral medication delivery [14]. Chitosan-based transdermal formulations enable the delivery of hydrophilic drugs by facilitating the alteration of secondary structure of stratum corneum proteins, the opening of tight junctions in the granular layer through the redistribution of transmembrane proteins, and the enhancement of the stratum corneum's water content [14]. Owing to the spherical shape of the chitosan nanoparticles, they can accommodate hydrophobic drug inside the inner core and hydrophilic drug directly with its surface [18]. Chitosan-based scaffolds can release the drug based on its pH-sensitive and temperature-sensitive nature, and thus, it can find multiple applications like wound healing and tissue regeneration. The other biomedical applications include biosensing, bioimaging, anti-plaque agent in dentistry, and as an antibacterial agent in wound dressings [16]. Thus, its versatile biological properties, biodegradability, biocompatibility, and less toxic nature seek the attention of researchers to explore its nutraceutical potential. This chapter, therefore, deals with biological properties of chitosan as a potential nutraceutical.

3 Nutraceutical Values of Chitosan

3.1 Antimicrobial Properties

Chitosan is reported to have promising antimicrobial activity against different microorganisms such as gram-negative, gram-positive bacteria, and fungi. It demonstrates antimicrobial activity through multiple mechanisms, as depicted in Fig. 1. Positively charged chitosan molecules interact with negatively charged cell wall components of microorganisms, which may increase cell membrane permeability and lead to microorganism destruction [19]. A higher degree of interaction was observed between the gram-negative bacteria compared to gram-positive organisms [20]. Numerous factors like the extent of deacetylation, molecular mass, pH, concentration, and bacterial type also have a great impact on its antimicrobial activity [21]. Facchinatto et al. demonstrated the antibacterial activity of chitosan nanostructures with varying levels of conjugation and degree of deacetylation under the blue light activation, and it was found to have a promising bactericidal property against gram-negative bacteria; *Escherichia coli* as well as gram-positive *Staphylococcus aureus* species [22]. Wang et al. conducted a comparative antimicrobial study between the chitosan dressing and common gauze dressing on wounds after 8 days of surgery. Chitosan-based wound dressing decreased the growth of the most abundant species of bacteria (*Enterococcaceae*) in a notable range of up to 31.6% compared to common gauze dressing, 23.1%. It was concluded that pure

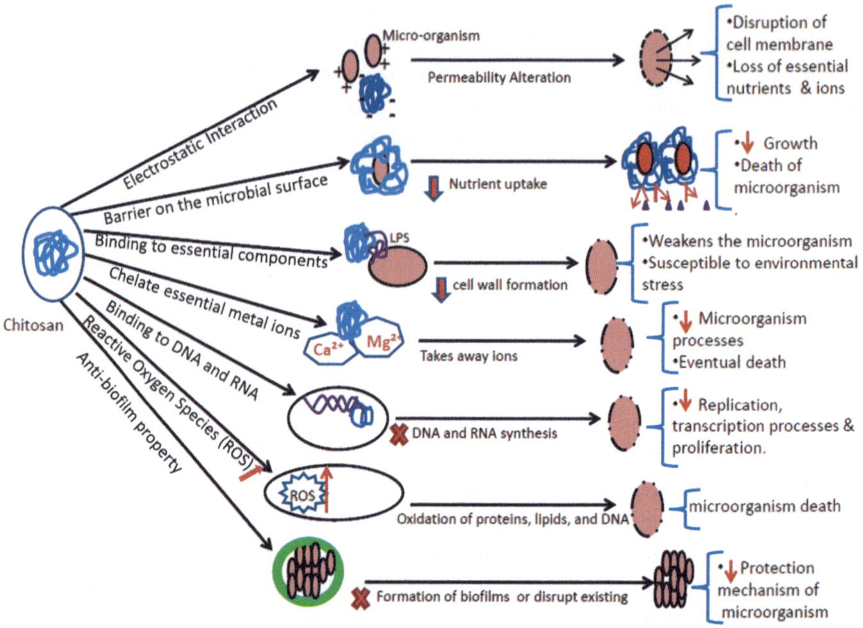

Fig. 1 Schematic depiction of mode of antimicrobial actions of chitosan. LPS: lipopolysaccharide; ROS: reactive oxygen species

chitosan exhibits a significant reduction in the microbial population with an effective wound-healing property [23]. Brilhante et al. investigated the antifungal ability of low molecular mass chitosan against the dimorphic fungus *Histoplasma capsulatum*. The results indicated that chitosan combined with classical antifungal drugs (amphotericin B or itraconazole) has a synergic effect, reducing chitosan's MICs by 32 times. Chitosan also reduced biomass and metabolic activity by about 40% at 512 μg/mL concentration, showing its promising effect *against Histoplasma capsulatum* [24].

Ahmed et al. conducted an antibacterial study of dietary chitosan nanostructures in the management of enteric redmouth disease (ERM) caused by the gram-negative bacteria *Yersinia ruckeri* strain in *Oncorhynchus mykiss* species of fish. In vitro antimicrobial ability of chitosan nanoparticles was evaluated against CSF007-82 strains of *Yersinia ruckeri* collected from the fish skin mucus. After 21 days of feeding, the minimum inhibitory concentration was found to be 4.28 ± 0.67 mm, two times greater than that of the untreated group, 2.54 ± 0.16 mm. It was also observed that the expression of the anti-inflammatory cytokine transforming growth factor-β and interleukin 1-β increased in ERM fish when treated with dietary chitosan nanoparticles compared to the untreated group. They concluded that chitosan nanoparticles activated the immunity responses of ERM fish compared to the untreated group of ERM fish [25].

Chi et al. developed chitosan microneedle array (CSMNA) patches integrated with smart responsive drug delivery for effective antibacterial activities and for promoting wound healing. In vitro antibacterial properties have been evaluated against gram-negative *Escherichia coli* and gram-positive *Staphylococcus aureus* bacterial strains. It showed that about 99% of both gram-negative and gram-positive bacteria populations were killed after 24 h of incubation with 4% concentration of chitosan microneedle arrays. This indicated the potential role of chitosan in the treatment of bacterial infections and enhanced wound-healing ability [26].

3.2 Anti-Inflammatory and Immunomodulatory Properties

Chitosan and its derivatives exhibit anti-inflammatory and immunomodulatory activities through various mechanisms, as shown in Fig. 2. Primarily, they suppress proinflammatory cytokines and mediators; interleukin-1β (IL-1β), tumor necrosis factor-alpha (TNF-α), interleukin-6 (IL-6), and nuclear factor kappa-B (NK-κB). Immunomodulatory and anti-inflammatory properties of chitosan mainly depend on the extent of deacetylation and the molecular mass [27–30]. Mohyuddin et al. studied the anti-inflammatory functions of chitosan (300–1,000 mg/kg/day) in

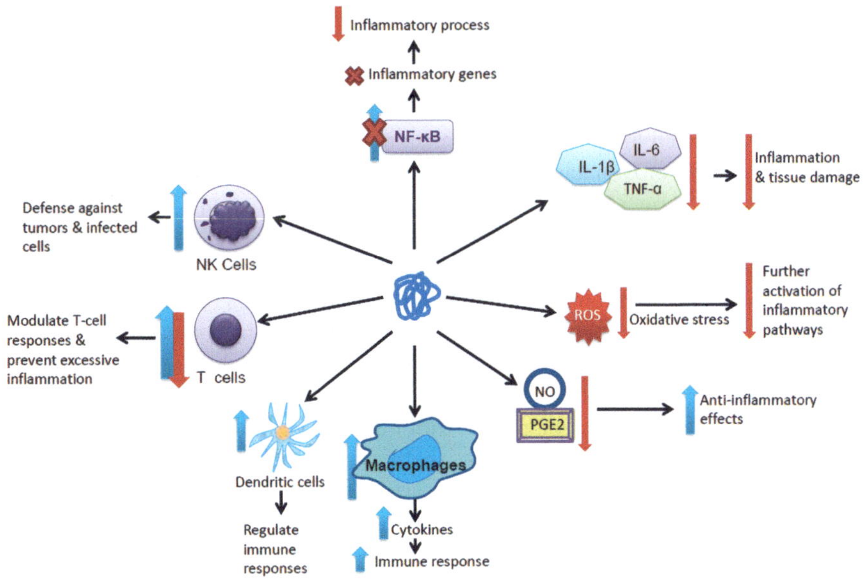

Fig. 2 Anti-inflammatory and immunomodulatory mode of actions of chitosan. Tumor necrosis factor-alpha (TNF-α), interleukin-6 (IL-6), nuclear factor kappa-B (NK-κB), and interleukin-1β (IL-1β). ROS: reactive oxygen species; PGE2: prostaglandin E2; NK cells: natural killer cells

heat-stressed male mice and compared it with control (heat-stressed animal without chitosan administration). Results demonstrated significant improvement in the colon histopathology, white blood cells, and lymphocyte count in the chitosan-treated animals. Plasma concentrations of TNF-α, interleukin-10, and heat shock protein-70 are substantially reduced in the treatment group, which indicates the potential anti-inflammatory activity of chitosan [31]. Davydova et al. investigated the immunotropic properties of chitosan and its polyelectrolyte complexes (PEC) with κ-carrageenan (κ-CGN) in male mice. Mice were exposed to inflammatory conditions with histamine, concanavalin A, and sheep erythrocyte immunization. The result showed a significant reduction in the inflammatory edema index (IEI) and enhanced anti-inflammatory activity in different test groups of chitosan, κ-carrageenan (κ-CGN), and their polyelectrolyte complexes (PEC) compared with the control group. PEC, κ-CGN, and CH showed the highest activity during histamine-induced exudative inflammation, which was linked to the stimulation of phagocytic cells like macrophages and neutrophils [32]. Tsvetkov et al. investigated biotin-modified chitosan oligosaccharide with 3,5, and 7 glucosamine components for immunomodulatory and anti-inflammatory properties. Chitosan oligosaccharides at various concentrations were incubated with the RAW 264.7 macrophage cell line for 24 and 48 h, and there was a significant increase in the major anti-inflammatory cytokine interleukin-10 and tumor necrosis factor-alpha. It was concluded that synthesized chito-oligosaccharides confirmed immunoregulatory activities [33]. Jhundoo et al. investigated chitosan's role in effectively managing inflammatory bowel disease. Pure chitosan with different molecular weights (MW) and degree of deacetylation (DD) and chitosan combined with 5-amino salicylic acid (5-ASA) were tested for pharmacological effects in a colitis model using CD-1 male Swiss mice. Both chitosan and 5-ASA at a dose of 30 mg/kg administered intrarectally significantly reduced the release of proinflammatory cytokines interleukin-6, interleukin-1β, tumor necrosis factor-α, as well as the effects of alkaline phosphate (ALP) and myeloperoxidase (MPO) compared to the control group. This confirmed chitosan's ability to suppress inflammatory conditions, independent of its structural properties (DD and MW) [34].

Immunomodulatory properties of chitosan oligosaccharides (COSs) with varying molecular mass have been investigated by Wu et al. in the head kidney macrophages isolated from blunt snout bream (*Megalobrama amblycephala*). These macrophages were stimulated by different molecular weights (~500 Da, ~1,000 Da, and 2,000–3,000 Da) COSs and examined with quantitative real-time PCR, transcriptional analysis, and RNA sequencing studies for immunopotentiation effect of COSs. The findings revealed that COSs activate the immune system to protect against pathogen invasion. Nevertheless, prolonged stimulation and increased molecular weight (MW) trigger macrophage apoptosis, possibly due to abnormal DNA replication and heightened inflammation [35].

3.3 *Antioxidant Effect*

The antioxidant activity of chitosan is widely recognized, making the biopolymer and its derivatives a subject of interest for researchers across various fields [36]. Numerous in vitro and in vivo studies have demonstrated that chitosan exerts redox-regulatory activity by inhibiting ROS production, preventing lipid oxidation through a significant reduction in serum-free fatty acids and malondialdehyde levels, and enhancing intracellular antioxidant enzymes in biological systems as represented in Fig. 3 [37, 38]. Marine chitosan is commonly incorporated into various dietary supplements due to its content of dietary fibers, nontoxicity, and high biocompatibility. Many research have shown that marine chitosan products enhance antioxidant activity or reduce oxidative stress. Xu et al. conducted a study in which sturgeons were fed a chitosan-supplemented diet (3.00 g chitosan/kg) for 8 weeks, followed by infection with *Aeromonas hydrophila*, to evaluate the effects of dietary chitosan on growth, antioxidant activity, immunity, intestinal morphology, and disease resistance. The results showed significant improvements in body weight gain, specific growth rate, and feed conversion ratio in chitosan-fed sturgeons compared to those on a control diet. Oral chitosan administration also significantly enhanced the activities of acid phosphatase, alkaline phosphatase, lysozyme, myeloperoxidase, superoxide dismutase, glutathione peroxidase, and catalase, along with increased resistance to *A. hydrophila* [39]. Abdel-Tawwab et al. studied the protective effects of dietary garlic (30 g/kg diet) and/or chitosan (10 g/kg diet) administered three times daily for 4 weeks against the toxic impact of feed-borne zearalenone (ZEN) on European seabass. Antioxidative status, immunity, and transcriptomic responses were measured during the investigation. Dietary

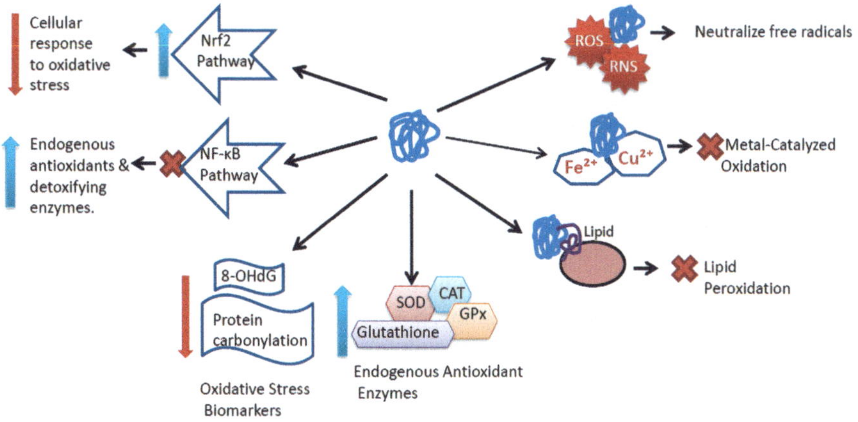

Fig. 3 Antioxidant mode of actions of chitosan. ROS: reactive oxygen species; RNS: reactive nitrogen species; SOD: superoxide dismutase; CAT: catalase; GPx: glutathione peroxidase; 8-OHdG: 8-hydroxy-2′-deoxyguanosine; NK-κB: nuclear factor kappa-B; Nrf2: nuclear factor erythroid 2-related factor 2

supplementation with garlic and chitosan significantly reduced ZEN-induced oxidative stress, restored total immunoglobulin levels that had been reduced during ZEN-induced immunosuppression, and altered the transcriptomic responses of ZEN-exposed fish. Furthermore, dietary supplementation with both feed additives produced better results than either alone [40]. Abdel et al. studied the effects of chitosan-*Ocimum basilicum* nanocomposite (COBN) on Nile tilapia (*Oreochromis niloticus*) growth, stress and antioxidant status, immune-related parameters, and gene expression. For 60 days, fish were fed diets with varying concentrations of COBN. After the trial, the fish were exposed to pathogenic bacteria (*Aeromonas sobria*) and yeast (*Candida albicans*) infections. The findings showed that in comparison to the control (COBN0), a dietary intervention with COBN promoted growth, intestinal architecture, immunity, and antioxidant markers, as well as protect *O. niloticus* against *A. sobria* and *C. albicans* infection. It was also determined that COBN, at a dose of 2 g/kg, could be used as a food additive in the sustainable aquaculture industry [41].

Lochi et al. investigated the effect of selenium nanoparticles and chitosan on production performance and antioxidant integrity of heat-stressed broilers. The broilers given nano selenium with chitosan (0.3 mg/kg) had the highest live body weight, feed intake, intestinal histomorphology, relative organ weight, and antioxidant parameters by the 42nd day of the study, whereas broilers fed a basal diet were the lowest. The study concluded that heat stress leads to significant decreases in performance, poor intestinal morphology, and reduced body weight in poultry, but regular supplementation with selenium nanoparticles and chitosan improves body weight, antioxidant status, and intestinal health [42].

3.4 Hypolipidemic and Hypocholesterolemic

Chitosan shows beneficial effects as a dietary supplement by lowering plasma cholesterol and triglyceride levels in both the blood and liver by various mechanisms as represented in Fig. 4. This effect is attributed to its ability to bind dietary lipids, which reduces intestinal lipid absorption and increases the excretion of total cholesterol and triglycerides through feces [43]. The hypolipidemic effect of chitosan may also be attributed to its disruption of enterohepatic bile acid circulation [44].

Zalaqi et al. investigated the interventional effects of marine chitosan (CS), swimming training (ST), and their combination (CS + ST) in mice fed high-fat diets (HFD). After an 8-week study, the CS + ST group outperformed the ST or CS alone in controlling body weight and inhibiting OS by increasing antioxidant capacity in HFD-fed mice [45]. Al-Okbi et al. evaluated the effectiveness of nutraceuticals containing chitosan, beta-sitosterol, and/or ferulic acid, as well as their nanostructures, in treating metabolic syndrome (MS) in rats and their interaction with atorvastatin. Rats were fed a high-fat, fructose-rich diet (HFFD) and then treated with various nutraceuticals, atorvastatin, or combinations of atorvastatin and nutraceuticals. Control groups included rats on a balanced diet and those fed HFFD

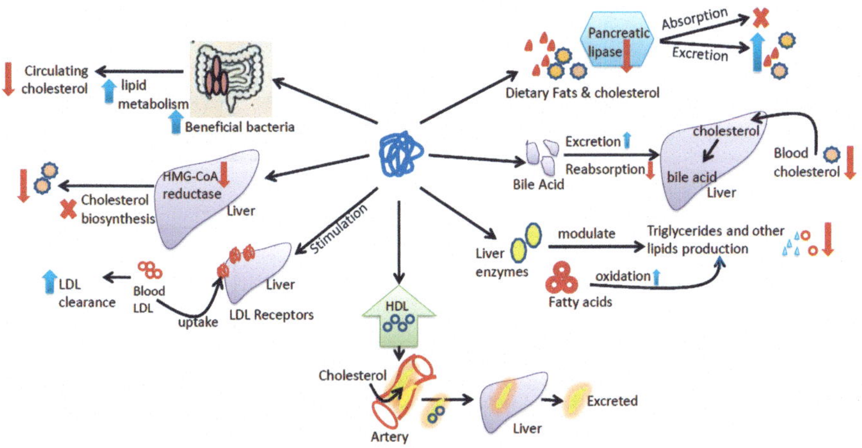

Fig. 4 Hypolipidemic and hypocholesterolemic mode of actions of chitosan. HDL: high-density lipoprotein; LDL: low-density lipoprotein

without treatment. Key metrics such as plasma glucose, lipid profile, aminotransferase activity, total antioxidant capacity, malondialdehyde, urea, creatinine, insulin, high-sensitivity C-reactive protein, adiponectin, and insulin resistance were measured. Liver fat content and histopathology were also analyzed. The study found that nutraceuticals containing chitosan and ferulic acid, in both standard and nanostructured forms, demonstrated superior efficacy [46]. Wang et al. studied the effects of various molecular weight chitosan (CS) and its quaternary ammonium salt (HACC) on rats fed high-fat diets. The results revealed that CS and HACC could reduce dietary fat and cholesterol absorption in vivo, thus effectively treating hyperlipidemia and fatty liver in rats [47]. Chiu et al. studied the effects of high and low molecular weight chitosan and chitosan-oligosaccharide on cholesterol regulation in rats fed a high-fat (HF) diet. In rats fed the HF diet, an 8-week 5% diet supplementation of each compound resulted in hypocholesterolemia. However, chitosan oligosaccharide was found to cause liver damage, as evidenced by increased hepatic cholesterol accumulation and intestinal cholesterol uptake. Both high and low molecular weight chitosan were found to reduce hypercholesterolemia and improve cholesterol homeostasis by modulating the hepatic and intestinal cholesterol regulators [48].

3.5 *Antitumor Activity*

Chitosan is a biocompatible polysaccharide with a low toxicity profile. The selective nature of chitosan exhibits different cytotoxic mechanisms in cancer cells compared with normal cells. This might be attributed by the electrostatic interaction between highly negatively charged cancer cells and the positively charged amino group of

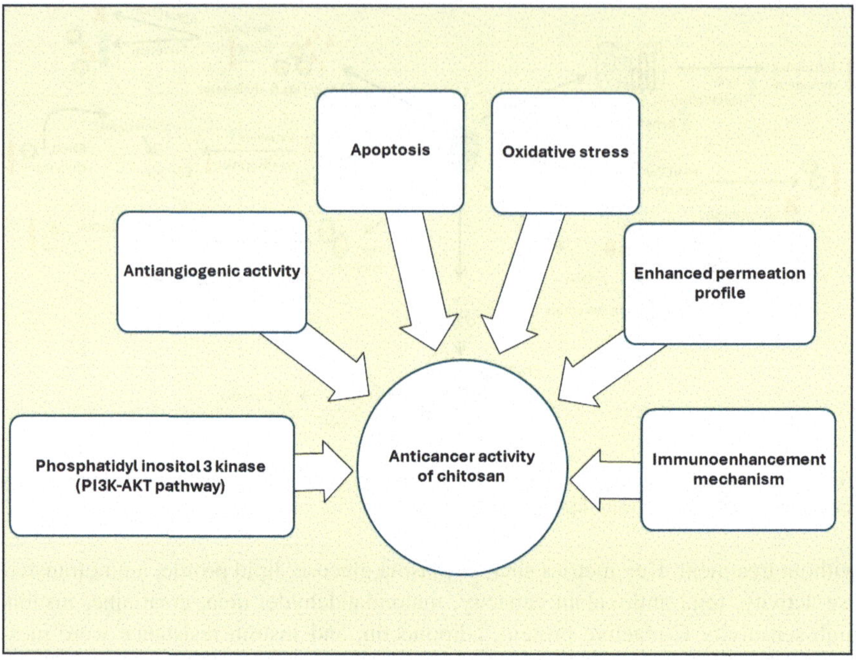

Fig. 5 Various anticancer mechanisms of chitosan

chitosan. Chitosan directly destroys the cancer cells by attacking tumor cell membranes or extracellularly by receptor-mediated endocytosis [49]. The anticancer activity of chitosan is mediated by various mechanisms like permeation-enhancing mechanism, enzymatic, antioxidant, immune enhancement mechanism, antiangiogenic, and apoptosis [50] as shown in Fig. 5. In different cancer cell lines, chitosan exhibits different mechanisms as shown in Table 1. In primary melanoma A375, it affects the adhesion and, in the case of primary melanoma SKMEL28 cell line, affects the proliferation. In RPMI7951, a metastatic melanoma cell line, chitosan activates the proapoptotic molecule like Bax and reduces antiapoptotic molecules like Bcl-2 and Bcl-XL, resulting in the apoptosis through the mitochondrial pathway [51]. In clinical studies, it has been proven that oral administration of water-soluble chitosan can improve the immune function by reducing the negative effect of radiotherapy on the immune function and has a good antitumor adjuvant effect in patients with radiotherapy suffering from lung cancer [52].

Factors including size, molecular weight, different modifications, and charge can influence the properties and biological activities of chitosan and its derivatives [53]. The hydrolysis of chitosan by various methods like enzymatic or chemical methods can result in the synthesis of a functional derivative of chitosan, that is, chitooligosaccharides (COS). Chitosan is water-insoluble, but COS exhibits more solubility in water, lower viscosity, and probably higher absorption rate. The effects

Table 1 Anticancer activity of chitosan in various preclinical studies

Chitosan type	Cancer model	Outcome	References
Chitosan nanoparticle isolated from squid, *Sepioteuthis lessoniana*	Hepatocellular carcinoma in rats	The protective effect on liver cells might be due to its robust antioxidant and antilipidemic properties	[70]
Chitosan nanoparticles from *Artemia salina*	Hepatocellular carcinoma in in vitro (HepG2) and in vivo (diethylnitrosamine-induced HCC in rats)	Cytotoxicity occurs through necrosis rather than apoptosis. In vivo studies revealed that treatment with chitosan nanoparticles reduces oxidative stress indicators in rat liver. In contrast, in the in vitro model, cytotoxicity is mediated by increased ROS generation	[71]
Chitosan nanoparticle	Diethyl nitrosamine (DEN)-induced hepatocellular carcinoma	The treatment with chitosan nanoparticle exhibited anticancer activity by reducing oxidative stress, inflammation, and apoptosis	[72]
Chitosan nanoparticles	Human gastric carcinoma cell line MGC803	The antitumor activity of chitosan nanoparticles is mediated by membrane-disrupting and apoptosis-inducing activities	
Chitosan nanoparticles	Human lymphoma SUDHL-4 cell line	Antiproliferative action mediated through apoptosis	[73]
Chitosan nanoparticle (CNP)	Human hepatocellular carcinoma (BEL7402) and mice model (BEL7402 induced)	CNP exhibit antitumor activity in vivo and in vitro by various mechanisms like neutralization of cell surface charge, decrease of mitochondrial membrane potential, and induction of lipid peroxidation	[74]
Chitosan nanoparticle	Esophageal cancer	The antimetastatic action of chitosan in cancer-associated fibroblast results in the downregulation of MMP1 and MMP9 and chemokines such as CXCR-4, CXCR-7, CCR-5, and SDF-1	[75]
Chitosan oligosaccharide	Mouse model of colitis-associated colorectal cancer	Inhibited the growth of colorectal cancer by activating AMPK and suppressing NF-κB and mTOR signaling pathways. COS has the potential to be useful in preventing colorectal cancer (CRC)	[69]

(continued)

Table 1 (continued)

Chitosan type	Cancer model	Outcome	References
Oligochitosan	Human acute leukemia HL-60	The apoptotic action of oligochitosan is linked to the activation of both the extrinsic and intrinsic pathways. In HL-60 cells, the expression of the antiapoptotic protein Bcl-2 was markedly downregulated, while the pro-apoptotic molecules; Fas, FADD, and Bax were upregulated. Moreover, the oligochitosan boosted the expression of galectin-9 mRNA	[76]
Water-soluble chitosan	Human leukemia cells (U937, K562, HL60 and THP-1)	Water-soluble chitosan initiates reduction in proliferation with cell death in leukemia cells by suppressing the Bcl-2 and protein kinase B (Akt)-dependent signaling pathways	[77]

of COS in lung cancer by in vitro (in human non-small-cell lung cancer A549 cells model) and in vivo model (C57BL/6 mice bearing lung cancer model) were evaluated. COS showed inhibition on the viability and proliferation of lung carcinoma cells (A549) without affecting the viability of human liver cells (HL-7702). In vivo study results showed that the administration of COS effectively suppressed the growth of Lewis tissues and induced necrosis of tumor cells. Additionally, COS has a role in regulating the expression of programmed cell death ligand 1 (PD-L1), which is important in reducing the therapeutic efficacy of anticancer agents. A study reported that COS is effective in downregulating programmed cell death ligand (PD-L1) mediated by interferon γ (IFN-γ) via AMPK activation and STAT1 inhibition promotes the T-cell-mediated immune killing in tumors and thus facilitate the anticancer action [54].

Various studies have shown that the use of chitosan or oligo chitosans can protect the body from cancer-induced oxidative stress [55]. Several studies investigated the role of the molecular weight of chitosan on anticancer activity. The cytotoxic activity of high molecular weight (HMWC) and low molecular weight chitosan (LMWC) on three cancerous cell lines—MCF-7 (human breast cancer), HeLa (human epithelial cervical cancer), and Saos-2 (osteosarcoma)—were evaluated. The study results showed that both types of chitosan exhibit anticancer activity in these three cell lines without affecting the viability of normal diploid fibroblast cells [49]. In another study, the administration of 21-kDa water-soluble chitosan and oligochitosan (at doses of 100 and 300 mg/kg body weight) resulted in the reduction in tumor growth and final weight in sarcoma 180-bearing mice model. The administration of

Table 2 Bioactivity of chitosan and relative mode of action

S. No.	Bioactivity	Mode of action	Advantages of chitosan
1	Antimicrobial	Chitosan's positively charged amino groups interact with negatively charged microbial cell membranes, resulting in increased permeability, leakage, functional disruption, and cell death	Biocompatible polysaccharide with low toxicity profile It can be processed into various forms such as membranes, sponges, gels, scaffolds, microparticles, nanoparticles, nanofibers, and 3D-printed structures Chitosan has a strong positive ionic charge and attracts negatively charged fats, cholesterol, toxic substances, and microorganisms Modification in the extent of deacetylation and the molecular mass of chitosan can modify its biopotential It is the most abundant natural biopolymer on earth after cellulose Chitosan is emerging as a very important raw material The potential for batch-to-bedside transformation in chitosan-based research is vast
2	Anti-inflammatory and immunomodulatory properties	Suppression of proinflammatory cytokines and mediators like interleukin-1β (IL-1β), tumor necrosis factor-alpha (TNF-α), interleukin-6 (IL-6), and nuclear factor kappa-B (NK-κB)	
3	Antioxidant	Inhibition of ROS production, prevention of lipid oxidation by significantly reducing serum free fatty acids, and malondialdehyde concentrations and it also increases intracellular antioxidant enzymes	
4	Hypolipidemic and hypocholesterolemic	Bind with dietary lipids and reduce intestinal lipid absorption. Interruption of the enterohepatic bile acid circulation	
5	Antitumor	Electrostatic interaction between highly negatively charged cancer cells and the positively charged amino group of chitosan The anticancer activity of chitosan is mediated by various mechanisms like permeation enhancing ability, enzymatic, antioxidant, immunoenhancement mechanism, antiangiogenic, and apoptotic potential	
6	Antihypertensive	Inhibiting the angiotensin-converting enzyme 1 and inhibiting rennin	
7	Antihypersensitive	Stabilize **mast cells**, reduce excessive Th2-mediated responses, and reduce IgE levels	

water-soluble chitosan (molecular weight of 46-kDa) at 100 mg/kg exhibited a decelerating effect on tumor growth and body weight. However, when the dosage was increased to 300 mg/kg, the chitosan (46KDa) did not provide any anticancer activity. Tumor growth was unaffected by a 130-kDa water-soluble chitosan. Thus, the study concluded that low molecular weight chitosan mediates antitumor activity through natural killer (NK) activity in intestinal intraepithelial lymphocytes (IELs) or splenic lymphocytes [56].

3.5.1 Antiangiogenic Activity

Angiogenesis is a process involving the growth of blood vessels from the preexisting one, resulting in a tumor condition. Downregulation of angiogenesis is considered a suitable therapeutic strategy for cancer treatment [57]. The angiogenic activity of chitosan is evident from the study conducted by Yinglei Xu et al. in nude mice xenografted with BEL-7402 (human hepatocellular carcinoma) cells. The immunohistochemical and RT-PCR studies showed that the inhibition of angiogenesis is mediated through the vascular endothelial growth factor receptor 2 (VEGFR2). The binding of VEGF with the VEGF2 receptor results in dimerization, and autophosphorylation followed by the downstream process that activates angiogenesis, vascular permeability enhancement, proliferation, and migration [58]. A study demonstrated that chitosan sulfate can be used instead of heparin to prevent angiogenesis and reduce negative effects by blocking the VEGF/VEGFR2 signaling pathway. In a study conducted in female mice with Ehrlich carcinoma on the neck region, the oral administration of chitosan nanoparticle at a dose of 0.5 mg/kg reduced the tumor size with downregulated levels of TNF-α, PDGE, and VEGF [57]. In comparison to chitosan, COS could decrease the expression of matrix metalloproteinase-9 (MMP-9) in human fibrosarcoma cells. Downregulation of MMP-9 expression may lead to the inhibition of vascular endothelial growth factor (VEGF) followed by antiangiogenic activity [59].

3.5.2 Apoptosis

Chitosan can modulate the expression of proteins linked with apoptosis, leading to the activation of caspase-9 and caspase-3 cleavage. This ultimately triggers apoptosis through the mitochondrial pathway [55].

Chitosan can trigger apoptosis by elevating the concentration of calcium ions and reactive oxygen species (ROS), and enhance the mitochondrial membrane potential. A study compared the anticancer activity of cisplatin and low molecular weight chitosan (LMWC) in SCC Ca9-22 (oral carcinoma cell line) and keratinocyte HaCaT cell lines (normal cell line). LMWC exhibits the cytotoxicity in the SCC Ca9-22 cell line; meanwhile, cisplatin exhibits toxicity in both cell lines. Thus, the study results indicate that cytotoxicity of LMWC is mediated via apoptosis and G1/S cell cycle arrest with minimal effect on the viability of normal cell lines [60]. R. Resmi et al. synthesized chitosan by demineralization and deprotonation from shrimp shell waste. The anticancer effect of chitosan was evaluated in MCF-7 breast cancer cells. The direct microscopic observation and MTT assay showed that the chitosan exhibited dose-dependent toxicity in MCF-7 cell line. Additionally, its cytocompatibility in L929 fibroblast indicates its nontoxicity in normal cells [61].

3.5.3 Oxidative Stress

Chitosan nanoparticle induces apoptosis in cancer cells by creating oxidative stress by reducing glutathione concentration and increasing ROS production in leukemia cells [62] Carolina et al. developed gold-modified chitosan nanoparticles to assess their antitumor efficacy against T-acute lymphocytic leukemia (CEM cell line) and chronic myeloid leukemia (K562 cell line). Treatment with modified chitosan nanoparticles exhibits apoptotic cell death in CEM cell line and necrotic cell death in the K562 cell without affecting the noncancerous cell line [63]. Additionally, the supplementation of chitosan was found to decrease the ROS production and lipid oxidation while elevating the plasma antioxidant capacity [36]. This action is mediated through the reaction of unstable radicals with the reactive amino and hydroxyl groups in the chitosan. In human bladder tumor cell line (human bladder tumor cells 5,637), chitosan induces apoptosis mediated by activating caspase -3 [64]. Takimoto et al. proved that the water-soluble chitosan triggers caspase 8 activity without affecting caspase 9, indicating that its antitumor activity is mediated through death receptor by TNF-alpha or TNF-R1 [65].

3.5.4 Enhanced Permeation Through Cell Membranes

The amino group of chitosan gets protonated in acidic and neutral environments. Thus, the positively charged chitosan favors the permeation through negatively charged basement and mucosal membrane. Various studies have reported that the antimetastatic action of chitosan is mediated by its enhanced permeation [50].

3.5.5 Immune-Mediated Mechanism

Chitosan accumulates in the tumor tissue, triggers the polarization of M1 macrophages, and modulates the tumor microenvironment into an immunosupportive state and mediates the anticancer activity. Chitosan activates the dendritic cell to promote the antitumor activity of NK cells by upregulating the IFN-gamma production [55]. Moreover, it is reported that chitosan enhances NK cell activity to kill leukemia cells. Chitosan suspension/microparticles have immune-stimulating activity such as cytokine release, promoting the accumulation and activation of macrophages and polymorphonuclear cells. Additionally, chitosan activates macrophages for anticancer activity and release of interleukin 1 [66].

3.5.6 Phosphatidyl Inositol 3 Kinase (PI3K-AKT Pathway)

Phosphatidyl inositol 3 kinase-protein kinase B (PI3K-AKT pathway) is important in carcinogenesis. Chitosan has a significant role in protein kinase B (Akt) downregulation. Abnormal activation of the pathway results in altered functions of cell proliferation, apoptosis, drug resistance, tumor recurrence, and occurrences. Chitosan interacts with the PI3K p85 subunit, resulting in the inhibition of the PI3K/Akt pathway, increasing the production of FAM172A. This effect proceeds by reducing the phosphorylation of MEK1/2 and blocking the activity of the MAPK/ERK signaling pathway [67]. This further leads to the inhibition of cell proliferation and promotes cell apoptosis. IlJin Sim et al. evaluated the effect of COS in osteosarcoma cells (MG63 and U2OS) and mice model of osteosarcoma. The study results showed that COS treatment significantly suppresses the osteosarcoma mediated through the PI3K-AKT/mTOR pathway (phosphorylated phosphatidylinositol 3-kinase-RAC-alpha serine/threonine-protein kinase-mammalian target of rapamycin) by downregulating cell migration-inducing protein (CMIP) [68]. The oral administration of COS (500 mg/kg/day) resulted in an ~60% reduction of tumor size in mouse model of colorectal cancer with AMPK activation and the suppression of NF-κB and mTOR signaling [69].

3.6 *Antihypertensive Action*

Hypertension is one of the most important risk factors in triggering cardiovascular diseases (CVD). The antihypertensive action of chitosan is mediated mainly by two mechanisms: inhibiting the angiotensin-converting enzyme 1 and inhibiting renin [78], as shown in Fig. 6. The administration of chitosan was found to be effective in reducing hypertension in both humans and animals. A study was carried out to evaluate the effect of chitosan supplementation combined with exercise in women with hypertension. The study results revealed that treatment with chitosan supplementation combined with exercise results in a significant reduction in raised levels of blood pressure [79]. The antihypertensive action of chitosan is additionally mediated by the protection against endothelial dysfunction facilitated by the anti-inflammatory and antioxidant action of COS [80]. Based on the meta-analysis study, chitosan consumption was found to effectively inhibit diastolic blood pressure with no effect on systolic blood pressure [81]. Treatment with water-soluble chitosan was found to be effective in reducing elevated blood pressure and membrane thickness/lumen diameter ratio in a spontaneously hypertensive rat model. A study on Wistar Kyoto rats proved that chitosan is effective in mediating vascular remodeling by reducing the expression of nuclear factor c1 of activated T cells (NFATc1) in abdominal aorta [82]. NaCl +3% chitosan's efficacy in preventing hypertension in a controlled, randomized, double-blind trial compared it to standard salt within the context of a salt reduction. In prehypertensive patients, NaCl +3% chitosan reduces the rise in blood pressure [83]. It has been discovered that terpolymer chitosan oligosaccharide

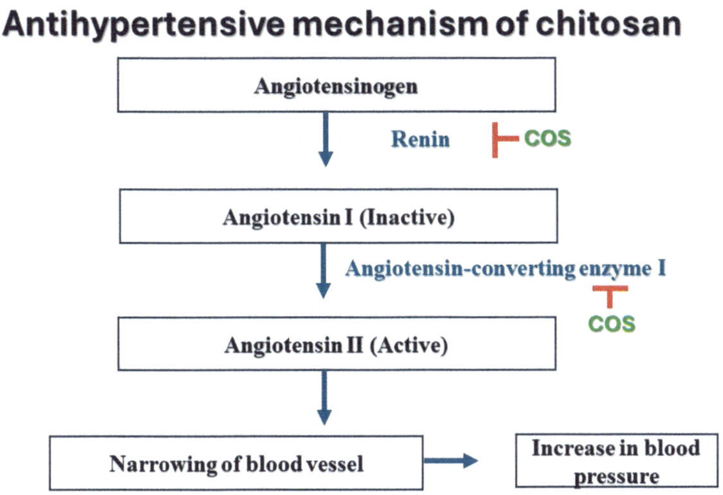

Fig. 6 Antihypertensive mechanism of chitosan

is a very good natural Angiotensin converting enzyme (ACE) inhibitor. Among all chitosan salts, 3% exhibited the biggest ACE inhibitory impact of 40.01 ± 0.14%, while captopril, a chemical ACE inhibitor, recorded the highest level of ACE inhibition (57.94 ± 1.33%) [84].

3.6.1 Inhibition of Angiotensin-Converting Enzyme 1

Angiotensin-converting enzyme 1 (ACE-1) plays a crucial role in converting inactive angiotensin I into angiotensin II, which leads to the constriction of blood vessels and an increase in blood pressure [78]. The effects of chitosan oligosaccharide (COS) on ACE inhibition and its antihypertensive effects were evaluated in spontaneously hypertensive rats. In vitro studies showed that among the various COS chitosans, the trimer exhibited better inhibitory action with IC50 value of 0.9 μM. In vivo study results showed that a single oral dose of chitosan trimer was found to be effective in reducing the blood pressure similar to that of standard drug, captopril [85]. The chemically modified COS like sulfated COS, carboxyl COS, and amino ethyl COS exhibited better ACE inhibition owing to its more negative charge density, which facilitates the binding into the ACE obligatory active site compared with unmodified COS [80]. Park et al. evaluated the role of molecular weight and deacetylation on ACE inhibitory action. In this study, they used different deacetylated (90, 75, and 50%) COS and different molecular weights like high molecular weight (5,000–10,000), medium molecular weight (1,000–5,000 Da), and low molecular weight (below 1,000 Da) COS. The medium molecular weight COSs with low deacetylation (50%) exhibited prominent ACE inhibition [86]. The role of molecular weight of COS in ACE inhibition was studied by Qian et al. [87]. From the study results, it is evident that low and medium molecular weight

COS exhibited better competition with natural substrate. Additionally, the C6 hydrogen substitution with sulfated group improves the ACE inhibition action [87]. In another study, DN Ngo et al. developed aminoethyl COS (MW: 800.79-4,765 Da) from COS (MW: 800-3,000 Da) with 90% deacetylation, and the results showed that the substitution of aminoethyl at C6 position promotes the ACE inhibition action [88].

3.6.2 Inhibition of Renin

The second main mechanism in regulating blood pressure is by acting on the renin–angiotensin system. Renin is involved in the cleavage of angiotensinogen into angiotensin I, which further gets converted into active form, that is, angiotensin II [78]. The degree of deacetylation and molecular weight of chitosan are critical parameters that influence renin inhibition. Park et al. evaluated the role of molecular weight and degree of deacetylation on renin inhibition. From the study, it is evident that renin inhibition was dose-dependent. COS with medium molecular weight COS with 90% deacetylation showed a high renin inhibition activity [89].

3.7 Antihypersensitive

Allergic reactions are disorders of the immune system due to exacerbation of the immune system to harmless environmental substances like animal dandruff, food, pollen, house dust, and mites. COS was found to be effective in mast RBL2H3 cell activation, indicating the protective action of COS against mast cell-mediated allergic reaction [90].

Dinitrophenyl bovine serum albumin antigen and DNP-specific immunoglobulin antibody are used to activate the mast cells; RBL-2H3 cells. The COS mediated the inhibition of mast cell activation by reducing the release of β-hexosamidase and histamine. Treatment with COS reduces the elevated intracellular calcium levels along with downregulating the expression of immunoglobulin Fc receptors. Varied molecular weight ranges can result in varying absorption, which correspond to differing inhibitory effects in mast cells [91]. Another study discovered that gallate oligomer chitosan (G-COS) effectively decreases calcium levels and histamine. Thus, the treatment with G-COS considerably prevents the degradation of IκB-α protein, the translocation of NF-κB (nuclear factor κB), and the activation of MAPK (mitogen-activated protein kinases), which results in the inhibition of cytokine generation. G-COS inhibited allergic reactions by suppressing FcεRI expression in antigen-stimulated cells [92]. In mouse model of shrimp allergen-induced hypersensitivity model, COS was found to effectively reduce the Th2 cytokines (IL4, IL5, and IL-13) by elevating the levels of Th1 cytokines (IFNγ) [93].

Allergic asthma is an inflammatory condition of the airways characterized by increased degranulation and cytokine production. Chung et al evaluated the role of low molecular weight chitosan (LM-COS) against allergic reactions and allergic asthma by in vivo and in vitro methods. The results of the in vivo study show that the ovalbumin-induced lung inflammation in asthma mice model increases the expression of several inflammatory mediators, such as interleukin (IL)-4, IL-5, IL-13, and tumor necrosis factor (TNF)-α, in both the lung tissue and bronchoalveolar lavage fluid. The oral treatment of LM-COS significantly reduced the expression of IL-4, IL-5, IL-13, and TNF-α in lung tissue and bronchoalveolar lavage fluid. From in vitro studies, it has been evident that COS is effective in reducing the antigen-stimulated degranulation and cytokine generation in rat basophilic leukemia RBL-2H3 cells. Thus, the study concluded that the oral administration of COS is found to be effective in mast cell-mediated allergic reaction by downregulating Th2 cytokines and proinflammatory cytokine [94].

Peanut allergy is IgE-mediated type I hypersensitivity, mediated by T helper 2 cytokines. The protective effect of chitin and chitosan was evaluated in peanut-induced anaphylaxis reactions in mice model. The administration of β-chitosan resulted in about 47% reduction in the production of peanut-specific IgE with considerable suppression of IL-5, IL-13, and IL-10 levels [95].

4 Current Status of Chitosan-Based Nutraceuticals

In the market, chitosan-based products are available as powder and capsules (350 mg) for weight management, heart health, immunity, liver health, and blood pressure regulation. LipoSan Ultra® is recognized as a safe and effective patented product for weight management (US Patent No. 6130321). It is reported that it intricately complexes and traps the fats and oils ingested, thereby diminishing the absorption of dietary fat (https://www.primex.is/products-services/; accessed on 11/9/2024). Furthermore, chitosan-based products are available as prebiotics, which promote the growth of beneficial gut bacteria. This enhances gut health, which may benefit the immune system and metabolism. Chitosan-based nutraceuticals represent a promising area of study with applications in weight control, cardiovascular health, and gut microbiota regulation. However, more clinical trials are required to completely evaluate these benefits and solve existing issues.

5 Future Prospective

Chitosan's versatility, bio-derived nature, broad range of health benefits, biocompatibility, biodegradability, and nontoxicity make it an appealing candidate for future nutraceuticals. Previous research indicates that chitosan exhibits a variety of biological activities, including antimicrobial, anti-inflammatory, immunomodulatory properties, antioxidant, hypolipidemic, hypocholesterolemic, antitumor, and antihypersensitive activities. These properties also depend on the molecular weight and degree of acetylation, which give them the potential to be highly effective nutraceuticals and a scope of modification in properties. As research continues to uncover its full potential, we can expect chitosan-based nutraceuticals to become increasingly popular in the global market, contributing to better health and wellness. With rising demand for sustainable and eco-friendly products, chitosan offers a viable alternative to synthetic materials, increasing its appeal in the nutraceutical industry. This could open up new markets and increase consumer acceptance.

6 Summary

Chitosan, a natural polysaccharide derived from chitin, has gained popularity as a nutraceutical due to its several health benefits and biocompatibility. Chitosan is extensively recognized for its antimicrobial, anti-inflammatory, immunomodulatory properties, antioxidant, hypolipidemic, hypocholesterolemic, antitumor, and antihypersensitive activities. The antimicrobial properties of chitosan contribute to defending against microbial infection and improve the immune system. Chitosan is known for its potent anti-inflammatory properties. It can inhibit the production of pro-inflammatory cytokines, lowering inflammation. Chitosan also has immunomodulatory properties because it regulates the activity of immune cells like macrophages and T-cells, which are important in maintaining immune balance and preventing chronic inflammatory diseases. Chitosan also has strong antioxidant properties that help neutralize free radicals and shield cells from oxidative stress. This activity is critical for preventing cellular damage, which can lead to a variety of chronic diseases, including heart disease and cancer. One of the major advantages of chitosan as a nutraceutical is its ability to help with weight management. Chitosan has a unique ability to bind dietary fats and prevent their absorption in the gastrointestinal tract. This fat-binding ability can lower total caloric intake, making it a popular ingredient in weight loss supplements. Furthermore, chitosan's ability to lower cholesterol levels by binding bile acids and promoting their excretion is a significant advantage that potentially lowers the risk of atherosclerosis and other cardiovascular conditions. Chitosan has shown promise as an antitumor agent, with studies indicating that it can inhibit the growth of cancer cells and induce apoptosis. This activity is partly due to its ability to boost the immune response to tumor cells and its direct effects on cancer cell metabolism. Chitosan also has antihypersensitive properties, which are useful for managing allergic reactions and other hypersensitivity disorders. It can control immune responses that cause hypersensitivity, reducing the severity of allergic symptoms. Chitosan's ability to form films also makes it an excellent candidate for drug delivery systems, offering controlled release and enhanced bioavailability of nutrients. Finally, chitosan's wide range of biological activities makes it an important nutraceutical. Its potential to promote health and prevent disease highlights its significance in developing functional foods and dietary supplements.

References

1. Santini A, Cammarata SM, Capone G, Ianaro A, Tenore GC, Pani L, Novellino E (2018) Nutraceuticals: opening the debate for a regulatory framework. Br J Clin Pharmacol 84:659–672. https://doi.org/10.1111/BCP.13496
2. Sharma M, Dwivedi P, Singh Rawat AK, Dwivedi AK (2016) Nutrition nutraceuticals: a proactive approach for healthcare. Forum Nutr:79–116. https://doi.org/10.1016/B978-0-12-804305-9.00003-8
3. Rajalakshmi R, Melians MA, Pon FF, Cosio DS, Buvarahamurthy V, Jayakumar AR, Paidas MJ, Rajalakshmi R, Pon FF, Cosio DS, Jayakumar AR, Paidas MJ, Melians MA (2023) Advantages and disadvantages of nutraceuticals. pp 245–286. https://doi.org/10.1007/978-981-99-0677-2_15
4. Siddiqui RA, Moghadasian MH (2020) Nutraceuticals and nutrition supplements: challenges and opportunities. Nutrients 12. https://doi.org/10.3390/NU12061593
5. Paniagua-Michel J (2015) Microalgal nutraceuticals. handbook of marine microalgae: biotechnology advances, pp 255–267. https://doi.org/10.1016/B978-0-12-800776-1.00016-9
6. Catalano A, Iacopetta D, Ceramella J, De Maio AC, Basile G, Giuzio F, Bonomo MG, Aquaro S, Walsh TJ, Sinicropi MS, Saturnino C, Geronikaki A, Salzano G (2022) Are nutraceuticals effective in COVID-19 and post-COVID prevention and treatment? Food Secur 11. https://doi.org/10.3390/FOODS11182884
7. Tripathi C, Girme A, Champaneri S, Patel RJ, Hingorani L (2020) Nutraceutical regulations: an opportunity in ASEAN countries. Nutrition 74:110729. https://doi.org/10.1016/J.NUT.2020.110729
8. Jadhav HB, Sablani S, Gogate P, Annapure U, Casanova F, Nayik GA, Alaskar K, Sarwar N, Raina IA, Ramniwas S, Mousavi Khaneghah A (2023) Factors governing consumers buying behavior concerning nutraceutical product. Food Sci Nutr 11:4988–5003. https://doi.org/10.1002/FSN3.3518
9. Parisi GF, Carota G, Castracani CC, Spampinato M, Manti S, Papale M, Di Rosa M, Barbagallo I, Leonardi S (2021) Nutraceuticals in the prevention of viral infections, including COVID-19, among the pediatric population: a review of the literature. Int J Mol Sci 22:1–26. https://doi.org/10.3390/IJMS22052465
10. Alesci A, Aragona M, Cicero N, Lauriano ER (2022) Can nutraceuticals assist treatment and improve covid-19 symptoms? Nat Prod Res 36:2672–2691. https://doi.org/10.1080/14786419.2021.1914032
11. Fratta Pasini AM, Stranieri C, Cominacini L, Mozzini C (2021) Potential role of antioxidant and anti-inflammatory therapies to prevent severe SARS-Cov-2 complications. Antioxidants (Basel) 10:1–22. https://doi.org/10.3390/ANTIOX10020272
12. Suleria HAR, Osborne S, Masci P, Gobe G (2015) Marine-based nutraceuticals: an innovative trend in the food and supplement industries. Mar Drugs 13:6336. https://doi.org/10.3390/MD13106336
13. Mawazi SM, Kumar M, Ahmad N, Ge Y, Mahmood S (2024) Recent applications of Chitosan and its derivatives in antibacterial, anticancer, wound healing, and tissue engineering fields. Polymers (Basel) 16. https://doi.org/10.3390/POLYM16101351
14. Desai N, Rana D, Salave S, Gupta R, Patel P, Karunakaran B, Sharma A, Giri J, Benival D, Kommineni N (2023) Chitosan: a potential biopolymer in drug delivery and biomedical applications. Pharmaceutics 15:1313. https://doi.org/10.3390/PHARMACEUTICS15041313
15. Kim Y, Zharkinbekov Z, Raziyeva K, Tabyldiyeva L, Berikova K, Zhumagul D, Temirkhanova K, Saparov A (2023) Chitosan-based biomaterials for tissue regeneration. Pharmaceutics 15. https://doi.org/10.3390/PHARMACEUTICS15030807
16. Parhi R (2020) Drug delivery applications of chitin and chitosan: a review. Environ Chem Lett 18:577–594. https://doi.org/10.1007/S10311-020-00963-5
17. Illum L (1998) Chitosan and its use as a pharmaceutical excipient. Pharm Res 15:1326–1331. https://doi.org/10.1023/A:1011929016601

18. Baharlouei P, Rahman A (2022) Chitin and Chitosan: prospective biomedical applications in drug delivery, cancer treatment, and wound healing. Mar Drugs 20. https://doi.org/10.3390/MD20070460
19. Chung Y, Su Y, Chen C, Jia G, Wang H, Wu JCG, Lin J (2016) Relationship between antibacterial activity of chitosan and surface characteristics of cell wall. Acta Pharmacol Sin 25:93236–93936
20. Li Z, Yang F, Yang R (2015) Synthesis and characterization of chitosan derivatives with dual-antibacterial functional groups. Int J Biol Macromol 75:378–387. https://doi.org/10.1016/J.IJBIOMAC.2015.01.056
21. Zheng LY, Zhu JF (2003) Study on antimicrobial activity of chitosan with different molecular weights. Carbohydr Polym 54:527–530. https://doi.org/10.1016/J.CARBPOL.2003.07.009
22. Facchinatto WM, Araujo LO, Moraes TB, Abelha TF, Lima THN, Santos DM dos, Campana-Filho SP, Colnago LA, Caires ARL (2022) Antimicrobial and photoantimicrobial activities of Chitosan/CNPPV nanocomposites. Int J Mol Sci 23:12519. https://doi.org/10.3390/IJMS232012519/S1
23. Wang CH, Cherng JH, Liu CC, Fang TJ, Hong ZJ, Chang SJ, Fan GY, Der HS (2021) Procoagulant and antimicrobial effects of Chitosan in wound healing. Int J Mol Sci 22:7067. https://doi.org/10.3390/IJMS22137067
24. Brilhante RSN, Costa A d C, Mesquita JRL d, dos Santos Araújo G, Freire RS, Nunes JVS, Nobre AFD, Fernandes MR, Rocha MFG, Pereira Neto W d A, Crouzier T, Schimpf U, Viera RS (2023) Antifungal activity of Chitosan against Histoplasma capsulatum in planktonic and biofilm forms: a therapeutic strategy in the future? J Fungi 9:1201. https://doi.org/10.3390/JOF9121201
25. Ahmed F, Soliman FM, Adly MA, Soliman HAM, El-Matbouli M, Saleh M (2021) Dietary Chitosan nanoparticles: potential role in modulation of rainbow trout (Oncorhynchus mykiss) antibacterial defense and intestinal immunity against enteric Redmouth disease. Mar Drugs 19: 72. https://doi.org/10.3390/MD19020072
26. Chi J, Zhang X, Chen C, Shao C, Zhao Y, Wang Y (2020) Antibacterial and angiogenic chitosan microneedle array patch for promoting wound healing. Bioact Mater 5:253–259. https://doi.org/10.1016/J.BIOACTMAT.2020.02.004
27. Bueter CL, Lee CK, Wang JP, Ostroff GR, Specht CA, Levitz SM (2014) Spectrum and mechanisms of inflammasome activation by chitosan. J Immunol 192:5943–5951. https://doi.org/10.4049/JIMMUNOL.1301695
28. Azuma K, Osaki T, Minami S, Okamoto Y, Puoci F (2015) Anticancer and anti-inflammatory properties of chitin and Chitosan oligosaccharides. J Funct Biomater 6:33–49. https://doi.org/10.3390/JFB6010033
29. Rabišková M, Bautzová T, Gajdziok J, Dvořáčková K, Lamprecht A, Pellequer Y, Spilková J (2012) Coated chitosan pellets containing rutin intended for the treatment of inflammatory bowel disease: in vitro characteristics and in vivo evaluation. Int J Pharm 422:151–159. https://doi.org/10.1016/J.IJPHARM.2011.10.045
30. Chang SH, Lin YY, Wu GJ, Huang CH, Tsai GJ (2019) Effect of chitosan molecular weight on anti-inflammatory activity in the RAW 264.7 macrophage model. Int J Biol Macromol 131: 167–175. https://doi.org/10.1016/J.IJBIOMAC.2019.02.066
31. Mohyuddin SG, Qamar A, Ying HC, Chen SW, Ying WJ, Xi LX, Bin MX, Chao YZ, Hong YY, Wu LY, Bao ML, Ju XH (2021) Effect of chitosan on blood profile, inflammatory cytokines by activating TLR4/NF-κB signaling pathway in intestine of heat stressed mice. Sci Rep 11. https://doi.org/10.1038/S41598-021-98931-8
32. Davydova VN, Sorokina IV, Volod'ko AV, Sokolova EV, Borisova MS, Yermak IM (2020) The comparative immunotropic activity of Carrageenan, Chitosan and their complexes. Mar Drugs 18:458. https://doi.org/10.3390/MD18090458
33. Tsvetkov YE, Paulovičová E, Paulovičová L, Farkaš P, Nifantiev NE (2020) Synthesis of biotin-tagged Chitosan oligosaccharides and assessment of their immunomodulatory activity. Front Chem 8:554732. https://doi.org/10.3389/FCHEM.2020.554732/BIBTEX

34. Jhundoo HD, Siefen T, Liang A, Schmidt C, Lokhnauth J, Béduneau A, Pellequer Y, Larsen CC, Lamprecht A (2020) Anti-inflammatory activity of Chitosan and 5-amino salicylic acid combinations in experimental colitis. Pharmaceutics 12:1–16. https://doi.org/10.3390/PHARMACEUTICS12111038
35. Wu C, Dai Y, Yuan G, Su J, Liu X (2019) Immunomodulatory effects and induction of apoptosis by different molecular weight chitosan oligosaccharides in head kidney macrophages from blunt snout bream (megalobrama amblycephala). Front Immunol 10:455516. https://doi.org/10.3389/FIMMU.2019.00869/BIBTEX
36. Ivanova DG, Yaneva ZL (2020) Antioxidant properties and redox-modulating activity of Chitosan and its derivatives: biomaterials with application in cancer therapy. Biores Open Access 9:64. https://doi.org/10.1089/BIORES.2019.0028
37. Kim KW, Thomas RL (2007) Antioxidative activity of chitosans with varying molecular weights. Food Chem 101:308–313. https://doi.org/10.1016/J.FOODCHEM.2006.01.038
38. Jayachitra A, Rajalakshmi A, Krithiga N (2013) Antioxidant activity of the chitosan extracted from shrimp exoskeleton. Middle East J Sci Res 16:1446–1451. https://doi.org/10.5829/idosi.mejsr.2013.16.10.12033
39. Xu H, Wang X, Liang Q, Xu R, Liu J, Yu D (2023) Dietary chitosan moderates the growth rate, antioxidant activity, immunity, intestinal morphology and resistance against Aeromonas hydrophila of juvenile hybrid sturgeon (Acipenser baerii♀ × Acipenser schrenckii♂). Int J Biol Macromol 224:1012–1024. https://doi.org/10.1016/J.IJBIOMAC.2022.10.185
40. Abdel-Tawwab M, Khalil RH, Diab AM, Khallaf MA, Abdel-Razek N, Abdel-Latif HMR, Khalifa E (2021) Dietary garlic and chitosan enhanced the antioxidant capacity, immunity, and modulated the transcription of HSP70 and Cytokine genes in Zearalenone-intoxicated European seabass. Fish Shellfish Immunol 113:35–41. https://doi.org/10.1016/J.FSI.2021.03.012
41. Abdel Rahman AN, Elshopakey GE, Behairy A, Altohamy DE, Ahmed AI, Farroh KY, Alkafafy M, Shahin SA, Ibrahim RE (2022) Chitosan-Ocimum basilicum nanocomposite as a dietary additive in Oreochromis niloticus: effects on immune-antioxidant response, head kidney gene expression, intestinal architecture, and growth. Fish Shellfish Immunol 128:425–435. https://doi.org/10.1016/J.FSI.2022.08.020
42. Lochi GM, Shah MG, Gandahi JA, Gadahi JA, Hadi SA, Farooq T, Vistro WA, Rahmani MM (2023) Effect of selenium nanoparticles and Chitosan on production performance and antioxidant integrity of heat-stressed broiler. Biol Trace Elem Res 201:1977–1986. https://doi.org/10.1007/S12011-022-03262-Y
43. Il AS, Cho S, Choi NJ (2021) Effectiveness of Chitosan as a dietary supplement in lowering cholesterol in murine models: a meta-analysis. Mar Drugs 19. https://doi.org/10.3390/MD19010026
44. Koide SS (1998) Chitin-chitosan: properties, benefits and risks. Nutr Res 18:1091–1101. https://doi.org/10.1016/S0271-5317(98)00091-8
45. Zalaqi Z, Ghazalian F, Khodayar MJ, Raesi Vanani A, Khorsandi L, Shushizadeh MR (2022) Swimming training combined with chitosan supplementation reduces the development of obesity and oxidative stress in high-fat diet-fed mice. Br J Nutr 127:837–846. https://doi.org/10.1017/S0007114521001379
46. Al-Okbi SY, Ali O, Aly AS, Refaat D, Esmail RSH, Elbakry HFH (2023) Management of metabolic syndrome by nutraceuticals prepared from chitosan and ferulic acid with or without beta-sitosterol and their nanoforms. Sci Rep 13:12176. https://doi.org/10.1038/S41598-023-38837-9
47. Wang Z, Zhang F, Yan Y, Zhang Z, Wang L, Qin C (2019) Lipid-lowering activities of chitosan and its quaternary ammonium salt for the hyperlipidemia rats induced by high-fat diets. Int J Biol Macromol 132:922–928. https://doi.org/10.1016/J.IJBIOMAC.2019.03.214
48. Chiu CY, Yen TE, Liu SH, Chiang MT (2020) Comparative effects and mechanisms of Chitosan and its derivatives on hypercholesterolemia in high-fat diet-fed rats. Int J Mol Sci 21. https://doi.org/10.3390/IJMS21010092

49. Abedia Z, Moghadamnia AA, Zabihi E, Pourbagher R, Ghasemi M, Nouri HR, Tashakorian H, Jenabian N (2019) Anticancer properties of chitosan against osteosarcoma, breast cancer and cervical cancer cell lines. Caspian J Intern Med 10:439. https://doi.org/10.22088/CJIM.10.4.439
50. Adhikari HS, Yadav PN (2018) Anticancer activity of Chitosan, Chitosan derivatives, and their mechanism of action. Int J Biomater 2018. https://doi.org/10.1155/2018/2952085
51. Gibot L, Chabaud S, Bouhout S, Bolduc S, Auger FA, Moulin VJ (2015) Anticancer properties of chitosan on human melanoma are cell line dependent. Int J Biol Macromol 72:370–379. https://doi.org/10.1016/J.IJBIOMAC.2014.08.033
52. Ma JX, Qian L, Zhou Y (2015) Stimulation effect of chitosan on the immunity of radiotherapy patients suffered from lung cancer. Int J Biol Macromol 72:195–198. https://doi.org/10.1016/J.IJBIOMAC.2014.08.027
53. Amirani E, Hallajzadeh J, Asemi Z, Mansournia MA, Yousefi B (2020) Effects of chitosan and oligochitosans on the phosphatidylinositol 3-kinase-AKT pathway in cancer therapy. Int J Biol Macromol 164:456–467. https://doi.org/10.1016/J.IJBIOMAC.2020.07.137
54. Chen J, Zhou Z, Zheng C, Liu Y, Hao R, Ji X, Xi Q, Shen J, Li Z (2022) Chitosan oligosaccharide regulates AMPK and STAT1 pathways synergistically to mediate PD-L1 expression for cancer chemoimmunotherapy. Carbohydr Polym 277:118869. https://doi.org/10.1016/J.CARBPOL.2021.118869
55. Ding J, Guo Y (2022) Recent advances in Chitosan and its derivatives in cancer treatment. Front Pharmacol 13:888740. https://doi.org/10.3389/FPHAR.2022.888740/BIBTEX
56. Maeda Y, Kimura Y (2004) Antitumor effects of various low-molecular-weight chitosans are due to increased natural killer activity of intestinal intraepithelial lymphocytes in sarcoma 180-bearing mice. J Nutr 134:945–950. https://doi.org/10.1093/JN/134.4.945
57. Ahmed NH, Said UZ, Meky NH, Mohamed MA (2018) Role of chitosan nanoparticles as anti-angiogenic in mice bearing Ehrlich carcinoma. Oncol Res Rev 1. https://doi.org/10.15761/ORR.1000117
58. Xu Y, Wen Z, Research ZX-A (2009) Chitosan nanoparticles inhibit the growth of human hepatocellular carcinoma xenografts through an antiangiogenic mechanism. Anticancer Res
59. Huang X, Jiao Y, Zhou C (2021) Impacts of chitosan oligosaccharide (COS) on angiogenic activities. Microvasc Res 134:104114. https://doi.org/10.1016/J.MVR.2020.104114
60. Wimardhani YS, Suniarti DF, Freisleben HJ, Wanandi SI, Siregar NC, Ikeda MA (2014) Chitosan exerts anticancer activity through induction of apoptosis and cell cycle arrest in oral cancer cells. J Oral Sci 56:119–126. https://doi.org/10.2334/JOSNUSD.56.119
61. Resmi R, Yoonus J, Beena B (2021) Anticancer and antibacterial activity of chitosan extracted from shrimp shell waste. Mater Today Proc 41:570–576. https://doi.org/10.1016/J.MATPR.2020.05.251
62. Sarangapani S, Patil A, Ngeow YK, Elsa Mohan R, Asundi A, Lang MJ (2018) Chitosan nanoparticles' functionality as redox active drugs through cytotoxicity, radical scavenging and cellular behaviour. Integr Biol (Camb) 10:313–324. https://doi.org/10.1039/C8IB00038G
63. Carolina A, Martínez-Torres YH, Lorenzo-Anota GM, García-Juárez Z-TDG, Rodríguez-Padilla C (2019) Chitosan gold nanoparticles induce different ROS-dependent cell death modalities in leukemic cells. Int J Nanomedicine 14:7173. https://doi.org/10.2147/IJN.S221021
64. Hasegawa M, Yagi K, Iwakawa S, Hirai M (2001) Chitosan induces apoptosis via Caspase-3 activation in bladder tumor cells. Jpn J Cancer Res 92:459. https://doi.org/10.1111/J.1349-7006.2001.TB01116.X
65. Takimoto H, Hasegawa M, Yagi K, Nakamura T, Sakaeda T, Hirai M (2004) Proapoptotic effect of a dietary supplement: water soluble chitosan activates caspase-8 and modulating death receptor expression. Drug Metab Pharmacokinet 19:76–82. https://doi.org/10.2133/DMPK.19.76
66. Nandgude T, Pagar R (2021) Plausible role of chitosan in drug and gene delivery against resistant breast cancer cells. Carbohydr Res 506:108357. https://doi.org/10.1016/J.CARRES.2021.108357

67. Ren Z, Xiao W, He M, Bai L (2023) Chitosan targets PI3K/Akt/FoxO3a axis to up-regulate FAM172A and suppress MAPK/ERK pathway to exert anti-tumor effect in osteosarcoma. Chem Biol Interact 373. https://doi.org/10.1016/J.CBI.2023.110354
68. Sim IJ, Choe WG, Ri JJ, Su H, Moqbel SAA, Yan WQ (2023) Chitosan oligosaccharide suppresses osteosarcoma malignancy by inhibiting CEMIP via the PI3K/AKT/mTOR pathway. Med Oncol 40:294. https://doi.org/10.1007/S12032-023-02165-9
69. Mattaveewong T, Wongkrasant P, Chanchai S, Pichyangkura R, Chatsudthipong V, Muanprasat C (2016) Chitosan oligosaccharide suppresses tumor progression in a mouse model of colitis-associated colorectal cancer through AMPK activation and suppression of NF-κB and mTOR signaling. Carbohydr Polym 145:30–36. https://doi.org/10.1016/J.CARBPOL.2016.02.077
70. Subhapradha N, Shanmugam V, Shanmugam A (2017) Chitosan nanoparticles from marine squid protect liver cells against N-diethylnitrosoamine-induced hepatocellular carcinoma. Carbohydr Polym 171:18–26. https://doi.org/10.1016/j.carbpol.2017.04.097
71. Elkeiy MM, Khamis AA, El-Gamal MM, Abo Gazia MM, Zalat ZA, El-Magd MA (2020) Chitosan nanoparticles from Artemia salina inhibit progression of hepatocellular carcinoma in vitro and in vivo. Environ Sci Pollut Res 27:19016–19028. https://doi.org/10.1007/S11356-018-3339-6/METRICS
72. Kadry MO, Abdel-Megeed RM, El-Meliegy E, Abdel-Hamid AHZ (2018) Crosstalk between GSK-3, c-Fos, NFκB and TNF-α signaling pathways play an ambitious role in Chitosan nanoparticles cancer therapy. Toxicol Rep 5:723–727. https://doi.org/10.1016/J.TOXREP.2018.06.002
73. Parida UK, Rout N, Bindhani BK, Parida UK, Rout N, Bindhani BK (2013) In vitro properties of chitosan nanoparticles induce apoptosis in human lymphoma SUDHL-4 cell line. Adv Biosci Biotechnol 4:1118–1127. https://doi.org/10.4236/ABB.2013.412148
74. Qi L, Xu Z, Chen M (2007) In vitro and in vivo suppression of hepatocellular carcinoma growth by chitosan nanoparticles. Eur J Cancer 43:184–193. https://doi.org/10.1016/J.EJCA.2006.08.029
75. Potdar PD, Shetti AU (2016) Evaluation of anti-metastatic effect of chitosan nanoparticles on esophageal cancer-associated fibroblasts. J Cancer Metastasis Treat 2:259
76. Dou J, Ma P, Xiong C, Tan C, Du Y (2011) Induction of apoptosis in human acute leukemia HL-60 cells by oligochitosan through extrinsic and intrinsic pathway. Carbohydr Polym 86:19–24. https://doi.org/10.1016/J.CARBPOL.2011.03.008
77. Kim MO, Moon DO, Kang CH, Choi YH, Lee JD, Kim GY (2013) Water-soluble chitosan sensitizes apoptosis in human leukemia cells via the downregulation of BCL-2 and dephosphorylation of AKT. J Food Biochem 37:270–277. https://doi.org/10.1111/J.1745-4514.2011.00628.X
78. Ngo DH, Vo TS, Ngo DN, Kang KH, Je JY, Pham HND, Byun HG, Kim SK (2015) Biological effects of chitosan and its derivatives. Food Hydrocoll 51:200–216. https://doi.org/10.1016/J.FOODHYD.2015.05.023
79. Rasouli Jokar E, Shamlou Kazemi S (2023) Effect of eight-weeks Chitosan supplementation and combined exercise on the blood pressure and lipid profile of women with hypertension. J Sport Biosci 15:55–70. https://doi.org/10.22059/JSB.2023.354853.1574
80. Tao W, Wang G, Wei J (2021) The role of Chitosan oligosaccharide in metabolic syndrome: a review of possible mechanisms. Mar Drugs 19. https://doi.org/10.3390/MD19090501
81. Huang H, Zou Y, Chi H (2018) Quantitative assessment of the effects of chitosan intervention on blood pressure control. Drug Des Devel Ther 12:67. https://doi.org/10.2147/DDDT.S148064
82. Du M, Kou L, Li S (2016) Water-soluble chitosan regulates vascular remodeling in hypertension via NFATc1. Eur Heart J Suppl 18:F38–F38. https://doi.org/10.1093/EURHEARTJ/SUW037

83. Allaert F andré (2017) Effect of NaCl + Chitosan 3% vs. NaCl on high blood pressure parameters of healthy volunteers with prehypertension. Minerva Cardioangiol 65:563–576. https://doi.org/10.23736/S0026-4725.17.04451-6
84. Kim HL, Park SM, Cho GS, Kim KY, Kim IC (2010) Physicochemical characteristics, antimicrobial activity, ACE inhibitory activity of chitosan-salt, and its antihypertensive effect. Food Sci Biotechnol 19:777–784. https://doi.org/10.1007/S10068-010-0109-8/METRICS
85. Hong S, Kim M, Oh S, Han C-K (1998) ACE inhibitory and antihypertensive effect of chitosan oligosaccharides in SHR. Korean J Food Sci Technol
86. Park PJ, Je JY, Kim SK (2003) Angiotensin I converting enzyme (ACE) inhibitory activity of hetero-chitooligosaccharides prepared from partially different deacetylated chitosans. J Agric Food Chem 51:4930–4934. https://doi.org/10.1021/JF0340557
87. Qian Z, Eom T, Ryu B, Chitosan SK-JC (2010) Angiotensin I-converting enzyme inhibitory activity of sulfated chitooligosaccharides with different molecular weights 15:75–79
88. Ngo D, Qian Z, Je J, Kim M, Biochemistry SK-P (2008) Aminoethyl chitooligosaccharides inhibit the activity of angiotensin converting enzyme. Elsevier
89. Park PJ, Ahn CB, Jeon YJ, Je JY (2008) Renin inhibition activity by chitooligosaccharides. Bioorg Med Chem Lett 18:2471–2474. https://doi.org/10.1016/J.BMCL.2008.02.041
90. Riaz Rajoka MS, Zhao L, Mehwish HM, Wu Y, Mahmood S (2019) Chitosan and its derivatives: synthesis, biotechnological applications, and future challenges. Appl Microbiol Biotechnol 103:1557–1571. https://doi.org/10.1007/S00253-018-9550-Z
91. Vo TS, Kim JA, Ngo DH, Kong CS, Kim SK (2012) Protective effect of chitosan oligosaccharides against FcεRI-mediated RBL-2H3 mast cell activation. Process Biochem 47:327–330. https://doi.org/10.1016/J.PROCBIO.2011.10.036
92. Vo TS, Ngo DH, Kim SK (2012) Gallic acid-grafted chitooligosaccharides suppress antigen-induced allergic reactions in RBL-2H3 mast cells. Eur J Pharm Sci 47:527–533. https://doi.org/10.1016/J.EJPS.2012.07.010
93. Jiang T, Ji H, Zhang L, Wang Y, Zhou H (2019) Chitosan oligosaccharide exerts anti-allergic effect against shrimp tropomyosin-induced food allergy by affecting Th1 and Th2 cytokines. Int Arch Allergy Immunol 180:10–16. https://doi.org/10.1159/000500720
94. Chung MJ, Park JK, Il PY (2012) Anti-inflammatory effects of low-molecular weight chitosan oligosaccharides in IgE–antigen complex-stimulated RBL-2H3 cells and asthma model mice. Int Immunopharmacol 12:453–459. https://doi.org/10.1016/J.INTIMP.2011.12.027
95. Bae MJ, Shin HS, Kim EK, Kim J, Shon DH (2013) Oral administration of chitin and chitosan prevents peanut-induced anaphylaxis in a murine food allergy model. Int J Biol Macromol 61: 164–168. https://doi.org/10.1016/J.IJBIOMAC.2013.06.017

Adv Polym Sci (2025) 295: 367–402
https://doi.org/10.1007/12_2024_179

Published online: 14 November 2024

Chitosan-Based Biomaterials for Hemostasis and Wound Healing

Matías Araneda Cisternas and Kokkarachedu Varaprasad

Contents

1 Introduction 369
2 Hemostasis Stages 373
3 Regulation of Hemostasis 374
4 Chitosan Interaction with Drugs for Biomedicine 376
4.1 Mucoadhesion 376
4.2 Ionotropic Gelation 376
4.3 Hydrogen Bonding 376
4.4 pH-Dependent Swelling and Dissolution 377
4.5 Enhanced Permeability 377
4.6 Targeted Drug Delivery 377
4.7 Biodegradability and Biocompatibility 378
5 Chitosan Interaction with Nanomaterials 378
5.1 Nanoparticle Formation 378
5.2 Surface Modification 378
5.3 Drug Encapsulation and Entrapment 379
5.4 pH-Responsive Drug Release 379
5.5 Active Targeting 379
5.6 Biocompatibility and Biodegradability 379
5.7 Multifunctional Platforms 379
6 Chitosan-Based Biomaterials for Hemostasis 380
6.1 Chitosan Sponges or Dressings 380
6.2 Chitosan Nanofibers 380
6.3 Chitosan Composites as Hemostatic Agents 383
6.4 Chitosan-Hemostatic Sprays or Sealants 383
6.5 Chitosan-Hemostatic Patches or Films 383
6.6 Chitosan-Coated Hemostatic Agents 384
6.7 Chitosan Hydrogels 385
6.8 Commercially Available Chitosan-Based Hemostatic Dressing 385
7 Recent Applications in Wound and Biomedicine 386

M. Araneda Cisternas and K. Varaprasad (✉)
Facultad de Ingeniería, Arquitectura y Diseño, Universidad San Sebastian, Concepción, Chile
e-mail: prasad.kokkarachedu@uss.cl

7.1 Advanced Wound Dressings ... 386
7.2 Drug Delivery Systems ... 386
7.3 Tissue Engineering Scaffolds ... 388
7.4 Dental Materials ... 388
7.5 Biomedical Coatings ... 388
7.6 Antimicrobial Textiles ... 390
8 Chitosan Biomaterials for Miscellaneous Applications ... 390
8.1 Cholesterol Detectors ... 390
8.2 Environmental Remediation ... 391
8.3 Heavy Metal Absorption ... 391
8.4 Biodegradable Bioplastics ... 391
8.5 Pharmaceutical Industry ... 391
8.6 Hydrogels ... 392
8.7 Food Packaging ... 392
8.8 Dietary Supplements ... 392
9 Future Aspects ... 393
9.1 Enhanced Hemostatic Performance ... 393
9.2 Controlled Drug Delivery ... 393
9.3 Bioactive Wound Healing ... 393
9.4 Antimicrobial Activity ... 393
9.5 Biocompatibility and Immunomodulation ... 394
9.6 Personalized Medicine and Theragnostic ... 394
9.7 Regulatory Approval and Clinical Translation ... 394
10 Conclusion ... 394
References ... 395

Abstract This chapter explores the versatile applications of biopolymer chitosan and its derivatives in hemostasis and wound healing. The cationic chitosan, a natural polysaccharide derived from chitin, has gained considerable attention in the biomedical field due to its unique properties, including biocompatibility, biodegradability, and hemostatic efficacy. The discussion encompasses the role of chitosan-based biomaterials in promoting hemostasis, facilitating clot formation, and accelerating the wound-healing process. Furthermore, the chapter delves into the various strategies employed to enhance the chitosan derivatives' hemostatic and wound-healing properties through chemical modifications and biocomposite formulations. Additionally, it highlights recent advancements and promising prospects in utilizing chitosan-based biomaterials for addressing bleeding disorders and promoting tissue regeneration. Overall, this chapter provides valuable insights into the multifaceted applications of chitosan in hemostasis and wound management, paving the way for innovative therapeutic interventions in the field of regenerative medicine.

Keywords Biomolecules · Chitosan biomaterials · Hemostasis · Wound healing

1 Introduction

Skin is the largest organ in the human and animal systems [1]. Damage to this organ refers to any injury that disrupts its normal structure and function [2]. These wounds can vary widely in severity, size, and depth, and they can be caused by a multitude of factors, including trauma, burns, surgical incisions, and infections, or underlying medical conditions. Skin wounds are classified based on their depth and extent of tissue damage [3], which influences their healing process and potential complications. The superficial wounds affect only the outermost layer of the skin, known as the epidermis. These are considered acute wounds [2]. Examples include minor cuts, scratches, or abrasions. Superficial or acute wounds of this nature heal within 1–12 weeks, depending on the nature of the wound, without significant scarring. The partial-thickness (chronic) wounds extend through the epidermis and into the underlying dermis. They may result from deeper cuts, burns, or abrasions. Partial-thickness wounds can be painful and may take longer to heal, often leaving behind scars. Finally, the full-thickness wounds penetrate through the entire skin thickness, extending into the subcutaneous tissue or even deeper into muscles, tendons, or bones. Full-thickness wounds are often caused by severe trauma, surgical procedures, or chronic conditions like ulcers. In addition, when skin damage occurs, external bacteria can enter the body, cause infections, and convert the wound into a chronic, non-healing state [2].

These types of deep, full-thickness wounds typically require medical intervention and may result in significant scarring and tissue loss if not properly managed. The process of wound healing involves a complex series of biological events aimed at restoring the integrity and function of the injured skin. This process typically occurs in several overlapping phases, including hemostasis, inflammation, proliferation, and remodeling [4].

Immediately after injury, the hemostasis process starts with blood vessels constricting to minimize bleeding [6], followed by the formation of a blood clot to seal the wound and initiate the inflammatory response (Fig. 1). The hemostasis phase starts to attract white blood cells, such as neutrophils and macrophages, to migrate to the wound site to remove debris, foreign particles, and pathogens. Inflammatory cytokines and growth factors are released to stimulate the proliferation of new cells and blood vessels. During the proliferation phase, fibroblasts produce collagen, the main structural protein in the skin, to rebuild the damaged tissue. Epithelial cells at the wound edges multiply and migrate to cover the wound surface, forming a new layer of skin in the remodeling step. This final phase of wound healing involves remodeling of the newly formed tissue, including the rearrangement and maturation of collagen fibers. This phase can last for months or even years and aims to strengthen the scar tissue and improve its appearance and function.

While most skin wounds heal without complications, certain factors can impede the healing process or lead to complications with infections and excessive scarring, which leads to chronic wounds. The initial process caused by bacterial contamination of the wound can lead to infection, delaying healing and increasing the risk of

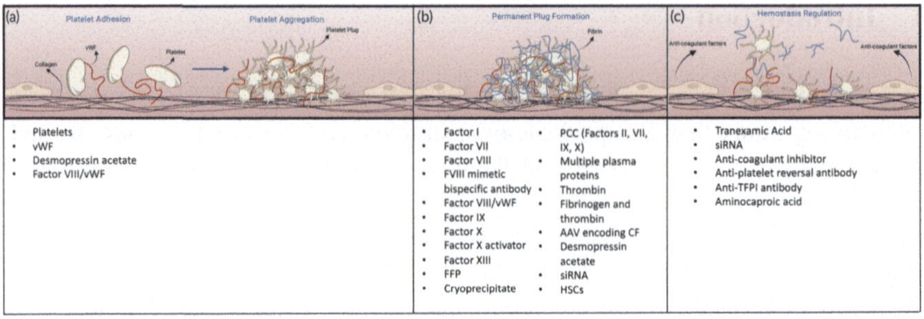

Fig. 1 Stages of hemostasis and corresponding agents are used to enhance the respective stage. (**a**) Primary hemostasis involves platelet adhesion and platelet aggregation, which eventually leads to platelet plug formation. (**b**) Secondary hemostasis involves the coagulation cascade, which terminates with the formation of a permanent plug characterized by a fibrin mesh. (**c**) Regulation of hemostasis consists of a delicate balance between pro-coagulant and anti-coagulant factors. Anti-coagulant factors released by the endothelial cells limit hemostasis to the site of vascular injury. In cases of elevated anti-coagulant activity, agents that inhibit anti-coagulant are employed as hemostats. FFP, fresh frozen plasma; HSC, hematopoietic stem cell; siRNA, small interfering RNA; TFPI, tissue-factor pathway inhibitor; vWF, von Willebrand [5] (Open Access This article is licensed under a Creative Commons Attribution License)

tissue damage or systemic infection. On the other hand, in some cases, wounds may heal with excessive scar tissue formation, resulting in hypertrophic scars or keloids that are raised, thickened, and cosmetically undesirable. Wounds that fail to recover within a reasonable timeframe (usually 4–6 weeks) are considered chronic [7]. They often result from underlying medical conditions such as diabetes, vascular disease, or immune deficiencies and require specialized treatment to facilitate healing.

The treatment of skin wounds depends on various factors [6], including the type, size, and location of the wound and the patient's overall health status. Common treatment modalities include: (a) Thorough cleaning of the wound to remove debris, foreign material, and bacteria, usually with saline or antiseptic solutions. (b) Dressings should be applied to protect the wound, promote a moist environment conducive to healing, and absorb excess exudate. Dressings may vary depending on the type of wound and its stage of healing. (c) Application of topical ointments, creams, or antimicrobial agents to prevent infection, reduce inflammation and promote tissue regeneration. (d) Surgical procedures may be necessary for the debridement of necrotic tissue, closure of large complex wounds, and reconstruction of damaged tissues. (e) Advanced wound care modalities such as negative pressure wound therapy, bioengineered skin substitutes, growth factors, and hyperbaric oxygen therapy may be indicated for non-healing or chronic wounds.

Currently, there are various factors that can damage the skin depending on how the injury occurs. There are three main classifications: (i) mechanical injuries resulting from physical damage by an external agent, (ii) thermal or chemical injuries due to the possible action of a corrosive reagent, wich can be exposure to temperatures unsuitable for the skin [8], and finally, (iii) ulcerative injuries inherent to the organism, which can be caused by pathologies, such as infections and dermatitis.

However, different approaches are currently proposed to address skin injuries and achieve wound healing depending on the context in which the injury occurs and its severity. Initially, the solution to these injuries was the use of gauze or bandages to stop bleeding physically. Nowadays, due to advances in medicine, therapies for wound healing have been developed using stem cells [9], growth factors [10], adhesive patches [11], and even hydrogels [12]. As mentioned earlier, wound dressings have been seen as an innovative solution for skin injuries, being used to moisturize the affected area, transport substances, drugs, cells, or simply to protect the wound from external factors such as infections. Within this area of biomedicine, advances have been observed in the synthesis of wound dressing materials from natural polymers, derived from natural sources such as plants, animals, and microorganisms. Due to their characteristics, such as biocompatibility, biodegradability, versatility, sustainability, and functionalization, they are used in various techniques of current biomedicine, such as the development of scaffolds for drug delivery, implants, and tissue engineering [13].

The main advantages of using natural polymers are well documented in the literature, with mainly 5 advantages are described. Most natural polymers do not cause harm to living organisms when used at an appropriate concentration, being harmless to living tissue and without generating an immune response, thus providing the first quality of being biocompatible [14]. It is important to highlight that this property precisely allows the use of these polymers in biomedical materials, such as implants or extended-release drug delivery systems, meaning that the presence of the material will not interrupt or pause the wound-healing process [15]. As a second advantage, natural polymers are biodegradable, meaning that when they decompose into their respective metabolites and are non-toxic to living tissue, they are absorbed and integrated into the different metabolic pathways of the organism in question. The third advantage is their great versatility; they have a wide range of uses, are widely diverse, and come from different origins and organic processes [16]. The fourth advantage is the possibility of being sustainable, since a large part of these types of polymeric materials originate from renewable sources. The last advantage present in natural polymers, directly connected to their versatility, and functionalization. As seen in different studies, hydrogels composed of natural polymers can be modified through the addition or incorporation of specific functional groups, providing the ability to enhance each of their advantages and adding new ones at the same time.

From its beginnigs in the 1990s, marked by its use in dietary supplements for human health, it has evolived significantly. French physiologist Charles Marie Benjamin Rouget (1824–1904) discovered chitosan by conducting experiments with the natural polymer chitin [17]. Among the natural polymers, chitosan attracted the attention of researchers to use in wound and hemostatic applications [18]. Through chemical and temperature modifications of chitin from animal sources, it can generate the biopolymer natural polysaccharide and it known as amino polysaccharide. This biopolymer originates from insects, crustaceans, and fungi. Due to its composition, it is known for not presenting toxicity, being highly biocompatible, and possessing preservative qualities. Without modifications, it is not soluble in water and organic solvents [19], although when added to a slightly acidic

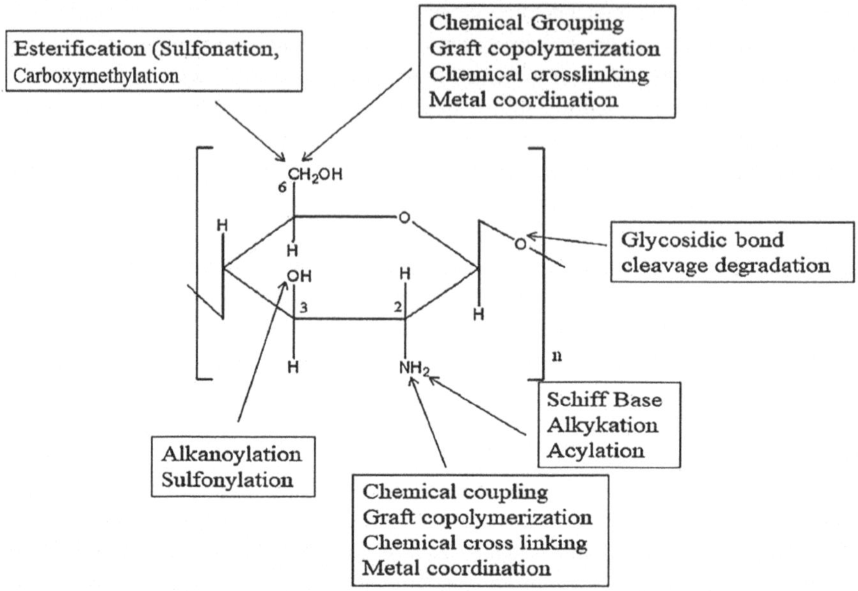

Fig. 2 Diagram of the chitosan modification process. Reproduced with permission [17], Copyright 2024, Elsevier

medium (<6.5), it can be dissolved, at lower pH levels and it can form gels. Conversely, in basic solutions, precipitation occurs. Among these are the primary amine group and the two primary and secondary hydroxyl groups, on carbons C2, C3, and C6, respectively. Carbon 2, with the primary amine group, is the leading actor in biological interaction.

The chitosan biomolecule is composed of variable proportions of carbon, nitrogen, hydrogen, and oxygen [20]. The presence of primary and secondary hydroxyl groups and the amino group results in the high chemical reactivity of chitosan, unlike its precursor chitin. It is important to remember that it is precisely the groups present at carbons 6 and 2 (amino and hydroxyl, respectively) that allow for modifications to be incorporated into the chitosan molecule. Different modification sites of chitosan have been described, as shown in Fig. 2 [17].

The conversion from chitin to chitosan is carried out from the exoskeletons of crustaceans and insects. The main processes for obtaining it involve bleaching with potassium permanganate, deproteinization, and demineralization with hydrochloric acid. After the latter process, chitin is obtained, which can be converted to chitosan through a deacetylation process carried out by chitin deacetylase. In this regard, it is important to remember that when the degree of deacetylation is greater than 90%, chitosan becomes slightly cytotoxic and non-degradable [21], so the degree of deacetylation of chitosan can be chosen in favor of its future use. Its molecular weight is divided into two categories: high molecular weight between 190 and 375 kDa and low molecular weight between 20 and 190 kDa. Both the degree of deacetylation and the molecular weight are important as they are inversely proportional to the physicochemical properties of biopolymer chitosan. Lately, chitosan has been used to enhance the biomaterial properties for hemostasis, infection control, and wound healing [18, 22]. This chapter discusses hemostasis stages and regulations, and chitosan-based biomaterials and their application in hemostasis, wound healing, and recent biomedicine applications. In addition, we discuss hemostasis phases and the ways to improve wound healing.

2 Hemostasis Stages

Hemostasis is a complex process carried out by cells, extracellular matrix elements, and signaling components such as factors, cytokines, and others [23]. The entire physiological process is divided into three main parts. It begins with vasoconstriction, which is the blocking of affected blood vessels, followed by platelet aggregation until coagulation occurs in the final stage. However, there are several hemostasis phases, such as vascular, platelet, coagulation, clot retraction, and fibrinolysis phases [24] (Fig. 1). Hemostasis begins immediately after injury to a blood vessel with vasoconstriction (vascular phase), which narrows the vessel lumen to reduce blood flow and minimize blood loss. Endothelial cells lining the blood vessels contract and release von Willebrand factor (vWF), which helps platelets adhere to the damaged vessel wall and start with the platelet phase. Platelets are small cell fragments in the blood that adhere to the exposed collagen fibers at the injury site, facilitated by vWF and other adhesive proteins. Once adhered, platelets become activated and change shape, releasing granules containing clotting factors such as thromboxane A2 and ADP [25]. This activation stimulates further platelet aggregation, forming a platelet plug or primary hemostatic plug.

The platelet phase must be continued with the coagulation phase in secondary hemostasis. The coagulation cascade, a complex series of enzymatic reactions, is initiated to reinforce the platelet plug and form a stable blood clot. Intrinsic and extrinsic pathways converge to activate factor X, which converts prothrombin to thrombin. Thrombin then converts fibrinogen into insoluble fibrin strands, which crosslink to form a mesh-like network that traps blood cells and platelets, consolidating the platelet plug into a fibrin clot. Moreover, once the fibrin clot is formed, platelets contract, pulling on the fibrin strands and compacting the clot. This process, known as clot retraction, reduces the size of the clot and stabilizes the wound site. In addition, when the wound heals, the clot is gradually dissolved by fibrinolysis, mediated by the enzyme plasmin, which breaks down fibrin into soluble fragments called fibrin degradation products (FDPs). Fibrinolysis prevents excessive clot formation and facilitates tissue repair.

3 Regulation of Hemostasis

Hemostasis is tightly regulated by a balance between procoagulant (promoting clot formation) and anticoagulant (inhibiting clot formation) factors to prevent inappropriate clotting or bleeding [26].

For sequential understanding, it is detailed that the procoagulant process begins when a blood vessel injury occurs, leading to the activation of platelets that adhere to the injury site, forming a temporary plug. Activated platelets release several chemicals, including adenosine diphosphate (ADP) and thromboxane A2, which activate additional platelets. Simultaneously, tissue factor (TF) is exposed at the injury site and binds with factor VII to form the TF-VIIa complex, which activates factor X to factor Xa. Factor Xa, in conjunction with factor Va, converts prothrombin into thrombin. Thrombin then converts fibrinogen into fibrin, which forms a mesh that stabilizes the platelet plug, creating a stable clot. This clot is further reinforced by the activation of factor XIII, which cross-links the fibrin mesh.

Once the necessary clot is formed, the anticoagulant process ensures that coagulation does not become excessive. Endothelial cells release anticoagulant factors such as tissue factor pathway inhibitor (TFPI), which inhibits the TF-VIIa complex, and thrombomodulin, which binds to thrombin and activates protein C. Activated protein C, along with its cofactor protein S, degrades factors Va and VIIIa. Additionally, antithrombin inhibits factors IIa (thrombin), IXa, Xa, XIa, and XIIa. The body maintains a delicate balance between procoagulant and anticoagulant factors to ensure proper coagulation only at the site of injury. Finally, the body eliminates the clot through fibrinolysis, where plasmin, activated from plasminogen by tissue plasminogen activator (tPA), degrades fibrin, restoring normal blood flow in the vessel.

The main regulatory mechanisms that are part of hemostasis are:

Endothelial Cells Endothelial cells produce anticoagulant factors, such as prostacyclin and nitric oxide, which inhibit platelet activation and coagulation [27]. They also express tissue factor pathway inhibitors to regulate the coagulation cascade.

Anticoagulant Proteins Circulating anticoagulant proteins, including antithrombin [28], protein C [29], and protein S [30], inhibit coagulation factors and limit thrombin formation.

Fibrinolytic System Plasminogen activators convert plasminogen into plasmin, which degrades fibrin clots. Tissue plasminogen activator and urokinase-type plasminogen activator are key regulators of fibrinolysis [31].

The series of processes involving hemostatic control occurs after the injury. When blood vessels rupture, thromboxane and epinephrine are released, respectively, causing the reduction of blood loss and the contraction of blood vessels [32]. Simultaneously, within the injury, collagen released from the ruptured blood vessel is surrounded by free-floating platelets in the blood. These release factors

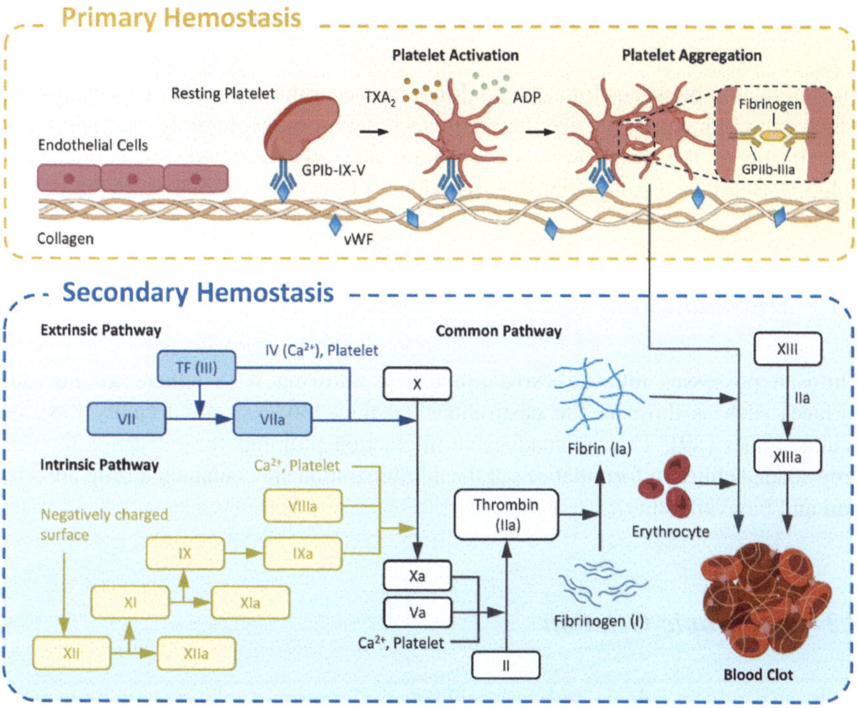

Fig. 3 Schematic illustration of the physiologic hemostasis process: primary and secondary hemostasis. Reproduced with permission [33], Copyright 2022, Elsevier

enhance platelet aggregation within the injury, increasing vasoconstriction and forming a platelet plug to block blood loss.

The activation of endogenous and exogenous coagulation pathways is also shown in Fig. 3 [33]. Specifically, the endogenous pathway begins, and factor XI is converted to activated factor XI (factor XIa). This factor, along with activated factor VII and tissue factor, converts factor IX to factor IXa. Factor IXa, when combined with factor VIIIa, activates factor X. Activated factor X (factor Xa) binds with factor Va and converts prothrombin to thrombin. The extrinsic pathway begins with the tissue factor binding to factor VII. Once activated, factors X and IX are activated along with VIIIa through proteolysis. Activated factor IX (IXa) binds to its cofactor, factor VIIIa, leading to the activation of factor X, producing Xa, which bound to activated factor V and calcium, generates a prothrombinase complex that cleaves prothrombin into thrombin. Finally, thrombin acts as a cofactor and catalyst for the involved proteolytic pathways [26].

4 Chitosan Interaction with Drugs for Biomedicine

The interaction between chitosan and drugs can occur through various mechanisms, which can ultimately influence the pharmacokinetics, bioavailability, and therapeutic efficacy of the drugs in wound healing and other biomedicine applications [34, 35]. Below is a comprehensive overview of how chitosan interacts with drugs,

4.1 Mucoadhesion

Chitosan possesses mucoadhesive properties, allowing it to adhere to mucosal surfaces such as those in the gastrointestinal tract [36, 37], nasal cavity [38], or ocular tissues [39]. This mucoadhesive interaction prolongs the residence time of drug-loaded chitosan formulations at the administration site, enhancing drug absorption and bioavailability.

4.2 Ionotropic Gelation

Chitosan can form polyelectrolyte complexes via ionotropic gelation with oppositely charged molecules, such as drugs or other polymers. This interaction occurs between the positively charged amino groups of chitosan and negatively charged functional groups of drugs, leading to the formation of stable complexes. The ionotropic gelation technique is one of the most studied methods for drug delivery. This is due to the ability to generate chitosan nanoparticles for drug transport using the ionotropic gelation method [40]. Ionotropic gelation is commonly employed to encapsulate drugs within chitosan nanoparticles or microparticles for controlled drug release.

4.3 Hydrogen Bonding

Chitosan molecules contain numerous hydroxyl and amino groups that can participate in hydrogen bonding with drug molecules. The presence of these groups has allowed for research development in various fields, such as wastewater remediation [41]. Hydrogen bonding interactions between chitosan and drugs contribute to the entrapment or encapsulation of drugs within chitosan matrices, enhancing drug stability and solubility. This phenomenon can improve the bioactivity of drug-encapsulated chitosan biomaterials in wound and other biomedical applications.

4.4 pH-Dependent Swelling and Dissolution

Chitosan has poor bioactivity at higher pH (more than 6), which can be improved by chitosan chemical modifications (carboxymethylation, hydroxypropylation, sulfation), complex formation with polyelectrolytes, metal ion complexation, and functionalization with nanomaterials and polymers [35]. These modifications can make chitosan water-soluble at higher pH, which is useful for wound applications. Chitosan exhibits pH-dependent swelling behavior, swelling in acidic environments due to the protonation of amino groups and forming a gel-like structure [42]. This pH-sensitive swelling property of chitosan can be exploited to control drug release, with faster release in acidic environments (e.g., stomach) and sustained release in neutral or basic environments (e.g., intestine). The ability to release drugs depending on the pH surrounding the chitosan matrix has initiated research in various areas aiming for targeted delivery. Specifically, the study presented involves a pH-sensitive hydrogel of gallic acid conjugated with chitosan that increases its water absorption capacity by 5,000% at pH 4 and simultaneously releases the chosen drug at the same pH [43], demonstrating the effectiveness and potential of chitosan for wound and biomedical applications.

4.5 Enhanced Permeability

Chitosan has been shown to enhance the permeability of drugs across biological barriers, such as the intestinal epithelium or blood-brain barrier. This enhancement in drug permeability is attributed to the reversible opening of tight junctions between epithelial cells, facilitated by interactions between chitosan and cell membrane proteins. The improvement in permeability due to chitosan has been utilized in transdermal hydrogels [44], chemotherapy [45], and intestinal absorption [46], among others.

4.6 Targeted Drug Delivery

Chitosan can be modified or conjugated with targeting ligands (e.g., antibodies, peptides) to deliver targeted drugs to specific cells or tissues and infected wounds. The interaction between chitosan and targeting ligands enables selective drug delivery, minimizing off-target effects and improving therapeutic outcomes. An example of targeted delivery conducted in recent years relates chitosan's mucoadhesive and pH-sensitive properties with drug delivery in breast cancer therapy [47].

4.7 *Biodegradability and Biocompatibility*

Chitosan is biodegradable and biocompatible, making it suitable for various drug delivery applications without eliciting significant adverse effects [48]. The interaction between chitosan and drugs occurs in a biocompatible environment, ensuring safe and effective drug delivery to the target site. These qualities have been utilized in various areas of current biomedicine. As mentioned throughout this text, biocompatibility and biodegradability are two of the most important qualities of chitosan and the main reasons for its current use. It is useful in compounds used within living organisms, such as implants, implant coatings, patches, and hydrogels that adhere to mucous membranes of the body.

5 Chitosan Interaction with Nanomaterials

The interaction between chitosan and nanomaterials plays a crucial role in the design and development of advanced drug delivery systems [49]. Here's a comprehensive overview of how chitosan interacts with nanomaterials in drug delivery and wound applications:

5.1 *Nanoparticle Formation*

Chitosan can self-assemble or be co-assembled with other materials to form nanoparticles with controlled size, shape, and surface properties [50]. Nanoparticles or nanomaterials composed of chitosan and various nanomaterials, such as polymers and inorganic nanoparticles, exhibit unique drug loading [51] and release [52] characteristics.

5.2 *Surface Modification*

Chitosan-coated or chitosan-modified nanomaterials can be synthesized to improve stability [53], biocompatibility [53, 54], and targetability of drug-loaded nanoparticles [55]. Surface modification of nanomaterials with chitosan enhances their colloidal stability, prolongs circulation time in vivo, and facilitates cellular uptake at the target site.

5.3 Drug Encapsulation and Entrapment

Chitosan interacts with drug molecules through various non-covalent interactions [56], including hydrogen bonding, electrostatic, and hydrophobic interactions. However, nanomaterials composed of chitosan can encapsulate or entrap hydrophobic, hydrophilic, or amphiphilic drugs within their matrix, providing protection and sustained release of the encapsulated drug [57].

5.4 pH-Responsive Drug Release

Chitosan-based nanomaterials can exhibit pH-responsive drug release behavior, with accelerated release in acidic environments (e.g., tumor microenvironment, endosomes) due to chitosan protonation. pH-responsive drug release from chitosan nanomaterials enables targeted and controlled drug delivery to specific tissues or cells, minimizing systemic side effects.

5.5 Active Targeting

Chitosan-functionalized nanomaterials can be conjugated with targeting ligands (e.g., antibodies, peptides, aptamers) to target specific cells or tissues actively. The interaction between chitosan and targeting ligands enhances the specificity and affinity of drug-loaded nanoparticles to the target site, improving therapeutic efficacy and reducing off-target effects [58].

5.6 Biocompatibility and Biodegradability

Chitosan-based nanomaterials exhibit excellent biocompatibility and biodegradability, making them suitable for biomedical applications without eliciting significant cytotoxic or immunogenic responses [59]. The interaction between chitosan and nanomaterials ensures the development of safe and effective drug delivery systems for clinical use [60].

5.7 Multifunctional Platforms

Chitosan nanomaterials can serve as multifunctional platforms for drug delivery, combining therapeutic agents, imaging agents, and targeting moieties within a single nanoparticle system. The integration of multiple functionalities into chitosan-based nanomaterials enables synergistic effects, personalized medicine approaches, and theranostic applications for precision medicine [61].

Overall, the chitosan and inorganic materials complex can improve the bioactivity of the nanoparticles, which are useful for wound and clinical applications.

6 Chitosan-Based Biomaterials for Hemostasis

Several chitosan-based materials have been developed for hemostasis, leveraging the hemostatic properties of chitosan to control bleeding (Fig. 4 I) in various clinical scenarios effectively [62]. Here are some examples:

6.1 Chitosan Sponges or Dressings

Chitosan sponges are porous materials that absorb blood and adhere to the wound site, promoting hemostasis through direct contact [62]. These dressings are biocompatible, biodegradable, and easily applied to wounds of different shapes and sizes [63]. In addition, chitosan sponges can be combined with other hemostatic or antimicrobial agents for enhanced efficacy. The positive charges within the chitosan molecule allow it to interact with different types of blood cells. This is why, upon contact with platelets, they adhere to the chitosan molecule, leading to platelet activation, increased platelet aggregation, and acceleration of the coagulation cascade [18].

Figure 4II shows the development of a super hydrophilic and high-strength chitosan hemostatic sponge prepared by freeze-drying and alkali treatment for rapid hemostasis. As a result, a sponge with strong absorption capacity, good mechanical properties, and rapid and superior hemostatic ability in vivo and in vitro was obtained [64].

6.2 Chitosan Nanofibers

Electrospun chitosan nanofibers offer a high surface area-to-volume ratio and enhanced mechanical strength, making them suitable for hemostatic applications [65]. Chitosan nanofiber mats can be applied directly to bleeding wounds to promote clot formation and accelerate hemostasis. The nanofiber morphology provides a scaffold for platelet adhesion and aggregation, facilitating the clotting process. The fabrication of nanofibers that include chitosan in their components has been successfully employed for wound healing in animal experiments, demonstrating its high biocompatibility and hemostatic action (Fig. 5) [66]. Figure 5 shows how researchers

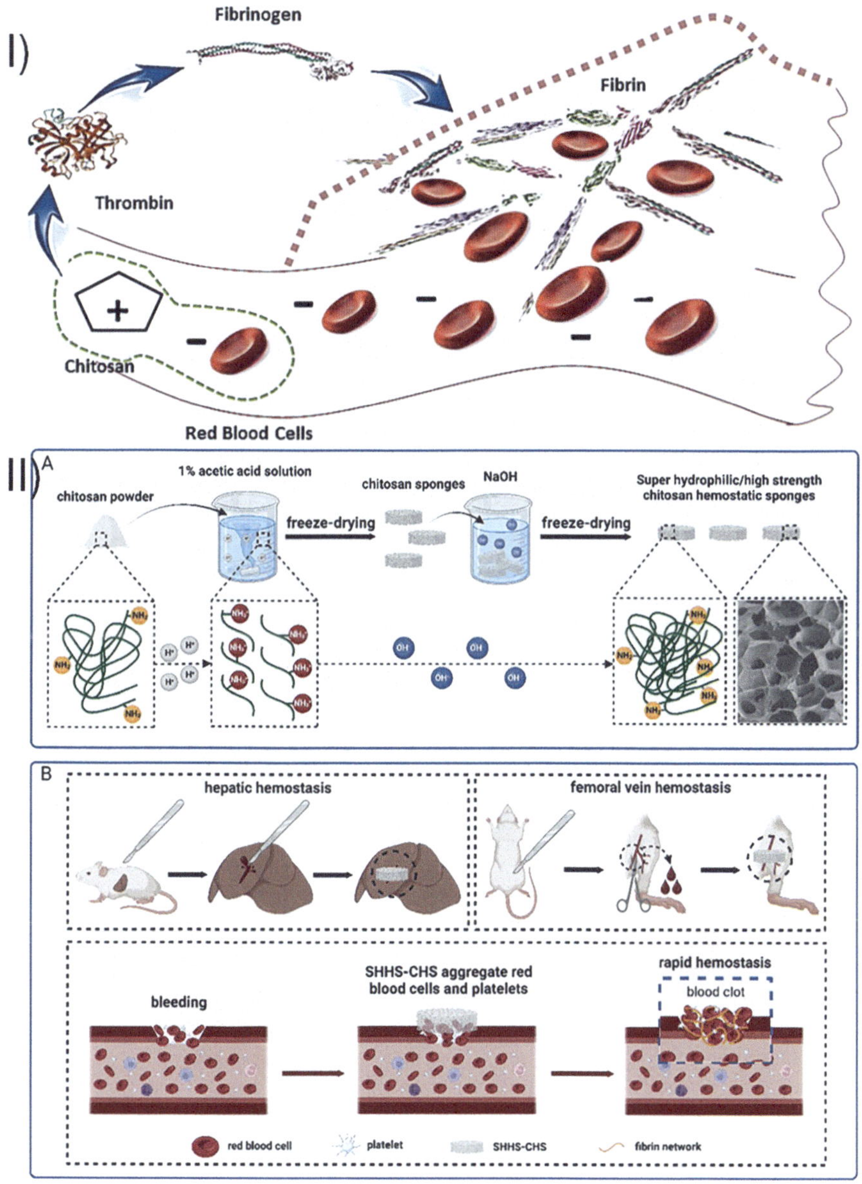

Fig. 4 (I) Hemostatic mechanism of chitosan in hemostasis [62] (Open Access This article is licensed under a Creative Commons Attribution License) and (II) The preparation of super hydrophilic and high strength chitosan hemostatic sponge (SHHS-CHS) (**a**). The application and hemostatic process of SHHS-CHS (**b**), Reproduced with permission [64], Copyright 2024, Elsevier

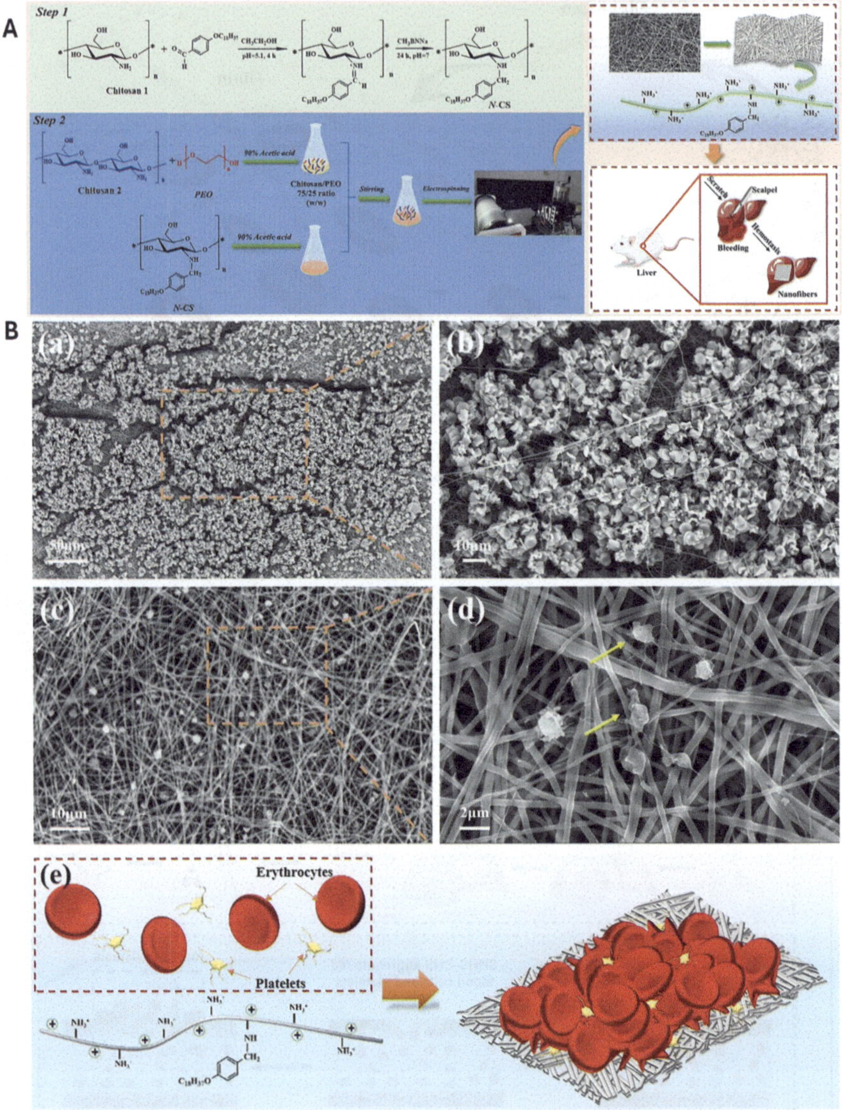

Fig. 5 Schematic diagram of the preparation and application for hemostasis in the liver injury of CPNs nanofibers (**A**): Step 1: The synthesis mechanism of N-alkylated chitosan (N-CS); (**B**) Step 2: The nanofibers preparation process of N-alkylated chitosan (N-CS) grafted with octadecyl into a chitosan/polyethylene oxide (PEO) matrix (CPNs). (**a**) and (**b**) SEM images of erythrocytes aggregation on the surface of nanofibers with a small amount of N-CS (CPN31); (**c**) and (**d**) SEM images of platelets adhesion on the surface of CPN31 (the yellow arrows represent activated platelets); (**e**) schematic diagram of erythrocytes and platelets adhesion on the surface of nanofibers. Reproduced with permission [66], copyright 2024, Elsevier

aimed to create an ideal hemostatic agent to stop bleeding and quickly prevent secondary bleeding during bandage changes. They developed a hydrophobic electrospun nanofiber membrane by doping N-alkylated chitosan (N-CS) grafted with octadecyl into a chitosan/polyethylene oxide (PEO) matrix. The principal findings include good membrane effectiveness, reduced blood loss, and easy removal. Mainly, it results in a significantly reduced hemostasis time and promotes stable and strong blood clot formation compared to traditional chitosan/PEO membranes and medical gauze.

6.3 Chitosan Composites as Hemostatic Agents

Chitosan can be combined with other hemostatic agents such as zeolites [67], clay minerals [68], or calcium salts [69] to enhance its hemostatic properties. These composite materials exhibit synergistic effects, with chitosan promoting clot formation and the additional hemostatic agents accelerating coagulation cascade reactions. Combining chitosan with these hemostatic agents results in more effective bleeding control than chitosan alone, making them valuable for medical applications, such as wound dressing and surgical sealants. In Fig. 6a, the use of a clay mineral, attapulgite, is described. When added to an alkylated chitosan matrix, it generates significant hemostatic effects by increasing blood absorption capacity, decreasing blood coagulation time, and stimulating platelet adhesion and aggregation [67].

6.4 Chitosan-Hemostatic Sprays or Sealants

Chitosan-based hemostatic sprays or sealants are liquid formulations that can be sprayed directly onto bleeding surfaces to achieve rapid hemostasis. These formulations adhere to the tissue surface, forming a seal or barrier that stops bleeding and promotes clot formation. In Fig. 6b, research shows that a hemostatic membrane system was created based on spray-assisted layer-by-layer electrostatic assemblies of oppositely charged polyphosphate (polyP) and chitosan (Cs). The membrane produced a microporous structure with high flexibility, and blood coagulation tests and a rodent hepatic severe hemorrhage model revealed the excellent hemostatic performance of the membrane system [70].

6.5 Chitosan-Hemostatic Patches or Films

Chitosan-based hemostatic patches or films provide a flexible and conformable platform for wound coverage and hemostasis. These patches or films can be applied to surgical incisions, vascular puncture sites, or traumatic injuries to control bleeding

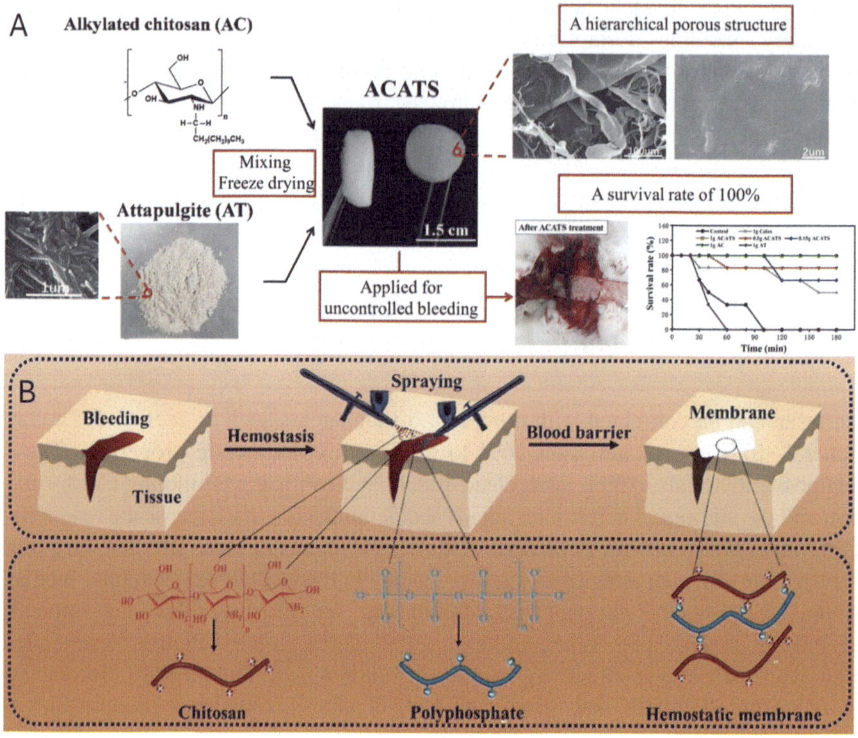

Fig. 6 (**a**) Graphical abstract from "Alkylated chitosan-attapulgite composite sponge for rapid hemostasis". Reproduced from ref. [67] with permission. This study reports the development of a composite sponge (ACATS) based on alkylated chitosan (AC) and attapulgite (AT) for rapid hemostasis. The principal highlights include a hierarchical porous structure with favorable biocompatibility and a survival rate of 100% in rabbit carotid artery injury model in vivo. (**b**) Graphical abstract extracted from the article: "Sprayable surface-adaptive biocompatible membranes for efficient hemostasis via assembly of chitosan and polyphosphate". In this work, the researchers produce a hemostatic membrane system based on spray-assisted layer by layer. The results are high flexibility, robust micro-stratified porous structure, excellent hemostatic performance, and good biocompatibility and biodegradability of the system. Reproduced from Ref. [70] with permission from Elsevier

and minimize blood loss. Chitosan-hemostatic patches may be impregnated with additional hemostatic agents or growth factors to promote wound healing [71].

6.6 *Chitosan-Coated Hemostatic Agents*

Chitosan coatings can be applied to existing hemostatic agents such as gauzes, sponges, or powders to improve their hemostatic efficacy [71]. The chitosan coating enhances the adhesion of the hemostatic agent to the wound site and provides additional hemostatic activity through direct contact with blood [72].

6.7 Chitosan Hydrogels

Chitosan has been used in the development of hydrogels for hemostatic applications due to its interaction with negatively charged erythrocytes and platelet cell membranes [73]. In addition, they are biocompatible, injectable, and easy to handle [18]. However, chitosan has been chemically modified to improve the homeostasis properties of hydrogels and other components have been used [71]. Recently, soft hydrogels have been developed from chitosan, sodium tripolyphosphate, aluminum chloride, and silica nanoparticles [74]. This soft hydrogel has a significantly higher blood clotting capacity (3.5-fold) than commercial HemoStat clear gel. Similarly, flexible hemostatic bio-hydrogels developed from modified chitosan (3-3,4-dihydroxyphenyl)propionic acid-modified chitosan), sebacic acid-terminated polyethylene glycol (PEGS) and p-hydroxybenzaldehyde terminated PEGS [75]. These hydrogels have self-adhesive, self-healing, stretchability, antibacterial performance, and biocompatibility properties with blood clotting.

In order to improve the hemostatic properties of chitosan based hydrogels, they were prepared with other materials such as gelatin, collagen, hyaluronic acid, fibrin, calcium derivatives (CaO, CaO_2, $CaCl_2$), nano-bioglass, nano whitlockite ($Ca_{18}Mg_2(HPO_4)_2(PO_4)_{12}$), MXene, copper and potassium aluminum sulfate [76–81]. According to the reports, chitosan/potassium aluminum sulfate/calcium chloride hydrogels can stop bleeding within a short time compared to commercially available Fibrin sealants and Floseal agents [81]. Overall, these chitosan-based hydrogels are a promising system that can stop bleeding and potentially save lives.

6.8 Commercially Available Chitosan-Based Hemostatic Dressing

Chitosan interacts with red blood cells and platelets to promote clotting, as it carries a positive charge that attracts negatively charged red blood cells and platelets. This helps to accelerate the clotting cascade and form a stable clot to stop bleeding. Therefore, it has been used to develop commercially available hemostatic dressing materials. Additionally, chitosan-based hemostatic dressing materials have received FDA approval for use in clinical applications, including use in the battlefield [77]. These chitosan-based hemostatic dressings have been used in emergency medical services in both the military and civilian hospital settings. Specifically, the commercially available ChitoFlex exhibits better hemostatic properties than other polymer-based hemostatic agents [82]. Other chitosan-based hemostatic materials used for hemostatic applications include Trauma Gauze, HemCon, Celox, Mini sponge dressing, Celox Gauze, and ChitoGauze [83]. These materials work through the exact chitosan-mediated clotting mechanism. Some of these chitosan-based hemostatic dressings, such as Celox trauma gauze, ChitoGauze, Mini-sponge, Celox, TraumaStat, Celox Gauze, Celox Rapid, and Celox Trauma Gauze, are used explicitly for arterial hemorrhage of the extremities. Overall, chitosan-based hemorrhage agents help accelerate the strengthening of blood clots and wound healing in clinical applications.

7 Recent Applications in Wound and Biomedicine

7.1 Advanced Wound Dressings

Chitosan-based wound dressings have been developed with enhanced properties, such as improved antimicrobial activity, promotion of tissue regeneration, and controlled drug release for effective wound healing [84–86]. Lately, dual-layer dermal substitutes have been generated to protect against external pathogens in a wound [87]. As a result, the dermal substitute made from polyvinylalcohol/chitosan/sildenafil citrate (PVA/CS/SC) electrospun onto the polycaprolactone (PCL) membrane showed a clear acceleration in wound healing (Fig. 7A) in rat models. As shown in Fig. 7B(a, b) this model increases wound closure and in histological sections, the development of a more mature dermis than the control sample and even the growth of hair follicles was demonstrated.

7.2 Drug Delivery Systems

The transport of substances with pharmacological functions into the body is an increasingly sophisticated technique that aims to deliver the drug to the correct location with the appropriate timing and dosage. This approach helps to prevent the drug from being degraded during digestion before reaching the target area and simultaneously avoids damaging healthy cells with the drug. To this end, various methods have been devised to transport substances into the body, including liposomes, nanoparticles, and micelles. Linked to these transport methods, it is important to note the matrix or substances through which they can be transported. Degradable biopolymers, such as chitosan, have become significant in this area. The main uses of chitosan in substance transport are due to its high mucoadhesion, which allows it to adhere to mucous membranes within the body and release the desired drug in a prolonged manner [88]. Chitosan nanoparticles and microparticles are being used as carriers for targeted drug delivery, allowing for the controlled release of therapeutic agents to specific tissues or organs, leading to improved treatment efficacy and reduced side effects.

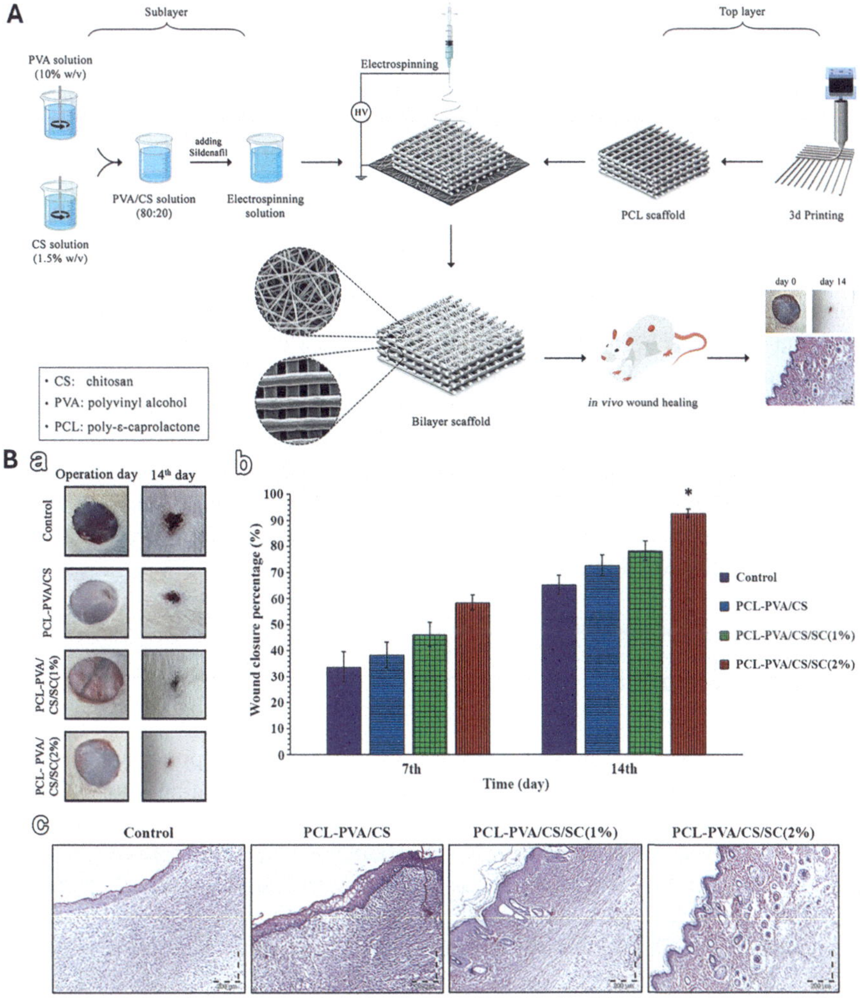

Fig. 7 Study in which double-layer dermal substitutes for wound healing were developed from polyvinyl alcohol, chitosan, and sildenafil citrate with a polycaprolactone membrane (**A**). In vivo evaluation of the wound dressings therapeutic efficacy on Winstar rats skin wounds model (**B**): macroscopic appearances of the treated wound on 1st and 14th days post-operation (**a**), histograms comparing the wound closure percentages in the control group, PCL-PVA/CS, PCL-PVA/CS/SC (1%w/v), and PCL-PVA/CS/SC (2%w/v) treated groups on 7th and 14th days post-operation (**b**). H&E-stained sections of skin samples from the wound site of different study groups. Reproduced with permission [87], Copyright 2024, Elsevier

7.3 *Tissue Engineering Scaffolds*

Chitosan scaffolds are utilized as three-dimensional structures to support cell growth and tissue regeneration in tissue engineering applications, particularly in the repair and regeneration of bone, cartilage, and skin tissues [87, 89]. Chitosan has been used for multiple applications in tissue engineering, either as matrices composed solely of chitosan or as hybrids with components such as polyvinylalcohol [90], silk fibroin [91], and collagen [92], among others. Types of technologies created from chitosan in this area include 3D cell culture matrices, regeneration of nervous system components [93], and promotion of osteogenic differentiation [91], among many others. Chitosan alone has been implemented in recent years in cell cultures that require slightly more care, such as neural stem cells [94]. The resulting product was a cost-effective, easily producible 3D culture system that allowed for quality control, standardization, and scalability. Chitosan in this product contributes to stem cell adhesion and metabolic activity, enrichment of the cell population, and suppression of the survival of the differentiated population [94]. Chitosan functionalized bioactive glass scaffolds have significant biocompatibility, non-toxicity, anti-inflammatory, and diabetic wound healing capacity (Fig. 8A).

7.4 *Dental Materials*

Chitosan-based materials are being explored for various dental applications, including oral drug delivery systems [95], dental implant coatings to enhance osseointegration [96], and antimicrobial dental composites [97]. An example of the use of chitosan in the dental area is the production of hybrids between chitosan and bioglass. Through these hybrids, drug delivery with antibacterial effects in the target area can be achieved, along with the capability to generate osteoinduction in rat models [95]. Chitosan gel was used to achieve hemostasis in dental extractions for patients on antiplatelet therapy, who are at a higher risk of bleeding [98]. According to recent reports, chitosan gel has been found to be more effective in stopping bleeding within 1.182 min than platelet-rich Fibrin (2.64 min) (Fig. 8B, C).

7.5 *Biomedical Coatings*

Chitosan coatings are used in medical devices and implants to improve biocompatibility, reduce infection risk [100], and promote tissue integration, thereby enhancing the performance and longevity of the implants. The possibility of coating existing materials used for implants in biomedicine generates special attention towards biopolymers like chitosan. A study using hydroxyapatite nanocomposites with chitosan

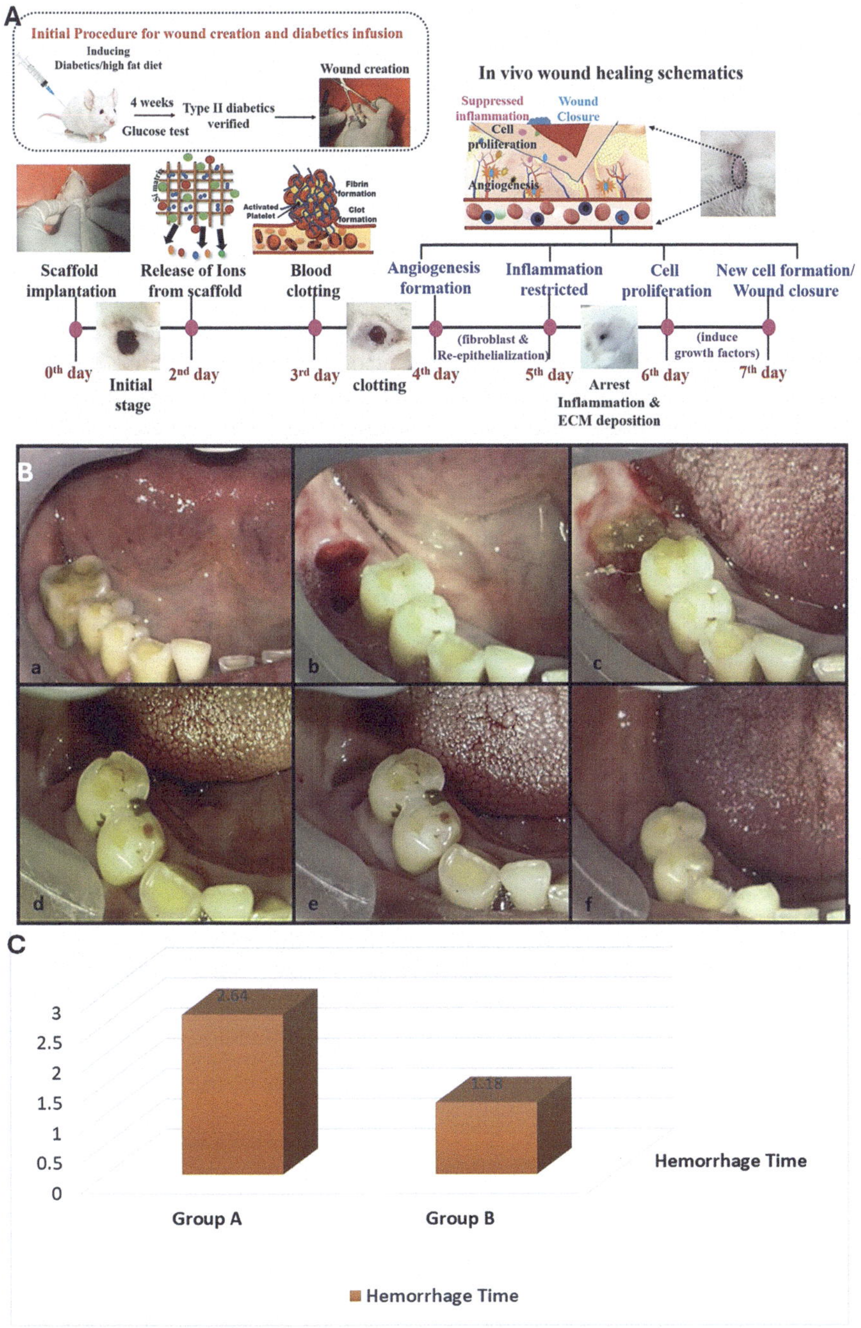

Fig. 8 (**A**) Experimental design of a study focused on the production of a subcutaneous chitosan-based implant for the healing of diabetic wounds in an animal model. The results showed good biocompatibility, cytocompatibility, antioxidant effect, and tissue regeneration. Reproduced with

to coat titanium bone implants resulted in the deposition of the nanocomposite on the titanium and an improvement in corrosion resistance compared to an uncoated implant [101].

7.6 Antimicrobial Textiles

Chitosan is incorporated into textile fibers and fabrics to impart antimicrobial properties, making them suitable for use in healthcare settings, personal protective equipment, and consumer products to prevent the spread of infections. In recent years, technologies have been developed for medical textiles with antibacterial and anti-SARS-CoV-2 action through smart nanoparticles in a chitosan matrix with gelatin as a crosslinker. This research yielded successful results with stable nanoparticles and a substantial and controlled release profile that allows for the antimicrobial effect [102].

8 Chitosan Biomaterials for Miscellaneous Applications

Chitosan-based materials have been used in several advanced applications due to their cationic nature and functionality. Due to this unique property, these biomaterials have been used for the following recent applications.

8.1 Cholesterol Detectors

Chitosan has been explored for biosensors to rapidly and accurately detect cholesterol levels in biological samples, such as blood or serum. It is a cationic polysaccharide and is used in the industry as a cholesterol detector due to its reactivity in forming hydrogen bonds between the nitrogen groups of its structure and the hydrogen groups of cholesterol [103].

Fig. 8 (continued) permission [99]. (**B**) Procedure in dental patients at different times with Chitosan hydrogel. (**a**) Patient without dental extraction, (**b**) immediate post extraction, (**c**) socket with Chitosan hydrogel, (**d**) first post-operation day, (**e**) third post-operation day (**f**) seventh post-operation day. (**C**) Correlation hemorrhage time in subjects of Group A (treated with platelet-rich fibrin) and Group B (treated with chitosan hydrogel). Reproduced **B** and **C** with permission [98]

8.2 Environmental Remediation

Chitosan-based materials are employed for wastewater treatment [104], heavy metal removal, and oil spill cleanup due to their adsorption capacity, biodegradability, and eco-friendly nature, contributing to environmental sustainability efforts. The use of chitosan in the absorption of heavy metals, such as Ni and Co, has been tested in magnetic nanocomposites in combination with iron oxide [105]. At the same time, the use of chitosan in absorbent matrices for compounds, such as oils, has been tested, showing high effectiveness in the absorption and removal of oil from water with an efficiency of 99% [106].

8.3 Heavy Metal Absorption

Chitosan has shown the ability to adsorb heavy metals in aqueous solutions, making it a promising material for the purification of contaminated water and environmental remediation. Due to the presence of chelating groups in chitosan, there is an attraction with metal ions through ionic exchange or coordination bridges. Research has described findings consistent with the absorption and removal of heavy metals through chitosan [107]. However, for adequate absorption of heavy metals, it is necessary to have a larger surface area, which is achieved by reducing its particle size.

8.4 Biodegradable Bioplastics

The ability of chitosan to form biodegradable films makes it suitable for use in the manufacturing of bioplastics, a more sustainable alternative to conventional plastics. An example of this is the hybrid plastic developed by Wei Liang and collaborators, which is a hybrid plastic based on polyvinyl alcohol (PVA), chitosan, and isolated soy protein. Compared to a common plastic, it exhibits less roughness on the surface, good water resistance, improved mechanical properties, and good degradation in soil [108].

8.5 Pharmaceutical Industry

Chitosan is used in a variety of biomedical applications in healthcare. Its use has been tested to improve drug survival; a clear example is probiotics, where its effectiveness in protecting the drug under gastrointestinal conditions has been demonstrated through encapsulation with chitosan [109]. It has even been studied in the field of agrochemistry for its use in the prolonged release of salicylic acid. In research, chitosan in nanoparticle form was combined with lignin nanoparticles to achieve prolonged drug release [110]. Several studies point to chitosan as an ideal macromolecule for wound healing, primarily used as a matrix for containing various substances that promote wound healing [111–113].

8.6 Hydrogels

Chitosan-based hydrogels have been extensively researched for biomedical applications such as tissue engineering, wound healing, controlled drug release, and organ regeneration [85]. The hydroxyl and amino functional groups of chitosan and its derivatives are important facilitators for the use of chitosan in the formation of hydrogels. These hydrogels are made using various physical methods, such as ionic interaction, electrostatic interaction, hydrogen bonding, and chemical methods, such as Schiff base reaction, and chemical crosslinking [15].

8.7 Food Packaging

Chitosan is used as an edible coating in the food industry due to its antimicrobial properties and ability to prolong fresh foods' shelf life. Recently, using chitosan, polyvinyl alcohol, and graphene oxide, a ternary bionanocomposite was generated for food packaging. In this research, a bionanocomposite capable of outperforming conventional packaging in terms of viability was obtained, and it was also demonstrated to have good food storage capacity. This resulted in the generation of a biocompatible, eco-friendly nanocomposite with promising applications in food packaging [114].

8.8 Dietary Supplements

Due to their natural origin and potential health benefits, chitosan is being investigated as a possible component in dietary supplements to promote weight loss and gastrointestinal health [115]. Regarding its role as a dietary supplement, experiments with fish have shown improvements in: protease and lipase activities, barrier function in the intestinal mucosa, antioxidant capacity, liver health, etc. However, in the same study, it was also discovered that an excessive intake of chitosan weakens its beneficial effects [115].

9 Future Aspects

The field of chitosan-based materials continues to evolve, with ongoing research focusing on addressing current challenges and exploring new opportunities. Here are some future aspects to consider for further advancement in chitosan-based materials:

9.1 Enhanced Hemostatic Performance

Develop significant chitosan-based materials with improved hemostatic properties, including faster onset of action, higher efficacy in controlling severe bleeding, and compatibility with various wound types and conditions. Explore novel strategies to enhance the adhesion, aggregation, and activation of platelets and clotting factors on chitosan surfaces for rapid and efficient hemostasis.

9.2 Controlled Drug Delivery

Design chitosan-based drug delivery systems with precise control over drug release kinetics, targeting specific tissues or cells, and minimizing systemic side effects. Investigate advanced formulations such as chitosan nanoparticles, microparticles, or hydrogels for sustained and controlled delivery of therapeutics, including small molecules, proteins, nucleic acids, and vaccines.

9.3 Bioactive Wound Healing

Develop chitosan-based materials that promote hemostasis and accelerate wound healing by releasing growth factors, cytokines, and other bioactive molecules. Explore the integration of chitosan with biomimetic scaffolds, extracellular matrix components, or stem cells to enhance tissue regeneration, angiogenesis, and re-epithelialization in chronic or non-healing wounds.

9.4 Antimicrobial Activity

Engineer chitosan-based materials with potent antimicrobial properties to prevent wound infections and promote aseptic wound healing. Investigate synergistic combinations of chitosan with antimicrobial agents, nanoparticles, or photodynamic therapy for combating multidrug-resistant pathogens and biofilm-associated infections.

9.5 Biocompatibility and Immunomodulation

Optimize chitosan-based materials' biocompatibility and immunomodulatory properties to minimize inflammation, foreign body reactions, and immune rejection. Explore surface modifications, functionalization techniques, or hybrid materials to tailor the immune response and promote tissue integration and regeneration.

9.6 Personalized Medicine and Theragnostic

Develop chitosan-based materials for personalized medicine applications, including patient-specific drug dosing, targeted therapy, and diagnostics. Explore theragnostic platforms combining therapeutic and diagnostic capabilities within a single chitosan-based nanomaterial for real-time monitoring and treatment of diseases.

9.7 Regulatory Approval and Clinical Translation

Conduct rigorous preclinical studies and clinical trials to evaluate chitosan-based materials' safety, efficacy, and long-term outcomes in various medical applications. Work closely with regulatory agencies to obtain approval for chitosan-based products and facilitate their translation from bench to bedside.

10 Conclusion

Chitosan and its derivatives present promising prospects for revolutionizing hemostasis and wound-healing strategies in the biomedical field. Chitosan-based materials effectively manage bleeding disorders and promote tissue repair through their inherent hemostatic properties and biocompatibility. The versatility of chitosan allows for tailored approaches, including the development of hemostatic dressings, scaffolds for tissue regeneration, and targeted drug delivery systems. Moreover, ongoing research and technological advancements expand the horizons of chitosan-based therapies, offering new avenues for addressing unmet clinical needs in hemostasis and wound care. By harnessing chitosan's unique properties, future endeavors can improve patient outcomes, enhance surgical procedures, and accelerate the healing process, ultimately contributing to regenerative medicine and patient care advancement.

References

1. Wang J et al (2024) Pharmaceutical applications of chitosan in skin regeneration: a review. Int J Biol Macromol 261:129064. https://doi.org/10.1016/j.ijbiomac.2023.129064
2. Varaprasad K, Jayaramudu T, Kanikireddy V, Toro C, Sadiku ER (2020) Alginate-based composite materials for wound dressing application: a mini review. Carbohydr Polym 236: 116025. https://doi.org/10.1016/j.carbpol.2020.116025
3. Patrulea V, Ostafe V, Borchard G, Jordan O (2015) Chitosan as a starting material for wound healing applications. Eur J Pharm Biopharm 97:417–426. https://doi.org/10.1016/j.ejpb.2015.08.004
4. Nandhini SN et al (2023) Recent advances in green synthesized nanoparticles for bactericidal and wound healing applications. Heliyon 9(2):e13128. https://doi.org/10.1016/j.heliyon.2023.e13128
5. Joshi M, Zhao Z, Mitragotri S (2024) Hemostats in the clinic. Bioeng Transl Med:e10673. https://doi.org/10.1002/btm2.10673
6. Guo S, DiPietro LA (2010) Factors affecting wound healing. J Dent Res 89(3):219–229. https://doi.org/10.1177/0022034509359125
7. Kanikireddy V, Varaprasad K, Jayaramudu T, Karthikeyan C, Sadiku R (2020) Carboxymethyl cellulose-based materials for infection control and wound healing: a review. Int J Biol Macromol 164:963–975. https://doi.org/10.1016/j.ijbiomac.2020.07.160
8. Chhabra S, Chhabra N, Kaur A, Gupta N (2017) Wound healing concepts in clinical practice of OMFS. J Maxillofac Oral Surg 16(4):403–423. https://doi.org/10.1007/s12663-016-0880-z
9. Yue Y et al (2024) A carboxymethyl chitosan/oxidized hyaluronic acid composite hydrogel dressing loading with stem cell exosome for chronic inflammation wounds healing. Int J Biol Macromol 257:128534. https://doi.org/10.1016/j.ijbiomac.2023.128534
10. Lin S et al (2021) Chitosan-poloxamer-based thermosensitive hydrogels containing zinc gluconate/recombinant human epidermal growth factor benefit for antibacterial and wound healing. Mater Sci Eng C 130:112450. https://doi.org/10.1016/j.msec.2021.112450
11. Wang L et al (2024) Tissue adhesives based on chitosan for skin wound healing: where do we stand in this era? A review. Int J Biol Macromol 258:129115. https://doi.org/10.1016/j.ijbiomac.2023.129115
12. Mottaghitalab F, Khodadadi Yazdi M, Reza Saeb M, Bączek T, Farokhi M (2024) Green and sustainable hydrogels based on quaternized chitosan to enhance wound healing. Chem Eng J 492:152288. https://doi.org/10.1016/j.cej.2024.152288
13. Varaprasad K, Karthikeyan C, Yallapu MM, Sadiku R (2022) The significance of biomacromolecule alginate for the 3D printing of hydrogels for biomedical applications. Int J Biol Macromol 212:561–578. https://doi.org/10.1016/j.ijbiomac.2022.05.157
14. Paradowska-Stolarz A, Wieckiewicz M, Owczarek A, Wezgowiec J (2021) Natural polymers for the maintenance of Oral health: review of recent advances and perspectives. Int J Mol Sci 22(19):10337. https://doi.org/10.3390/ijms221910337
15. Varaprasad K, Raghavendra GM, Jayaramudu T, Yallapu MM, Sadiku R (2017) A mini review on hydrogels classification and recent developments in miscellaneous applications. Mater Sci Eng C 79:958–971. https://doi.org/10.1016/j.msec.2017.05.096
16. Plotniece A et al (2023) Selected strategies to fight pathogenic bacteria. J Enzyme Inhib Med Chem 38(1):2155816. https://doi.org/10.1080/14756366.2022.2155816

17. Jagdale S, Agarwal B, Dixit A, Gaware S (2024) Chitosan as excellent bio-macromolecule with myriad of anti-activities in biomedical applications – a review. Int J Biol Macromol 257: 128697. https://doi.org/10.1016/j.ijbiomac.2023.128697
18. Fan P, Zeng Y, Zaldivar-Silva D, Agüero L, Wang S (2023) Chitosan-based hemostatic hydrogels: the concept, mechanism, application, and prospects. Molecules 28(3):1473. https://doi.org/10.3390/molecules28031473
19. Panda PK, Dash P, Chang Y-H, Yang J-M (2022) Improvement of chitosan water solubility by fumaric acid modification. Mater Lett 316:132046. https://doi.org/10.1016/j.matlet.2022.132046
20. Raghavendra GM, Jung J, Kim D, Varaprasad K, Seo J (2016) Identification of silver cubic structures during ultrasonication of chitosan $AgNO_3$ solution. Carbohydr Polym 152:558–565. https://doi.org/10.1016/j.carbpol.2016.07.045
21. Islam N, Dmour I, Taha MO (2019) Degradability of chitosan micro/nanoparticles for pulmonary drug delivery. Heliyon 5(5):e01684. https://doi.org/10.1016/j.heliyon.2019.e01684
22. Karthikeyan C et al (2023) Hybrid nanomaterial composed of chitosan, curcumin, ZnO and TiO_2 for antibacterial therapies. Int J Biol Macromol 242:124814. https://doi.org/10.1016/j.ijbiomac.2023.124814
23. Wilhelm G et al (2023) The crossroads of the coagulation system and the immune system: interactions and connections. Int J Mol Sci 24(16):12563. https://doi.org/10.3390/ijms241612563
24. Boon GD (1993) An overview of hemostasis. Toxicol Pathol 21(2):170–179. https://doi.org/10.1177/019262339302100209
25. Periayah MH, Halim AS, Saad AZM (2017) Mechanism action of platelets and crucial blood coagulation pathways in hemostasis. Int J Hematol-Oncol Stem Cell Res 11(4):319–327
26. Palta S, Saroa R, Palta A (2014) Overview of the coagulation system. Indian J Anaesth 58(5): 515. https://doi.org/10.4103/0019-5049.144643
27. Neubauer K, Zieger B (2022) Endothelial cells and coagulation. Cell Tissue Res 387(3): 391–398. https://doi.org/10.1007/s00441-021-03471-2
28. Loncar R, Kalina U, Stoldt V, Thomas V, Scharf RE, Vodovnik A (2006) Antithrombin significantly influences platelet adhesion onto immobilized fibrinogen in an in-vitro system simulating low flow. Thromb J 4:19. https://doi.org/10.1186/1477-9560-4-19
29. Rezaie AR (2010) Regulation of the protein C anticoagulant and antiinflammatory pathways. Curr Med Chem 17(19):2059–2069. https://doi.org/10.2174/092986710791233706
30. Gierula M, Ahnström J (2020) Anticoagulant protein S – new insights on interactions and functions. J Thromb Haemost 18(11):2801–2811. https://doi.org/10.1111/jth.15025
31. Hvas CL, Larsen JB (2023) The fibrinolytic system and its measurement: history, current uses and future directions for diagnosis and treatment. Int J Mol Sci 24(18):14179. https://doi.org/10.3390/ijms241814179
32. Rodrigues M, Kosaric N, Bonham CA, Gurtner GC (2019) Wound healing: a cellular perspective. Physiol Rev 99(1):665–706. https://doi.org/10.1152/physrev.00067.2017
33. Zou C-Y et al (2022) Design of biopolymer-based hemostatic material: starting from molecular structures and forms. Mater Today Biol 17:100468. https://doi.org/10.1016/j.mtbio.2022.100468
34. Guadarrama-Escobar OR et al (2023) Chitosan nanoparticles as oral drug carriers. Int J Mol Sci 24(5):4289. https://doi.org/10.3390/ijms24054289
35. Liu H et al (2018) A functional chitosan-based hydrogel as a wound dressing and drug delivery system in the treatment of wound healing. RSC Adv 8(14):7533–7549. https://doi.org/10.1039/C7RA13510F
36. Phuong Ta L, Bujna E, Kun S, Charalampopoulos D, Khutoryanskiy VV (2021) Electrosprayed mucoadhesive alginate-chitosan microcapsules for gastrointestinal delivery of probiotics. Int J Pharm 597:120342. https://doi.org/10.1016/j.ijpharm.2021.120342

37. Bian D, Chen Z, Ouyang Y, Wang S, Wang M, Chen W (2024) Ultrafast self-gelling, sprayable, and adhesive carboxymethyl chitosan/poly-γ-glutamic acid/oxidized dextran powder for effective gastric perforation hemostasis and wound healing. Int J Biol Macromol 254: 127960. https://doi.org/10.1016/j.ijbiomac.2023.127960
38. Mahajan S, Nangare S, Chaudhari A, Patil G (2024) Synthesis of chitosan-graphene oxide thermosensitive in situ hydrogel for nasal delivery of rasagiline mesylate: in-vitro-ex vivo characterization. J Drug Deliv Sci Technol 95:105549. https://doi.org/10.1016/j.jddst.2024.105549
39. Javed S et al (2023) Tobramycin-loaded nanoparticles of thiolated chitosan for ocular drug delivery: preparation, mucoadhesion and pharmacokinetic evaluation. Heliyon 9(9):e19877. https://doi.org/10.1016/j.heliyon.2023.e19877
40. Algharib SA et al (2022) Preparation of chitosan nanoparticles by ionotropic gelation technique: effects of formulation parameters and in vitro characterization. J Mol Struct 1252: 132129. https://doi.org/10.1016/j.molstruc.2021.132129
41. Arshia MH, Shahvelayati AS, Sheshmani S, Hajiaghababaei L, Ghasri MRA (2024) Hydrogen bond-mediated self-assembly of Tin (II) oxide wrapped with chitosan/[BzPy]Cl network: an effective bionanocomposite for textile wastewater remediation. Heliyon 10(3):e24771. https://doi.org/10.1016/j.heliyon.2024.e24771
42. Jiménez-Gómez CP, Cecilia JA (2020) Chitosan: a natural biopolymer with a wide and varied range of applications. Molecules 25(17):3981. https://doi.org/10.3390/molecules25173981
43. Park GR, Gwak MA, Choi YH, Park WH (2023) pH-sensitive gallol-rich chitosan hydrogel beads for on-off controlled drug delivery. Int J Biol Macromol 240:124346. https://doi.org/10.1016/j.ijbiomac.2023.124346
44. Faris TM et al (2023) Preparation and evaluation of transdermal hydrogel of chitosan coated nanocurcumin for enhanced stability and skin permeability. Arab J Chem 16(12):105302. https://doi.org/10.1016/j.arabjc.2023.105302
45. Cao J, Zhu C, Cao Z, Ke X (2024) CPPs-modified chitosan as permeability-enhancing chemotherapeutic combined with gene therapy nanosystem by thermosensitive hydrogel for the treatment of osteosarcoma. Int J Biol Macromol 267:130915. https://doi.org/10.1016/j.ijbiomac.2024.130915
46. Xiong W et al (2021) Brij-functionalized chitosan nanocarrier system enhances the intestinal permeability of P-glycoprotein substrate-like drugs. Carbohydr Polym 266:118112. https://doi.org/10.1016/j.carbpol.2021.118112
47. Zoghi M, Pourmadadi M, Yazdian F, Nigjeh MN, Rashedi H, Sahraeian R (2023) Synthesis and characterization of chitosan/carbon quantum dots/Fe2O3 nanocomposite comprising curcumin for targeted drug delivery in breast cancer therapy. Int J Biol Macromol 249: 125788. https://doi.org/10.1016/j.ijbiomac.2023.125788
48. Zhong C et al (2024) Biocompatibility and osteogenic potential of choline phosphate chitosan-coated biodegradable Zn1Mg. Acta Biomater 175:395–410. https://doi.org/10.1016/j.actbio.2023.12.014
49. Bai Q et al (2022) Current challenges and future applications of antibacterial nanomaterials and chitosan hydrogel in burn wound healing. Mater Adv 3(17):6707–6727. https://doi.org/10.1039/D2MA00695B
50. Yuan J et al (2024) Chitosan-taurine nanoparticles cross-linked carboxymethyl chitosan hydrogels facilitate both acute and chronic diabetic wound healing. Int J Biol Macromol 273:132762. https://doi.org/10.1016/j.ijbiomac.2024.132762
51. Wang X et al (2023) Stimuli-responsive chitosan-based injectable hydrogel for "on-demand" drug release. Mater Adv 4(23):6439–6448. https://doi.org/10.1039/D3MA00430A
52. Qiu J et al (2024) An injectable, self-healable, and antimicrobial hydroxypropyl chitosan/poly (vinyl alcohol) hydrogel for drug delivery systems. Colloids Surf Physicochem Eng Asp 698: 134587. https://doi.org/10.1016/j.colsurfa.2024.134587

53. Saberian M, Safari Roudsari R, Haghshenas N, Rousta A, Alizadeh S (2024) How the combination of alginate and chitosan can fabricate a hydrogel with favorable properties for wound healing. Heliyon 10(11):e32040. https://doi.org/10.1016/j.heliyon.2024.e32040
54. De La Mora-López DS et al (2024) Production and performance evaluation of chitosan/ collagen/honey nanofibrous membranes for wound dressing applications. Int J Biol Macromol:133809. https://doi.org/10.1016/j.ijbiomac.2024.133809
55. Unsoy G, Khodadust R, Yalcin S, Mutlu P, Gunduz U (2014) Synthesis of doxorubicin loaded magnetic chitosan nanoparticles for pH responsive targeted drug delivery. Eur J Pharm Sci 62: 243–250. https://doi.org/10.1016/j.ejps.2014.05.021
56. Ćirić A, Krajišnik D, Čalija B, Đekić L (2020) Biocompatible non-covalent complexes of chitosan and different polymers: characteristics and application in drug delivery. Arh Za Farm 70(4):173–197. https://doi.org/10.5937/arhfarm2004173Q
57. Peers S, Montembault A, Ladavière C (2020) Chitosan hydrogels for sustained drug delivery. J Control Release 326:150–163. https://doi.org/10.1016/j.jconrel.2020.06.012
58. Du H, Cai X, Zhai G (2013) Advances in the targeting molecules modified chitosan-based nanoformulations. Curr Drug Targets 14(9):1034–1052. https://doi.org/10.2174/ 1389450111314090012
59. Rizeq BR, Younes NN, Rasool K, Nasrallah GK (2019) Synthesis, bioapplications, and toxicity evaluation of chitosan-based nanoparticles. Int J Mol Sci 20(22):5776. https://doi. org/10.3390/ijms20225776
60. Li J et al (2018) Chitosan-based nanomaterials for drug delivery. Molecules 23(10):2661. https://doi.org/10.3390/molecules23102661
61. Fathi M, Majidi S, Zangabad PS, Barar J, Erfan-Niya H, Omidi Y (2018) Chitosan-based multifunctional nanomedicines and theranostics for targeted therapy of cancer. Med Res Rev 38(6):2110–2136. https://doi.org/10.1002/med.21506
62. Gozali D, Hudaya AR, Suharyani I (2022) A review on chitosan-based materials as potential wound dressing materials. Int J Appl Pharm:27–32. https://doi.org/10.22159/ijap.2022.v14s4. PP23
63. Abdul Fatah IF, Mydin RBSMN (2023) A review on mechanical properties of chitosan-based for wound dressing materials. J Health Transl Med sp2023(1):331–337. https://doi.org/10. 22452/jummec.sp2023no1.35
64. Hu Z et al (2024) A super hydrophilic and high strength chitosan hemostatic sponge prepared by freeze-drying and alkali treatment for rapid hemostasis. Mater Today Commun 39:108855. https://doi.org/10.1016/j.mtcomm.2024.108855
65. Cui C, Sun S, Wu S, Chen S, Ma J, Zhou F (2021) Electrospun chitosan nanofibers for wound healing application. Eng Regen 2:82–90. https://doi.org/10.1016/j.engreg.2021.08.001
66. Liu T et al (2024) Electrospun nanofiber membranes for rapid liver hemostasis via N-alkylated chitosan doped chitosan/PEO. Int J Biol Macromol 258:128948. https://doi.org/10.1016/j. ijbiomac.2023.128948
67. Chen X, Yan G, Chen M, Yang P, Xu B (2023) Alkylated chitosan-attapulgite composite sponge for rapid hemostasis. Biomater Adv 153:213569. https://doi.org/10.1016/j.bioadv. 2023.213569
68. Wang X, Mu B, Zhang H, Du Y, Yang F, Wang A (2023) Incorporation of mixed-dimensional palygorskite clay into chitosan/polyvinylpyrrolidone nanocomposite films for enhancing hemostatic activity. Int J Biol Macromol 237:124213. https://doi.org/10.1016/j.ijbiomac. 2023.124213
69. Lei J et al (2023) A bioactive nanocomposite sponge for simultaneous hemostasis and antimicrobial therapy. Nano Res 16(3):4004–4012. https://doi.org/10.1007/s12274-022-5226-1
70. Liu C et al (2023) Sprayable surface-adaptive biocompatible membranes for efficient hemostasis via assembly of chitosan and polyphosphate. Carbohydr Polym 302:120360. https://doi. org/10.1016/j.carbpol.2022.120360

71. Lunkov AP, Zubareva AA, Varlamov VP, Nechaeva AM, Drozd NN (2023) Chemical modification of chitosan for developing of new hemostatic materials: a review. Int J Biol Macromol 253:127608. https://doi.org/10.1016/j.ijbiomac.2023.127608
72. Yang Y et al (2023) Chitosan-based mussel-inspired hydrogel for rapid self-healing and high adhesion of tissue adhesion and wound dressings. Carbohydr Polym 316:121083. https://doi.org/10.1016/j.carbpol.2023.121083
73. Leonhardt EE, Kang N, Hamad MA, Wooley KL, Elsabahy M (2019) Absorbable hemostatic hydrogels comprising composites of sacrificial templates and honeycomb-like nanofibrous mats of chitosan. Nat Commun 10(1):2307. https://doi.org/10.1038/s41467-019-10290-1
74. Patil G, Pawar R, Jadhav S, Ghormade V (2022) A chitosan based multimodal "soft" hydrogel for rapid hemostasis of non-compressible hemorrhages and its mode of action. Carbohydr Polym Technol Appl 4:100237. https://doi.org/10.1016/j.carpta.2022.100237
75. Song F et al (2021) Chitosan-based multifunctional flexible hemostatic bio-hydrogel. Acta Biomater 136:170–183. https://doi.org/10.1016/j.actbio.2021.09.056
76. Gheorghiță D et al (2023) Chitosan-based biomaterials for hemostatic applications: a review of recent advances. Int J Mol Sci 24(13):10540. https://doi.org/10.3390/ijms241310540
77. Zhong Y, Hu H, Min N, Wei Y, Li X, Li X (2021) Application and outlook of topical hemostatic materials: a narrative review. Ann Transl Med 9(7):577–577. https://doi.org/10.21037/atm-20-7160
78. Huang Y et al (2024) MXene-NH_2/chitosan hemostatic sponges for rapid wound healing. Int J Biol Macromol 260:129489. https://doi.org/10.1016/j.ijbiomac.2024.129489
79. Sundaram MN, Amirthalingam S, Mony U, Varma PK, Jayakumar R (2019) Injectable chitosan-nano bioglass composite hemostatic hydrogel for effective bleeding control. Int J Biol Macromol 129:936–943. https://doi.org/10.1016/j.ijbiomac.2019.01.220
80. Muthiah Pillai NS, Eswar K, Amirthalingam S, Mony U, Kerala Varma P, Jayakumar R (2019) Injectable nano Whitlockite incorporated chitosan hydrogel for effective hemostasis. ACS Appl Bio Mater 2(2):865–873. https://doi.org/10.1021/acsabm.8b00710
81. Sundaram MN, Mony U, Varma PK, Rangasamy J (2021) Vasoconstrictor and coagulation activator entrapped chitosan based composite hydrogel for rapid bleeding control. Carbohydr Polym 258:117634. https://doi.org/10.1016/j.carbpol.2021.117634
82. Devlin JJ, Kircher S, Kozen BG, Littlejohn LF, Johnson AS (2009) Comparison of ChitoFlex®, CELOX™, and QuikClot® in control of hemorrhage. J Emerg Med: S073646790900105X. https://doi.org/10.1016/j.jemermed.2009.02.017
83. Khan MA, Mujahid M (2019) A review on recent advances in chitosan based composite for hemostatic dressings. Int J Biol Macromol 124:138–147. https://doi.org/10.1016/j.ijbiomac.2018.11.045
84. Vimala K et al (2010) Fabrication of porous chitosan films impregnated with silver nanoparticles: a facile approach for superior antibacterial application. Colloids Surf B Biointerfaces 76(1):248–258. https://doi.org/10.1016/j.colsurfb.2009.10.044
85. Kokkarachedu Varaprasad K, Vimala SR, Narayana Reddy N, Siva Mohana Reddy G, Mohana Raju K (2012) Biodegradable chitosan hydrogels for in vitro drug release studies of 5-flurouracil an anticancer drug. J Polym Environ 20(2):573–582. https://doi.org/10.1007/s10924-012-0412-y
86. Karthikeyan C, Varaprasad K, Akbari-Fakhrabadi A, Hameed ASH, Sadiku R (2020) Biomolecule chitosan, curcumin and ZnO-based antibacterial nanomaterial, via a one-pot process. Carbohydr Polym 249:116825. https://doi.org/10.1016/j.carbpol.2020.116825
87. Rezaei ES, Poursamar SA, Naeimi M, Taheri MM, Rafienia M (2024) An in vitro and in vivo study of electrospun polyvinyl alcohol/chitosan/sildenafil citrate mat on 3D-printed polycaprolactone membrane as a double layer wound dressing. Int J Biol Macromol 269:131859. https://doi.org/10.1016/j.ijbiomac.2024.131859
88. Haider A et al (2024) Advances in chitosan-based drug delivery systems: a comprehensive review for therapeutic applications. Eur Polym J 210:112983. https://doi.org/10.1016/j.eurpolymj.2024.112983

89. Sultankulov B, Berillo D, Sultankulova K, Tokay T, Saparov A (2019) Progress in the development of chitosan-based biomaterials for tissue engineering and regenerative medicine. Biomol Ther 9(9):470. https://doi.org/10.3390/biom9090470
90. Narmadha R, Sathish PB, Nagarajan R, Prateek W, Vijaygopal R, Selvakumar R (2024) Biomolecule functionalized PVA-chitosan composite nanofibrous scaffold for human endometrium stromal cells attachment and proliferation: an in-vitro study. Mater Lett 363:136326. https://doi.org/10.1016/j.matlet.2024.136326
91. Zhang X et al (2024) Cell-free chitosan/silk fibroin/bioactive glass scaffolds with radial pore for in situ inductive regeneration of critical-size bone defects. Carbohydr Polym 332:121945. https://doi.org/10.1016/j.carbpol.2024.121945
92. Dasgupta S, Reddy KP, Datta P, Barui A (2023) Vitamin D3-incorporated chitosan/collagen/fibrinogen scaffolds promote angiogenesis and endothelial transition via HIF-1/IGF-1/VEGF pathways in dental pulp stem cells. Int J Biol Macromol 253:127325. https://doi.org/10.1016/j.ijbiomac.2023.127325
93. Si J, Li T, Yang X, Lu X (2024) Regulating Schwann cells and endothelial cells by YIGSR functionalized porous chitosan scaffolds for nerve regeneration. Mater Today Commun 38:108031. https://doi.org/10.1016/j.mtcomm.2024.108031
94. Chang F-C et al (2023) 3D chitosan scaffolds support expansion of human neural stem cells in chemically defined condition. Matter 6(10):3631–3660. https://doi.org/10.1016/j.matt.2023.08.014
95. El-Sayed SAM, Mabrouk M, Khallaf ME, Abd El-Hady BM, El-Meliegy E, Shehata MR (2020) Antibacterial, drug delivery, and osteoinduction abilities of bioglass/chitosan scaffolds for dental applications. J Drug Deliv Sci Technol 57:101757. https://doi.org/10.1016/j.jddst.2020.101757
96. Roy TR, Nelogi S, Chowdhary R, Naik VV, Peter M, Santhosh VN (2024) Influence of chitosan and Cissus quandrangularis coating on osseointegration in titanium implants in rabbits: a preclinical in vivo study. J Stomatol Oral Maxillofac Surg 125(3):101727. https://doi.org/10.1016/j.jormas.2023.101727
97. Kikuchi LNT et al (2022) Effects of the crosslinking of chitosan/DCPA particles in the antimicrobial and mechanical properties of dental restorative composites. Dent Mater 38(9):1482–1491. https://doi.org/10.1016/j.dental.2022.06.024
98. Sarkar S, Prashanth NT, Shobha ES, Rangan V, Nikhila G (2019) Efficacy of platelet rich fibrin versus chitosan as a hemostatic agent following dental extraction in patients on antiplatelet therapy. J Oral Biol Craniofacial Res 9(4):336–339. https://doi.org/10.1016/j.jobcr.2019.07.003
99. Manjubaashini N, Bargavi P, Thomas NG, Krishnan N, Balakumar S (2024) Chitosan bioactive glass scaffolds for in vivo subcutaneous implantation, toxicity assessment, and diabetic wound healing upon animal model. Int J Biol Macromol 256:128291. https://doi.org/10.1016/j.ijbiomac.2023.128291
100. Andrade Del Olmo J et al (2023) Multifunctional antibacterial chitosan-based hydrogel coatings on Ti6Al4V biomaterial for biomedical implant applications. Int J Biol Macromol 231:123328. https://doi.org/10.1016/j.ijbiomac.2023.123328
101. Hassan MK, Abdelrehim SAA, Elkhooly TA, Elmezayyen AS, Mansour-Gabr MM, Ghany NAA (2024) Investigation of structure, morphology, and corrosion behavior of carboxylic acids/hydroxyapatite/chitosan coatings on Ti discs for implants. Thin Solid Films 798:140378. https://doi.org/10.1016/j.tsf.2024.140378
102. Ali IH, Elkashlan AM, Hammad MA, Hamdi M (2023) Antimicrobial and anti-SARS-CoV-2 activities of smart daclatasvir-chitosan/gelatin nanoparticles-in-PLLA nanofibrous medical textiles; in vitro, and in vivo study. Int J Biol Macromol 253:127350. https://doi.org/10.1016/j.ijbiomac.2023.127350
103. Nainggolan I et al (2024) Sensitivity of chitosan/reduced graphene oxide/manganese dioxide modified electrodes for cholesterol detection using cyclic voltammetry. South Afr J Chem Eng 48:329–336. https://doi.org/10.1016/j.sajce.2024.03.009

104. Das L, Das P, Bhowal A, Bhattachariee C (2020) Synthesis of hybrid hydrogel nano-polymer composite using graphene oxide, chitosan and PVA and its application in waste water treatment. Environ Technol Innov 18:100664. https://doi.org/10.1016/j.eti.2020.100664
105. Sharifi MJ, Nouralishahi A, Hallajisani A (2023) Fe_3O_4-chitosan nanocomposite as a magnetic biosorbent for removal of nickel and cobalt heavy metals from polluted water. Int J Biol Macromol 248:125984. https://doi.org/10.1016/j.ijbiomac.2023.125984
106. Su C, Yang H, Zhao H, Liu Y, Chen R (2017) Recyclable and biodegradable superhydrophobic and superoleophilic chitosan sponge for the effective removal of oily pollutants from water. Chem Eng J 330:423–432. https://doi.org/10.1016/j.cej.2017.07.157
107. Ashraf A, Dutta J, Farooq A, Rafatullah M, Pal K, Kyzas GZ (2024) Chitosan-based materials for heavy metal adsorption: recent advancements, challenges and limitations. J Mol Struct 1309:138225. https://doi.org/10.1016/j.molstruc.2024.138225
108. Liang W et al (2022) Insight into crosslinked chitosan/soy protein isolate /PVA plastics by revealing its structure, physicochemical properties, and biodegradability. Ind Crop Prod 187: 115548. https://doi.org/10.1016/j.indcrop.2022.115548
109. Kumar P, Dhanda S (2024) Chitosan encapsulation of Pediococcus acidilactici NCDC 252 improved its survival in simulated gastro-intestinal conditions. Process Biochem 138: 130–138. https://doi.org/10.1016/j.procbio.2024.01.013
110. Lotfy VF, Basta AH (2024) Performance effectiveness of nano-lignin in production of gel with nano-chitosan for controlling release of salicylic acid. Int J Biol Macromol 265:131098. https://doi.org/10.1016/j.ijbiomac.2024.131098
111. Chappidi S, Ankireddy SR, Sree CG, Rayalcheruvu U, Buddolla V (2024) Enhancing diabetic rat wound healing through chitosan-mediated nano-scaffolds loaded with quercetin-silver complex. Mater Lett 361:136167. https://doi.org/10.1016/j.matlet.2024.136167
112. Zhou C et al (2024) AgNPs loaded adenine-modified chitosan composite POSS-PEG hybrid hydrogel with enhanced antibacterial and cell proliferation properties for promotion of infected wound healing. Int J Biol Macromol 267:131575. https://doi.org/10.1016/j.ijbiomac.2024.131575
113. Kumar V et al (2023) Synthesis and characterization of chitosan nanofibers for wound healing and drug delivery application. J Drug Deliv Sci Technol 87:104858. https://doi.org/10.1016/j.jddst.2023.104858
114. Singh S, Kumar S, Koh J (2024) Physicochemical and optical activity of chitosan based ternary nanocomposites for food packaging applications. J Mol Struct 1311:138210. https://doi.org/10.1016/j.molstruc.2024.138210
115. Yu W et al (2023) Effects of dietary chitosan on the growth, health status and disease resistance of golden pompano (Trachinotus ovatus). Carbohydr Polym 300:120237. https://doi.org/10.1016/j.carbpol.2022.120237

Adv Polym Sci (2025) 295: 403–428
https://doi.org/10.1007/12_2024_188

Published online: 17 December 2024

Delivery of Biomolecules Using Chitosan for Tissue Engineering

Abhirami Dinesan, Mary Susan J. Anthraper, and R. Jayakumar

Contents

1 Introduction 404
2 Biomolecule Delivery Using Different Forms of Chitosan for Tissue Engineering 406
2.1 Chitosan Microparticles/Nanoparticles 407
2.2 Chitosan Hydrogel 414
2.3 Chitosan Scaffolds 419
3 Conclusions and Future Directions 424
References 425

Abstract Various advancements were made in the field of tissue engineering to address the critical need for functional tissue regeneration in different clinical settings. Biomolecule delivery systems are crucial for orchestrating the tissue repair process by modulating the spatiotemporal release of bioactive molecules. The successful delivery of bioactive agents can result in various cellular interactions to promote the repair of defective tissues. Chitosan derived from chitin is a promising multifunctional biopolymer utilized in regeneration applications due to its exceptional properties, such as biocompatibility, biodegradability and antimicrobial activities. It has a unique backbone structure containing free amine groups, which allows for versatile chemical modifications, enhancing its functionality in various biomaterial applications in tissue engineering. It can be processed alone or blended with other polymers to formulate hydrogels, nanoparticles, microparticles, films or scaffolds, depending on the required release kinetics for the delivery of biomolecules. This chapter covers the latent applications of chitosan as a flexible platform for the delivery of biomolecules in tissue engineering and discusses future approaches for its growth in regenerative medicine.

A. Dinesan, M. S. J. Anthraper, and R. Jayakumar (✉)
Polymeric Biomaterials Lab, School of Nanosciences and Molecular Medicine, Amrita Vishwa Vidyapeetham, Kochi, India
e-mail: rjayakumar@aims.amrita.edu

Keywords Biomolecule delivery · Hydrogels · Microparticles · Nanoparticles · Regenerative medicine · Tissue repair

1 Introduction

Tissue engineering (TE) is a rapidly growing frontier within regenerative medicine, which combines principles from both biology and engineering to establish substitutes for impaired tissues [1]. In tissue engineering, the accurate control and release of regulatory biomolecules are pivotal for directing cellular behavior and assisting tissue regeneration. Within this scenario, a variety of biomolecules, such as antigens, proteins, nucleic acids, peptides, and antibodies, must penetrate biological barriers in order to reach their target sites. This challenge emphasizes the utmost necessity for developing delivery mechanisms that can effectively navigate physiological barriers and deliver biomolecules precisely to their target sites, thus providing innovative tissue engineering solutions [2]. Controlled release strategies are employed in biomolecule delivery systems to spatially and temporally regulate the presentation of these biomolecules within engineered tissue constructs [3]. Both synthetic and natural polymers offer remarkable benefits for biomolecule delivery in tissue regeneration. Starch, hyaluronic acid (HA), collagen, alginate, cellulose, chitin, and chitosan (CS) are highly appealing natural polymers for delivering biomolecules in tissue regeneration.

Biopolymers have captured a lot of interest in the last few decades compared to synthetic polymers. Peptides and polysaccharides are increasingly recognized as versatile materials for drug delivery in TE due to their biocompatibility, biodegradability, and tunable properties. Chitin is a natural polysaccharide commonly found in the exoskeleton of crustaceans and nonanimal sources such as fungi (Fig. 1). It is the precursor of chitosan and is obtained through the deacetylation process (Fig. 2). This copolymer is made up of glucosamine and N-acetyl glucosamine, making it stand out as one of the most prevalent biopolymers, and its versatility has led to its widespread use in various biomedical fields.

Molecular weight is one of the most vital factors affecting bioactivity in chitosan. Bioactive characteristics are apparently better expressed in lower molecular weight (<20 kDa) of chitosan than in higher molecular weight (>120 kDa) of chitosan [4]. The divergence in bioactivity could be attributed to various factors, like increased solubility, better interaction with biological membranes, and higher ability to penetrate tissues. Lower MW chitosan is generally more efficient for clinical applications, such as antimicrobial treatments, wound-healing applications, and drug release systems [4]. Chitosan's hydrophilic surface significantly benefits cell adhesion and proliferation in tissue engineering by enhancing the interaction between cells and the scaffold material. It also improves water retention, creating a favorable microenvironment for cell attachment. Chitosan has found extensive application as a hemostatic agent in various forms, such as sponges, gauze, and granules [5]. As a

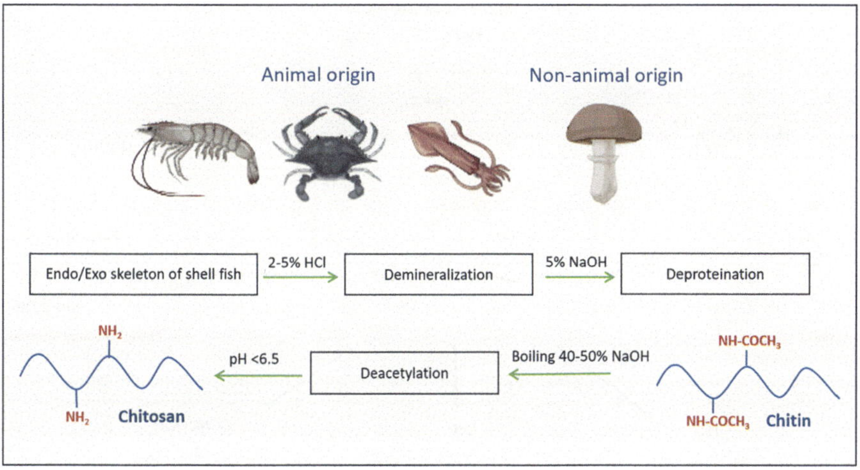

Fig. 1 Schematic diagram showing the chitosan production steps

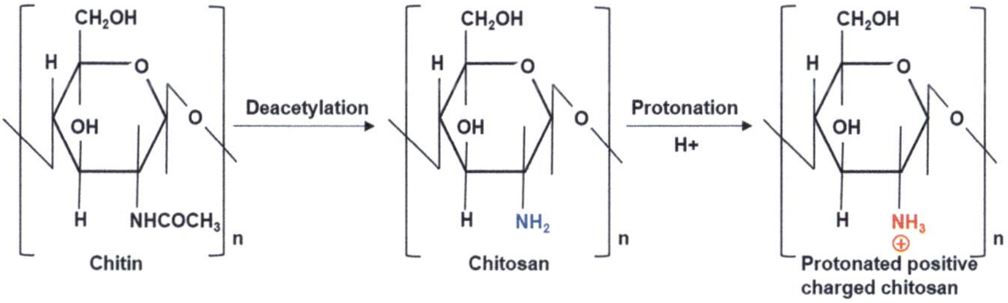

Fig. 2 Deacetylation of chitin to chitosan

polycationic polymer, chitosan contains multiple functional groups, which can be chemically modified with a variety of ligands [6]. Amine and hydroxyl functional groups in the backbone of polymers can be chemically modified to synthesize various chitosan derivatives, with enhanced affinity toward the drug functional group and high solubility. The conventional methods for modifying chitosan include alkylation, sulfation, silylation, acylation, and phosphorylation [7]. Chitosan can undergo chemical modifications at its amino and hydroxyl groups, resulting in derivatives that are N modified, O modified, or modified at both N and O sites (Fig. 3) [8].

The rising interest in chitosan for biomedical applications has spurred the development of tailored biomaterials through chemical and physical modifications, often involving combinations with other polymers and inorganic materials, to create composite materials with enhanced properties. The final product can be engineered through different procedures, such as three-dimensional (3D) printing, electrospinning, freeze-drying, and solvent casting [9]. Chitosan-based formulations have been widely explored for applications in bone, cartilage, nerve, and skin TE.

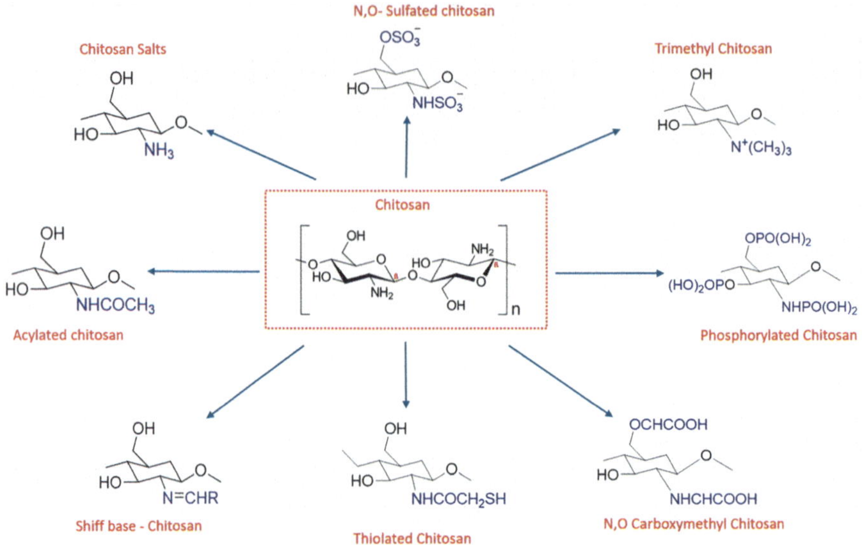

Fig. 3 Chemical modifications of chitosan

2 Biomolecule Delivery Using Different Forms of Chitosan for Tissue Engineering

Chitosan is the most widely applied biomolecule carrier for TE due to its remarkable properties, like biodegradability, low toxicity, and ease of modification. This biocompatible material has vast potential applications as it can be engineered into different forms, such as microparticles, nanoparticles, hydrogels, and scaffolds, each offering unique benefits for targeted delivery and the sustained release of biomolecules (Fig. 4). Microparticles allow a sustained local delivery, which is decisive for keeping local therapeutic levels of bioactive agents. Hydrogels offer huge potential for clinical translation with their high water content and tunable characteristics. They can produce a conducive environment for cell growth and differentiation. Meanwhile, scaffolds offer structural support and mimic the extracellular matrix to enhance tissue regeneration and integration. These various forms of chitosan-based delivery systems highlight the capacity to enhance the efficacy of tissue engineering applications by ensuring precise and effective biomolecule delivery [10]. On the other hand, incorporating antiseptic agents into chitosan formulations for wound healing can result in numerous advantages, such as biocompatibility, antimicrobial activity, stimuli responsiveness, moisture retention, controlled delivery, and ease of fabrication [11, 12].

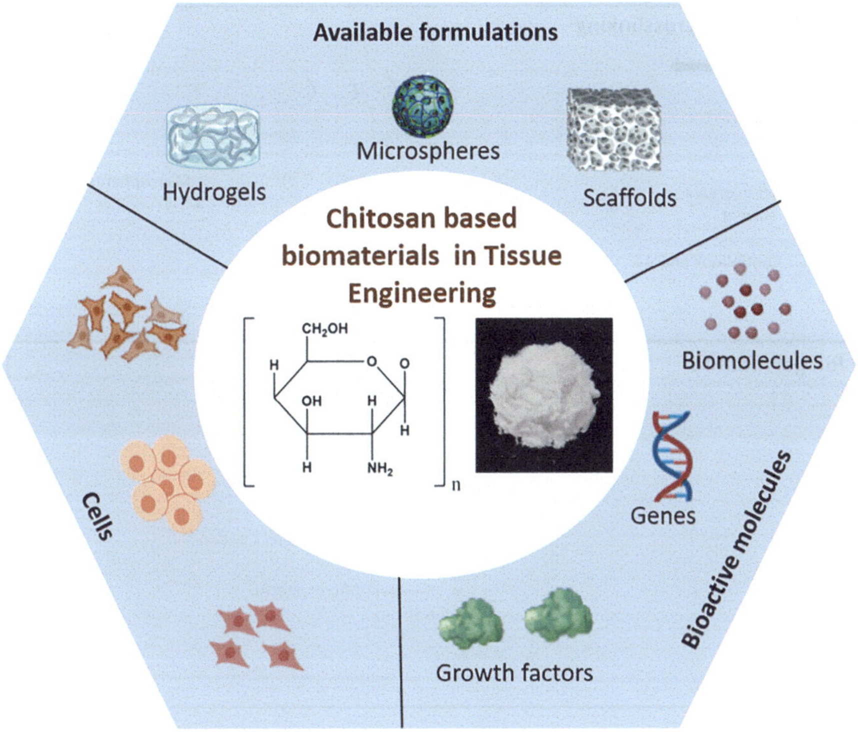

Fig. 4 Schematic diagram of chitosan-based biomaterials in TE utilizing a combination of cells and bioactive molecules

2.1 *Chitosan Microparticles/Nanoparticles*

Microspheres are small spherical particles with sizes ranging from 0.1 to 300 μm, made from synthetic or natural polymers [13]. Their reduced size enhances drug absorption and improves bioavailability. Currently, microencapsulated pharmaceutical formulations like chitosan microspheres serve as particulate systems capable of incorporating various bioactive molecules. Biomaterials are preferred for the synthesis of microspheres due to their biocompatibility, biodegradability, nontoxicity, safety, and cost-effectiveness. Microspheres are particularly advantageous for controlled and prolonged drug release, which often relies on the polymer's mucoadhesiveness. Their high surface area enhances cell attachment and proliferation [14]. Consequently, microspheres made with natural polymers, like chitosan (CS), enhance sustained drug release and reduce the need for frequent dosing [15]. The primary techniques for synthesizing chitosan microspheres include the ion gel method, the spray-drying method, and the emulsification cross-linking method (Fig. 5). In the ion gel method, an anionic solution will be utilized as a physical cross-linking agent, and the preparation medium will be acetic acid. During

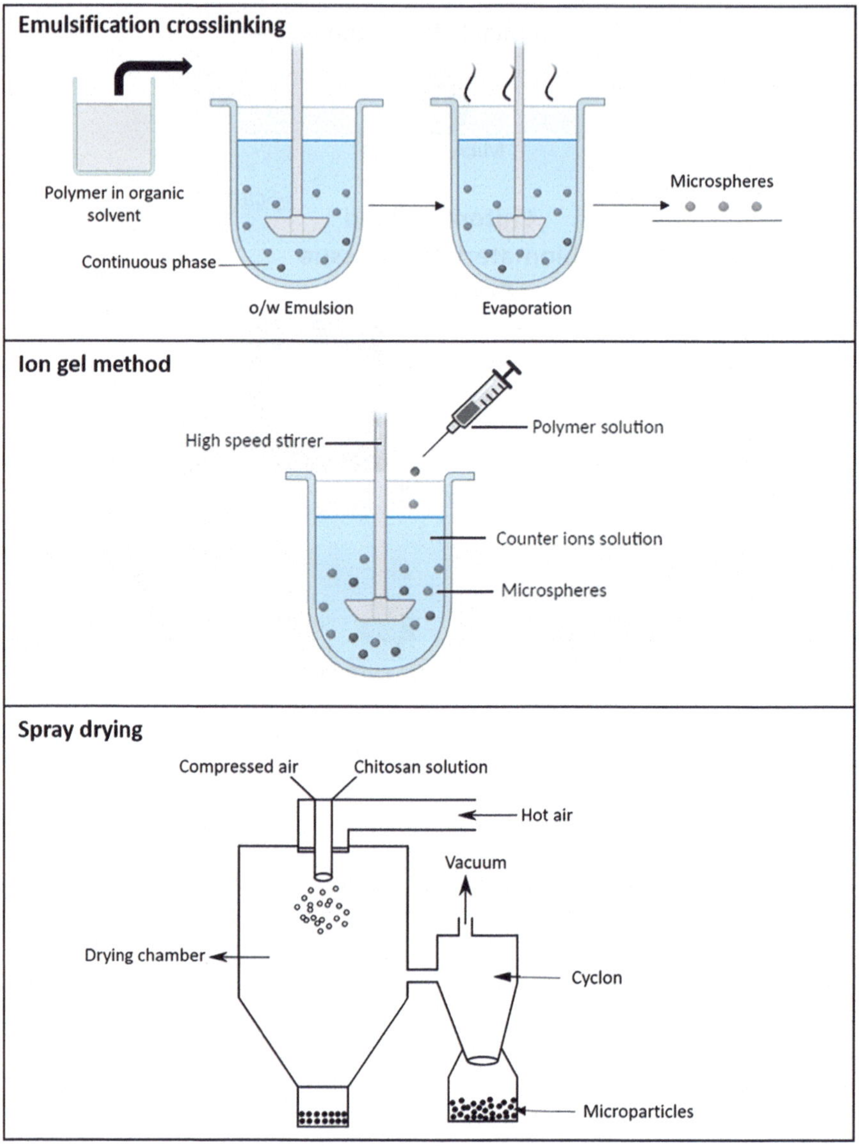

Fig. 5 Different methods used for the synthesis of chitosan microspheres

this process, the interaction between the negative charge of anionic gel and the positive charge of chitosan results in the formation of chitosan microspheres. Spray-drying technology is based on the dissolution and dispersion of chitosan in an acid solution, where the emulsion is sprayed into a hot air stream. Therefore, this process will quickly remove the solvent and obtain the chitosan microspheres. These microspheres after being formed, and excess solvent is evacuated from them by vacuum drying. This technique provides precise control over particle size and morphology, with the final microsphere size influenced by the nozzle size, atomization pressure, and air temperature. Another widely used method is emulsification cross-linking. This is a straightforward technique combining water and oil phases and involves emulsifying this mixture through constant stirring. Due to continuous stirring emulsion, droplets were formed and then cross-linked using a chemical cross-linker. The aldehyde groups in the cross-linking agent react with the amino groups in chitosan, resulting in the formation of solid microspheres [16].

Optical microscopy can be used to measure the microsphere particle size [17]. Additionally, drug loading efficiency is assessed by evaluating the percentage of drug incorporated in the material, which is measured by UV-visible spectrophotometer. Moreover, other analytical methods, such as scanning electron microscopy (SEM), can provide detailed surface morphology, and high-performance liquid chromatography can offer the precise quantification of drug content within microspheres.

Observations indicated that the dissolution rate of the drug improved with an increase in the surface area of both the drug and its carrier. Consequently, nanoparticles (size ranges from 0.1 to 100 nm) were employed to encapsulate the drug for targeted delivery toward the desired location. For the production of chitosan nanoparticles, there are five methods available, as given in Fig. 6. A widely used method is ionotropic gelation, in which a polyanion will be added to the chitosan solution, which is dissolved in an acetic acid medium. Hence, the negative-charged group of polyanion will electrostatically interact with the amine group of chitosan under high-stirring conditions [18].

In the microemulsion method, chitosan dissolved in an acetic acid solution will be mixed in a surfactant hexane mixture with the cross-linker glutaraldehyde. Under a constant stirring condition, nanoparticles will be formed overnight. The solvent will evaporate, and with the use of calcium chloride, surfactants will precipitate and be removed via centrifugation. Finally, nanoparticles will be dried under a vacuum or through freeze drying [19]. The reverse micellar method is another effective method for the synthesis of chitosan nanoparticles. It involves forming reverse micelles by adding a surfactant to an organic solvent mixed with a chitosan solution. It allows fine control over particle size, which makes it suitable for biomolecule delivery [20]. Similar to microparticle synthesis, the emulsification solvent diffusion method involves the addition of an organic solvent into chitosan solution with a stabilizing agent under high stirring, followed by homogenization. The final end product is the polyelectrolyte complex, which is formed when two or more oppositely charged polymers interact and bind together via electrostatic interaction. Chitosan nanoparticles can be formed by adding deoxyribonucleic acid (DNA) into the chitosan solution with constant stirring [18].

Having explored the unique properties and advantages of chitosan microspheres/nanoparticles for biomolecule delivery, we can now delve into various tissue engineering applications where these chitosan carriers are used. Table 1 depicts previous studies in which chitosan microspheres/nanoparticles were utilized for the delivery of biomolecules for various tissue engineering purposes.

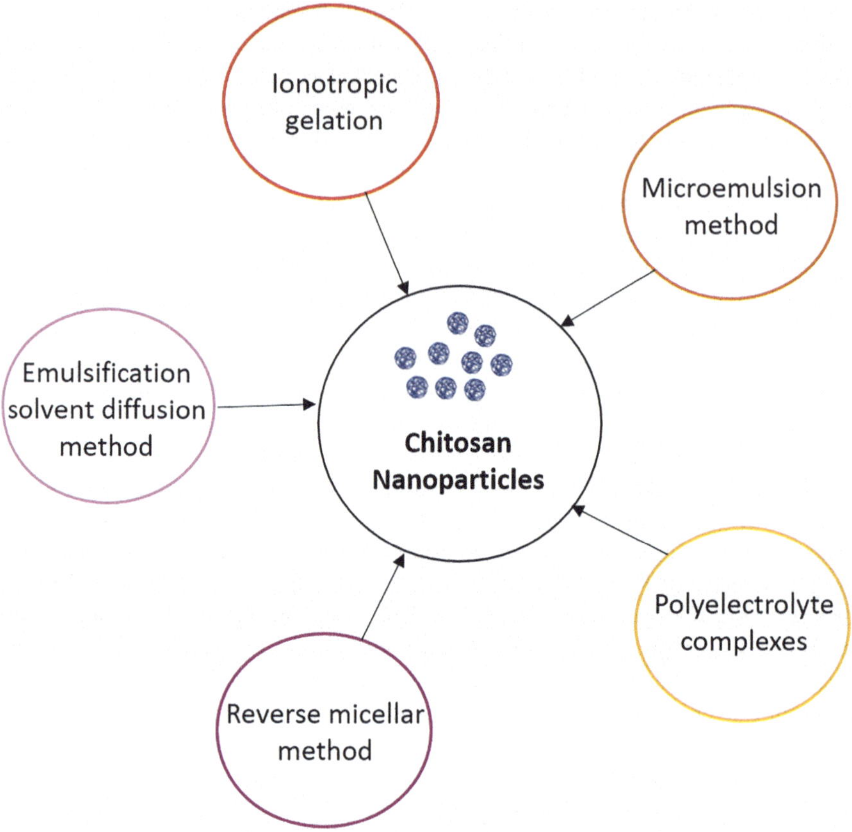

Fig. 6 Different methods for formulating chitosan nanoparticles

2.1.1 Adipose Tissue Engineering

Adipose-derived stem cells (ADSCs) are exceptionally valuable due to their high yield and ease of extraction through minimally invasive procedures, such as liposuction. These cells possess significant proliferative capacity and multipotency, enabling them to differentiate into various cell types. This versatility makes ADSCs highly suitable for a wide range of regenerative therapies, like TE and wound healing. The study conducted by C. A. Custódio et al. observed that cell-selective chitosan microparticles, developed as injectable carriers for tissue regeneration, demonstrated the ability to selectively isolate specific cell types from mixed populations and promote their expansion. Using human-adipose-derived stem cells and human umbilical vein endothelial cells as examples, the study showed that these bifunctionalized particles could effectively target and expand particular cells [21]. This finding highlights the potential of these microparticles in targeted cell therapy and tissue engineering applications.

Table 1 Chitosan microspheres for the delivery of biomolecules in tissue engineering

Targeted tissue	Biomaterial composition	Type of biomolecule delivery	Respective function	Reference
Adipose	Chitosan microspheres conjugated with biotinylated antibodies	Adipose stem cells	Bifunctionalized particles could effectively target and expand adipose stem cells.	[21]
Bone	CS-microsphere-loaded scaffold	BMP-2	Slow and sustained BMP2 release for bone regeneration.	[23]
Cartilage	Chitosan-microsphere- loaded oxidized alginate, carboxymethyl chitosan, and a polyethylene glycol hydrogel system	Melatonin and methylprednisolone	The dual-drug-releasing system induces chondrogenic differentiation and improved hyaline cartilage formation.	[26]
Cardiac	Alginate chitosan microspheres	Human adipose stem cells	Microencapsulation of stem cells to enhance cell retention in the infarcted myocardium.	[27]
Corneal	Chitosan-nanoparticle- incorporated chitosan/ polycaprolactone (PCL) membranes	Human corneal endothelial cells	Enhanced corneal endothelial regeneration.	[29]
Skin	Chitosan microspheres	Silver nanoparticles	Controlled release of silver nanoparticles, which leads to good cell growth, proliferation, and reinforcement in antibacterial ability.	[31]
	Chitosan nanoparticles	Curcumin	Wound dressing with enhanced antioxidant and antibacterial properties. Facilitated a more effective wound-healing process in MRSA-infected wounds.	[32]

2.1.2 Bone Tissue Engineering

Chitosan microspheres have garnered attention due to their biocompatibility and ability to degrade naturally, for applications such as injectable biomaterials for bone repair and drug delivery in bone tissue engineering, particularly for non-load-bearing situations [22]. A novel drug delivery system for bone regeneration has been developed, featuring a scaffold loaded with chitosan microspheres. This system effectively incorporates BMP-2, a bioactive molecule, into the microsphere-loaded scaffold, demonstrating excellent biocompatibility and efficient BMP-2 release kinetics. It is mainly aimed at a slow and sustained release of bioactive agents. Consequently, this new CS-microsphere-loaded scaffold shows great potential as a scaffold for bone regeneration [23]. In another study, chitosan microspheres with extracellular matrix mimicking nanofibers were developed, which exhibited excellent biocompatibility, suitable biodegradability, and strong cell adhesion properties. They also facilitated the osteogenic differentiation of rat mesenchymal stem cells (rMSCs). The sol-gel transition process was utilized for the synthesis of these chitosan microspheres. The size and morphology of microspheres varied according to the temperature during thermal induction and the concentration of the chitosan used. In further in vivo studies, the constructs of predifferentiated rMSCs and chitosan microspheres significantly improved bone healing in rat calvarial defects [24].

2.1.3 Cartilage Tissue Engineering

Articular cartilage has minimal self-repair capabilities because of its avascular and aneural nature. Additionally, its dense extracellular matrix hinders the rapid migration of chondrocytes to damaged areas for repair. Glucosamine sulfate and chondroitin sulfate are the major drugs utilized in cartilage regeneration, which helps replenish cartilage components [25]. In a recent study, researchers devised an innovative approach for cartilage TE by creating an injectable in situ forming microparticle system. This system featured a combination of two chondrogenic drugs (melatonin and methylprednisolone), aimed at exploring its potential for chondrogenesis, both in vitro and in vivo. Chitosan microparticles encapsulated with chondrogenic drugs were synthesized via the ion gelation method. The study demonstrated promising results, indicating that the developed microparticle system not only facilitated chondrogenic differentiation but also enhanced the formation of hyaline cartilage [26].

2.1.4 Cardiac Tissue Engineering

The introduction of viable cells directly into the infracted myocardium tissue is the traditional procedure followed in cardiac TE. Nevertheless, there are several drawbacks to this method, including lower cell survival rates and limited cell retention capacity. In order to meet these obstacles, researchers have introduced some innovative solutions, like chitosan microparticles, in cardiac tissue engineering. By successfully encapsulating cells within chitosan microparticles, it was observed that it could improve cell retention and protect cells from the harsh environment of injured heart tissue. To overcome the limitations of direct cell injection, the abovementioned strategy could serve as an effective therapy for heart diseases. Biocompatible microcapsules fabricated from alginate chitosan cross-linked with genipin have gained wide interest in encapsulating human adipose stem cells. This could enhance implant retention in an infracted myocardium and thereby improve results in clinical settings [27].

2.1.5 Corneal Tissue Engineering

Corneal endothelial keratoplasty (CEK) is a surgical procedure that aims to restore vision by replacing damaged endothelial cells with healthy donor tissue in order to address corneal endothelial disorders (CEDs). These CEK procedures can be refined by optimizing the survival of grafts, thereby reducing complications and offering a potential alternative therapy for corneal endothelial disorders. Growing research on these engineered artificial endothelial cell sheets, which are able to mimic native corneal endothelial tissue, provides a potential solution for addressing CEDs. The main aim of this innovative idea is to restore vision with improved outcomes in patients who find difficulty from corneal endothelial disorders [28].

Tayebi et al. developed a biodegradable scaffold for corneal endothelial cell culture by integrating chitosan nanoparticles into chitosan/polycaprolactone (PCL) membranes; the final product was completely transparent. Chitosan nanoparticles were synthesized using the ionotropic gelation method. The in vitro degradation study was assessed by suspending the scaffold in phosphate buffered saline (PBS) (pH 7.4). The degradation rate remained relatively stable for 21 days. The scaffold maintained its structure during the degradation study. It also exhibited good biocompatibility and an appropriate microenvironment for cell adhesion and proliferation [29].

2.1.6 Wound Healing and Skin Tissue Engineering

Managing the healing process of deep wounds presents a significant challenge for wound care professionals. In a previous study, the chitosan alginate polyelectrolyte complex (PEC) was fabricated as microparticles and beads in order to be utilized as controlled release systems in transdermal applications and as wound dressings [30]. This chitosan alginate PEC not only serves as an antimicrobial barrier but also provides a humid wound environment suitable for healing. The semi-rigid nature of the PEC provides excellent dress formability because it can conform to wound contours for full coverage and protection. In addition, the complex is slowly degraded at the wound site so that embedded therapeutic agents like antibiotics or growth factors can be continuously released to the wound bed, allowing for an improved healing outcome. In a recently published study, silver (Ag) nanoparticles (NPs) have been embedded into chitosan microspheres for the topical delivery of the Ag NPs to a wound site, which has potential applications as antibacterial-activity-reinforcing scaffolds in skin tissue engineering. CS microspheres encapsulating silver nanoparticles were fabricated using the emulsification cross-linking technique and were further embedded in CS scaffolds, followed by lyophilization, to obtain porous scaffolds. In vitro results demonstrated the sustained release of Ag microparticles, contributing significantly to excellent antibacterial properties. Additionally, the scaffolds' biocompatibility and mechanical strength make them ideal for skin tissue engineering, providing a dual benefit of infection control and tissue repair [31].

In another study, the wound-healing activity of chitosan/PCL nanofibers with curcumin-loaded chitosan nanoparticles was studied. This novel wound dressing with desirable mechanical strength demonstrated antioxidant properties, antibacterial properties, and improved cell proliferation in vitro. The in vivo study showed significant improvement in the wound-healing process of wounds affected by methicillin-resistant *Staphylococcus aureus* (MRSA). Curcumin encapsulation within chitosan nanoparticles could significantly improve bioavailability and sustained delivery [32].

2.2 Chitosan Hydrogel

A hydrogel consists of a three-dimensional, swellable polymer network capable of absorbing and holding a significant volume of water. In their swollen state, hydrogels exhibit key physical characteristics similar to living tissues, including low surface tension with biological fluids, a rubbery texture, and a flexible consistency. Moreover, their biocompatibility and ability to mimic the extracellular matrix make them highly suitable for various biomedical applications, such as drug delivery systems, tissue engineering, wound dressings, etc. [33].

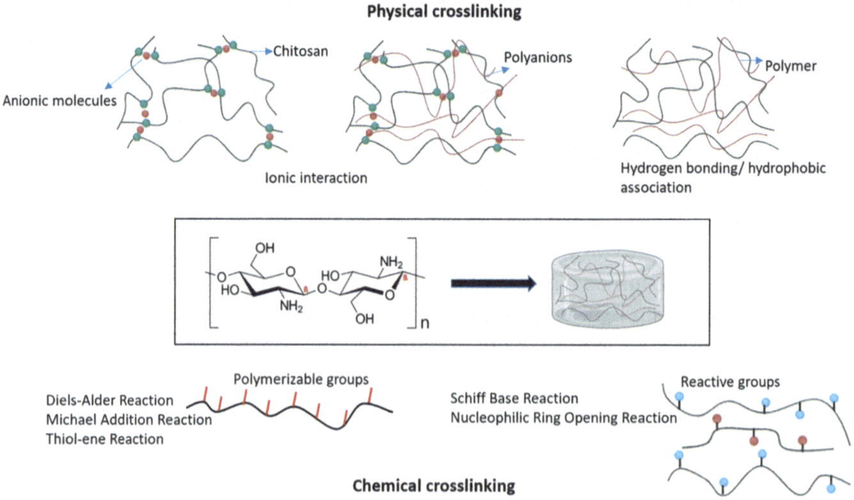

Fig. 7 Synthesis of chitosan-based hydrogels

Chitosan hydrogels can be fabricated through both physical and chemical cross-linking methods (Fig. 7). Physically cross-linked hydrogel fabrication depends on significant physical interactions, including ionic, hydrophobic, and polyelectrolytic interactions. Chitosan-based physical gels are particularly effective as drug delivery agents for short-term release. In the method involving hydrophobic interactions, gelation is based on sol-gel chemistry [34]. The incorporation of hydrophobic interactions in chitosan hydrogel formulations depends on the degree of N-acetylation with hydrophobic segments. Here, an amphiphile gel is produced by combining hydrophilic polymers and hydrophobic solvents. The hydrophobic surface of the resultant gel enhances drug entrapment efficiency and the stability of gels [35]. A stable and irreversible gel will be formed by chemical cross-linking with superior mechanical properties. Highly reactive hydroxyl and amino groups enable covalent bond formation across the polymer chains.

From bone to cardiac regeneration, the versatility of chitosan hydrogels allows for tailored therapeutic strategies that meet the specific needs of each tissue type. The upcoming sections will examine the diverse applications of chitosan hydrogels in

Table 2 Chitosan hydrogels for the delivery of biomolecules in tissue engineering

Targeted tissue	Biomaterial (composition)	Type of biomolecule delivery	Respective function	Reference
Bone	Chitosan/POSS hydrogel	Ketoprofen	Demonstrated controlled delivery of ketoprofen for bone regeneration.	[38]
Nucleus pulposus	Chitosan/HA hydrogel	Kartogenin	Showed controlled release of kartogenin and promoted chondrogenic differentiation of ADMSCs.	[53]
Cartilage	Chitosan hydrogel	Primary articular chondrocytes	This biomimetic cell carrier showed good cell adhesion and maintained chondrogenic properties in vivo.	[40]
Cardiac	Chitosan hydrogel with graphene quantum dots	Human mesenchymal stem cells	Hydrogel promoted angiogenesis and reduced cardiomyocyte necrosis in the infarcted heart region.	[45]
Corneal	Thermosensitive chitosan hydrogel	Ferulic acid	Demonstrated controlled delivery of ferulic acid in both in vitro and in vivo studies.	[47]
Skin	Composite alginate- chitosan hydrogel	Hesperidin	Exhibited anti-inflammatory, antibacterial, and angiogenic effects on the excisional wound model.	[51]
	Chitosan/HA hydrogel	Deferoxamine	In vivo model demonstrated sustained release of DFO for more than 1 week. Gel formulation was not cytotoxic and showed enhanced angiogenesis.	[52]

tissue engineering. Table 2 depicts previous studies where chitosan hydrogels were utilized for the delivery of biomolecules for various tissue engineering purposes.

2.2.1 Adipose Tissue Engineering

Soft tissue replacement is frequently necessary when conditions such as disease, trauma, or injury result in damage to the adipose tissue in different body regions. These soft tissue defects can significantly impact a patient's appearance, emotional health, and, sometimes, physical functionality. As a result, there is a substantial clinical demand for effective adipose tissue substitutes. In the context of breast tissue reconstruction, current medical practices involve the use of autologous tissue flaps or various commercial fillers. These fillers include substances like fibrin, hyaluronic acid, and viscoelastic hylan gels. However, these methods have limitations, and the quest for improved solutions continues to be a critical area of research and development in medical science [36].

2.2.2 Bone Tissue Engineering

The limited mechanical strength and nonspecific osteoinductive properties of chitosan hydrogel restrict its effectiveness in bone regeneration applications. These limitations hinder its ability to provide adequate structural support and targeted stimulation for bone regeneration, reducing its overall potential in medical treatments aimed at bone repair and regeneration [37]. However, chitosan-based hybrid hydrogels have emerged as potential candidates for drug delivery in bone TE and regeneration. Studies have shown that chitosan-based hybrid hydrogels, incorporating polyhedral oligomeric silsesquioxanes (POSS), offer improved properties over pure chitosan, including enhanced rheology and drug delivery capabilities. A Michael-type reaction was utilized to formulate a chitosan/POSS hydrogel, where genipin was used as a cross-linker. The study demonstrated its potential for the controlled release of drugs such as ketoprofen, making it valuable for precise therapeutic applications in bone regeneration. In vitro biocompatibility was assessed on human fetal osteoblastic cells, and negligible toxicity was shown [38].

2.2.3 Cartilage Tissue Engineering

The broad utilization of chitosan-based hydrogels is notable for cartilaginous tissue regeneration due to several advantageous characteristics. These include a structural similarity to glycosaminoglycan (GAG), a vital element in the cartilage matrix, along with biocompatibility, biodegradability, bioadhesion, and inherent antibacterial, chondro-conductive and chondro-integrative properties [39].

Researchers investigated an injectable and self-gelling chitosan-based carrier for cells to assess its effectiveness in delivering primary chondrocytes and its ability to adhere to and remain in defects in living animals [40]. This study explored the impact of adding microparticulate chitosan to lyophilized, porous silk scaffolds on cartilage formation using goat chondrocytes. The optimized blend of chitosan and silk notably improved the mechanical properties of the constructs. This underscores the importance of creating customized biomaterial scaffolds to guide cell differentiation, proliferation, and extracellular matrix deposition in cartilage tissue engineering [41]. In 2020, Beniani et al. reported on a thermosensitive injectable chitosan hydrogel incorporated with kartogenin to promote chondrogenesis for cartilage tissue regeneration This small molecule has been reported to promote the selective differentiation of MSCs toward chondrocytes. β-glycerophosphate was utilized to make the hydrogel thermosensitive. The hydrogel demonstrated a sustained kartogenin release for 40 days, and the chondrogenesis of MSCs were confirmed by PicoGreen assay and collagen expression analysis [42].

2.2.4 Cardiac Tissue Engineering

Injectable, preformed hydrogels represent a reasonable new therapeutic strategy for inducing cardiac regeneration in the context of progressive heart failure after myocardial infarction. In this case, the combination of chitosan and smart technologies indicates great progress in the field of cardiac therapy, which can provide novel approaches to treat cardiac diseases and eventually improve the life quality of patients [43]. An appropriate hydrogel that facilitates cell adhesion, proliferation, and vascularization has always been a challenge in the field of myocardial tissue engineering [44]. Researchers have discussed an engineered conductive hydrogel containing graphene quantum dots for cardiac regeneration. In myocardial infarction (MI) models, the hydrogel not only promoted angiogenesis and reduced cardiomyocyte necrosis in the infarcted heart area when combined with human mesenchymal stem cells but also significantly enhanced cardiac function. The material suitability for cardiac applications were shown to be appropriate, which was confirmed via physicochemical analytical methods. Compared to untreated cells, in vitro tests demonstrated higher cell viability with increased expression of pro-inflammatory, angiogenic, and cardiogenic markers. The in vivo result showing an increased ejection fraction, decreased fibrosis, increased vessel density, and decreased size in the infarcted area suggests significant heart regeneration post-MI [45].

2.2.5 Corneal Tissue Engineering

Hydrogel-based systems are attractive for use in corneal tissue engineering applications because their characteristics closely resembles the ECM of cornea. Their structural and mechanical characteristics align well with the natural corneal tissue, enhancing their potential to integrate seamlessly with the body's own tissues and promote healing.

Earlier research has explored the characteristics of innovative thin chitosan-poly (ethylene glycol) composite hydrogel films (CPHFs) for potential use in corneal TE. The study delves into the mechanical properties, biocompatibility, and structural integrity of these hydrogel films, aiming to evaluate their suitability for regenerative medicine applications, specifically for repairing and regenerating corneal tissue [46]. Similarly, Tsai et al. fabricated chitosan-based thermosensitive hydrogel for the controlled delivery of ferulic acid on corneal wound healing. Ferulic acid is an antioxidant and has been reported to enhance cell adhesion and proliferation by regulating cyclin D1 and VEGF pathways. The in vivo studies on the rabbit corneal burn model demonstrated reduced inflammation and apoptosis rates, which promoted corneal wound healing [47, 48].

2.2.6 Skin Tissue Engineering

In recent decades, chitosan-based biomaterials have found significant applications in skin TE, owing to their biocompatibility, hemostatic properties, antimicrobial activities, and biodegradability. Chitosan functions dually as a wound dressing and a drug delivery system (DDS), facilitating the precise delivery of antibacterial agents, growth factors, stem cells, and other bioactive substances, thus potentially accelerating the wound-healing process [49, 50]. A composite alginate-chitosan hydrogel fabricated with different concentrations of hesperidin was analyzed for wound healing on an excisional wound model. Hesperidin is a flavonoid that has anti-inflammatory, antibacterial, and angiogenic effects, which could potentially be used as a therapeutic agent for treating skin conditions. This combination was able to provide an appropriate porous structure for better cell attachment and migration. In this study, the combined effect of hesperidin and hydrogel was evaluated on skin wound healing. The material did not show any hemolysis, and the results of the in vivo evaluation demonstrated that the alginate-chitosan hydrogel with 10% hesperidin could heal almost 95% of the wound after 14 days [51]. In a related study, chitosan/HA hydrogel was developed with angiogenic molecule deferoxamine (DFO)-incorporated nanoparticles for angiogenesis. The in vivo study demonstrated that the chitosan/HA hydrogel system containing DFO achieved sustained release of the drug for 8 days and thereby enhanced blood vessel formation. The gel system also showed excellent cytocompatibility and cell proliferation [52].

2.2.7 Nucleus Pulposus Tissue Engineering

Hydrogels are three-dimensional hydrophilic polymer networks with high water-holding capacity that closely mimic the extracellular matrix of nucleus pulposus tissues. Chitosan hydrogels can be tailored to have the mechanical and biochemical properties required for nucleus pulposus tissue engineering. Injectable in situ forming hydrogels can overcome limitations associated with invasive surgeries by enabling minimally invasive delivery. These hydrogels are administered via syringe injection, allowing the aqueous solution to be introduced directly into the target sites. This method is particularly advantageous for filling irregularly shaped defects, providing a less invasive and more precise treatment option. For instance, Zhu et al. developed kartogenin-incorporated chitosan-hyaluronic acid hydrogel (CS/HA) that promoted nucleus pulposus regeneration. The developed CS/HA hydrogel was capable of establishing a microenvironment similar to that of the native nucleus pulposus tissue. The inclusion of chitosan, which is structurally similar to glycosaminoglycans, and HA, an ECM component, facilitated cellular growth and deposition. Additionally, CS/HA hydrogel demonstrated a controlled release of kartogenin over time. This combination potentially plays a critical role in regulating the differentiation and function of nucleus pulposus [53].

2.3 Chitosan Scaffolds

Structurally, a scaffold is a three-dimensional porous material that resembles biological tissues to help cell growth and differentiation. This advanced design is at the forefront of the customizable architecture for application-specific porosity, degradation rates, and mechanical properties, which can be modified to meet its intended target tissue. Moreover, its multifunctionality can permit the delivery of growth factors and different bioactive substances to improve their regeneration characteristics, which makes it suitable for tissue engineering and regenerative medicine (Fig. 8) [54]. An ideal scaffold utilized for defective tissue should approximately mimic the native extracellular matrix (ECM) and immensely induce the therapeutic regenerative responses of the injured tissue [55].

In the upcoming sections, we will look into the specific roles of chitosan scaffolds in enhancing tissue repair and regeneration. Table 3 depicts previous studies where chitosan scaffolds were utilized for the delivery of biomolecules for various tissue engineering purposes.

2.3.1 Adipose Tissue Engineering

Soft tissue reconstruction to repair deficits or pathology-related issues is a clinical necessity, which demands suitable biomaterials that can fill voids and foster constructive tissue growth [56]. An example of this is a chitosan/glycerol phosphate (C/GP), thermosensitive hydrogel, which was utilized to incorporate acellular adipose tissue that offers biomimetic properties and supports cell interactions similarly to the natural extracellular environment. Acellular adipose tissue was added to the bio-scaffold, which contains extracellular matrix components that include biochemical cues for cell signaling and the integration of surrounding tissues. The ASCs embedded within this bioactive environment preserved the stem cell attributes and improved their regeneration capabilities [57].

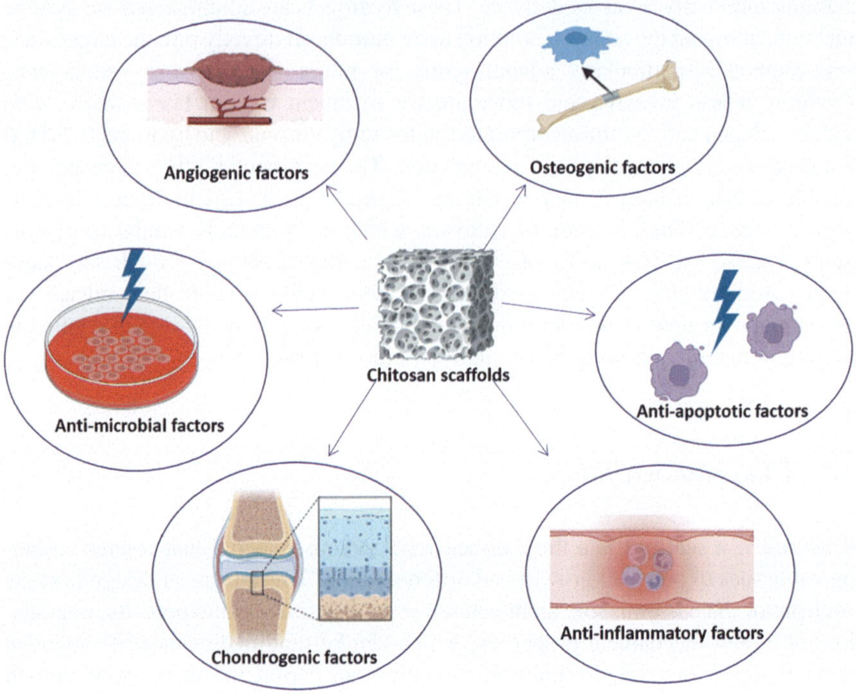

Fig. 8 Biomolecule delivery using chitosan scaffolds in tissue engineering

2.3.2 Bone Tissue Engineering

An ideal tissue-engineered scaffold for bone defect repair must have several key features. It should be biocompatible and biodegradable, and it should have a microstructured surface that promotes osteoconductivity and must possess appropriate mechanical properties. These are primarily to facilitate integration within the surrounding tissue, adequate degradation rates, as desired for tissue regeneration, promote good attachment of cells and allow migration. Apart from mechanical properties, the microstructure and osteoconductivity of the scaffold are crucial for guided neo-osteogenesis. All features should be similar in terms of strength and flexibility with natural living tissues to support normal physiological functions. Bioactive agents may also be included in advanced scaffolds to enhance bone healing and vascularization [58]. Ghaliah et al. studied the effect of simvastatin-impregnated chitosan (CS/SIM) scaffolds on the osteoblastic differentiation of human bone marrow MSCs. The study revealed that different concentrations of simvastatin did not significantly affect the osteogenic differentiation of human bone marrow MSCs, although there was an increase in alkaline phosphatase activity in the CS/SIM 0.05 mg group at day 14. The study concluded that the effect of simvastatin was not dose dependent [59].

Table 3 Chitosan scaffolds for the delivery of biomolecules in tissue engineering

Targeted tissue	Biomaterial (composition)	Type of biomolecule delivery	Respective function	Reference
Bone	Chitosan scaffold	Simvastatin	Showed structural similarity to bone ECM and promoted osteoblastic differentiation.	[59]
Cartilage	Chitosan/ polycaprolacton scaffold	Kartogenin	Showed similar mechanical properties to native cartilage tissue and chondrogenic differentiation of MSCs.	[61]
Skin	Chitosan scaffold	Amikacin	Demonstrated the sustained release of amikacin, promoting infectious wound healing.	[69]
	Chitosan membranes with glycerol	Glycerol and antibacterial agents	Enhanced mechanical properties and improved tensile strength; additionally, it can effectively load and release antibacterial drugs, significantly boosting their antibacterial performance for wound healing.	[70]
Neural tissue	Chitosan scaffold	Nerve growth factor	Biomimetic and biodegradable scaffold, enabling the long-term release of NGF (8 weeks) and enhancing nerve repair in a sciatic nerve injury model.	[71]
	Chitosan scaffold	Matrilin-2	Enhanced adhesion and the migration of Schwann cells via the controlled release of matrilin-2.	[72]
Periodontal tissue	Chitosan-calcium-hydroxide scaffold	Simvastatin	Demonstrated sustained delivery of simvastatin and improved tissue neogenesis in vivo.	[74]
	Carboxymethyl chitosan and poly-lysine (PLL-CMC) scaffold	Metformin	Dual-functioning scaffolds not only possess antioxidative characteristics but also exhibit antibacterial effects, which prevent infections commonly associated with periodontal diseases.	[75]

2.3.3 Cartilage Tissue Engineering

Cartilage plays a crucial role in biomechanics by evenly distributing the load onto the bones beneath it. Additionally, it offers a smooth, low-friction surface essential for the efficient mobility of joints. This tissue helps absorb impact and reduce wear on bones, thereby maintaining joint health and function over time. Garcia et al. have found that (CS)-based fiber mats with poly(ethylene oxide) (PEO), created through electrospinning, can be used as a scaffold for cartilage regeneration. The material showed good mechanical properties and a porous nature. The degradation study showed that the scaffold was stable for 21 days in a cell culture medium. This work mainly studied the in vitro cell viability, adhesion, and proliferation of chondrocytes on CS-PEO scaffolds. The results showed better adhesion and high proliferation of chondrocytes over the scaffold [60]. Chitosan can also be converted into multilayer composite scaffolds for the continuous release of therapeutic biomolecules. Researchers demonstrated that multilayer chitosan scaffolds with poly(caprolactone) mats and conjugated kartogenin enabled the gradual release of kartogenin, which promoted the chondrogenesis of MSCs [61].

2.3.4 Cardiac Tissue Engineering

The heart has a significantly lower regenerative capacity compared to other organs. Consequently, a key objective in cardiac TE is to enhance the repair and regeneration of damaged arteries, valves, and heart muscles. Hence, research on chitosan scaffolds aims to develop innovative solutions to improve heart function and recovery following injury, addressing the limitations of natural cardiac regeneration [62]. Chitosan possesses the capability to form scaffolds with interconnected structures and sufficient porosity, facilitating optimal cell viability through a steady supply of oxygen and nutrients. This aspect is vital for promoting cell growth, proliferation, and functionality in tissue engineering applications; ensuring successful regenerative therapies aimed at restoring damaged tissues. One extensively researched characteristic of scaffolds in cardiac tissue engineering is their pore diameter. Studies have indicated that pore sizes ranging from 60 to 200 μm facilitate effective cell-cell communication, growth, anchoring, and nutrient transport to cardiomyocytes and other cells [63].

Researchers have demonstrated the suitability of a composite collagen/chitosan scaffold for myocardial tissue engineering. The native myocardium is composed of an extracellular matrix, which contains collagen, elastin and glycosaminoglycans [64]. Accordingly, in myocardial tissue engineering, a collagen/chitosan composite scaffold was produced by using an acetic acid medium, and the cross-linking effects of glutaraldehyde (GTA), genipin (GP), and tripolyphosphate (TPP) were investigated. The results revealed that TPP cross-linked scaffolds exhibited better biocompatibility, swelling ratio, and porosity. Moreover, TPP scaffold also exhibited outstanding mechanical properties, featuring an elastic modulus that falls within the physiological range of native myocardial tissue (20–100 kPa) [65].

2.3.5 Corneal Tissue Engineering

The cornea is an avascular tissue. It protects the eye from infections by serving as a structural barrier [66]. Research has shown that chitosan scaffolds support the growth and differentiation of corneal epithelial cells, stromal cells, and endothelial cells, which are crucial for the successful regeneration of the multilayered corneal structure. Chitosan's transparency and optical properties can also be optimized to match those of the natural cornea, ensuring that it does not impair vision [67]. In a similar study, a chitosan scaffold was developed utilizing glutaraldehyde (GTA) as a cross-linker to the assess mechanical properties and transparency, aiming to mimic the corneal tissue [68]. While there are numerous studies investigating the use of chitosan in corneal tissue engineering, the overall number of studies specifically focused on this application is relatively limited, indicating that further research is needed for more robust applications and long term benefits.

2.3.6 Skin Tissue Engineering

Modern chitosan wound dressings have biocompatibility, nontoxicity, and antiseptic effects, which have led to their wide applications. Besides, they are hemostatic and accelerate the process of tissue regeneration. They may also be incorporated with a wide range of bioactive compounds to enhance their scavenging effect, thus increasing their therapeutic potential. In a recent study, an antibacterial agent and diopside nanoparticles were loaded into a chitosan scaffold for the effective control of wound infection and to promote wound healing. Utilizing the sol-gel method, diopside nanoparticles were synthesized and further mixed with 1% chitosan and 1% amikacin. The obtained gel was lyophilized in order to get a scaffold, which demonstrated the sustained release of amikacin for 7 days. Further cell culture studies proved that the material was biocompatible and promoted cell adhesion and migration, thereby facilitating faster wound healing. In the early stages of wound healing, wounds are often predominantly colonized by various Gram-positive and Gram-negative bacteria. The antibacterial activity of scaffolds was studied ex vivo in porcine skin infected with bacteria, S.aureus and E.coli, and the scaffold was found to have excellent antibacterial activity against both the bacteria [69]. Ye Ma et al. reported the fabrication of chitosan membranes infused with glycerol and antibacterial agents for burn wound treatment. Glycerol was incorporated to enhance the long-term stability of the membranes, while the antibacterial agents targeted *Escherichia coli* and *S. aureus* growth. In vitro experiments using dermal fibroblast cultures indicated the increased proliferation of fibroblast cells on the membranes. These findings imply that the created membrane holds promise for use as an antimicrobial dressing system in the treatment of skin burns [70].

2.3.7 Neural Tissue Engineering

Chitosan scaffolds are highly recognized for their potential in biomolecule delivery and nerve regeneration. In 2021, Liu et al. investigated chitosan scaffolds loaded with a nerve growth factor for long-term delivery to address sciatic nerve injuries in aged rats. The in vivo treatment significantly enhanced motor and sensory function recovery, facilitating the reconnection of the regenerated sciatic nerve with neural circuits in the sensory nervous system. This biomimetic scaffold was designed to completely degrade in 2 years, allowing for gradual replacement by regenerated nerve tissue as the scaffold breaks down over time [71].

A recent investigation focused on creating a biomimetic chitosan scaffold by incorporating matrilin-2 into a chitosan-derived porous scaffold. Matrilin-2 enhances the adhesion and migration of Schwann cells, which is crucial for nerve regeneration. This scaffold aimed to enhance peripheral nerve regeneration by providing a supportive environment for nerve cell growth and facilitating the controlled release of matrilin-2 [72].

2.3.8 Periodontal Tissue Engineering

Periodontitis is a gum disease distinguished by inflammation and infection of the periodontal tissue, primarily caused by the deposition of plaque. Recent advancements in the treatment strategy focus on the regeneration of periodontal tissue, including the periodontal ligament, cementum, and alveolar bone. To enhance periodontal regeneration, new carriers for localized drug delivery within the periodontal cavity are being developed [73].

A recent study developed a composite chitosan-calcium hydroxide scaffold incorporated with simvastatin, to induce the differentiation of dental pulp cells toward a highly mineralized phenotype, as confirmed by the overexpression of odontoblastic markers. Moreover, the interconnected porous nature of the scaffold enabled cell attachment, proliferation, and enhanced mineralized matrix deposition. The study demonstrated the synergistic effect of simvastatin and calcium hydroxide in cell differentiation. This process is facilitated by the release of simvastatin and calcium, which promotes the expression of genes associated with odontoblastic activity [74].

In 2021, another study by Gao et al. developed a bioactive coating utilizing a metformin-loaded multilayered carboxymethyl chitosan and poly(lysine) (PLL-CMC) scaffold that could enhance the regenerative potential of periodontal tissue while addressing the challenges posed by diabetes-related complications. In vitro cell culture studies in periodontal ligament stem cells (PDLSCs) demonstrated good biocompatibility of the scaffold. The scaffold facilitated the controlled release of metformin for 14 days and significantly inhibited bacterial adhesion and growth. Hence, metformin-loaded multilayered scaffolds could restore the impaired osteogenic potential of human PDLSCs affected by hyperglycemia [75].

3 Conclusions and Future Directions

In this chapter, we have given a concise summary of the current status and challenges associated with the delivery of biomolecules using chitosan in tissue engineering. In conclusion, while chitosan formulations initially face limitations in drug delivery and tissue engineering applications due to their inherent hydrophilicity and insolubility in specific physiological conditions, the potential for modifying the polymer offers promising avenues for overcoming these challenges. By strategically altering the properties of chitosan formulations, researchers can enhance their suitability for biomolecule delivery in tissue engineering, thereby unlocking new opportunities for advanced therapeutic interventions and regenerative medicine approaches.

References

1. Han F, Wang J, Ding L, Hu Y, Li W, Yuan Z, Guo Q, Zhu C, Yu L, Wang H, Zhao Z (2020) Tissue engineering and regenerative medicine: achievements, future and sustainability in Asia. Front Bioeng Biotechnol 8:83
2. Mohammadi Z, Eini M, Rastegari A, Tehrani MR (2021) Chitosan as a machine for biomolecule delivery: a review. Carbohydr Polym 256:117414
3. Guan J, Feng J, Lu F (2022) Recent advances in the hydrogel based biomolecule delivery system for cartilage tissue engineering. Adv Mater Sci Eng 1:1899400
4. Kołodziejska M, Jankowska K, Klak M, Wszoła M (2021) Chitosan as an underrated polymer in modern tissue engineering. Nanomaterials 11:3019
5. Sundaram MN, Mony U, Varma PK, Rangasamy J (2021) Vasoconstrictor and coagulation activator entrapped chitosan based composite hydrogel for rapid bleeding control. Carbohydr Polym 258:117634
6. Piekarska K, Sikora M, Owczarek M, Jóźwik-Pruska J, Wiśniewska-Wrona M (2023) Chitin and chitosan as polymers of the future-obtaining, modification, life cycle assessment and main directions of application. Polymers 15:793
7. Ways TMM, Lau WM, Khutoryanskiy VV (2018) Chitosan and its derivatives for application in mucoadhesive drug delivery systems. Polymers 10:267
8. Ibrahim HM, Klingner A (2020) A review on electrospun polymeric nanofibers: production parameters and potential applications. Polym Test 90:106647
9. Sidharthan DS, Abhinandan R, Adithya SP, Balagangadharan K, Selvamurugan N (2021) Chitosan and its potential use for the delivery of bioactive molecules in bone tissue engineering. In: Jayakumar R, Prabaharan M (eds) Chitosan for biomaterials IV. Springer, Cham, pp 117–162
10. Sultankulov B, Berillo D, Sultankulova K, Tokay T, Saparov A (2019) Progress in the development of chitosan-based biomaterials for tissue engineering and regenerative medicine. Biomolecules 9:470
11. Antony IR, Pradeep A, Pillai AV, Menon RR, Kumar VA, Jayakumar R (2023) Antiseptic chitosan-poly (hexamethylene) biguanide hydrogel for the treatment of infectious wounds. J Funct Biomater 14:528
12. Nithya S, Nimal TR, Baranwal G, Suresh MK, Anju CP, Kumar VA, Mohan CG, Jayakumar R, Biswas R (2018) Preparation, characterization and efficacy of lysostaphin-chitosan gel against *Staphylococcus aureus*. Int J Biol Macromol 110:157–166
13. Prajapati J, Misra SK, Upadhyay S, Kapoor A (2022) Microspheres: the efficient controlled drug delivery carriers. Acta Sci Pharm Sci 6:33–44
14. Huang L, Xiao L, Poudel AJ, Li J, Zhou P, Gauthier M, Liu H, Wu Z, Yang G (2018) Porous chitosan microspheres as microcarriers for 3D cell culture. Carbohydr Polym 202:611–620
15. Frenţ OD, Duteanu N, Teusdea AC, Ciocan S, Vicas L, Jurca T, Muresan M, Pallag A, Ianasi P, Marian E (2022) Preparation and characterization of chitosan-alginate microspheres loaded with quercetin. Polymers 14:490
16. Luo X, Tang K-Y, Wang Y-Y (2023) Preparation of drug-loaded chitosan microspheres repair materials. Eur Rev Med Pharmacol Sci 27:6489–6495

17. Das MK, Ahmed AB, Saha D (2019) Microsphere a drug delivery system–a review. Int J Curr Pharm Res 11:34–41
18. Divya K, Jisha MS (2018) Chitosan nanoparticles preparation and applications. Environ Chem Lett 16:101–112
19. Arredondo-Ochoa T, Silva-Martínez GA (2022) Microemulsion based nanostructures for drug delivery. Front Nanotechnol 3:753947
20. Yanat M, Schroën K (2021) Preparation methods and applications of chitosan nanoparticles; with an outlook toward reinforcement of biodegradable packaging. React Funct Polym 161: 104849
21. Custódio CA, Cerqueira MT, Marques AP, Reis RL, Mano JF (2015) Cell selective chitosan microparticles as injectable cell carriers for tissue regeneration. Biomaterials 43:23–31
22. Ding C-C, Teng S-H, Pan H (2012) In-situ generation of chitosan/hydroxyapatite composite microspheres for biomedical application. Mater Lett 79:72–74
23. Li Q, Zhou G, Yu X, Wang T, Xi Y, Tang Z (2015) Porous deproteinized bovine bone scaffold with three-dimensional localized drug delivery system using chitosan microspheres. Biomed Eng 14:33
24. Yang H, Wang S, Bian H, Xing X, Yu J, Wu X, Zhang L, Liang X, Lu A, Huang C (2022) Extracellular matrix-mimicking nanofibrous chitosan microspheres as cell micro-ark for tissue engineering. Carbohydr Polym 292:119693
25. Lin F, Li Y, Cui W (2023) Injectable hydrogel microspheres in cartilage repair. Biomed Tech 1: 18–29
26. Naghizadeh Z, Karkhaneh A, Nokhbatolfoghahaei H, Farzad MS, Rezai-Rad M, Dehghan MM, Aminishakib P, Khojasteh A (2021) Cartilage regeneration with dual-drug-releasing injectable hydrogel/microparticle system: in vitro and in vivo study. J Cell Physiol 236:2194–2204
27. Paul A, Chen G, Khan A, Rao VTS, Shum-Tim D, Prakash S (2012) Genipin-cross-linked microencapsulated human adipose stem cells augment transplant retention resulting in attenuation of chronically infarcted rat heart fibrosis and cardiac dysfunction. Cell Transplant 21: 2735–2751
28. Delaey J, De Vos L, Koppen C, Dubruel P, Van Vlierberghe S, Van Den Bogerd B (2022) Tissue engineered scaffolds for corneal endothelial regeneration: a material's perspective. Biomater Sci 10:2440–2461
29. Tayebi T, Baradaran-Rafii A, Hajifathali A, Rahimpour A, Zali H, Shaabani A, Niknejad H (2021) Biofabrication of chitosan/chitosan nanoparticles/polycaprolactone transparent membrane for corneal endothelial tissue engineering. Sci Rep 11:7060
30. Kim HJ, Lee HC, Oh JS, Shin BA, Oh CS, Park RD, Yang KS, Cho CS (1999) Polyelectrolyte complex composed of chitosan and sodium alginate for wound dressing application. J Biomater Sci Polym 10:543–556
31. Niu X, Wei Y, Liu Q, Yang B, Ma N, Li Z, Zhao L, Chen W, Huang D (2020) Silver-loaded microspheres reinforced chitosan scaffolds for skin tissue engineering. Eur Polym J 134:109861
32. Fahimirad S, Abtahi H, Satei P, Ghaznavi-Rad E, Moslehi M, Ganji A (2021) Wound healing performance of PCL/chitosan based electrospun nanofiber electrosprayed with curcumin loaded chitosan nanoparticles. Carbohydr Polym 259:117640
33. Hamedi H, Moradi S, Hudson SM, Tonelli AE (2018) Chitosan based hydrogels and their applications for drug delivery in wound dressings: a review. Carbohydr Polym 199:445–460
34. Pita-López ML, Fletes-Vargas G, Espinosa-Andrews H, Rodríguez-Rodríguez R (2021) Physically cross-linked chitosan-based hydrogels for tissue engineering applications: a state-of-the-art review. Eur Polym J 145:110176
35. Hong F, Qiu P, Wang Y, Ren P, Liu J, Zhao J, Gou D (2023) Chitosan-based hydrogels: from preparation to applications-a review. Food Chem: X 21:101095
36. Choi JH, Gimble JM, Lee K, Marra KG, Rubin JP, Yoo JJ, Vunjak-Novakovic G, Kaplan DL (2010) Adipose tissue engineering for soft tissue regeneration. Tissue Eng Part B Rev 16:413–426
37. Kazemi-Aghdam F, Jahed V, Dehghan-Niri M, Ganji F, Vasheghani-Farahani E (2021) Injectable chitosan hydrogel embedding modified halloysite nanotubes for bone tissue engineering. Carbohydr Polym 269:118311

38. Celesti C, Iannazzo D, Espro C, Visco A, Legnani L, Veltri L, Visalli G, Di Pietro A, Bottino P, Chiacchio MA (2022) Chitosan/POSS hybrid hydrogels for bone tissue engineering. Materials 15:8208
39. De Sousa Victor R, Santos AMDC, De Sousa BV, De AraújoNeves G, Santana LNDL, Menezes RR (2020) A review on Chitosan's uses as biomaterial: tissue engineering, drug delivery systems and cancer treatment. Materials 13:4995
40. Hoemann CD, Sun J, Légaré A, McKee MD, Buschmann MD (2005) Tissue engineering of cartilage using an injectable and adhesive chitosan-based cell-delivery vehicle. Osteoarthr Cartil 13:318–329
41. Chameettachal S, Murab S, Vaid R, Midha S, Ghosh S (2017) Effect of visco-elastic silk-chitosan microcomposite scaffolds on matrix deposition and biomechanical functionality for cartilage tissue engineering: viscoelastic silk-chitosan microcomposite scaffolds. J Tissue Eng Regen Med 11:1212–1229
42. Dehghan-Baniani D, Chen Y, Wang D, Bagheri R, Solouk A, Wu H (2020) Injectable in situ forming kartogenin-loaded chitosan hydrogel with tunable rheological properties for cartilage tissue engineering. Colloids Surf B Biointerfaces 192:111059
43. Beleño Acosta B, Advincula RC, Grande-Tovar CD (2023) Chitosan-based scaffolds for the treatment of myocardial infarction: a systematic review. Molecules 28:1920
44. Chang Y-C, Mirhaidari G, Kelly J, Breuer C (2021) Current challenges and solutions to tissue engineering of large-scale cardiac constructs. Curr Cardiol Rep 23:47
45. Si R, Gao C, Guo R, Lin C, Li J, Guo W (2020) Human mesenchymal stem cells encapsulated-coacervated photoluminescent nanodots layered bioactive chitosan/collagen hydrogel matrices to indorse cardiac healing after acute myocardial infarction. J Photochem Photobiol 206:111789
46. Ozcelik B, Brown KD, Blencowe A, Daniell M, Stevens GW, Qiao GG (2013) Ultrathin chitosan–poly (ethylene glycol) hydrogel films for corneal tissue engineering. Acta Biomater 9:6594–6605
47. Tsai CY, Woung LC, Yen JC, Tseng PC, Chiou SH, Sung YJ, Liu KT, Cheng YH (2016) Thermosensitive chitosan-based hydrogels for sustained release of ferulic acid on corneal wound healing. Carbohydr Polym 135:308–315
48. Wang J, Yuan Z, Zhao H, Ju D, Chen Y, Chen X, Zhang J (2011) Ferulic acid promotes endothelial cells proliferation through up-regulating cyclin D1 and VEGF. J Ethnopharmacol 137:992–997
49. Amirthalingam S, Rajendran AK, Mani P (2021) Perspectives and challenges of using chitosan in various biological applications. In: Rangasamy J (ed) Chitosan for biomaterials III: structure-property relationships. Springer, Cham, pp 1–22
50. Liu H, Wang C, Li C, Qin Y, Wang Z, Yang F, Li Z, Wang J (2018) A functional chitosan-based hydrogel as a wound dressing and drug delivery system in the treatment of wound healing. RSC Adv 8:7533–7549
51. Bagher Z, Ehterami A, Safdel MH, Khastar H, Semiari H, Asefnejad A, Davachi SM, Mirzaii M, Salehi M (2020) Wound healing with alginate/chitosan hydrogel containing hesperidin in rat model. Drug Deliv Technol 55:101379
52. Vignesh S, Sivashanmugam A, Annapoorna M, Janarthanan R, Subramania I, Jayakumar R (2018) Injectable deferoxamine nanoparticles loaded chitosan-hyaluronic acid coacervate hydrogel for therapeutic angiogenesis. Colloids Surf B Biointerfaces 161:129–138
53. Zhu Y, Tan J, Zhu H, Lin G, Yin F, Wang L, Song K, Wang Y, Zhou G, Yi W (2017) Development of kartogenin-conjugated chitosan–hyaluronic acid hydrogel for nucleus pulposus regeneration. Biomater Sci 5:784–791
54. Gupta P, Sharma S, Jabin S, Jadoun S (2024) Chitosan nanocomposite for tissue engineering and regenerative medicine: a review. Int J Biol Macromol 254:127660
55. Kazemi Asl S, Rahimzadegan M, Ostadrahimi R (2023) The recent advancement in the chitosan hybrid-based scaffolds for cardiac regeneration after myocardial infarction. Carbohydr Polym 300:120266

56. Mahoney CM, Imbarlina C, Yates CC, Marra KG (2018) Current therapeutic strategies for adipose tissue defects/repair using engineered biomaterials and biomolecule formulations. Front Pharmacol 9:507
57. Huang Y-B, Lin M-W, Liu M-Y, C.-L. (2015) Chen composite of decellular adipose tissue with chitosan-based scaffold for tissue engineering with adipose-derived stem cells. J Biomater Tissue Eng 5:56–63
58. Zhang L, Yang G, Johnson BN, Jia X (2019) Three-dimensional (3D) printed scaffold and material selection for bone repair. Acta Biomater 84:16–33
59. Alsawah GM, Al-Obaida MI, Al-Madi EM (2021) Effect of a simvastatin-impregnated chitosan scaffold on cell growth and osteoblastic differentiation. Appl Sci 11:5346
60. Garcia CE, Lardy B, Bossard F, Soltero Martínez FA, Rinaudo M (2021) Chitosan based biomaterials for cartilage tissue engineering: chondrocyte adhesion and proliferation. Food Hydrocoll Health 1:100018
61. Houreh AB, Masaeli E, Nasr-Esfahani MH (2021) Chitosan/polycaprolactone multilayer hydrogel: a sustained Kartogenin delivery model for cartilage regeneration. Int J Biol Macromol 177: 589–600
62. Gil-Cabrerizo P, Scacchetti I, Garbayo E, Blanco-Prieto MJ (2023) Cardiac tissue engineering for myocardial infarction treatment. Eur J Pharm Sci 185:106439
63. EsmaeiliPourfarhangi K, Mashayekhan S, Asl SG, Hajebrahimi Z (2018) Construction of scaffolds composed of acellular cardiac extracellular matrix for myocardial tissue engineering. Biologicals 53:10–18
64. Frangogiannis NG (2017) The extracellular matrix in myocardial injury, repair, and remodeling. J Clin Invest 127:1600–1612
65. Fang Y, Zhang T, Song Y, Sun W (2020) Assessment of various crosslinking agents on collagen/chitosan scaffolds for myocardial tissue engineering. Biomed Mater 15:045003
66. Bashir H, Seykora J, Lee V (2017) Invisible shield: review of the corneal epithelium as a barrier to UV radiation, pathogens, and other environmental stimuli. J Ophthalmic Vis Res 12:305
67. Li W, Long Y, Liu Y, Long K, Liu S, Wang Z, Wang Y, Ren L (2014) Fabrication and characterization of chitosan–collagen crosslinked membranes for corneal tissue engineering. J Biomater Sci Polym Ed 25:1962–1972
68. Rusdiputra S, Wibowo A, TresnaWulanAsri LA, Purwasasmita BS (2019) Fabrication of chitosan based-scaffold as potential cornea implant. IOP Conf Ser Mater Sci Eng 547:012062
69. Mothilal NP, Pradeep A, Arthi C, Gopal K, Kaliannagounder VK, Park CH, Kumar VA, Rangasamy J (2024) Amikacin sulphate loaded chitosan-diopside nanoparticles composite scaffold for infectious wound healing. Int J Biol Macromol 263:130217
70. Ma Y, Xin L, Tan H, Fan M, Li J, Jia Y, Ling Z, Chen Y, Hu X (2017) Chitosan membrane dressings toughened by glycerol to load antibacterial drugs for wound healing. Mater Sci Eng C 81:522–531
71. Liu FD, Duan HM, Hao F, Zhao W, Gao YD, Hao P, Yang ZY, Li XG (2022) Biomimetic chitosan scaffolds with long-term controlled release of nerve growth factor repairs 20-mm-long sciatic nerve defects in rats. Neural Regen Res 17:1146–1155
72. Li NY, Vorrius B, Ge J, Qiao Z, Zhu S, Katarincic J, Chen Q (2023) Matrilin-2 within a three-dimensional lysine-modified chitosan porous scaffold enhances Schwann cell migration and axonal outgrowth for peripheral nerve regeneration. Front Bioeng Biotechnol 11:1142610
73. Atia GA, Shalaby HK, Zehravi M, Ghobashy MM, Attia HA, Ahmad Z, Khan FS, Dey A, Mukerjee N, Alexiou A, Rahman MH (2022) Drug-loaded chitosan scaffolds for periodontal tissue regeneration. Polymers 14:3192
74. Soares DG, Bordini EA, Bronze-Uhle ES, Cassiano FB, Silva IS, Gallinari MD, Matheus HR, Almeida JM, Cintra LT, Hebling J, de Souza Costa CA (2021) Chitosan-calcium-simvastatin scaffold as an inductive cell-free platform. J Dent Res 100:1118–1126
75. Gao X, Cong X, Wang Y, Zhong W, Huang L, Wang X, Chen P, He Z, Hu B, Song J (2021) Fabrication of antioxidative and antibacterial surface coatings with metformin-loaded self-assembled multilayers for periodontal regeneration in diabetes mellitus patients. J Mater Sci 56:18668–18683

Adv Polym Sci (2025) 295: 429–462
https://doi.org/10.1007/12_2024_187

Published online: 9 January 2025

Chitosan-based Emulsions: Versatile Vesicular Capsules in Pharmaceutical Applications

Wanpen Tachaboonyakiat

Contents

1 Introduction 430
2 Role of CS on Emulsion Formation and Stabilization 432
2.1 Role of CS in Emulsion Formation 433
2.2 Role of CS on Emulsion Stabilization 434
3 Phenomena of Unstable Emulsion 435
4 Conventional vs Pickering Emulsions: Vesicular Capsules 437
4.1 Conventional Emulsions 439
4.2 Pickering Emulsions 439
4.3 Comparison Between Pickering and Conventional Emulsions 444
5 Emulsion: A Versatile Capsule in Pharmaceutical Applications 444
5.1 Drug Delivery 445
5.2 Targeted Drug Delivery 446
5.3 Cosmeceuticals and Topical Delivery 447
5.4 Nutraceuticals and Food Products 449
5.5 Antimicrobial Treatment 452
5.6 Wound Healing 453
6 Summary 455
References 456

Abstract This chapter provides a comprehensive overview of chitosan (CS)-based emulsions, encompassing both conventional and Pickering emulsions. The fundamental roles of CS that contribute to emulsion formation and stabilization are elucidated. The phenomena of emulsion instability, which should be of concern in formulating robust CS-based systems, are described. The versatility of CS-based

W. Tachaboonyakiat (✉)
Department of Materials Science, Faculty of Science, Chulalongkorn University, Bangkok, Thailand
e-mail: wanpen.ta@chula.ac.th

emulsions as capsules or for drug encapsulation allows their diverse applications in pharmaceutical and related fields. In drug delivery systems, these emulsions demonstrate a good potential for controlled release within suitable conditions for oral and nasal administration with enhanced targeting to mucosal areas. They can serve as stable vehicles for active ingredients in cosmeceutics. Moreover, CS emulsions facilitate the encapsulation of bioactive compounds, retard lipid oxidation, and modulate lipid digestion in nutraceutics. The inherent antimicrobial properties of CS are examined in the context of emulsion-based formulations, highlighting potential applications in food preservation and topical treatments to prevent infection, which can further enhance wound healing.

Keywords Antimicrobial · Chitosan emulsion · Cosmeceutics · Encapsulation · Nutraceutics · Pharmaceutics · Wound healing

1 Introduction

Chitin is a fibrous polysaccharide that is arranged along the exoskeleton of crustaceans. It can also be found in many aquatic, terrestrial, and some microorganisms. Chitosan (CS) is a cationic polymer obtained through the partial deacetylation of chitin (Fig. 1a). Chitin/CS is a ubiquitous and the second-most plentiful polysaccharide after cellulose. Chemically (Fig. 1b), chitin/CS, a linear polysaccharide, is a copolymer of randomly distributed β-(1,4)-linked D-glucosamine and *N*-acetyl-D-glucosamine units [1]. The chemical structure of chitin is predominantly *N*-acetyl-D-glucosamine units, while that of CS is predominantly D-glucosamine units. Thus, the

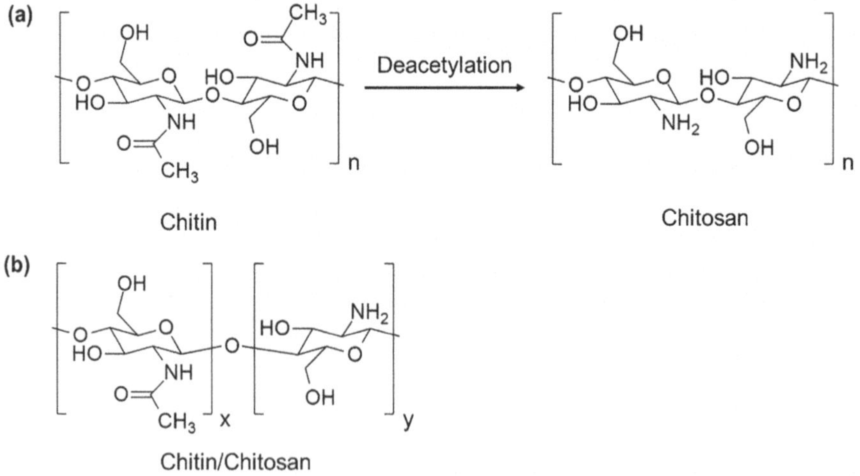

Fig. 1 (**a**) Deacetylation of chitin and (**b**) the chemical structure of chitin/CS

predominant number of *N*-acetyl-D-glucosamine units or D-glucosamine units is reported as the degree of acetylation (DA) or the degree of deacetylation (DD), respectively. Thus, CS with a higher DD contains a larger number of amino groups, which are protonated into positive charges in acidic conditions, allowing CS to become soluble in diluted acid and to possess a greater capacity to form complexes with anionic molecules [2]. In contrast, CS with a lower DD presents a larger number of acetylated residues, favoring the hydrogen bonding formation and so increasing the rigidity of polymeric chains [2]. Moreover, the molecular weight (M_W) of CS also influences its viscosity and solubility. The higher the M_W of CS, the lower the solubility and the higher the viscosity in solutions, whereas CS with a lower M_W has a better solubility and forms solutions with a lower viscosity [2]. Because CS consists of multiple functional groups (hydroxyl and amino groups), it can be readily chemically modified into numerous functional derivatives. Moreover, CS exhibits remarkable biocompatibility, low cytotoxicity, tissue regeneration, antibacterial, anti-inflammatory, and hemostatic properties. Importantly, chitin/CS is biodegradable, being partially broken down by enzymes (i.e., lysozymes) naturally present in the body [3, 4], and so it does not accumulate in the body, reducing the risk of adverse effects. In addition, the cell culture evaluation of its toxicity and interactions with cells has revealed that CS is nontoxic, even at high concentrations [3]. This makes it safe for various pharmaceutical applications [5]. Its biocompatibility, biodegradability, and nontoxicity make it highly valuable in numerous biomedical and pharmaceutical applications [6]. Moreover, the unique chemical structure of CS provides several key properties that make it an attractive and better choice than other synthetic polymers. The combination of these properties makes CS a versatile and promising biomaterial for developing innovative pharmaceutical products, including drug delivery systems, cosmeceutical and nutraceutical applications, and antimicrobial activities and wound healing.

Emulsion is a biphasic system of immiscible liquids consisting of a lipophilic phase dispersed in an aqueous phase, or vice versa, in the form of globules or droplets stabilized by the emulsifying agent [7]. Emulsifiers, which structurally contain both lipophilic and hydrophilic parts, align at the oil-water interface to minimize interfacial tension and facilitate emulsification. The liquid phase, which is dispersed as globules, is termed as the dispersed phase or discontinuous phase, while the dispersion medium is known as the continuous phase. The dispersion of an immiscible liquid can be formed via a mechanical energy input, resulting in metastable dispersion. Homogenization is a well-known process used to reduce the droplet size and mix two immiscible liquids into an emulsion [7]. Agitation is simple mechanical stirring or high-intensity mixing and is the main industrial process used for forming emulsions [8]. The stronger the agitation of a liquid phase is, the more it is dispersed in the form of smaller droplets. This technique is mostly applied in the food industry for the manufacturing process of butter, mayonnaise, and salad dressing and in cosmetic industries for the formulation of moisturizing creams and body balms. Emulsions in pharmaceutical and cosmeceutical applications are generally in the form of vitamin drops, lotions, creams, and so on.

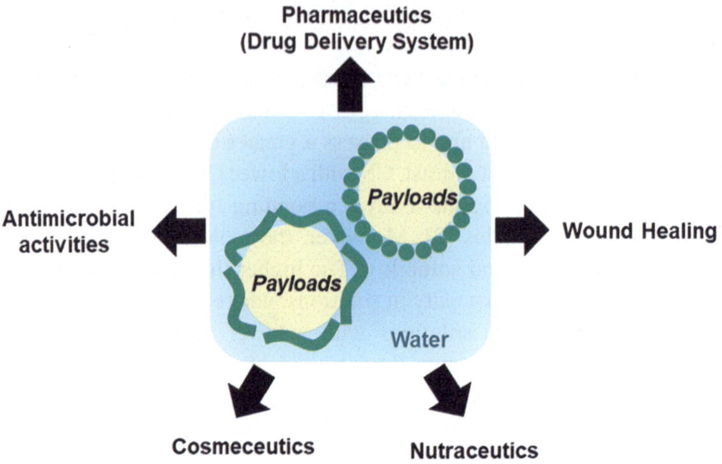

Fig. 2 Illustration of CS-based conventional and Pickering emulsions: versatile capsules in pharmaceutics and related applications

This chapter outlines the versatile applications of CS and its derivatives in forming emulsions, specifically highlighting their role in pharmaceutical applications. By exploring the distinctions between conventional and Pickering emulsions, this chapter intends to focus on the advantages and potential of CS-based emulsions in a controlled/targeted drug delivery system and related fields of cosmeceutical and nutraceutical applications as well as in antimicrobial activities and wound healing (Fig. 2).

2 Role of CS on Emulsion Formation and Stabilization

Emulsion systems are thermodynamically unstable because of the aggregation or flocculation of the emulsion droplets, followed by gravitational separation as a creaming layer. The phenomenon of an unstable emulsion is described later in Sect. 3. Emulsifiers, typically synthetic surfactants, are employed to minimize the surface tension at the oil-water interface and facilitate emulsification, while stabilizers are often added to develop the long-term stability of an emulsion by preventing droplets from coming in close contact via electrostatic repulsion and/or steric hindrance. The role of CS in emulsion formation and stabilization is summarized in this section.

2.1 Role of CS in Emulsion Formation

The solubility of the emulsifying agents in either the oil or water phase is often evaluated in terms of hydrophilic-lipophilic balance (HLB) [9]. This number is an empirical predictive model for determining the emulsifying characteristic of emulsifiers. A molecule with a greater proportion of hydrophilic groups compared to lipophilic groups presents a high HLB value. For CS, its emulsifying capacity may reflect its structural heterogeneity as it comprises of hydrophilic (D-glucosamine) residues and some hydrophobic (*N*-acetyl-D-glucosamine) residues. The hydrophobic residues are likely to turn themselves toward and into the oil phase, while the hydrophilic residues are likely to turn themselves toward and into the water phase. The molecular arrangement in between the oil and water interface reduces the contact area and lowers the interfacial tension [2]. The HLB value of CS was determined by measuring the diameter of a toluene drop on the surface of a 1% (w/v) CS solution in 1% (v/v) acetic acid and compared to that on the surface of several known emulsifying agent solutions. The CS solution was found to exhibit an HLB value in the range of 34–36.7, promoting the formation of oil-in-water (O/W) emulsions [10].

However, the emulsifying efficiency of CS may be affected by many factors, including the differences between the oils and the intrinsic and extrinsic features of CS, such as its DD, M_W, pH, and concentration. For example, when CS with a M_W of 100–300 kDa and a DD of 80 was used as an emulsifying agent for a paraffin oil-water interface, it had no adsorption at the paraffin oil interface when CS concentration was used below 0.1% (w/v) [11]. However, the interface tension decreased as the CS concentration increased up to 2% (w/v) [11]. On the other hand, in the sunflower oil-water system, the emulsifying capacity of CS was scaled directly with the CS concentration [12]. Increasing the CS concentration (0.25–0.75%) would probably promote the breakdown of oil into smaller droplets, whereas further increasing the CS concentration (0.75–1.25%) would promote electrostatic interactions among the molecules, leading to reduced chances for CS molecules to adsorb with the oil [12].

Lower M_W CS has a superior emulsifying activity than that with a higher M_W [10] since the lower Mw CS is absorbed more easily and densely onto the surface of the oil droplets than that for the higher Mw CS [12]. The effect of the DD of CS on its emulsifying efficiency has been issued previously [12], where CS with a DD of 60.5% and 86.1% had a better emulsifying activity than CS with a DD of 65.4%, 70.8%, and 77.3% [12]. The decreased DD reflected the increased level of hydrophobic *N*-acetylated residues, and so the CS molecule was promoted to absorb on the surface of oil droplets. Subsequently, CS with a DD of 60.5% showed the best emulsifying activity. On the other hand, CS with a high DD (86.1%) mostly contained protonated amino groups in acidic conditions, where the electrostatic repulsion of the protonated amino groups would extend the molecular structure, and so the hydrophobic residues of the *N*-acetylated residues had a greater chance to interact with the oil. In contrast, CS with a reduced DD (less than 80%) had a lower emulsifying capacity [13], whereas two optimum DD values (81% and 88%) achieved thorough emulsification of the sunflower oil without sedimentation.

2.2 Role of CS on Emulsion Stabilization

Chitosan is not only an emulsifying agent but also a stabilizer in the emulsification process in terms of increasing the viscosity of the emulsion system and creating electrostatic stabilization. The presence of CS increases the viscosity of the continuous phase of O/W emulsion and delays the agglomeration of droplets and flocculation, therefore it decreases the rate of creaming. The M_W and concentration of CS strongly affect the viscosity of the solution [12]. Rapeseed O/W emulsions stabilized by 4% (w/v) whey protein isolate (WPI) was described to be relatively unstable at pH 3, while emulsion systems containing 4% (w/v) WPI with different concentrations of CS [0.1–0.5% (w/v)] delayed the rate of emulsion creaming. Emulsions with 0.5% (w/v) CS showed stability of longer than 150 h [14].

Moreover, electrostatic interactions can significantly keep droplets from getting close enough to flocculate. The association of oppositely charged polymers, termed layer-by-layer (LBL) electrostatic deposition, was employed to fabricate multilayer ultrathin films of polyelectrolytes [15]. With this strategy, the coating of oppositely charged polymers to emulsifier-coated droplets can be applied with emulsion for an alternative approach to create electrosteric stabilization. For this, an ionic emulsifier, such as lecithin, was used to generate a primary emulsion, and the positively charged CS was subsequently added into the system, where it adsorbed to the surface of droplets and subsequently produced a secondary emulsion that presented a two-layer interface coating on the emulsion droplets [16]. The repetition of this LBL procedure can produce oil droplets coated by a multilayer interface emulsion that exhibits a better thermal and freeze-thaw stability than single-layer O/W emulsions [16].

Several environmental factors affect CS characteristics as well as electrostatic interactions, such as the pH and the presence of salt. Since CS has a pKa of around 6.3, the amino groups of CS can be protonated into positive charges; therefore, conventional emulsions stabilized by CS are stable at pH values below pKa via electrosteric interaction, in contrast, are not stable at pH values above pKa because of the relatively low electrostatic repulsion between the droplets, resulting in extensive droplet flocculation. Bridging flocculation probably happens from the partial desorption of CS molecules from the surface of the flocs [16]. Additionally, the presence of salt affects the ionic strength of the CS solution and can also affect electrosteric interactions and the amount of adsorbed CS at the interface. A minimum concentration of salt is required to neutralize the charges of CS due to the screening effect of salt. Higher concentrations of salt result in a weaker electrostatic attraction between CS and the anionic surface, and then molecular rearrangement after adsorption probably occurs by hydrophobic interactions, resulting in a higher quantity of

CS molecules adsorbed to the surface, thus a thicker coat than in the absence of salt [17].

However, the amino groups of CS are deprotonated at pH values above its pKa, so self-aggregating CS nanoparticles (NPs) can be formed at the protonation and deprotonation equilibrium and used as stabilizers for the formulation of Pickering emulsions [18]. The pH affects the type of emulsion, in terms of a conventional or a Pickering emulsion, and their stabilization is discussed in Sect. *4*. Pickering emulsions stabilized by self-aggregating CS particles are significantly affected by many factors, including the concentration, DD, and M_W of the CS. High CS concentrations, a low DD, and a low M_W facilitate the formation and stabilization of Pickering emulsions [19].

3 Phenomena of Unstable Emulsion

Emulsion instability is a dynamic phenomenon and is governed by the four key mechanisms of creaming (sedimentation), flocculation, Ostwald ripening, and coalescence (Fig. 3). Creaming and sedimentation respond directly to gravity, with no changes in emulsion droplet size or size distribution. These phenomena are induced by a difference in density between dispersed and continuous phases, leading to two layers of dispersion [20]. Creaming describes the upward movement of droplets because the dispersed phase has a lower density than the surrounding liquid and so is more concentrated as a cream layer, whereas sedimentation describes the downward movement of the droplets because the dispersed phase has a higher density than the surrounding liquid and may exhibit a clear upper serum. Therefore, O/W emulsions tend to show creaming, while water-in-oil (W/O) emulsions tend to show sedimentation [21].

Flocculation occurs when droplets get close enough to form aggregates (flocs), possibly touching at some points with no changes in their droplet size, size distribution, or total surface area [21]. Flocculation can be divided into two types: depletion and bridging. Depletion flocculation can be induced by the presence of excess unadsorbed polymeric surfactant micelles remaining in the continuous phase, which then merge the emulsion droplets to form an aggregate, a so-called submicelle, while bridging flocculation can be induced by the interacting or bridging of polymeric surfactant molecules between the droplets [22, 23]. Depletion flocculation leads to an increase in the low-shear viscosity of the emulsion because of the unadsorbed surfactant molecules remaining in the aqueous phase and exhibiting a shear-thinning behavior (the emulsion has a high viscosity at low shear rates where a further increase in the shear rate will reduce viscosity). On the other hand, bridging flocculation leads to a decrease in the overall low-shear viscosity of the emulsion because the surfactant molecules interact or bridge those droplets without contributing to the viscosity in an aqueous phase. Rather, those surfactant molecules are adsorbed onto the droplets and form an extended chain in the aqueous phase. Bridging flocculation frequently occurs with long polymeric surfactant molecules.

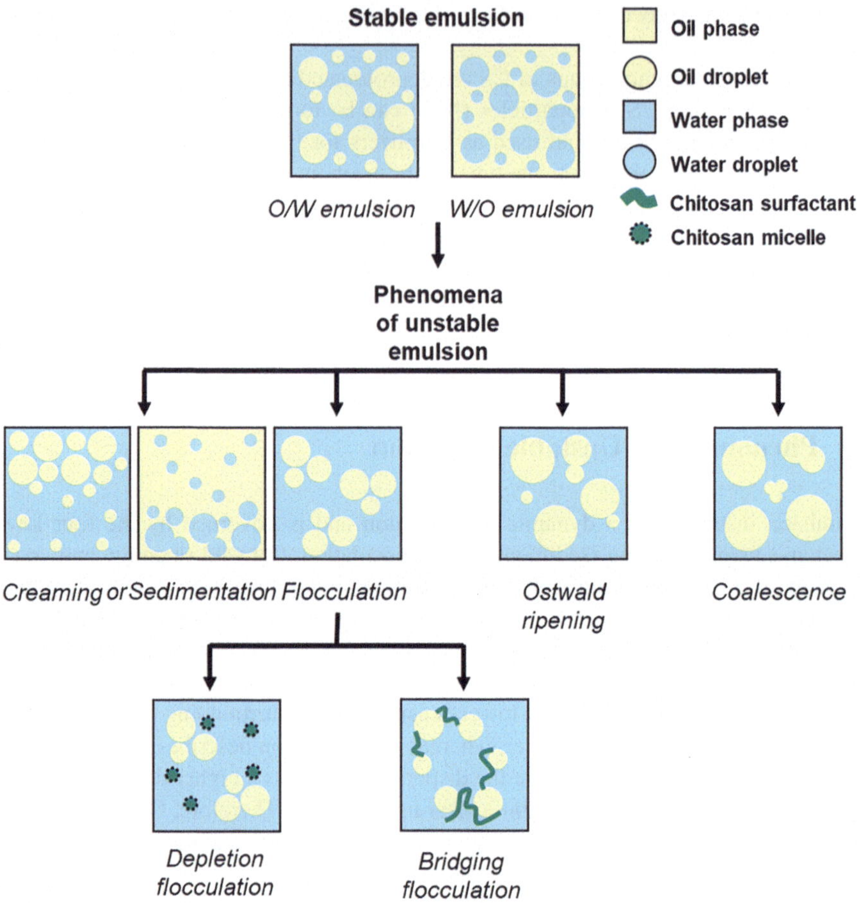

Fig. 3 Kinetic processes of emulsion instability through creaming (sedimentation), flocculation, Ostwald ripening, and coalescence

Depletion flocculation is generally more reversible upon dilution or shear than bridging flocculation.

Ostwald ripening occurs when the dispersed liquid has a restricted solubility in the continuous phase, causing smaller droplets to shrink, while larger droplets grow as the dispersed liquid moves from smaller droplets to larger ones against the continuous phase [21].

Coalescence happens when the small droplets merge to form larger droplets and frequently proceeds to creaming (sedimentation). Therefore, a two-phase separation is developed under Ostwald ripening or coalescence [21].

The physical mechanisms leading to unstable emulsions can be categorized into two types: reversible and irreversible processes. Flocculation is a reversible aggregation of droplets, while coalescence and Ostwald ripening are an irreversible fusion of droplets. Both instabilities are sensitive to the concentration of emulsifiers.

Flocculation predominates when the emulsifier is extensively water soluble, whereas coalescence predominates when the emulsifier has limited water solubility [24].

4 Conventional vs Pickering Emulsions: Vesicular Capsules

Encapsulation is a vesicular system having an inner core cavity that contains active substances (payload), which are enclosed by a polymeric membrane shell [25]. Emulsions, a biphasic heterogeneous system of immiscible liquids in which one is dispersed into the another in the form of droplets, can be prepared in different types based on the dispersed phase in either a simple W/O or O/W emulsion or as multiple water-in-oil-in-water (W/O/W) or oil-in-water-in-oil (O/W/O) emulsions [26]. Thus, an emulsion provides either lipophilic or hydrophilic cavities in which respectively lipophilic or hydrophilic substances (payloads) can be loaded (Fig. 4). In addition, the emulsion can be prepared in various droplet sizes, which can be sorted into a macroemulsion, microemulsion, or nanoemulsion based on their particle size, which can be larger than 400 nm (0.4 μm), in between 100 and 400 nm (0.1–0.4 μm), and less than 100 nm (0.1 μm), respectively [26], to yield macrocapsules, microcapsules, and nanocapsules, respectively. For this reason, emulsions are a simple and interesting system for encapsulation and a versatile capsule for pharmaceutical applications.

Emulsion formation involves two key steps: (i) breaking down liquid into tiny droplets through homogenization, which increases the system's surface area, and (ii) stabilizing these droplets with emulsifiers, which adsorb onto the droplet surfaces to reduce interfacial tension and form a protective barrier that prevents droplets from flocculation and coalescence [27, 28]. Surfactants and emulsifiers play crucial roles in stabilizing emulsions, though their mechanisms vary significantly depending on their molecular structure. Understanding these differences is essential for effective

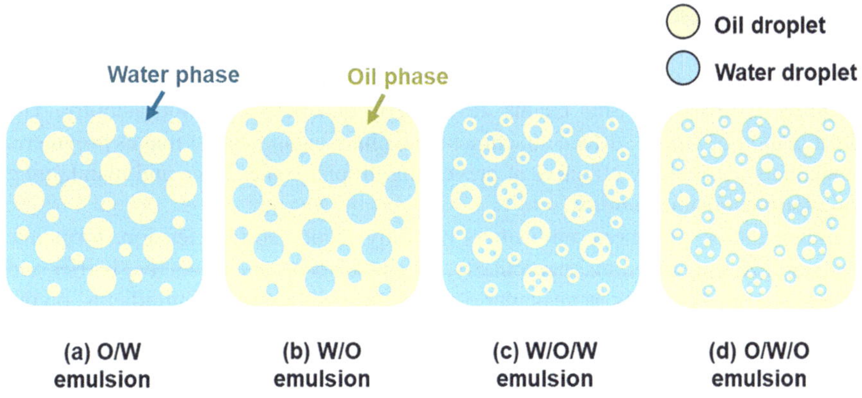

Fig. 4 Schematic illustration of (**a**, **b**) simple (**a**) O/W and (**b**) W/O emulsions and (**c**, **d**) multiple (**c**) W/O/W and (**d**) O/W/O emulsions

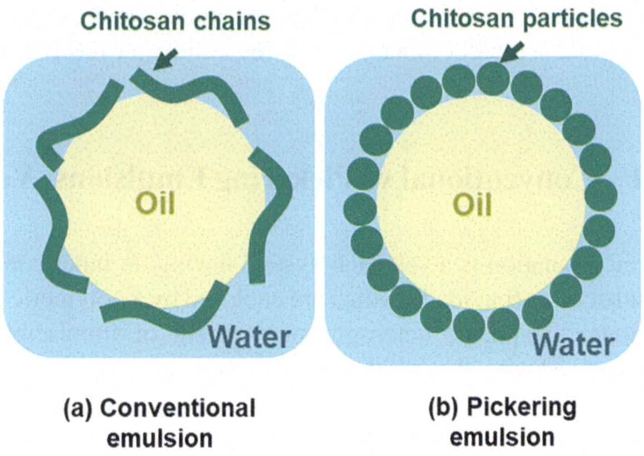

Fig. 5 Schematic representation of an O/W emulsion using (left) polymeric emulsifiers to form a conventional emulsion and (right) solid particles as the emulsifier to form a Pickering emulsion

emulsion formulation and stability control. Small molecule surfactants operate through a relatively simple mechanism. These molecules attach to interfaces through a single anchor point, while allowing them to reduce surface tension, also makes them somewhat unstable. This single-point attachment means they can easily detach from the interface, potentially leading to phase separation in the emulsion system [29]. Their effectiveness is heavily concentration-dependent, with optimal stabilization occurring only above their critical micelle concentration (CMC) [30]. Below this threshold, the emulsion may gradually coalesce. In contrast, polymeric emulsifiers, such as proteins and polysaccharides, offer more robust stabilization through multiple mechanisms. Unlike their smaller counterparts, these large molecules attach to interfaces through multiple segments, creating complex loop-train-tail structures [29]. This multi-point attachment results in the formation of thick protective layers around emulsion droplets, providing superior stability. One of their key advantages is their ability to remain firmly attached to interfaces even at relatively low concentrations, making them more efficient stabilizers. Polymeric emulsifiers demonstrate optimal performance at or above their overlapping concentration, (C*) [21]. At these concentration, the polymer chains in the bulk solution begin to entangle with each other, creating an even more effective stabilizing network. This entanglement occurs not only between free polymer chains in solution, but also between adsorbed and free chains, resulting in enhanced emulsion stability. The combination of strong surface attachment and chain entanglement makes polymeric emulsifiers particularly effective at preventing droplet coalescence and maintaining long-term emulsion stability.

Interestingly, polymeric emulsifiers not only reduce surface tension, formulate and stabilize the emulsions, particularly, they also serve as a protective barrier and create dispersed droplets as vesicular capsules for loading active ingredients, owing to their alignment at the oil-water interface [25]. Emulsions stabilized by polymeric emulsifiers, such as CS, can be categorized into two main groups – conventional and Pickering emulsions – based on the conformation of the emulsifiers, i.e., polymer chains and solid particles, respectively [21], as illustrated in Fig. 5.

4.1 Conventional Emulsions

Conventional emulsions are formed from the simple mixing of two immiscible liquids (typically oil and water), with one being dispersed as small spherical droplets within the other, with some type of surfactant at the interface. They are thermodynamically unstable due to the positive Gibbs free energy change (ΔG) associated with the increased interfacial area (ΔA) during emulsification, and ΔH is almost zero. Since ΔA is much greater than the existing entropy change (ΔS), then ΔG is mainly ascribed to $\gamma_{ow} \cdot \Delta A$. However, the addition of emulsifiers or surfactants can stabilize emulsions by minimizing the interfacial tension (γ_{ow}) between the oil and water phases [21].

$$\Delta G = \Delta H - T\Delta S + \gamma_{ow} \cdot \Delta A$$

where ΔG is the Gibbs free energy change, ΔH is the enthalpy change, ΔS is the entropy change, T is the temperature (°K), γ_{ow} is the interfacial tension between oil and water, and ΔA is the interfacial area change.

The stabilization mechanism in conventional emulsions primarily involves the formation of a thin interfacial film of surfactant molecules. This film creates a repulsive force between droplets, preventing them from merging and maintaining the stability of the emulsion. Conventional emulsions are widely used due to their versatility and ease of preparation. However, they may have limitations in terms of long-term stability.

That CS plays an important role in conventional emulsions as an emulsifying agent, due to its emulsifying capacity, is attributed to the heterogeneity of its structure, with hydrophilic D-glucosamine residues and hydrophobic *N*-acetylglucosamine residues. The effectiveness of CS as an emulsifier depends on many factors, such as its concentration, DD, and M_W. CS provides dual stabilization in O/W emulsions by both encapsulating oil droplets with a protective layer (independently or in combination with surfactants) and enhancing the continuous phase viscosity [31].

4.2 Pickering Emulsions

The first systemic stabilization of paraffin oil droplets in water by several types of solid particles in a surfactant-free system was discovered by Pickering in 1907 [32], and such emulsions have been widely referred to as Pickering emulsions. Pickering emulsions are emulsions stabilized by solid particles that accumulate at the oil-water interface. These particles can be of various origins, including inorganic materials (e.g., silica, clay) or organic materials (e.g., CS, starch) [33]. The stabilization mechanism in Pickering emulsions is established by the irreversible adsorption of solid particles at the oil-water interface. These particles form a physical barrier that prevents droplet coalescence, leading to an increase in stability compared to conventional emulsions [34]. Therefore, Pickering emulsions offer several advantages over conventional emulsions, including an increased stability against coalescence, the ability to incorporate particles with specific functionalities (e.g., magnetic, stimuli-responsive), and an often-improved biocompatibility due to the use of nontoxic and biodegradable particles.

4.2.1 Factors Influencing Pickering Emulsion Formation

Wettability

Particle wettability plays a fundamental role in the formation and stability of Pickering emulsions, which are unique systems stabilized by solid particles rather than traditional surfactants. The key measure of this wettability is the contact angle (θ_{ow}) formed where the particle adsorbed at the oil-water interface (Fig. 6). The stabilization of Pickering emulsions is achieved when solid particles adsorb onto the surface of emulsion droplets, effectively reducing the interfacial tensions (γ_{ow}) between the two immiscible phases [31]. This stabilization mechanism is strongly dependent on the contact angle of the particles, which can be fine-tuned through various physical and chemical surface modifications. The contact angle also determines the type of emulsion that will preferentially form. When solid particles exhibit a contact angle less than 90° (hydrophilic), they tend to stabilize O/W emulsions. Conversely, particles with contact angles greater than 90° (hydrophobic) typically

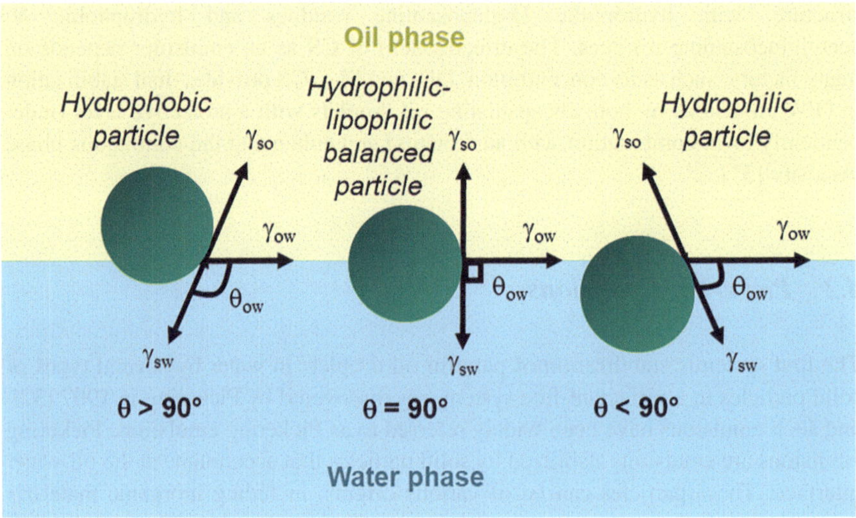

Fig. 6 Schematic illustration of the adsorption behavior of solid particles at the oil-water interface based on wettability as determined by the contact angle (θ_{ow}) for a (left) hydrophobic particle, (right) hydrophilic particle, and (center) hydrophilic and lipophilic balanced particle

produce W/O emulsions. For optimal Pickering emulsion stabilization, solid particles should align with a contact angle below or above but close to 90° [35]. This optimal angle ensures strong solid particles attachment at the interface while maintaining the desired emulsion types. Understanding and controlling particle wettability is therefore crucial for developing stable Pickering emulsions with specific characteristics for various applications. It should be noted that the θ_{ow} varies with different oils and water systems, resulting in different interfacial tensions (γ_{ow}) [36].

Particle Size and Surface Charge

The stability of Pickering emulsions is significantly influenced by both particle size and surface charge characteristics. Regarding particle size, the fundamental rule is that solid particles must be smaller than the stabilized emulsion droplets, ensuring proper adsorption at the interface [37]. However, extremely tiny particles behave similarly to a small molecular surfactant, making them easily detach from the oil-water interface. While larger particles demonstrate greater stability due to their higher desorption energy [38], through the particles must remain smaller than the stabilized emulsion droplets. Regarding surface charge, CS-based polyelectrolyte complex solid particles can be prepared with the variation of its surface charge by varying the pH and ionic strength, as well as by adjusting the mass or charge ratios of CS relative to counter complexing polymer. Highly charged particles should be ineffective at formulating Pickering emulsions due to strong electrostatic repulsion among the particles preventing interfacial adsorption [39, 40]. For gliadin/CS NP, as a representative example, the lower surface charge of gliadin/CS NP correlates with the higher interfacial absorption [41]. However, the impact of surface charge is complex, with both highly charged and less charged particles successfully stabilizing emulsions [31]. Increasing the pH from 3 to 7 in gliadin/CS NP systems reduces surface charge due to weakening electrostatic repulsion between particles. This reduction in repulsive forces allows more particles to accumulate at the oil-water interface, potentially enhancing emulsion stability [41].

Particle Concentration

Higher particle concentrations lead to an increase in coverage of the interfacial area and develop the particle networks formation in the continuous phase, both of which prevent droplet coalescence [31].

4.2.2 Role of CS in Pickering Emulsions

Chitosan-based particles can be prepared through several methods, including self-aggregation, ionic gelation, polyelectrolyte complexation, and hydrophobic modification. Self-aggregating CS particles form when the polymer is deprotonated under alkaline condition, causing it to self-assemble into colloidal particles [42]. Ionic gelation involves crosslinking the positively charged amino groups of CS with the negatively charged groups of tripolyphosphate (TPP) [43]. Polyelectrolyte complexation happens when CS interacts electrostatically with another oppositely charged polymer to form a complex [31]. Hydrophobic modification introduces covalently bonded hydrophobic groups to the CS, changing its self-assembly properties [44]. These diverse preparation techniques allow for the tailoring of CS-based particle properties for various applications. Such CS particles create the potential to function as a stabilizer for Pickering emulsions without using a surfactant. The choice of particle types and preparation method influences the wettability, particle sizes, and surface charges of the obtained particles and consequently influences the qualities and stability of the emulsion [31]. By adjusting the pH or through chemical modifications, CS can self-aggregate or form complexes with other molecules, creating particles that adsorb irreversibly at the oil-water interface as well as stabilize the emulsion [42], and form a physical barrier that prevents droplet coalescence.

Note that CS is a pH-responsive polymer that is sensitive to changes in the pH of the solution [31]. At pH below its pKa, the amino groups are protonated, allowing chitosan to dissolve in dilute acetic acid. At pH above the pKa, the amino groups deprotonate, causing the CS to aggregate into particles and precipitate [42]. Interestingly, the pH of the CS solution influences the type and stability of the obtained emulsions. High intensity ultrasonication can produce both conventional and Pickering emulsions with long-term stability, without requiring additional surfactants or crosslinking agents [45]. At pH values below the pKa of CS (< 6.5), conventional emulsions were formulated, while at pH 6.5 (above the pKa of CS), Pickering emulsions were formed. The conformational transition of CS molecules will change from extended chains to more flexible folded chains as pH is increased in the range below the pKa of CS, while the CS molecules will self-assemble into well-dispersed NPs at pH 6.5 (Scheme 1). As a consequence of the pH-influenced conformational transition, increasing the pH from 3.5 to 5.5 (below the pKa) decreased the droplet sizes of the corresponding conventional emulsions and enhanced emulsion stability from days to months. However, the corresponding Pickering emulsion formed at pH 6.5 (above the pKa) had the smallest droplet size and exhibited the longest stability of up to 5 months, owing to the adsorption of solid CS NPs at the oil-water interface.

The rate of creaming is proportional to the square of the droplet diameter according to Stokes's law, so that reducing the droplet size can effectively enhance the emulsion stability [46]. This suggests that the ability of CS to form emulsions is mainly affected by pH and ultrasonication treatment. Furthermore, the initial pH of the aqueous phase probably affects the conformation of self-aggregated CS particles, and subsequently affects the stability of the corresponding Pickering emulsions [47]. The self-aggregated CS NP assembled at pH 6–6.5 presents a highly hydrophilic surface because of the partial deprotonation of the amino groups, allowing them to be weak emulsifiers for O/W Pickering emulsion. In contrast, the self-

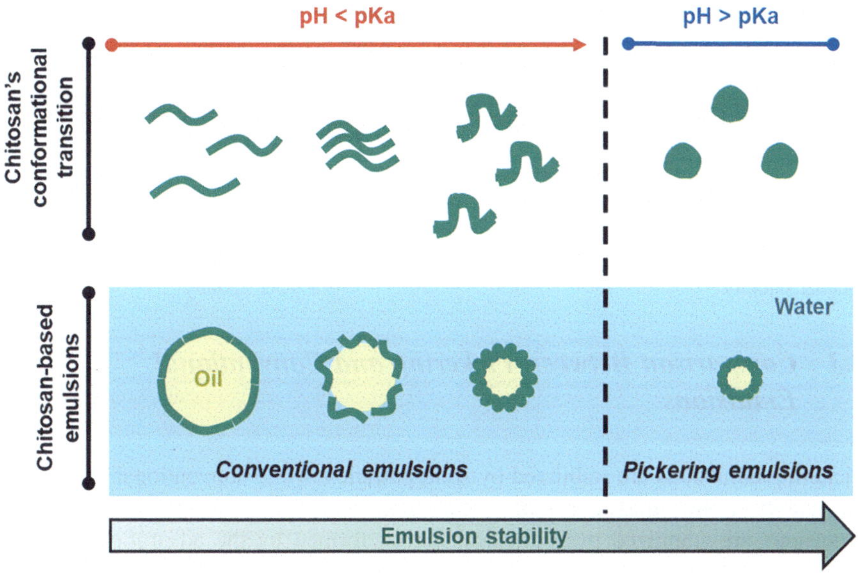

Scheme 1 Influence of the pH of the CS solution on the conformational transition, emulsion type, and stability of CS-based emulsions

aggregated CS NP assembled at pH > 9 presents a relatively hydrophobic surface because of the complete deprotonation of the amino groups, allowing them to be suitable emulsifiers for O/W Pickering emulsions.

Ultrasound-assisted emulsification is useful for generating both conventional and Pickering emulsions with CS. Ultrasonication helps homogenize the system and enhance the emulsifying properties of CS by breaking down particle agglomerates. However, the intense mechanical forces can break through the glycosidic linkages and cause some depolymerization [45]. Ultrasonication treatment influences both the size and the polydispersity of CS particles used to stabilize Pickering emulsion. Untreated particles are larger (~ 1000 nm) with a unimodal size distribution, while the ultrasonic-treated particles are smaller with a bimodal size distribution. The smaller, more uniform layer of particles adsorb and align better at the droplet interface, improving emulsion stability [48].

The emulsifying properties of CS are highly influenced by the pH as this impacts charge density and chain conformation. So far, CS has shown significant potential as a pH-controlled emulsifier or stabilizer, offering a promising approach for creating biodegradable, surfactant-free emulsions with long-term stability. This is of high importance for various pharmaceutical and related applications.

Table 1 Comparative advantages of Pickering emulsions over conventional emulsions in terms of their stability

Pickering emulsion	Conventional emulsion
Irreversible adsorption of solid particles at the oil-water interface, preventing emulsion droplet aggregation	Reversible adsorption of surfactant, causing flocculation
Usually unaffected by external factors (pH, temperature, and oil phase composition)	Sensitive to external factors
Requires lower or no concentration of additional emulsifiers	Often requires higher concentrations for stability
Higher stability	Lower stability

4.3 Comparison Between Pickering and Conventional Emulsions

Pickering emulsions are stabilized by solid particles, while conventional emulsions are stabilized by emulsifier molecules (often surfactants). Therefore, Pickering emulsions are stabilized by a physical barrier formed by the adsorption of solid particles, while conventional emulsions are stabilized by the reduction of the interfacial tension caused by emulsifier molecules [49]. In addition, Pickering emulsions typically have a dense, packed layer of particles at the interface, while conventional emulsions have a less dense monolayer of emulsifier molecules. Pickering emulsions offer several advantages over conventional emulsions [35, 50], as summarized in Table 1.

This comparison highlights the greater potential benefits of using CS-based particle-stabilized Pickering emulsions than utilizing CS or its derivatives to form conventional emulsions to encapsulate and enhance the stability of loaded bioactive substances for various applications, particularly in the pharmaceutical, cosmeceuticals and nutraceuticals where safety, stability, and environmental impacts are critical considerations.

5 Emulsion: A Versatile Capsule in Pharmaceutical Applications

This chapter gives detailed information on the emulsions prepared by CS or its solid particles as an emulsifier or stabilizer. The alignment of the polymers at the oil-water interface functions as a protective barrier or a capsular for the dispersed payload droplet, making the emulsion a versatile vesicular capsule for pharmaceutical applications. The encapsulation of drugs or bioactive substances within the capsules will protect active drugs, increase their life, and control drug concentration in the bloodstream through controlled release rates [51]. Accordingly, emulsions of CS and its derivatives are used as versatile capsules for pharmaceutical applications, particularly in drug delivery systems, and cosmeceutical and nutraceutical applications, as well as antimicrobial agents and wound healing. These are described below in Sects. 5.1–5.6.

5.1 Drug Delivery

Emulsions formed with CS exhibit outstanding biodegradable and biocompatible properties and have been studied extensively for the encapsulation of active substances or drug carriers. They can be administered through nasal, oral, and transdermal routes and protect the payload drugs from degradation. Conventional emulsions formulated with synthetic surfactants can potentially cause various adverse effects in humans [52]. Thus, instead of using synthetic surfactants, polysaccharides and proteins are alternative groups of interest. Casein, a hydrophobic protein, was conjugated to CS via the Maillard reaction using grafting degrees of 3.10–15.08% to obtain a group of emulsifiers with a wide range of HLB values, i.e., from 6 to 14. The curcumin was successfully encapsulated, protected, and gradually released from these stable emulsion formulations, which maintained their integrity over extended periods [53].

Moreover, phosphorylated CS (PCTS) was developed as a new bio-based water-soluble emulsifier with an HLB value of around 19, and so it was possible to formulate an O/W emulsion with a lipophilic cavity for the encapsulation of naproxen, a model lipophilic payload [25]. Besides, PCTS has been reported to be cytocompatible and so is suitable for biomedical applications and pharmaceutical drug delivery systems [54]. The PCTS forms a thickly adsorbed barrier at the oil and water interface, which serves a role as a good emulsifier as well as a protective layer for payloads. The naproxen at concentration of 6 mg/mL of oil was loaded within the emulsion stabilized by 1% (w/v) PCTS solution, revealing an encapsulation efficiency (EE) of $91.2 \pm 4.7\%$, indicating the high potential of emulsion-based drug capsules. The naproxen exhibited a slower release rate and at a lower level in an acidic condition (pH 1.2) than in phosphate-buffered saline (PBS, pH 7.4), owing to its lower solubility in an acidic media, making it preferable for release in a neutral environment.

On the other hand, a Pickering emulsion stabilized by CS-TPP NP was developed for the encapsulation and delivery of curcumin. Curcumin demonstrates diverse therapeutic properties, including anti-inflammation, antimicrobial, antioxidant, and anticancer. The results revealed that the CS-TPP-stabilized Pickering emulsions were both stable and effective in encapsulating curcumin. The curcumin showed a faster release rate from CS-TPP-stabilized Pickering emulsions in gastric conditions (pH 2.0) than in a physiological condition (pH 7.4) [43], indicating a potential and preferable delivery system for bioactive compounds in the gastric environment.

Oral administration is the common administration route, where drug capsules are taken through the mouth and processed via the gastrointestinal tract (GIT). The design of the drug capsules for oral administration needs to take into account where the drug is to be absorbed and its controlled release along the GIT, which has different pH conditions. The absorption of drugs will be enhanced in the acidic environment of the stomach or in the nearly neutral environment of the intestine, depending on the ionic properties of the compound. However, due to the significantly larger surface area of the intestinal villi, adsorption in the intestine tends to be more prominent [55]. Herein, hydrophobic sodium alginate-CS (HSA-CS) nanocomposite-stabilized Pickering emulsions were proposed as drug-loading capsules for oral administration [56]. The integration of the pH-triggered performance of HSA-CS with a Pickering emulsion gave emulsion drug capsules a high loading capacity within the solid particles layer of the Pickering emulsion. Their pH response allowed for a pH-triggered drug release behavior in the continuous simulated GIT using simulated gastric fluid (SGF, pH 1.5) and simulated intestinal fluid (SIF, pH 6.8). Pickering emulsions loaded of Ibuprofen at concentration of 30 mg/mL of oil released essentially no ibuprofen in 3 h in SGF but nearly 100% in 12 h in SIF, revealing their great potential as a pH-responsive vehicle for oral drug delivery.

5.2 Targeted Drug Delivery

The protonated positive charge of the amino groups of CS can interact with the negative charges of mucus, giving CS mucoadhesive properties [57]. The adhesion to mucus membranes is significantly influenced by hydrophobic interaction, where the polymer chains physically entangle with mucin molecules [58]. Therefore, to improve the mucoadhesive properties of CS so as to increase drug accumulation on the mucosa by prolonging retention time at the target area, the incorporation of acyl-CS exhibiting positive charges, including octanoyl CS (OC), lauroyl CS (LC), and palmitoyl CS (PC), were introduced to the self-nanoemulsifying drug delivery systems (SNEDDS) [59]. The SNEDDS were formulated by dispersing a mixture of oil (solvent), low M_W polyethylene glycol or propylene glycol (cosolvent), and surfactant (Tween 20 or Tween 80) at a ratio of 1mL in 100 mL of 0.1 M PBS (pH 6.8). Modifying the SNEDDS surface with acyl-CS created effective mucus interactions. Incorporation of 2% acyl-CS doubled the SNEDDS retention time on intestinal mucus surfaces. The mucoadhesive properties were significantly enhanced as the acyl chain length increased from octanoyl (C8) to lauroyl (C12) to palmitoyl (C16). Further research reported on loading cefixime (CFX), a broad-spectrum antibacterial drug, into SNEDDS incorporated with mucoadhesive acyl-CS [60]. An in vitro release study conducted at a physiological pH revealed that SNEDDS incorporated with acyl-CS were able to release the drug at more than 60%. Subsequently, the SNEDDS incorporated with acyl-CS was given to albino rabbits (average body weight of 1.2 kg) at 5 mg/kg for in vivo oral administration, and then the bioavailability of the drug in plasma was examined. The SNEDDS incorporated with acyl-CS showed an enhanced oral performance for CFX administration with an increased plasma concentration (15 ± 3 μg/mL) compared to the control CFX suspension (4 ± 1 μg/mL).

Nasal vaccine delivery offers key advantages over traditional injections, including reduced infection risk and better patient acceptance. While mucosal vaccines can trigger immune responses at mucosal surfaces, their effectiveness is often limited by mucociliary clearance and epithelial barriers [61]. To address these challenges, researchers developed a novel vaccine incorporating a mucoadhesive biopolymeric system of CS/oligoCS/Tween 80 (CS-COS-T80) costabilized emulsion loaded with ovalbumin (OVA) antigen for a nasal adjuvant [62]. This formulation significantly extended the antigen's nasal retention time beyond 12 h. Intranasal immunization with CS-COS-T80 co-stabilized emulsion-adjuvanted antigen successfully stimulated both local mucosal and systemic immune responses, while also generating memory T cells for long-term immunity. These results suggest that CS-based emulsions hold promise as effective enhanced immune responsive vaccine delivery systems.

5.3 *Cosmeceuticals and Topical Delivery*

Cosmeceuticals represent a fusion of cosmetics and pharmaceuticals, offering beauty products with therapeutic benefits. Pickering emulsions have emerged as valuable formulations in cosmetic science, particularly when incorporating active ingredients for skin benefits [63]. CS has gained significant attention in this field due to its multiple advantageous properties, including moisturizing effects, anti-inflammatory and antioxidant activities, and UV protection [64, 65]. As a particulate emulsifier, chitosan's positive charge enables it to enhance skin penetration through interactions with the *stratum corneum*'s tight junctions [66, 67]. Combined with its biocompatibility and biodegradability, CS-based Pickering emulsions show great promise for developing green cosmetic products for skin applications [31, 68].

Cosmetic formulations such as creams, lotions, and gels contain surfactants, which facilitate the homogenization of the oil phase with the water phase. The most commonly used surfactants in cosmetic emulsions are sodium lauryl sulfate, polyethylene glycol ethers, polysorbate 80, and sorbitan oleate (Span 80), among others [69]. Recent studies have highlighted natural-based cosmetics as a solution to the side effects and toxicity associated with synthetic ingredients in cosmetics and skincare products. Additionally, there is growing dedication from the cosmetic and pharmaceutical industry sectors to address consumers' demands for safer and healthier products through green chemistry principles.

The greater stability and surfactant-free nature of Pickering emulsions make them a preferable alternative to conventional emulsions for skin applications and also a promising option for topical formulations to avoid the adverse effects of synthetic emulsifiers on the skin, such as irritation and eczema [70]. Resveratrol-loaded Pickering emulsions stabilized by CS/gum Arabic (CS/GA) NPs had been developed for cosmetic purposes [68].

Resveratrol, a plant-derived polyphenol, exhibits multiple therapeutic properties including antimicrobial, anti-inflammatory, chemopreventive, and strong antioxidant activities [71]. Topical delivery of resveratrol via porcine skin adsorption, Pickering emulsions stabilized by CS/GA NP effectively delivered resveratrol. The formulations of Pickering emulsions stabilized by 0.5% and 1.5% (w/v) of CS/GA NP releasing 53.7% and 38.5% of the total applied resveratrol respectively over 24 h. The controlled release rate is governed by its diffusion through the NP shell layering on the surface of the oil droplets. Therefore, the higher the CS/GA NP concentration, the more CS/GA NP adsorbed at the surface of oil droplets and so reduced the rate of drug release from the core olive oil droplets. For cosmetic ingredients, optimal delivery systems should maximize skin retention while minimizing systemic permeation [68]. The Pickering emulsions stabilized by 0.5% and 1.5% (w/v) of CS/GA NP achieved this goal, showing high skin retention rates at 11.60% and 10.82% with minimal permeation at 2.20% and 3.16% of total applied dose of resveratrol, respectively [68]. This formulation successfully concentrated resveratrol in the epidermis and dermis layers without requiring chemical penetration enhancers. The cationic CS-based NP effectively facilitated penetration through the *stratum corneum* while also protecting resveratrol from UV degradation, making it an ideal delivery system for cosmetic applications.

Furthermore, CS/collagen peptide (CS/CP) NPs were used to stabilize a Pickering emulsion aiming for cosmetic applications [72]. The CS/CP NPs gave better wettability and lower interfacial tension. Since the interaction between the CS and CP exposes more hydrophobic moieties, the CS/CP NP-stabilized Pickering emulsions demonstrated excellent storage stability. The emulsion stability increased with an increase in CS/CP NP concentrations. Pickering emulsions stabilized by CS/CP NP demonstrated effective skin penetration capabilities. These emulsion droplets successfully traversed the *stratum corneum* and reached deeper skin layers, with penetration depth varying based on CS/CP NP concentration and skin contact duration. The polysaccharide-protein complex NP enabled efficient deep-layer deposition of the emulsion droplets, making them suitable as green carriers. The CS/CP NP-stabilized Pickering emulsions proved to be exceptional carriers for topical CBD delivery, achieving over 99% EE [73]. This high EE resulted from CBD's strong lipophilicity and the robust barrier formed by CS/CP NP around oil droplets, preventing CBD leakage. While the emulsions demonstrated biocompatibility with human keratinocytes (HaCaT cells), CBD's lipophilic nature caused it to concentrate primarily in the *stratum corneum* rather than penetrating the viable epidermis and dermis. Overall, these biocompatible, surfactant-free Pickering emulsions represent an effective platform for topical therapeutic delivery.

Oil-in-water (O/W) nanoemulsions also serve as effective capsules for the delivery of hydrophobic bioactive compounds in cosmeceuticals. The CS has emerged as a promising natural emulsifier for sunscreen formulations, where skin retention rather than bloodstream penetration is crucial [74]. The macromolecular structure of CS enables it to form a stable interfacial film around oil droplets, creating lipophilic cavities that can incorporate photoprotective agents and antioxidant plant extracts like pomegranate. The CS-stabilized interface demonstrates strong skin affinity, extending the formulation's retention time in the epidermis while maintaining safety. This makes CS-stabilized O/W nanoemulsions particularly suitable for topical sunscreen applications requiring enhanced skin retention.

To this end, Pickering emulsions based on CS particles not only protect active substances from degradation, prolong their stability, and extend the shelf life of the cosmetic product but also control the release profile after application to the skin. Pickering formulations show promise for green cosmetics due to their biodegradability, biocompatibility and non-toxicity. Moreover, CS and CS-based NP are particularly effective in topical drug delivery systems. Their positive charges facilitate adhesion to the negatively charged skin surface, while their ability to maintain *stratum corneum* hydration enhances cell membrane fluidity, collectively improving transdermal drug absorption [75]. The ability of CS to electrostatically interact with the negative charges of the skin or damaged hair and to form polymeric films, which are required in the conditioning and moisturizing of cosmetic formulation, particularly makes CS an attractive ingredient for the development of skin and hair care products [76].

5.4 Nutraceuticals and Food Products

Since CS can help maintain normal blood low-density lipoprotein-cholesterol concentrations [77] due to its hypocholesterolemic and fat-binding properties, it is commonly included in many nutritional supplements designed to prevent fat absorption in the digestive tract [78]. Moreover, CS is generally considered safe for oral administration, although it is not specifically degraded in the human intestine because chitin/CS degradable enzymes (chitinase and chitosanase) are not present in the human body [79]. Thus, consuming several grams of CS daily may lead to diarrhea as CS can trap water and lipids in the intestine [80]. Although CS is clinically well safe, it is still not recommended for people who are allergic to crustaceans.

The major food applications of CS include encapsulating nutraceuticals, antimicrobial agents, dietary fiber, edible supplements, enzyme immobilization, thickener, gelling, and emulsifying agents [81]. Emulsion formation is an intermediate step in food processing, such as in cakes and ice-cream mixing. Most fat-containing foods, such as milk, cream, beverages, dressing, and sauces, are mostly in the form of O/W emulsions [82].

The NP-stabilized Pickering emulsions, particularly those utilizing food-grade polysaccharides and proteins, present a viable solution for reducing synthetic surfactant usage in food systems [83]. Since CS has been reported as a powerful bioactive molecule with antimicrobial and mucoadhesive properties [31] and is generally recognized as safe (GRAS) by the US Food and Drug Administration (US FDA) [84–86], CS-based Pickering emulsions have found diverse applications in nutraceuticals and food products development.

5.4.1 Encapsulation of Nutrients and Bioactive Compounds

Pickering emulsions stabilized by CS-based NPs can encapsulate nutraceuticals or bioactive compounds. Encapsulation in Pickering emulsions provides protection for sensitive compounds against environmental degradation factors including light, pH, oxygen, and temperature, thereby enhancing their stability and bioavailability [87]. This approach proved particularly valuable for hesperidin, a poorly water-soluble flavonoid prone to degradation and loss of antioxidant properties when complexed with food proteins. A solution was developed using chitosan CS NP-stabilized, lecithin-enhanced O/W Pickering emulsions, created through homogenization, which successfully improved its water dispensability and protected it against oxidation [88]. In addition, CS/guar gum (CS/GG) NPs were developed to stabilize Pickering emulsions for astaxanthin encapsulation. The CS/GG-stabilized Pickering emulsions exhibited superior emulsion stability with a higher retention rate of astaxanthin than the O/W emulsion stabilized by Tween 80 [89]. The effectiveness of CS in encapsulating and protecting sensitive bioactive compounds highlights its potential benefit in the development of functional foods and nutraceuticals.

Microcapsules derived from Pickering emulsion templates offer an innovative approach to lipid encapsulation. In a study of roasted coffee oil encapsulation using CS NP-stabilized Pickering emulsions, freeze-dried microcapsules demonstrated high oil retention but exhibited irregular morphology and porosity. Spray-dried alternatives yielded smaller (11 μm), uniformly spherical particles that provided superior protection and antioxidant activity for coffee oil phenolics during in vitro digestion [90]. Similarly, oregano essential oil (OEO) was successfully encapsulated using CS/GA NP-stabilized Pickering emulsions, achieving 2.5 μm capsules with 60% EE through spray-drying [91]. The optimized CS/GA hybrid complexes enhanced emulsion stability. Therefore, these microcapsules derived from Pickering emulsion templates demonstrate a potential as versatile carriers for active ingredients across medical, agricultural, and other industrial applications.

5.4.2 Controlling Lipid Digestibility

Pickering emulsions based on CSs provide an effective approach for controlling lipid digestibility. Their ability to stabilize emulsions, form barriers to digestive enzymes, and respond to pH changes make them versatile tools in the development of nutraceuticals, functional foods, and medical nutrition products aimed at the improvement of health and the management of diseases related to lipid metabolism [92]. A comparative in vitro lipid digestion study examined lipid hydrolysis rates in Pickering emulsions stabilized by three food-grade polysaccharide particles (CS, cellulose crystals (CCrys), and cellulose nanofibers (CNFs)) by sequential incubation in SGF and SIF [93]. Cellulose-based and CS-based Pickering emulsions demonstrated contrasting behaviors during gastrointestinal digestion, offering distinct applications in food and pharmaceutical systems. CCrys and CNFs created robust interfacial barriers that effectively inhibited lipid digestion through strong particle adsorption at the oil-water interface. Their stability maintained consistent droplet sizes throughout digestion, successfully preventing lipase access to the interface. In contrast, CS NPs exhibited pH-responsive behavior that modulated lipid digestion. Under gastric conditions, CS chains partially extended, promoting bridging flocculation and reducing droplet sizes, which increased the surface area available for enzyme interaction. During intestinal digestion, CS NP aggregated and detached from the interface of the emulsion droplets, enabling controlled lipase access and gradual lipid hydrolysis. [93]. These distinct characteristics suggested specific applications: cellulose-based Pickering emulsions showed promise as effective lipid digestion inhibitors, potentially beneficial for weight management applications or low lipid adsorption products development. Meanwhile, CS-based Pickering emulsions offered controlled release properties ideal for nutraceutical delivery, providing delayed lipid hydrolysis that could enhance bioactive compound absorption.

Research demonstrates that CS coating of non-CS particle-stabilized Pickering emulsions significantly affects the lipid digestion pattern. For instance, CS coating on emulsions stabilized by waxy maize starch nanocrystals improved its stability under acidic conditions, offering a crucial strategic approach for enhancing food hydrocolloid stability and controlling GIT digestion. The CS coating thickness, determined by concentration, proved crucial in modulating lipolysis in SIF, with higher CS concentrations both slowing lipid digestion and reducing fatty acid release [94].

5.4.3 Lipid Oxidation Retardation in Food Formulations

Lipids, essential components in both food and biological systems, provide critical nutritional value through energy provision, essential fatty acids, and fat-soluble vitamins delivery in human nutrition. However, lipid oxidation in food products leads to off-flavors, nutrient loss, color and texture deterioration, potential health risks, and economic losses. These adverse effects make it crucial to control oxidation to maintain food quality, safety, and shelf life [95].

β-carotene is a lipid-soluble bioactive component that is sensitive to oxidation when exposed to an acidic condition, heat, light, and oxygen. Thus, CS hydrochloride/carboxymethyl starch (CSHC-CMS) complex nanogel-stabilized O/W Pickering emulsions were investigated for the oral delivery of β-carotene, focusing on their protective effects against lipid oxidation and in vitro release in a simulated GIT. Lipid oxidation was determined by the primary production of lipid hydroperoxide (LH). Bulk corn oil revealed an increase in LH from 15.31 to 230.11 mmol/kg of oil over 15-day storage, while the β-carotene-loaded Pickering emulsion demonstrated delayed lipid oxidation [96] due to the CSHC-CMS nanogels forming an interfacial barrier of lipid droplets. This interfacial barrier not only inhibited the contact of lipase with the lipid droplets, slowing down the digestion, but it also retarded the lipid oxidation.

A study examined the oxidation resistance of sunflower O/W Pickering emulsions using two different stabilizers: CS-stearic acid (SA) nanogels versus the conventional emulsifier Tween 80. The CS-SA nanogels demonstrated superior stabilizing properties, particularly when using higher SA to CS ratios and under alkaline conditions. The nanogel-stabilized emulsions showed better resistance to lipid oxidation compared to those made with Tween 80, suggesting that CS-SA nanogels could serve as effective alternative emulsifiers for the retardation of lipid oxidation [97].

Noted that the free amino groups of CS could remove the transition metals involved in the lipid oxidation process, such as ferrous ion (Fe^{2+}), which can induce and accelerate lipid peroxidation [97]. Building on this property, researchers developed a new dual-purpose stabilizer by grafting caffeic acid (CA) onto CS (CA-g-CS) for used in food-grade Pickering emulsions with its antioxidant properties. In the presence of Fe^{2+} as an oxidative accelerator, CA-g-CS exhibited superior Fe^{2+} chelating properties than CA alone and retained a comparable radical scavenging activity as CA. The CA-g-CS compound effectively prevented emulsion oxidation in acidic to near-neutral conditions (pH 3-6), demonstrating its effectiveness as both an emulsion stabilizer and antioxidant [98].

5.5 *Antimicrobial Treatment*

Chitosan exhibits a broad antimicrobial activity against various bacteria, fungi, and viruses [99, 100], which is attributed to its cationic nature, enabling it to interact with negatively charged microbial cell membranes, which leads to cell disruption and death. Moreover, several intrinsic and extrinsic factors related to CS (its DD, M_W, and structural modification) and the composition of bacterial cell walls can affect the antimicrobial activities of CS [101, 102]. The antimicrobial properties of CS are crucial for preventing wound infections, a common complication that can significantly delay healing and lead to serious health issues. A combination of CS and other active substances may have synergistic effects to improve the antibacterial activities of CS. A spray of tea-tree-oil-loaded Pickering emulsion stabilized by CS NPs was found to have high antibacterial activity and an excellent wound-healing effect [103]. The tea-tree-oil-loaded Pickering emulsion provided a sustained release and further strengthened the tea tree oil's antibacterial activity toward Gram-positive *Staphylococcus aureus* (*S. aureus*), Gram-positive methicillin-resistant *S. aureus* (MRSA), and Gram-negative *Escherichia coli* (*E. coli*). Higher concentrations of CS particles increased the viscosity of emulsion, allowing better wound coverage and longer contact time when sprayed. Unlike conventional surfactant-based emulsions, this CS NP-stabilized Pickering emulsion showed improved safety and reduced skin irritation, making it particularly suitable for application on damaged skin.

The antimicrobial activity of CS-based emulsions is not only crucial for preventing wound infections, but it is also promising as an alternative food preservative. Synthetic preservatives and antimicrobials can pose health risks, including toxic residues and potential links to cancer and chronic diseases, while essential oils (EO) provide a natural alternative [104]. These plant-derived compounds, which contain phenolic compounds and terpenes, effectively inhibit microbial growth and help preserve food products without the health concerns associated with synthetic additives [104]. Therefore, O/W emulsions offer an oil droplet whereby EO can be presented and dispersed in water as a continuous phase. However, conventional emulsions prepared with synthetic surfactants can potentially cause various adverse effects in humans [52]. The Pickering emulsion (O/W) stabilized by chitin nanofibers (ChNFs) containing *Cinnamon cassia* oil (*C. cassia* oil, antimicrobial essential oil) exhibited prolonged antibacterial activities against *E. coli* with the controlled release of antimicrobial *C. cassia* oil, promising prolonged food preservation [105]. Moreover, the Pickering emulsion stabilized by zwitterionic Z-ChNF containing the EO of citrus exhibited antimicrobial properties against *S. aureus* and *E. coli*. The amphiphilic nature of Z-ChNF, which stabilized the Pickering emulsion, also had a synergistic effect for enhancing antimicrobial activities in a concentration-dependent manner [106].

5.6 *Wound Healing*

5.6.1 Mechanisms of CS-Mediated Wound Healing

Wound healing occurs through four interconnected phases that overlap during the recovery process: (i) hemostasis and coagulation, (ii) inflammation, (iii) proliferation, and (iv) remodeling. Several factors can influence the progression and success of these healing stages, as previously documented in some literature [107]. The mechanisms of action of CS involved in wound-healing phases are briefly summarized in the section below.

(i) ***Hemostasis and Coagulation Phase***

The hemostatic properties of CS are derived from its ability to enhance blood clotting by activating platelets, which are essential for blood clot formation, and to interact with red blood cells [99, 100]. This property is crucial in the early stages of wound healing as it helps control bleeding and create a stable environment for the subsequent healing processes.

(ii) ***Inflammation***

The wound-healing ability of CS is mediated through its ability to modulate the inflammatory response, reduce oxidative stress, and provide a moist environment for wound healing. These further support its therapeutic potential. Pei et al. [108] demonstrated the anti-inflammatory effects of a CS-based Pickering emulsion in a mouse model, while Wu et al. [109] showed that a gel-like CS-based Pickering emulsion could regulate the wound-healing environment and inhibit inflammatory reactions.

(iii) ***Cell Proliferation and Migration***

The proliferation and migration of cells essential for wound repair, such as fibroblasts, keratinocytes, and endothelial cells, have been shown to be promoted by CS through its interaction with cell surface receptors and the activation of intracellular signaling pathways [109, 110]. For instance, CS oligosaccharide (COS) was found to significantly promote the proliferation of human umbilical cord mesenchymal stem cells in an injectable hydrogel system [111]. Similarly, modified CS hydrogels containing platelet-derived growth factor-BB demonstrated increased fibroblast proliferation in vitro [112]. The enhanced cell proliferation and migration contribute to faster wound closure and tissue regeneration.

(iv) ***Remodeling***

Extracellular matrix remodeling involves both matrix reorganization and the development of new blood vessels (angiogenesis). This vascularization process is crucial for wound healing, as it delivers essential oxygen and nutrients to regenerating tissues. Research has shown that certain CS derivatives, particularly those that can bind heparin, promote blood vessel formation [113]. This was demonstrated when heparin-containing chitosan hydrogels successfully stimulated angiogenesis. Collagen, a major structural protein in the skin, is crucial for wound healing, and CS stimulates collagen production by fibroblasts [114, 115]. This is evident in studies where CS-based hydrogels significantly increased collagen deposition and the hydroxyproline content in wound tissue [114]. Increased collagen deposition provides structural support, promotes wound contraction, and ultimately leads to faster healing.

Overall, the modes of action of CS in wound healing are multifaceted, involving hemostasis, the modulation of the inflammatory response, cell proliferation and migration, angiogenesis, and collagen synthesis. These mechanisms contribute to the effectiveness of CS-based materials in promoting wound healing as well as their role in forming emulsions.

5.6.2 Chitosan-Based Emulsions in Wound Healing

CS-based Pickering emulsions, which use solid CS NPs as stabilizers, have become increasingly popular over conventional emulsions due to their enhanced stability, biocompatibility, and simple preparation methods. These formulations are particularly valuable in topical pharmaceutical applications. In O/W emulsions, the water-based continuous phase allows CS to create a moist environment, which is beneficial for wound healing [116, 117].

Asfour et al. [110] developed a CS-based Pickering emulsions for the topical delivery of rutin, a hydrophobic drug with wound-healing properties. The formulation combined rutin dissolved in a propylene glycol-oleic acid mixture as the oil phase, stabilized by self-aggregated CS NPs. This O/W Pickering emulsion showed remarkable stability for 8 months and achieved nearly 100% rutin release within 24 h, more effectively released than a plain drug suspension. When tested on Wistar rats, the formulation successfully delivered rutin to wound sites, resulting in complete healing within 10 days, demonstrating its effectiveness as a delivery system for hydrophobic therapeutic compounds. The synergistic effects of the sustained release of rutin as well as the emulsion components (CS and oleic acid) in the CSNPs-stabilized Pickering emulsion advanced the wound-healing process.

Pei et al. [108] formulated a Pickering emulsion stabilized by CS/GA NPs incorporating tea tree oil and vitamin E. This showed an effective inhibition against *S. aureus*, *E. coli*, *Pseudomonas aeruginosa*, and *Candida albicans*. The antibacterial effect was greater than that of tea tree oil alone. In vivo tests showed that the vitamin-E-loaded Pickering emulsion stabilized by CS/GA NPs significantly reduced peritoneal vascular permeability in mice, indicating good acute anti-inflammatory effects. In a mouse wound model, this Pickering emulsion showed a high wound-healing rate and healed the wound completely by day 10, while the sustained release of vitamin E from the solid particle shell of the Pickering emulsion as well as the combination of CS/GA NPs and tea tree oil appeared to have synergistic beneficial effects on wound healing compared to the individual components. Therefore, the developed Pickering emulsion stabilized by CS/GA NPs demonstrated multiple beneficial effects, which promoted faster and more effective wound healing in mice through its antibacterial, anti-inflammatory, and tissue regeneration properties.

6 Summary

Both conventional and Pickering emulsions based on CS have demonstrated remarkable versatility as vesicular capsules for pharmaceutical and related applications. This chapter has explored their diverse roles in drug delivery systems, cosmeceutics, nutraceutics, antimicrobial activities, and wound healing.

Overall, CS emulsions offer a promising platform for the controlled release and enhanced bioavailability of various active pharmaceutical ingredients. The unique properties of CS, including its biocompatibility, biodegradability, and mucoadhesive nature, contribute to improved drug absorption and targeting. The applications of CS-based emulsions extend beyond traditional pharmaceuticals into the fields of cosmeceutics and nutraceutics. These systems have shown efficacy in encapsulating sensitive active ingredients, protecting them from degradation and facilitating their controlled release. Furthermore, CS-based emulsions have demonstrated the capacity to modulate lipid digestion and retard lipid oxidation, offering potential benefits for both nutraceutical formulations and functional food products. The inherent antimicrobial properties of CS, combined with its emulsification capabilities, present a powerful tool for combating pathogenic microorganisms. This dual functionality makes CS-based emulsions particularly attractive for developing antimicrobial formulations with enhanced efficacy and stability. In wound-healing applications, CS-based emulsions have shown promise in creating moist environments while simultaneously offering antimicrobial protection. The ability to incorporate growth factors and other bioactive compounds into these emulsions further enhances their potential in advanced wound care.

In conclusion, both conventional and Pickering emulsions based on CS represent versatile and promising capsules for a diverse range of pharmaceutical and related applications. Exploring more sophisticated and effective CS emulsion-based formulations in the future may further expand their role in improving human health and well-being.

References

1. Kumari S, Kishor R (2020) Chitin and chitosan: origin, properties, and applications. In: Kumari S, Kishor R (eds) Handbook of chitin and chitosan. Elsevier, Amsterdam, pp 1–33
2. Klinkesorn U (2013) The role of chitosan in emulsion formation and stabilization. Food Rev Intl 29:371–393
3. Kurita K (2006) Chitin and chitosan: functional biopolymers from marine crustaceans. Mar Biotechnol 8:203–226
4. Rinaudo M (2006) Chitin and chitosan: properties and applications. Prog Polym Sci 31:603–632
5. Ravi Kumar MNV (2000) A review of chitin and chitosan applications. React Funct Polym 46: 1–27
6. Azmana M, Mahmood S, Hilles AR et al (2021) A review on chitosan and chitosan-based bionanocomposites: promising material for combatting global issues and its applications. Int J Biol Macromol 185:832–848
7. Håkansson A (2019) Emulsion formation by homogenization: current understanding and future perspectives. Annu Rev Food Sci Technol 10:239–258
8. Matsumoto S, Makino H, Ueno A (1987) Formation of W/O/W emulsions through the simple agitation of oil-water mixtures. J Japan Oil Chem Soc 36:320–327
9. Davis HT (1994) Factors determining emulsion type: Hydrophile – lipophile balance and beyond. Colloids Surf A Physicochem Eng Asp 91:9–24
10. Schulz PC, Rodríguez MS, Del Blanco LF et al (1998) Emulsification properties of chitosan. Colloid Polym Sci 276:1159–1165
11. Payet L, Terentjev EM (2008) Emulsification and stabilization mechanisms of O/W emulsions in the presence of chitosan. Langmuir 24:12247–12252
12. Li X, Xia W (2011) Effects of concentration, degree of deacetylation and molecular weight on emulsifying properties of chitosan. Int J Biol Macromol 48:768–772
13. Del Blanco LF, Rodriguez MS, Schulz PC et al (1999) Influence of the deacetylation degree on chitosan emulsification properties. Colloid Polym Sci 277:1087–1092
14. Speiciene V, Guilmineau F, Kulozik U et al (2007) The effect of chitosan on the properties of emulsions stabilized by whey proteins. Food Chem 102:1048–1054
15. Tachaboonyakiat W, Serizawa T, Endo T et al (2000) The influence of molecular weight over the ultrathin films of biodegradable Polyion complexes between chitosan and poly(-glutamic acid). Polym J 32:481–485

16. Klinkesorn U, Sophanodora P, Chinachoti P et al (2005) Encapsulation of emulsified tuna oil in two-layered interfacial membranes prepared using electrostatic layer-by-layer deposition. Food Hydrocoll 19:1044–1053
17. Laplante S, Turgeon SL, Paquin P (2006) Emulsion-stabilizing properties of chitosan in the presence of whey protein isolate: effect of the mixture ratio, ionic strength and pH. Carbohydr Polym 65:479–487
18. Meng W, Sun H, Mu T et al (2023) Effects of environmental stimuli on the physicochemical and rheological properties of chitosan-macroalgal polyphenol stabilized Pickering emulsion. Int J Biol Macromol 227:1245–1257
19. Zhang F, Cai X, Ding L et al (2020) Effect of pH, ionic strength, chitosan deacetylation on the stability and rheological properties of O/W emulsions formulated with chitosan/casein complexes. Food Hydrocoll 111:106211
20. Frank SG (2002) Emulsions: theory and practice. Third Edition by Paul Becher (Wilmington, DE). Oxford University Press (In Partnership with American Chemical Society): Oxford. J Am Chem Soc 124:7876–7877
21. Kawaguchi M (2016) Silicone oil emulsions stabilized by polymers and solid particles. Adv Colloid Interf Sci 233:186–199
22. Dickinson E, Golding M (1997) Depletion flocculation of emulsions containing unadsorbed sodium caseinate. Food Hydrocoll 11:13–18
23. Zhai H, Gunness P, Gidley MJ (2021) Depletion and bridging flocculation of oil droplets in the presence of β-glucan, arabinoxylan and pectin polymers: effects on lipolysis. Carbohydr Polym 255:117491
24. Aronson MP (1989) The role of free surfactant in destabilizing oil-in-water emulsions. Langmuir 5:494–501
25. Chongprakobkit S, Maniratanachote R, Tachaboonyakiat W (2013) Oil-in-water emulsions stabilized by sodium phosphorylated chitosan. Carbohydr Polym 96:82–90
26. Debnath BK, Saha UK, Sahoo N (2015) A comprehensive review on the application of emulsions as an alternative fuel for diesel engines. Renew Sust Energ Rev 42:196–211
27. McClements DJ (2005) Food emulsions: principles, practices, and techniques. CRC Press, Boca Raton
28. Dickinson E (2009) Hydrocolloids as emulsifiers and emulsion stabilizers. Food Hydrocoll 23: 1473–1482
29. Fleer GJ, Cohen Stuart MA, Scheutjens JMHM, Cosgrove T, Vincent B (1993) Polymers at interfaces. Chapman & Hall, London
30. Binks BP (1998) Modern aspects of emulsion science. The Royal Society of Chemistry, Cambridge
31. Sharkawy A, Barreiro MF, Rodrigues AE (2020) Chitosan-based Pickering emulsions and their applications: a review. Carbohydr Polym 250:116885
32. Pickering SU (1907) Emulsions. J Chem Soc Trans 91:2001–2021
33. Yang Y, Fang Z, Chen X et al (2017) An overview of Pickering emulsions: solid-particle materials, classification, morphology, and applications. Front Pharmacol 8:287
34. Dickinson E (2010) Food emulsions and foams: stabilization by particles. Curr Opin Colloid Interface Sci 15:40–49
35. Muiz A, Klojdová I, Stathopoulos C (2023) Utilization of by-products for preparation of Pickering particles. Eur Food Res Technol 249:1–15
36. Inoue M, Hashizaki K, Taguchi H et al (2008) Emulsion preparation using beta-cyclodextrin and its derivatives acting as an emulsifier. Chem Pharm Bull 56(9):1335–1337
37. Tavernier I, Wijaya W, Van Der Meeren P et al (2016) Food-grade particles for emulsion stabilization. Trends Food Sci Technol 50:159–174
38. Binks BP (2002) Particles as surfactants – similarities and differences. Curr Opin Colloid Interface Sci 7:21–41

39. Low LE, Siva SP, Ho YK et al (2020) Recent advances of characterization techniques for the formation, physical properties and stability of Pickering emulsion. Adv Colloid Interf Sci 277: 102117
40. Wang H, Singh V, Behrens SH (2012) Image charge effects on the formation of Pickering emulsions. J Phys Chem Lett 3:2986–2990
41. Li M-F, He Z-Y, Li G-Y et al (2019) The formation and characterization of antioxidant Pickering emulsions: effect of the interactions between gliadin and chitosan. Food Hydrocoll 90:482–489
42. Meng W, Sun H, Mu T et al (2023) Chitosan-based Pickering emulsion: a comprehensive review on their stabilizers, bioavailability, applications and regulations. Carbohydr Polym 304:120491
43. Shah BR, Li Y, Jin W et al (2016) Preparation and optimization of Pickering emulsion stabilized by chitosan-tripolyphosphate nanoparticles for curcumin encapsulation. Food Hydrocoll 52:369–377
44. Tabatabaei M, Rajaei A, Hosseini E et al (2022) Effect of type of fatty acid attached to chitosan on walnut oil-in-water Pickering emulsion properties. Carbohydr Polym 291:119566
45. Wang X-Y, Heuzey M-C (2016) Chitosan-based conventional and Pickering emulsions with long-term stability. Langmuir 32:929–936
46. Robins M (2000) Emulsions - creaming phenomena. Curr Opin Colloid Interface Sci 5:265–272
47. Liu H, Wang C, Zou S et al (2012) Simple, reversible emulsion system switched by pH on the basis of chitosan without any hydrophobic modification. Langmuir 28:11017–11024
48. Costa ALR, Gomes A, Cunha RL (2018) One-step ultrasound producing O/W emulsions stabilized by chitosan particles. Food Res Int 107:717–725
49. Deng W, Li Y, Wu L et al (2022) Pickering emulsions stabilized by polysaccharides particles and their applications: a review. Food Sci Technol 42:e24722. https://doi.org/10.1590/fst.24722
50. Adewunmi AA, Mahboob A, Kamal MS et al (2022) Pickering emulsions stabilized by chitosan/natural acacia gum biopolymers: effects of pH and salt concentrations. Polymers 14(23):5270
51. Sinha VR, Singla AK, Wadhawan S et al (2004) Chitosan microspheres as a potential carrier for drugs. Int J Pharm 274:1–33
52. Halmos EP, Mack A, Gibson PR (2019) Review article: emulsifiers in the food supply and implications for gastrointestinal disease. Aliment Pharmacol Ther 49:41–50
53. Meng X, Liu H, Xia Y et al (2021) A family of chitosan-peptide conjugates provides broad HLB values, enhancing emulsion's stability, antioxidant and drug release capacity. Carbohydr Polym 258:117653
54. Srakaew V, Ruangsri P, Suthin K et al (2011) Sodium-phosphorylated chitosan/zinc oxide complexes and evaluation of their cytocompatibility: an approach for periodontal dressing. J Biomater Appl 27(4):403–412
55. Claassen V (1994) 6- Oral drug administration. In: Huston JP (ed) Techniques in the behavioral and neural sciences. Elsevier, pp 59–115
56. Mao Q, Li M, Zhang S et al (2020) Chitosan-hydrophobic alginate nanocomposites stabilized pH-triggered Pickering emulsion for drug controlled-release. Int J Biol Macromol 162:1888–1896
57. Bernkop-Schnürch A, Dünnhaupt S (2012) Chitosan-based drug delivery systems. Eur J Pharm Biopharm 81:463–469
58. Takeuchi H, Yamamoto H, Kawashima Y (2001) Mucoadhesive nanoparticulate systems for peptide drug delivery. Adv Drug Deliv Rev 47:39–54
59. Efiana NA, Mahmood A, Lam HT et al (2017) Improved mucoadhesive properties of self-nanoemulsifying drug delivery systems (SNEDDS) by introducing acyl chitosan. Int J Pharm 519:206–212

60. Saifullah S, Kanwal T, Ullah S et al (2021) Design and development of lipid modified chitosan containing muco-adhesive self-emulsifying drug delivery systems for cefixime oral delivery. Chem Phys Lipids 235:105052
61. Teng Z, Meng L-Y, Yang J-K et al (2022) Bridging nanoplatform and vaccine delivery, a landscape of strategy to enhance nasal immunity. J Control Release 351:456–475
62. Li D, Li W, Li S et al (2024) Fabrication of chitosan-based emulsion as an adjuvant to enhance nasal mucosal immune responses. Int J Biol Macromol 272:132913
63. Marto J, Ascenso A, Simões S et al (2016) Pickering emulsions: challenges and opportunities in topical delivery. Expert Opin Drug Deliv 13:1093–1107
64. Aranaz I, Acosta N, Civera C et al (2018) Cosmetics and cosmeceutical applications of chitin, chitosan and their derivatives. Polymers 10:213
65. Afonso CR, Hirano RS, Gaspar AL et al (2019) Biodegradable antioxidant chitosan films useful as an anti-aging skin mask. Int J Biol Macromol 132:1262–1273
66. Ammala A (2013) Biodegradable polymers as encapsulation materials for cosmetics and personal care markets. Int J Cosmet Sci 35:113–124
67. He W, Guo X, Xiao L et al (2009) Study on the mechanisms of chitosan and its derivatives used as transdermal penetration enhancers. Int J Pharm 382:234–243
68. Sharkawy A, Casimiro FM, Barreiro MF et al (2020) Enhancing trans-resveratrol topical delivery and photostability through entrapment in chitosan/gum Arabic Pickering emulsions. Int J Biol Macromol 147:150–159
69. Venkataramani D, Tsulaia A, Amin S (2020) Fundamentals and applications of particle stabilized emulsions in cosmetic formulations. Adv Colloid Interf Sci 283:102234
70. Seweryn A (2018) Interactions between surfactants and the skin – theory and practice. Adv Colloid Interf Sci 256:242–255
71. Chedea VS, Vicaş SI, Sticozzi C et al (2017) Resveratrol: from diet to topical usage. Food Funct 8:3879–3892
72. Sharkawy A, Barreiro MF, Rodrigues AE (2021) New Pickering emulsions stabilized with chitosan/collagen peptides nanoparticles: synthesis, characterization and tracking of the nanoparticles after skin application. Colloids Surf A Physicochem Eng Asp 616:126327
73. Sharkawy A, Silva AM, Rodrigues F et al (2021) Pickering emulsions stabilized with chitosan/collagen peptides nanoparticles as green topical delivery vehicles for cannabidiol (CBD). Colloids Surf A Physicochem Eng Asp 631:127677
74. Cerqueira-Coutinho C, Santos-Oliveira R, Dos Santos E et al (2015) Development of a photoprotective and antioxidant nanoemulsion containing chitosan as an agent for improving skin retention. Eng Life Sci 15:593–604
75. Ta Q, Ting J, Harwood S et al (2021) Chitosan nanoparticles for enhancing drugs and cosmetic components penetration through the skin. Eur J Pharm Sci 160:105765
76. Morin-Crini N, Lichtfouse E, Torri G et al (2019) Applications of chitosan in food, pharmaceuticals, medicine, cosmetics, agriculture, textiles, pulp and paper, biotechnology, and environmental chemistry. Environ Chem Lett 17:1667–1692
77. Efsa Panel on Dietetic Products N, Allergies (2011) Scientific opinion on the substantiation of health claims related to chitosan and reduction in body weight (ID 679, 1499), maintenance of normal blood LDL-cholesterol concentrations (ID 4663), reduction of intestinal transit time (ID 4664) and reduction of inflammation (ID 1985) pursuant to Article 13(1) of Regulation (EC) No 1924/2006. EFSA J 9:2214
78. Muzzarelli RA (1999) Clinical and biochemical evaluation of chitosan for hypercholesterolemia and overweight control. EXS 87:293–304
79. Shahidi F, Arachchi JKV, Jeon Y-J (1999) Food applications of chitin and chitosans. Trends Food Sci Technol 10:37–51
80. Koide SS (1998) Chitin-chitosan: properties, benefits and risks. Nutr Res 18:1091–1101
81. Agulló E, Rodríguez MS, Ramos V et al (2003) Present and future role of chitin and chitosan in food. Macromol Biosci 3:521–530

82. Zhu X, Chen J, Hu Y et al (2021) Tuning complexation of carboxymethyl cellulose/cationic chitosan to stabilize Pickering emulsion for curcumin encapsulation. Food Hydrocoll 110: 106135
83. Jo M, Ban C, Goh K et al (2020) Enhancement of the gut-retention time of resveratrol using waxy maize starch nanocrystal-stabilized and chitosan-coated Pickering emulsions. Food Hydrocoll 112:106291
84. FDA, Chitosan derived from A. Niger, GRAS Notice 000397. Accessed October 17, 2017, https://www.fda.gov/downloads/food/ingredientspackaginglabeling/gras/noticeinventory/ucm277279
85. FDA, Shrimp-derived chitosan, GRAS Notice 000170. Accessed October 17, 2017, https://www.fda.gov/downloads/Food/IngredientsPackagingLabeling/GRAS/NoticeInventory/ucm268859
86. FDA, Shrimp-derived chitosan,GRAS Notice 000073. Accessed October 17, 2017, https://www.fda.gov/downloads/Food/IngredientsPackagingLabeling/GRAS/NoticeInventory/ucm266601
87. Maleki G, Woltering EJ, Mozafari MR (2022) Applications of chitosan-based carrier as an encapsulating agent in food industry. Trends Food Sci Technol 120:88–99
88. Dammak I, José Do Amaral Sobral P (2018) Formulation optimization of lecithin-enhanced Pickering emulsions stabilized by chitosan nanoparticles for hesperidin encapsulation. J Food Eng 229:2–11
89. Yang L, Cao X, Gai A et al (2022) Chitosan/guar gum nanoparticles to stabilize Pickering emulsion for astaxanthin encapsulation. LWT 165:113727
90. Franco Ribeiro E, Carregari Polachini T, Dutra Alvim I et al (2022) Microencapsulation of roasted coffee oil Pickering emulsions using spray- and freeze-drying: physical, structural and in vitro bioaccessibility studies. Int J Food Sci Technol 57:145–153
91. Lobato-Guarnido I, Luzón G, Ríos F et al (2023) Synthesis and characterization of environmentally friendly chitosan–Arabic gum nanoparticles for encapsulation of oregano essential oil in Pickering emulsion. Nanomaterials 13:2651
92. Hu S, Li W, Cai Z et al (2023) Research progress on chitin/chitosan-based emulsion delivery systems and their application in lipid digestion regulation. Crit Rev Food Sci Nutr:1–23. https://doi.org/10.1080/10408398.2023.2264392
93. Costa ALR, Gomes A, Furtado GDF et al (2020) Modulating in vitro digestibility of Pickering emulsions stabilized by food-grade polysaccharides particles. Carbohydr Polym 227:115344
94. Jo M, Ban C, Goh KKT et al (2019) Influence of chitosan-coating on the stability and digestion of emulsions stabilized by waxy maize starch crystals. Food Hydrocoll 94:603–612
95. Wang D, Xiao H, Lyu X et al (2023) Lipid oxidation in food science and nutritional health: a comprehensive review. Oil Crop Science 8:35–44
96. Li X-M, Li X, Wu Z et al (2020) Chitosan hydrochloride/carboxymethyl starch complex nanogels stabilized Pickering emulsions for oral delivery of β-carotene: protection effect and in vitro digestion study. Food Chem 315:126288
97. Atarian M, Rajaei A, Tabatabaei M et al (2019) Formulation of Pickering sunflower oil-in-water emulsion stabilized by chitosan-stearic acid nanogel and studying its oxidative stability. Carbohydr Polym 210:47–55
98. İlyasoğlu H, Nadzieja M, Guo Z (2019) Caffeic acid grafted chitosan as a novel dual-functional stabilizer for food-grade emulsions and additive antioxidant property. Food Hydrocoll 95:168–176
99. Mouro C, Simões M, Gouveia IC (2019) Emulsion electrospun fiber mats of PCL/PVA/chitosan and eugenol for wound dressing applications. Adv Polym Technol 2019:9859506
100. Gutha Y, Pathak JL, Zhang W et al (2017) Antibacterial and wound healing properties of chitosan/poly(vinyl alcohol)/zinc oxide beads (CS/PVA/ZnO). Int J Biol Macromol 103:234–241
101. Tachaboonyakiat W (2017) 9 - Antimicrobial applications of chitosan. In: Jennings JA, Bumgardner JD (eds) Chitosan based biomaterials, vol 2. Woodhead Publishing, pp 245–274

102. Morkaew T, Pinyakong O, Tachaboonyakiat W (2017) Structural effect of quaternary ammonium chitin derivatives on their bactericidal activity and specificity. Int J Biol Macromol 101: 719–728
103. Bao X, Wu J, Ma G (2020) Sprayed Pickering emulsion with high antibacterial activity for wound healing. Prog Nat Sci Mater Int 30:669–676
104. Lima RC, Carvalho AP, Vieira CP et al (2021) Green and healthier alternatives to chemical additives as cheese preservative: natural antimicrobials in active nanopackaging/coatings. Polymers 13:2675
105. Huang Y, Liu H, Liu S et al (2020) Cinnamon Cassia oil emulsions stabilized by chitin nanofibrils: physicochemical properties and antibacterial activities. J Agric Food Chem 68: 14620–14631
106. Zhong W, Li D, Li L et al (2024) pH-responsive Pickering emulsion containing citrus essential oil stabilized by zwitterionically charged chitin nanofibers: physicochemical properties and antimicrobial activity. Food Chem 433:137388
107. Tachaboonyakiat W (2021) Physical and chemical modification of chitin/chitosan for functional wound dressings. In: Jayakumar R, Prabaharan M (eds) Chitosan for biomaterials III: structure-property relationships. Springer, Cham, pp 257–299
108. Pei J, Wei N, Cui C et al (2022) Preparation of Pickering emulsion for antibacterial, anti-inflammatory and wound healing. J Biosci Med 10:134–150
109. Wu X, Zhang Q, Wang Z et al (2022) Investigation of construction and characterization of carboxymethyl chitosan - sodium alginate nanoparticles to stabilize Pickering emulsion hydrogels for curcumin encapsulation and accelerating wound healing. Int J Biol Macromol 209:1837–1847
110. Asfour MH, Elmotasem H, Mostafa DM et al (2017) Chitosan based Pickering emulsion as a promising approach for topical application of rutin in a solubilized form intended for wound healing: in vitro and in vivo study. Int J Pharm 534:325–338
111. Lv X, Liu Y, Song S et al (2019) Influence of chitosan oligosaccharide on the gelling and wound healing properties of injectable hydrogels based on carboxymethyl chitosan/alginate polyelectrolyte complexes. Carbohydr Polym 205:312–321
112. Yang DH, Seo DI, Lee D-W et al (2017) Preparation and evaluation of visible-light cured glycol chitosan hydrogel dressing containing dual growth factors for accelerated wound healing. J Ind Eng Chem 53:360–370
113. Yar M, Shahzad S, Shahzadi L et al (2017) Heparin binding chitosan derivatives for production of pro-angiogenic hydrogels for promoting tissue healing. Mater Sci Eng C 74:347–356
114. Chatterjee NS, Sukumaran HG, Dara PK et al (2022) Nano-encapsulation of curcumin in fish collagen grafted succinyl chitosan hydrogel accelerates wound healing process in experimental rats. Food Hydrocolloids Health 2:100061
115. Xu N, Wang L, Guan J et al (2018) Wound healing effects of a Curcuma zedoaria polysaccharide with platelet-rich plasma exosomes assembled on chitosan/silk hydrogel sponge in a diabetic rat model. Int J Biol Macromol 117:102–107
116. Boateng JS, Matthews KH, Stevens HNE et al (2008) Wound healing dressings and drug delivery systems: a review. J Pharm Sci 97:2892–2923
117. Junker J, Kamel R, Caterson E et al (2013) Clinical impact upon wound healing and inflammation in moist, wet, and dry environments. Adv Wound Care 2:348–356